Janek von Byern • Ingo Grunwald
Editors

Biological Adhesive Systems

From Nature to Technical and Medical Application

SpringerWienNewYork

Dr. Dipl.-Biol. Janek von Byern
Core Facility Cell Imaging and Ultrastructure Research
Faculty of Life Science
University of Vienna
Vienna, Austria

Dr. Dipl.-Biol. Ingo Grunwald
Fraunhofer Institute for Manufacturing Technology
and Advanced Materials (IFAM)
Department of Adhesive Bonding Technology
and Surfaces, Adhesives and Polymer Chemistry
Bremen, Germany

Printed in Austria

SpringerWienNewYork is part of
Springer Science+Business Media
springer.at

Cover Illustrations: Background © Fraunhofer Institute for Manufacturing Technology
and Advanced Materials (IFAM), 28359 Bremen, Germany
Image of *Mytilus edulis* © APA-PictureDesk GmbH, Laimgrubengasse 10, 1060 Wien, Austria
Image of *Drosera affinis* © Petr Soural www.naturfoto.cz
Typesetting: Thomson Press (India) Ltd., Chennai
Printing: Druckerei Theiss GmbH, 9431 St. Stefan im Lavanttal, Austria

Printed on acid-free and chlorine-free bleached paper
SPIN: 12722556

With 156 (partly coloured) Figures

Library of Congress Control Number: 2010938616

ISBN 978-3-7091-0141-4 SpringerWienNewYork

Foreword

J. Herbert Waite

Like many graduate students before and after me I was mesmerized by a proposition expressed years earlier by Krogh (1929) – namely that "for many problems there is an animal on which it can be most conveniently studied". This opinion became known as the August Krogh Principle and remains much discussed to this day, particularly among comparative physiologists (Krebs, 1975). The words "problems" and "animal" are key because they highlight the two fundamental and complementary foci of biological research: (1) expertise about an animal (zoo-centric), which is mostly observational and (2) a mechanistic analysis of some problem in the animal's life history or physiology (problem-centric), which is usually a hypothesis-driven investigation.

Among the biologists of my acquaintance, there are few if any who are only one or the other; instead, most are different blends of the two with a slight polarization towards one of the two foci. If she loved her animal beyond anything else, the zoo-centric biologist might undertake to investigate what this organism was specifically well adapted to do. On the other hand, if a specific problem rather than the organism was his *idée fixe*, then the problem-centric biologist might undertake a systematic search into research about this problem in many different organisms before selecting one from among them. Both pathways are valid, although the routines, challenges and thrills are different. If I had my druthers today and research funding was no obstacle, I would follow a zoo-centric pathway; that is, get familiar with an organism first and then investigate its most fascinating adaptations. Why? At the current level of human activity, 40% of all marine species are predicted to become functionally extinct in ten years (Worm et al., 2006). As a member of *Homo sapiens*, I feel a measure of responsibility for the stewardship of biological diversity on this planet. Consequently, there is an increasing urgency for acquaintance with this diversity without first asking about convenience. There are so many species about which nothing is known, and the curse of not knowing is apathy.

Bioadhesion is the adaptation featured in this book, and biology has many adhesive practitioners. Indeed, every living organism is adhesively assembled in the most exquisite way. Clearly, specific adhesion needs to be distinguished from the opportunistic variety. I think of specific adhesion as the adhesion between cells in the same tissue, whereas opportunistic adhesion might be the adhesion between pathogenic microbes and the urinary tract, or between a slug and the garden path. If opportunistic bioadhesion is our theme, then there are still many practitioners but the subset is somewhat more select than before. How is bioadhesion best studied in one's favorite organism? Which organism is the best model for adhesion research? These are questions many of us have agonized over and continue to agonize over.

The August Krogh Principle can be handily applied to investigating opportunistic bioadhesion. However, the insights about adhesion obtained are probably less transcendent than those deduced from the model organisms used to investigate "core concepts" in cell biology such as ion transport (toad skin), nerve conductance (squid axon) or neural networks and behavior (sea hare) to mention a few (Krebs, 1975). Based on the limited number of well-characterized bioadhesive strategies, I would hazard to guess that the sheer magnitude of evolutionary pressure on opportunistic adhesive strategies vis-à-vis what goes on inside cells and tissues is a powerful driver for diversification. However, until more is known about opportunistic bioadhesion, this remains conjecture. Perhaps comparison to another highly diversified and well-studied process such as coloration is appropriate here. Living organisms have evolved a plethora of physical and chemical strategies for making color. Two different insects may be perceived as red but the mechanisms producing their color are likely as not to be totally unrelated (Shawkey et al., 2009).

I have studied invertebrates for the past 35 years, beginning as a lowly undergraduate assistant to the curator of a shell collection at the Museum for Comparative Zoology in Cambridge, MA. I was well drilled in invertebrate zoology and intrigued by the distinctions for classifying organisms. But to be perfectly honest, I found taxonomical details quite mind numbing and for that reason was gravitating towards biochemistry as my major. Everything unexpectedly fell into place after graduate school in the late 1970s, when two different granting agencies – US National Institute of Dental Research and the US Office of Naval Research – announced research initiatives urging investigators to explore the use of model organisms as a way (i) of discovering a new generation of wet adhesives (biomimetics) and (ii) of keeping surfaces clean (antifouling), respectively. Thus, I was stoked by the science of adhesion before selecting marine mussels as my model organism (problem-centric). The selection was actually the outcome of performing some simple tests on a line-up of sticky culprits that included ivy, kelp, barnacles, oysters, sea stars, sea anemones and ascidians among others based on the following simple criteria: (1) how robust are the organisms in captivity, (2) how quick is adhesion, (3) how much adhesive material is secreted, (4) how much adhesive precursor is stockpiled by the organism, and (5) can the molecules responsible for adhesion be isolated and specifically assayed. Needless to say marine mussels particularly in the genus *Mytilus* scored well in all criteria.

Due largely to the existence of specific assays for Dopa – a key functional group in byssal adhesive proteins – mussels have been excellent experimental models for revealing the adhesive tricks that one group of sessile organisms is capable of Waite et al. (2005). Mussels do not, however, have any monopoly on effective bioadhesion to solid surfaces in seawater, nor is their solution unique – at least one group of sabellariid polychaete tubeworms has independently evolved a very similar biochemistry (Zhao et al., 2005). In many known cases, bioadhesion, like coloration, appears to be the result of fundamentally different molecular adaptations. Apart from an inherent clinginess, the adhesion of barnacles, ascidians, limpets, geckos, ants, and orchids, etc. to surfaces is distinctive with great and subtle variations in performance. Adhesion that is permanent or temporary, gliding or step-wise, wet or dry, textured or chemical – all these types have biomimetic potential.

One popular definition of biomimetics is the abstraction of useful design from biology (Vincent, 1994). It is no longer absurd to predict that a compilation of bio-inspired adhesive designs will someday fill a tome with more recipes than the *Handbook for Adhesives* (Skeist, 1977) currently lists for technological adhesive options. Alas, whether or not our model organisms survive the poisoned environments that our biomimetic technologies help create for them is a problem rather greater than adhesion. There is no knowing whether we will rise to that challenge.

References

Krebs HA (1975) The August Krogh principle: "For many problems there is an animal on which it can be most conveniently studied". Journal of Experimental Zoology 194(1): 221–226.

Krogh A (1929) The progress of physiology. The American Journal of Physiology 90(2): 243–251.

Shawkey MD, Morehouse NI, and Vukusic P (2009) A protean palette: colour materials and mixing in birds and butterflies. Journal of the Royal Society Interface 6(Suppl 2): S221–S231.

Skeist I (1977) Handbook of Adhesives, 2nd Ed. Van Nostrand Reinhold Company, New York.

Vincent JFV (1994) Borrowing the best from nature. Encyclopaedia Britannica, Yearbook of Science and the Future, 1995: pp 168–187.

Waite JH, Holten-Andersen N, Jewhurst SA, and Sun CJ (2005) Mussel adhesion: finding the tricks worth mimicking. Journal of Adhesion 81: 297–317.

Worm B, Barbier EB, Beaumont N, Duffy JE, Folke C, Halpern BS, Jackson JB, Lotze HK, Micheli F, Palumbi SR, Sala E, Selkoe KA, Stachowicz JJ, and Watson R (2006) Impacts of biodiversity loss on ocean ecosystem services. Science 314(5800): 787–790.

Zhao H, Sun C, Stewart RJ, and Waite JH (2005) Cement proteins of the tube-building polychaete *Phragmatopoma californica*. Journal of Biological Chemistry 280(52): 42938–42944.

Contents

Part A

Each individual is singular. This is reflected in the following chapters, which show that each plant and animal has re-innovated adhesion.

No described adhesive system resembles any other one – neither in development, form and occurrence, nor in terms of the composition and function of the chemical components.

In order to understand and use these adhesives practically, a topic of Part B of this book, it is essential to examine the biological system and understand its glue synthesis and bonding function.

One underlying problem, however, is the search for a "common language". Most of the cell types introduced here have several different designations, depending on the respective researcher. This complicates within-genus comparisons, not to mention establishing relationships at higher taxonomic levels.

The present book offers, for the first time, the opportunity to directly compare well-known adhesive systems in the animal and plant kingdom and to correlate similarities between the different systems.

1 Bonding Single Pollen Grains Together: How and Why?

Michael Hesse

Contents

1.1 The Anther Tapetum as a Glandular Tissue in Seed Plants

In their early developmental stages, the anthers (the pollen-producing organs of a male flower) form a tapetum between the sporogeneous tissue and the anther wall; both the tapetal cells and the sporogeneous cells have developed originally from the same subepidermal tissue. The tapetum is of considerable physiologic significance because all the nutritional material entering the microspores and later on the pollen grains passes or originates from it. In addition, during certain periods of pollen development, it accumulates substantial quantities of reserve compounds (e.g., starch and/or protein crystals in plastids, lipid droplets inside and outside the plastids, soluble polysaccharides in the vacuoles). These stored substances successively disappear during and after the tapetum degeneration, but several characters of mature pollen grains, which are of considerable interest in pollination, depend just on these substances. For details of tapetum development and ultrastructure the reader is referred to Hesse (1993); Halbritter et al. (1997); and Pacini and Hesse (2005).

The tapetum is a specialized tissue concerned with the nutrition of the developing spores and pollen grains, and is found in the sporangia of lower plants and anthers of higher plants (Pacini, 1997). Tapetal cells exhibit a variety of developmental pathways, especially in terms of the behavior of the cell walls and the synthesis of pollen wall precursors (Barnes and Blackmore, 1992).

1.1.1 The Tapetum Types

Usually two main tapetum types are distinguished in the angiosperms:

The **secretory** (or parietal or glandular or cellular) tapetum, and 2. the **amoeboid** (or invasive, or plasmodial, or periplasmodial) tapetum. The secretory tapetum is characterized by persistent cell walls and is also referred to as glandular, parietal, or cellular. In contrast, the cell walls of the amoeboid tapetum break down during microspore development. This second kind of tapetum has been attributed at least in the past to ± aquatic taxa (a recent review was presented by Furness (2008), demonstrating that the plasmodial tapetum types are not rare in eudicots, as suspected before). It should be stressed that in later developmental (i.e., degeneration) stages, the resulting substances do not allow a distinction between the former amoeboid or secretory tapetum.

In secretory tapeta, the cells are polarized both in structure and function, while the cells of the amoeboid tapeta are not. The inner and the outer tangential faces of the secretory tapetum differ in their capacity to release certain substances, e.g., the outer tangential face is apparently not concerned with material secretion (with the possible exception of orbicules). The polarity of the tapetum cells is also expressed by the intercellular tapetal space (radial cell faces) through which secretion also takes place: a gradient from the outer to the inner face of the tapetum cells exists with respect to the timing of cell wall solution, and the production and the release of the various stored substances. Conversely, no polarity is found on the cell surface of the amoeboid tapeta, even if (e.g., in *Arum italicum)* two zones are temporarily evident in the periplasmodium: one zone just surrounds the microspores (with polyribosomes, microtubules, vesicles, and few dilated ER cisternae), while the other zone (with nuclei, mitochondria, plastids, small vacuoles, and ribosomes) does not.

In both tapetum types, the produced substances are either "readymade" (the locular fluid, the callase, the PAS-positive content, partly the sporophytic proteins) or become polymerized after their release (the exine precursors, the viscin threads, the culture sac, the orbicules, the pollenkitt/tryphine substances, and partly the sporophytic proteins).

The role of the tapetum types with respect to the produced material is different: All the substances produced by the secretory tapetum reach the microspores/pollen grains *via* the locular fluid. In contrast, in the amoeboid tapeta, the cytoplasm adheres closely to the microspores. The locular fluid, which mostly occurs only in the secretory tapetum, represents an infiltrating medium between the sporophyte and the developing gametophytes. This intermediate fluid, which transports the nutrients, is extremely reduced in volume or even absent in amoeboid tapeta and in some particular parietal ones. The nutrition of meiocytes/microspores/pollen grains therefore takes place indirectly in secretory tapeta and directly in amoeboid tapeta.

The advantages or disadvantages of the two main tapetum types can be seen as follows: The main advantage of amoeboid tapeta is that nutrition can take place without the locular fluid as an intermediate, while the main disadvantage is that this can only take place in small anthers with few microspores/pollen grains per locule. In contrast, the main advantage of the secretory tapetum is that the anthers may be large with many microspores/pollen grains per locule, while its main disadvantage is the need of water inside the locule to protect against dehydration and to transfer nutrients via the locular fluid.

Although the nourishing function of the tapetal cells probably starts after the differentiation of the anther tissues, it has at the beginning a low rate because the microspore mother cells/meiocytes are encased in a callosic wall and the tapetum cells are also enveloped by a mostly cellulosic wall. Later on, however, the flow of nutrients increases, as all the tapetum cells loose their walls and the callose disappears. The eallosic wall is formed during the prophase between the pectocellulosic wall and the plasmalemma. Due to its very low permeability it acts as a molecular filter between the sporophyte and the gametophyte. In the "successive" type of microspore ontogenesis a new callosic wall is formed after the first meiotic division separating the dyads and after the second division the spores from one another. In the "simultaneous" type of microspore ontogenesis, no walls are laid down until the second meiotic division has been completed. However, more recent investigations indicate that indeed new callosic walls are laid down (Ressayre et al., 2005; also for review).

The most important roles commonly ascribed to gymnosperm and angiosperm tapeta, aside from their essential role in nutrition, are the formation of pollen-connecting agents (with four types: formation of viscin threads, i.e., long, flexible, fragile sporopollenin ropes, or strands on the surface of pollen tetrads or single pollen grains, coalesced with the exine itself; the formation of pollenkitt, a hydrophobic, oily layer containing mainly lipids and carotenoids; the formation of tryphine, which in contrast to pollenkitt is comprised of a mixture of hydrophobic and hydrophilous substances often containing cytoplasmic elements (degenerated organelles); the formation of slimy/sticky rope-like derivates from the tapetal ground plasm).

1.1.2 Pollen-connecting Agents: Nature, Function, and Systematic Distribution

Anemophilous seed plants shed their dry pollen always as single grains, transferred by wind, any pollen association would be counterproductive. The simultaneous transfer of larger pollen associations by insects or vertebrates between flowers certainly increases the probability that ovules will be fertilized (Knox and McConchie, 1986; Rose and Barthlott, 1995). Only Angiosperm pollen grains may clump together, and this takes place in three different ways: (1) by means of fluids as, e.g., pollenkitt and other forms of pollen coatings, (2) by means of ropes, strings, or fibers of different nature, and (3) by means of special sporoderm walls as, e.g., in permanent tetrads. These types are listed in Table 1.1, with special reference to terminology, origin, and main components of the pollen-clumping agents.

1.1.2.1 Pollenkitt and Tryphine, the Principal Forms of Pollen Coatings

Pollen coatings are represented not only by pollenkitt (Knoll, 1930), but also by proteins and cytoplasmic remnants (see below). There is no evidence to suggest that the pollenkitt material would play any part in recognition reactions. **Pollenkitt** means the deposited lipid fraction of the tapetum organelles plastids and ER only, and **tryphine** is the mixed fraction of degenerated tapetal organelles with lipid and cytoplasmic remnants.

An extensive review of chemical properties of the pollen coatings, and of pertinent literature was given by Boavida (2005) and Murphy (2006).

1.1.2.2 Tryphine

Tryphine formation was first described by Dickinson and Lewis (1973) in *Raphanus*. They also coined this term to distinguish the specific Brassicaceae features from pollenkitt known since Knoll's papers. It has only been studied and described for Brassicaceae (a precise definition is given by Dickinson et al. (2000), but is probably also present in other groups (Pacini, 1997)). Pollenkitt and tryphine, irrespective of their mode of formation, composition, and so on, are similar in many respects because both are formed by fusion of tapetal elaiosomes and spherosomes (Platt et al., 1998).

Table 1.1: Typology of pollen-clumping agents (adapted from Pacini and Hesse, 2005)

Origin	Pollen-clumping agents (quoted are their common terms and – so far known – their main components)	Angiosperm taxa
of tapetal origin	pollenkitt (plastids involved in lipid formation) elastoviscin (plastids not involved in the formation of lipid droplets)	most entomophilous angiosperms *Raphionacme* (Aristolociaceae) many Orchidaceae
-„-	thecal slime/mucilage rafts	*Halophila*, *Thalassia* (Hydrocharitaceae)
-„-	tryphine	Brassicaceae
-„-	slimy, highly viscous mixtures of probably various lipidic and cytoplasmic components	e.g. *Porcelia* (Annonaceae) Aristolochia (*Aristolochiaceae*) some Caesalpiniaceae, Araceae, and many other examples[a]
both of tapetal and non-tapetal origin	(sporopollenin) ektexinous viscin threads, respectively exinal connections	Onagraceae, Ericaceae/Rhododendroideae; *Jacqueshuberia*
-„-	(ekt- and end-)exine cohesions forming compound pollen (pollen units: tetrads, polyads, massulae)	in (at least) 56 angiosperm families; exine-"bridges" in some Onagraceae

[a]Occurrence includes Annonaceae: *Porcelia* (Morawetz and Waha, 1991); the slimy strands wrapping pollen grains in some Araceae (Troll, 1928; Richter, 1929; Hesse, 2009). Aristolochiaceae: *Aristolochia* (M. Wolter: pers. comm.); Asclepiadaceae: *Mondia* (H. Kunze, Minden: pers. comm.); *Raphionacme* (Dannenbaum and Schill, 1991); Caesalpiniaceae: *Bauhinia*, *Caesalpinia*, *Cercis*, and *Delonix* (all in Hesse, 1986); Heliconiaceae: *Heliconia* (Rose and Barthlott, 1995); Hydrocharitaceae: *Halophila* and *Thalassia* (Pettitt, 1981; Cox and Tomlinson, 1988); Marcgraviaceae: *Norantea* (Sazima et al., 1993; Pinheiro et al., 1995); Orchidaceae: *Cypripedium* (Burns-Balogh and Hesse, 1988), *Disa* (Vogel, 1959), *Doritis* (Wolter et al., 1988), *Habenaria* (Hesse and Burns-Balogh, 1984), *Zeuxine* (Vijayaraghavan and Shukla, 1980; Shukla, 1984), orchids in general (cf. Schill and Wolter, 1986; Dressler, 1993); Passifloraceae: *Tetrastylis* (Buzato and Franco, 1992); Strelitziaceae: *Strelitzia* (Kronestedt-Robards, 1996); Zannichelliaceae: *Lepilaena* (Cox and Knox, 1989).

A report of biosynthesis and processing of proteins, forming main pollen-coating components, was published by Piffanelli et al. (1998).

1.1.3 Pollenkitt: Function and Origin

Mature pollen grains of zoophilous (and amphiphilous) angiosperms are generally coated with an oily, sticky material, commonly called pollenkitt since Knoll (1930). Sometimes pollenkitt may assume a filiform habit, see Chapter 1.1.4 "Filiform Pollen-connecting Structures", p. 9.

The abundant presence of pollenkitt is mostly a sign for entomophily, but in some cases (e.g., Ericaceae: *Calluna*, *Erica*; Primulaceae: *Cyclamen*) the pollenkitt gets dry after some hours, and whitish, and the pollination mode may shift toward anemophily.

This may be in some taxa related to the end of nectar production (Franchi and Pacini, 1996) and to the presence of pending flowers, when the release of dry pollen is facilitated. At the end of anthesis, pollenkitt changes its physicochemical properties, such as color and viscosity, probably to avoid further visits by insects. This was observed in *Erica* (Hesse, 1979), which, like *Calluna*, is visited first by insects, mainly bees, and then become wind-pollinated.

Pollenkitt was recently discovered very well preserved in fossil flower buds on and between fossil pollen grains. Amorphous, highly electron-opaque substances (pollenkitt) and orbicules both very similar to those of recent *Tilia*, were observed on and between pollen grains of fossil flower buds of *Craigia bronnii* (Tilioideae, Malvaceae). Staining and sectioning of this amorphous substance demonstrated that it was indeed pollenkitt (Zetter et al. (2002), also for review).

1.1.3.1 Function

Listed below are **the many functions of pollenkitt** during the period from anther opening to pollen hydration on the stigma [adapted from Pacini and Hesse (2005)]:

(1) to hold pollen in the anther until dispersal;
(2) to enable secondary pollen presentation;
(3) to facilitate pollen dispersal;
(4) to protect pollen from water loss;
(5) to protect pollen from ultra-violet radiation;
(6) to maintain sporophytic proteins responsible for pollen–stigma recognition inside exine cavities;
(7) to protect pollen protoplasts from fungi and bacteria;
(8) to keep together pollen grains during transport[1];
(9) to protect pollen from hydrolysis;
(10) to render pollen attractive to animals;
(11) to render pollen visible to animal eyes;
(12) to hide pollen from animal eyes;
(13) to avoid predation of pollen through smell;
(14) to enable adhesion to insect bodies;
(15) to enable pollen packaging by bees and to form corbicules;
(16) to provide a digestible reward for pollinators;
(17) to enable pollen clumps to reach the stigma;
(18) to allow self-pollination;
(19) to facilitate adhesion to the stigma;
(20) to facilitate pollen hydration.

1.1.3.2 Pollenkitt Ontogenesis (Adapted from Hesse, 1993, with Additions)

Pollenkitt is produced in the anther tapetum. Mostly the plastids – and often also the tapetal cytoplasm/other organelles – are involved in its formation (Weber, 1992). It forms the mostly dense and structurally homogeneous exine coating as seen by light microscopy and also by TEM. Investigations of mature/degenerating tapetal cells by TEM (Reznickova and Dickinson, 1982), and especially by thin-layer chromatography (Dobson, 1988) have shown that pollenkitt is a complex mixture of lipids, carotenoids, etc. It varies considerably between species and consists of many unsaturated and saturated lipids, of carotenoids, and sometimes also of proteins (Dobson, 1988).

There is good evidence for pollenkitt production by usually more than a single tapetum organelle. A well-studied example is *Tilia platyphyllos*, where two different types of pollenkitt (pk) precursors, showing separate cytologic/compartimental origins, occur within the tapetum cells. The first – initially not membrane-bound – lipid droplets within the cytoplasm can be found early in the late tetrad stage with the young microspores still enclosed in callose (fig. 1a in Hesse, 1993). Later on, in the free microspore stage, during exine formation, the number of these globules increases, and they become tightly enclosed by SER-profiles (figs. 1b, c in Hesse, 1993).

[1]A rare functional aspect of pollenkitt was described by Wang et al. (2004). The pollenkitt in *Caulokaempferia coenobialis* is clear and rich in unsaturated lipids. It forms an oily film in which the pollen grains are suspended. This was the first report of pollination in angiosperms by pollen that is conveyed in a mobile secreted medium. The lateral flow of the film of pollen along the style seems to be due only to the spreading properties of the oily emulsion and not to gravity.

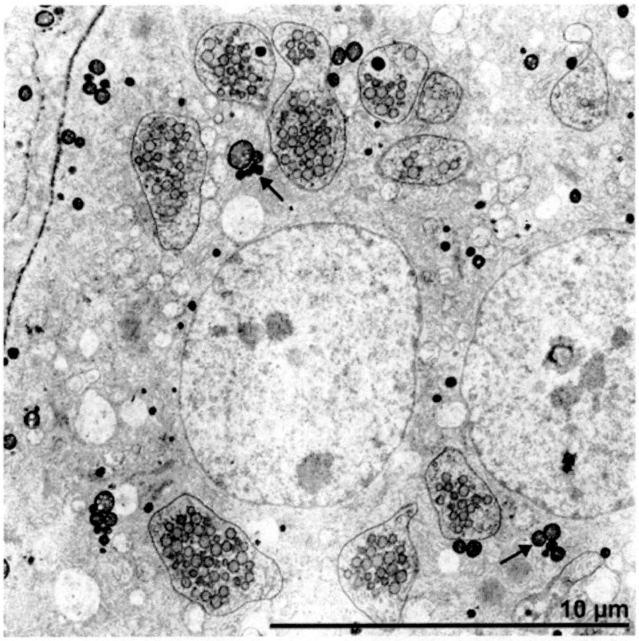

Fig. 1.1 *Tilia platyphyllos* (Malvaceae – Tilioideae). Detail of a young tapetal cell. Two pollenkitt components: one component is represented by free-floating electron-dense lipid droplets (arrows), the other by electron-transparent droplets in the modified plastids (Photo by M. Hesse)

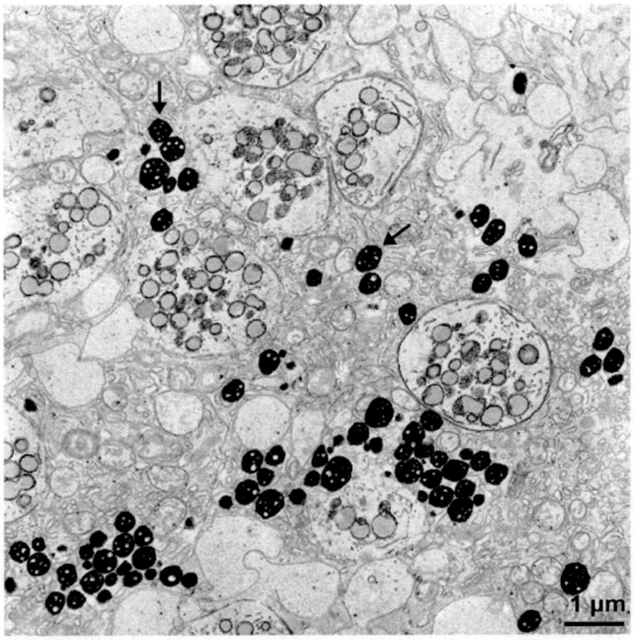

Fig. 1.2 *Tilia platyphyllos* (Malvaceae – Tilioideae). Detail of a fully developed tapetal cell in its maximum stage of secretion. Abundant electron-dense free-floating lipid droplets (arrows), and electron-transparent droplets within modified plastids (Photo by M. Hesse)

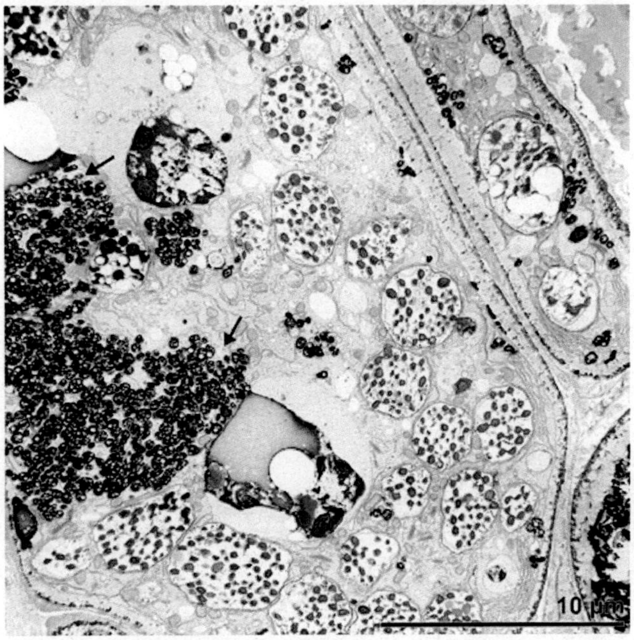

Fig. 1.3 *Tilia platyphyllos* (Malvaceae – Tilioideae). Detail of a mature tapetal cell, with masses of electron-dense lipid droplets (arrows), and modified plastids with the plastidal component of pollenkitt (Photo by M. Hesse)

After exine formation is completed, the globules are released into the cytoplasm, and are no longer enclosed by ER. Tapetal plastids do not show lipid droplets at this stage. Before the first microspore mitosis the number and dimension of the formerly ER-bound globules (the first pollenkitt precursor type) are greatly enlarged; only few connections to ER profiles can be seen at this stage (figs. 2a and 3 in Hesse (1993)). After the first microspore mitosis the abundant non-plastidal, "cytoplasmic" lipid droplets often appear to be again connected with or even enclosed by a single (SER-) membrane (figs. 4 and 5 in Hesse (1993)).

The second pk component is formed independently from the first one within modified tapetal plastids (elaioplasts). Up to exine formation stage the plastids do not show modifications and do not produce any lipid droplets. Shortly before the first pollen mitosis the plastids within the tapetal cells undergo remarkable modifications. Only very few thylakoids exist, but many proplastid-like tubules (Fig. 1.1).

The plastids (elaioplasts) enlarge greatly and form a lot of widely electron-transparent lipid droplets. Shortly after the first pollen mitosis, the plastidal droplets undergo some significant changes. In this developmental stage the plastidal droplets show bubbles (fig. 2c in Hesse (1993)) and a striking difference in electron-density. Some elaioplasts are filled with highly electron-transparent globules, some with electron-dense globules, while others contain both sorts simultaneously (fig. 2b in Hesse (1993)). Shortly after this stage the plastidal droplets show an electron-transparent core and a dense outer zone (fig. 3 in Hesse (1993)). The pk globules are composed of a granular component also, which derives from the globular bubbles seen in an earlier stage (Fig. 1.2).

Until late stages of tapetum development in *Tilia platyphyllos* both pk precursor types are well separated. Only during the final degeneration stages of the tapetum cells the second pk precursor type becomes released from the plastids, since now the plastid membranes degenerate (Fig. 1.3).

Both precursor types independently float into the loculus. First they can be distinguished as blistered lumps occurring from the SER-born component and as small, mostly dense droplets resulting from the plastidal pk precursor type[2]. In late developmental stages of *Tilia* both components may become mixed.

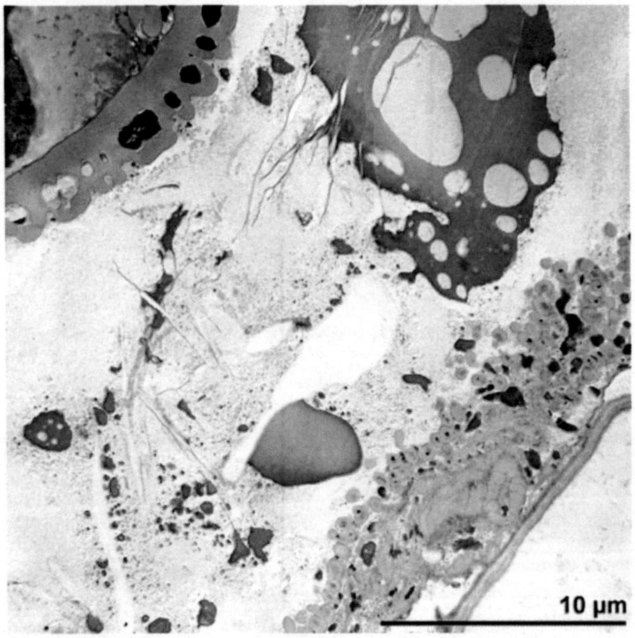

Fig. 1.4 *Tilia platyphyllos* (Malvaceae – Tilioideae). After release of both pollenkitt components they mix and form electron-dense masses, either free in the loculus, or filling the exine in the mature pollen grains (Photo by M. Hesse)

The components are finally deposited onto the pollen exine (Fig. 1.4). The pollenkitt is placed in part on top of the exine tectum or is filling in part the intercolumellar spaces either as small droplets or – at the latest stage just before dehiscence – as a widely compact, sometimes lamellate, but generally extremely dense substance, forming the proper pollenkitt.

1.1.3.3 Pollen-gluing Agents not Formed by Pollenkitt

Two rare types of pollen-gluing material, different from pollenkitt and tryphine, occur in some Araceae. The first one is a somewhat combination of tapetal and non-tapetal material, e.g., in *Anubias*, *Calla*, *Cryptocoryne*, *Culcasia*, *Schismatoglottis*, *Xanthosoma*, or *Zantedeschia* (Richter, 1929[3]). This type of glue originates not only from the tapetum, but also from thecal material already from the PMC stage onward (i.e., extremely soon), forming, e.g., the so-called pollen droplets in *Cryptocoryne*, or the so-called pollen-strands in, e.g., *Schismatoglottis* (Hesse et al. 2009, unpubl. obs.).

Another quite dissimilar type of a pollen-gluing agent, occurring also in an aroid, namely in *Philodendron*, derives curiously from the spathe, without any contributing role of the tapetum: pollen becomes extruded in strands, and is embedded in the resin produced by the spathe, becoming very sticky from the moment of thecae dehiscence (Croat, 1999; see also Armbruster, 1984). The latter author (Armbruster, 1984), studying the role of resin in angiosperm pollination, has questioned the efficacy of floral resin in the transport of pollen, citing its possible toxicity and the difficulty of transporting pollen embedded in resin. While he stresses the role of resin for other purposes, mainly in nest building by bees, Croat (1999) and http://www.aroid.org/genera/philodendron/pollibiol.php pointed out that bees which use resin for nest building play no role whatever in *Philodendron* pollination. In

[2]In contrast to the *Tilia* events Weber (1992) has shown that in *Apium nodiflorum* one of the two pollenkitt precursors accumulate within membrane-bound domains continuous with SER. Sometimes, curiously, e.g., in *Rosmarinus*, there is evidence that only a single organelle is responsible for pollenkitt production, namely the ER (Ubera Jimenez et al., 1996). However, *Tilia* may be seen as the common example for pollenkitt production modes.

[3]S. Richter (originally in German): „abholbereiter Pollen liegt eingebettet in einer schleimigen Masse; sie besteht aus den plasmodialen Tapetumablagerungen und den Zersetzungsprodukten der Ausgussöffnung.“ … „Das reichlich vorhandene Material von *Zantedeschia aethiopica* ermöglichte genauere Untersuchungen. Schon im Jugendstadium der Anthere, während die Pollenmutterzellen noch vorhanden sind, beginnt eine Zersetzung des Tapetums … Dieses dringt nach Auflösung des Tetradenverbandes zwischen die einzelnen Pollenkörner als Periplasmodium ein. Es wird aber nicht vollständig zur Ernährung der heranwachsenden Pollenkörner aufgebraucht. Die letzten Reste bleiben an der Exine haften und halten die reifen Pollen zusammen“.

contrast, the near universal availability of resin, its close association with pollen delivery, the non-tacky nature of *Philodendron* pollen and its availability only at anthesis of flowers all point toward a strong role in *Philodendron* pollination. In those species with resiniferous spadices the pollen is shed with and incorporated in the resin from the moment of thecae dehiscence. Alternatively those species which lack staminal resin and instead have resin only on the inner spathe surface have pollen presented as slender filaments.

1.1.4 Filiform Pollen-connecting Structures

Two different types of filiform (thread-shaped) structures are involved in pollen grain connection agents. In the first part we will consider the nature, occurrence, and putative function of the various thread- or rope-like structures; the second part is devoted to a conspectus on the respective origin.

1.1.5 Acetolysis-resistant, Sporopollenin Pollen-connecting Threads

These curious structures are not formed by components of the lipid/protein tapetum fraction, but consist of a highly resistant organic polymer, the sporopollenin, which is the main component of the exine, the outer pollen wall (for details see Hesse et al., 2009).

Occurrence: Acetolysis-resistant threads ("viscin threads") are rare and may be regarded as restricted to angiosperms. They are not only derived from the ektexine, but are in fact part of the ektexine itself, and thus their ornamentation is often quite similar to the ektexine sculpturing. Their mode of origin is up to now poorly known. Viscin threads occur in almost all recent Onagraceae and Ericaceae-Rhododendroideae (Skvarla et al., 1975, 1978; Waha, 1984).

Nixon and Crepet (1993) and Zetter and Hesse (1996) presented reports on fossil Ericaceae pollen with viscin threads. The more prominent and more numerous viscin threads of fossil Onagraceae are well known (Keri and Zetter, 1992). The so-called exinal connections in *Jacqueshuberia*, Caesalpiniaceae, are chemically, structurally, and most probably functionally very closely related to viscin threads (Patel et al., 1985; Hesse, 1986).

Figures 1.5–1.11 demonstrate the most important characters of viscin threads in Onagraceae and in Ericaceae-Rhododendroideae. One important difference between Onagraceae and *Rhododendron* viscin threads is that the threads emerge in Onagraceae from the inner side of the tetrad (i.e., proximally), while in *Rhododendron* the threads have their offspring distally.

Function: Any pollen material with viscin threads is related to a highly specialized pollination mode (Waha, 1984; Hesse, 1986; Nixon and Crepet, 1993; Crepat, 1996; Zetter and Hesse, 1996).

The viscin threads also play a role in pollen presentation. In Onagraceae pollination often takes place by long-

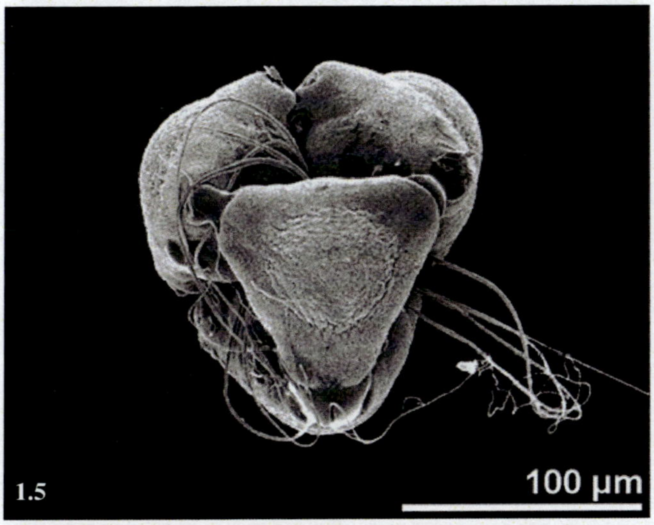

Fig. 1.5 *Epilobium hirsutum* (Onagraceae), pollen tetrad, the individual pollen grains are connected by viscin threads. Note proximal emergence of viscin threads (PalDat, Photo by H. Halbritter)

Fig. 1.6 *Epilobium hirsutum* (Onagraceae), several pollen tetrads, all connected by viscin threads (PalDat, Photo by H. Halbritter)

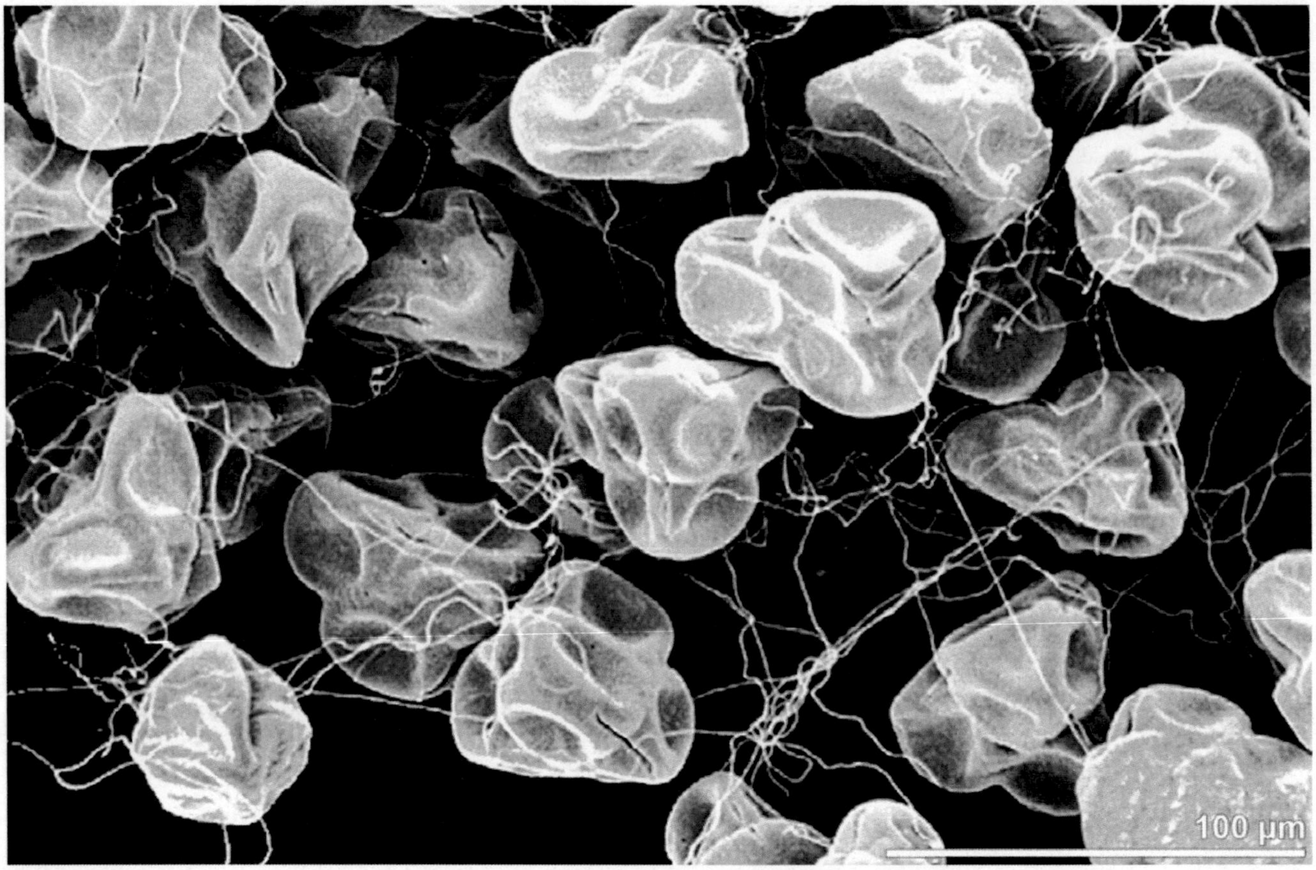

Fig. 1.7 *Rhododendron luteum* (Ericaceae), many pollen tetrads in dry stage, connected with viscin threads (PalDat, Photo by H. Halbritter)

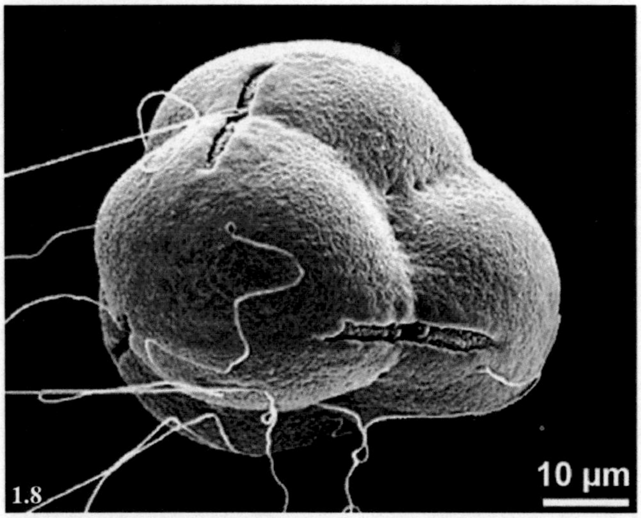

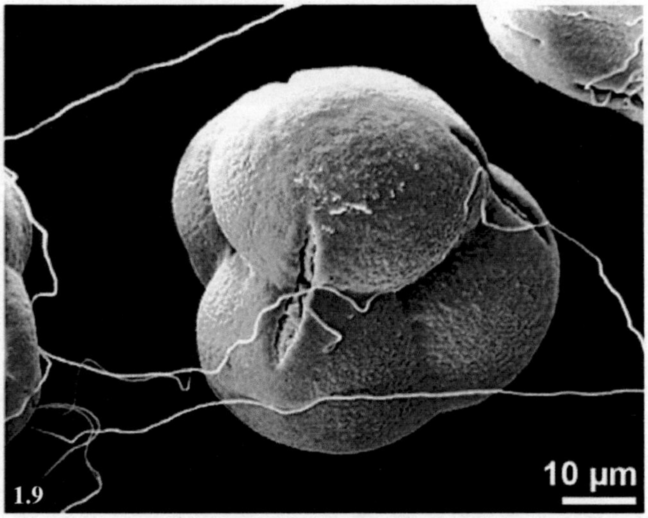

Fig. 1.8 *Rhododendron luteum* (Ericaceae), single pollen tetrad with many viscin threads, note viscin threads occurring from one, i.e., the uppermost pollen grain at its distal side (PalDat, Photo by H. Halbritter)

Fig. 1.9 *Rhododendron luteum* (Ericaceae), three pollen tetrads connected by some viscin threads, note that some threads emerge from the apertures (PalDat, Photo by H. Halbritter)

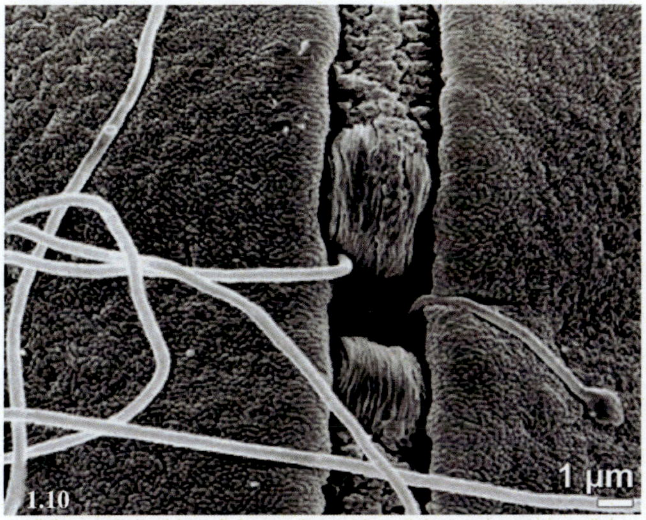

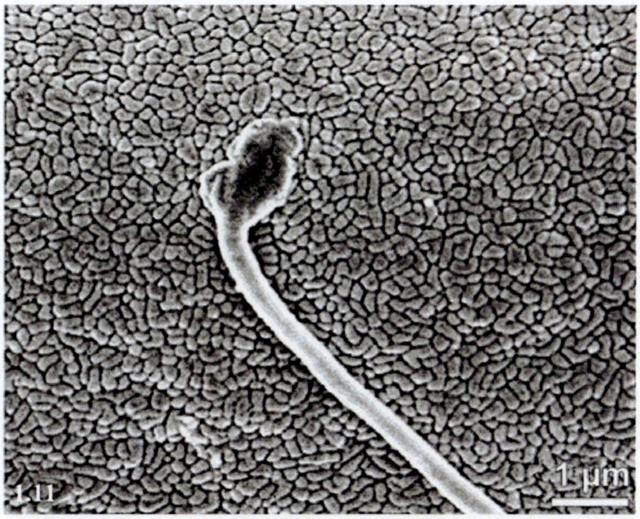

Fig. 1.10 *Rhododendron luteum* (Ericaceae), detail of an aperture, viscin threads emerge from the aperture (PalDat, Photo by H. Halbritter)

Fig. 1.11 *Rhododendron luteum* (Ericaceae), close-up of pollen ornamentation, detail of viscin thread (PalDat, Photo by H. Halbritter)

tongued insects (Willemstein, 1987) or birds (*Fuchsia*). The heavy and robust flower visiting animals may require greater viscin thread strength, thus the thick, often complex and generally sculptured viscin threads of Onagraceae should have more tensile strength ("stiffness") than their thin, smooth counterparts in the Ericaceae-Rhododendroideae. It should be noted that all viscin threads are highly flexible, but not elastic like a rubber band.

The viscin threads did not break apart even under the most vigorous vibration in a shaker, which points toward an astonishing flexibility and simultaneous stability of the seemingly fragile viscin threads.

1.1.6 Pollen-connecting Threads not Consisting of Sporopollenin

The nature and mode of origin of sporopollenin-lacking threads, which are known to arise in various ways, are much better known (Halbritter et al., 1997).

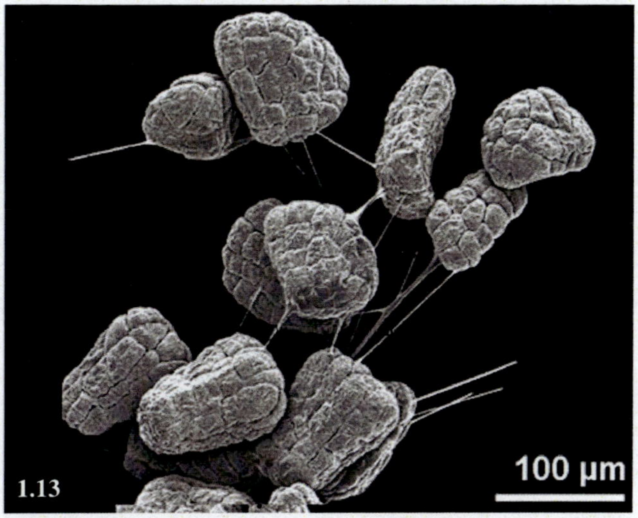

Fig. 1.12 *Restrepia dodsonii* (Orchidaceae), pollinia connected by elastoviscin threads (PalDat, Photo by M. Svojtka)

Fig. 1.13 *Nigritella rhellicani* (Orchidaceae), dry massulae connected by elastoviscin threads (PalDat, Photo by M. Svojtka)

Five modes of origin have been recognized to date:

(A) Ordinary pollenkitt may sometimes assume a rope-like habit (Halbritter and Hesse, pers. obs.). The pollenkitt viscosity probably depends on its specific chemical composition: pollenkitt composed predominantly of unsaturated lipids with a higher number of bonds is of higher viscosity than pollenkitt mainly composed of saturated lipids.

(B) In some unrelated angiosperm taxa, as in *Porcelia* (Annonaceae), in *Aristolochia*, and in all mentioned Caesalpiniaceae, a slimy mixture of lipids and cytoplasmic components (both most probably deriving from tapetal cells only) may form thread-like structures.

(C) In other angiosperm taxa, such as in *Raphionacme* (Asclepiadaceae) and in many Orchidaceae, the "elastoviscin" is composed of lipid products of the tapetal cells' ground plasm and endoplasmic reticulum (ER) only (Wolter et al., 1988; Dannenbaum and Schill, 1991; Dressler, 1993). Elastoviscin is regarded as in part homologous to pollenkitt: It occurs between pollen tetrads or pollinia and their appendicular structures as a clear, highly viscous, and elastic substance (Schill and Wolter, 1986). Either it holds the orchid pollinia together (Dressler, 1993), or it may act as its own viscidium (Burns-Balogh and Hesse, 1988). Elastoviscin assumes a thread-like appearance only when pollen grains are physically separated.

(D) In other orchids cellular and/or sporopollenin parts of the caudicle and pollen-forming tissues may be involved in the formation of thread-forming elastoviscin (*Disa*, Vogel, 1959; *Habenaria*, Hesse and Burns-Balogh, 1984) (Figs. 1.12 and 1.13).

(E) A quite different manner of thread formation occurs in *Echium*, *Esterhazya*, *Gymnocalycium*, *Heliconia*, *Impatiens*, and *Strelitzia*. Their long and robust threads derive from cells along or near the stomial dehiscence line, and, contrary to (A–D), they are formed before pollen is physically separated. These threads are either of cellular origin, as in *Strelitzia*, or only parts of distinct cells, as in *Impatiens* and in *Heliconia*. In *Gymnocalycium*, and likewise in *Echium* and *Esterhazya*, they are made up of modified stomium and/or septal cell walls together with distinct derivates of tapetal cells. However, these threads have significantly different functions. In *Heliconia* and *Strelitzia* they act – being sticky – as a pollen-clumping agent. In *Echium*, *Esterhazya*, or *Gymnocalycium* the threads entangle pollen grains, where pollen baskets are formed.

Acknowledgments

The scanning electron micrographs of thread-forming pollen-connecting agents are copyright by PalDat (http://www.paldat.org/, see Figure Legends). The author is grateful to Mrs. Andrea Frosch-Radivo for excellent technical assistance.

References

Armbruster WS (1984) The role of resin in angiosperm pollination: ecological and chemical considerations. American Journal of Botany 71: 1149–1160.

Barnes SH and Blackmore S (1992) Ultrastructural organization of two tapetal types in angiosperms. Archives of Histology and Cytology 55(Suppl): 217–224.

Boavida (2005) The making of gametes in higher plants. International Journal of Developmental Biology 49: 595–614.

Burns-Balogh P and Hesse M (1988) Pollen morphology of the cypripedioid orchids. Plant Systematic 158: 165–182.

Buzato S and Franco ALM (1992) *Tetrastylis ovalis*: a second case of bat-pollinated passionflower (Passifloraceae). Plant Systematics and Evolution 181: 261–267.

Cox PA and Knox RB (1989) Two-dimensional pollination in hydrophilous plants: convergent evolution in the genera *Halodule* (Cymodoceaceae), *Halophila* (Hydrocharitaceae), *Ruppia* (Ruppiaceae), and *Lepilaena* (Zannichelliaceae). American Journal of Botany 76: 164–175.

Cox PA and Tomlinson PB (1988) Pollination ecology of a seagrass, *Thalassia testudinum* (Hydrocharitaceae), in St. Croix. American Journal of Botany 75: 958–965.

Crepat WL (1996) Timing in the evolution of derived floral characters: Upper Cretaceous (Turonian) taxa with tricolpate and tricolpate-derived pollen. Review of Palaeobotany and Palynology 90: 339–359.

Croat TB (1999) Pollination Biology. A Revision of *Philodendron* subgen. *Philodendron* (Araceae) of Central America. http://www.aroid.org/genera/philodendron/Contents.php and http://www.aroid.org/genera/philodendron/pollibiol.php.

Dannenbaum C and Schill R (1991) Die Entwicklung der Pollentetraden und Pollinien bei den Asclepiadaceae. Bibliotheca Botanica 141: 1–138.

Dickinson HG and Lewis FRS (1973) The formation of the tryphine coating the pollen grains of *Raphanus*, and its properties relating to the self-incompatibility system. Proceedings of the Royal Society London B: Biological Sciences 184: 149–165.

Dickinson HG, Elleman CJ, and Doughty J (2000) Pollen coatings – chimaeric genetics and new functions. Sexual Plant Reproduction 12: 302–309.

Dobson HEM (1988) Survey of pollen and pollenkitt lipids – chemical cues to flower visitors? American Journal of Botany 75: 170–182.

Dressler RL (1993) Phylogeny and classification of the orchid family. Cambridge University Press, Cambridge.

Franchi GG and Pacini E (1996) Types of pollination and seed dispersal in mediterranean plants. Giornale botanico italiano 130: 579–585.

Furness CA (2008) A review of the distribution of plasmodial and invasive tapeta in eudicots. International Journal of Plant Sciences 169: 207–223.

Halbritter H, Hesse M, and Buchner R (1997) Pollen-connecting threads in *Gymnocalycium* (Cactaceae): their origin, function, and systematic relevance, with a review on pollen-clumping modes. Grana 36: 1–10.

Hesse M (1979) Entwicklungsgeschichte und Ultrastruktur von Pollenkitt und Exine bei nahe verwandten entomophilen und anemophilen Sippen der Salicaceae, Tiliaceae und Ericaceae. Flora 168: 540–557.

Hesse M (1986) Nature, form and function of pollen-connecting threads in angiosperms. Academic Press, London.

Hesse M (1993) Pollenkitt development and composition in *Tilia platyphyllos* (Tiliaceae) analysed by conventional and energy filtering TEM. Plant Systematics and Evolution Suppl 7: 39–52.

Hesse M (2009) Pollen of *Anubias*, *Culcasia*, *Lagenandra* and *Piptospatha* (Aroideae, Araceae): Functional and systematic relevance. Aroideana 32: 147–158.

Hesse M and Burns-Balogh P (1984) Pollen and pollinarium morphology of *Habenaria* (Orchidaceae). Pollen Spores 26: 385–400.

Hesse M, Halbritter H, Zetter R, Weber M, Buchner R, Frosch-Radivo A, and Ulrich S (2009) Pollen Terminology. An Illustrated Handbook. Springer-Verlag, Wien.

Keri C and Zetter R (1992) Notes on the exine ultrastructure of Onagraceae and Rhododendroideae (Ericaceae). Grana 31: 119–123.

Knoll F (1930) Über Pollenkitt und Bestäubungsart. Zeitschrift für Botanik 23: 610–675.

Knox RB and McConchie CA (1986) Structure and function of compound pollen. In: Blackmore S and Ferguson IK (eds) Pollen and Spores. Form and Function. Academic Press, London: pp 265–282.

Kronestedt-Robards E (1996) Formation of the pollen-aggregating threads in *Strelitzia reginae*. Annals of Botany 77: 243–250.

Morawetz W and Waha M (1991) Zur Entstehung und Funktion pollenverbindender Fäden bei *Porcelia* (Annonaceae). Beiträge zur Biologie der Pflanzen 66: 145–154.

Murphy DJ (2006) The extracellular pollen coat in members of the Brassicaceae: composition, biosynthesis, and functions in pollination. Protoplasma 228: 31–39.

Nixon KE and Crepet WL (1993) Late Cretaceous fossil flowers of ericalean affinity. American Journal of Botany 80: 616–623.

Pacini E (1997) Tapetum character states: analytical keys for tapetum types and activities. Canadian Journal of Botany 75: 1448–1459.

Pacini E and Hesse M (2005) Pollenkitt: its composition, forms and function. Flora 200: 399–415.

Patel V, Skvarla JJ, Ferguson IK, Graham A, and Raven PH (1985) The nature of threadlike structures and other morphological characters in *Jacqueshuberia* pollen (Leguminosae: Caesalpinioideae). American Journal of Botany 72: 407–413.

Pettitt JM (1981) Reproduction in seagrasses: pollen development in *Thalassia hemprichii*, *Halophila stipulacea* and *Thalassodendron ciliatum*. Annals of Botany 48: 609–622.

Piffanelli P, Ross JHE, and Murphy DJ (1998) Biogenesis and function of the lipidic structures of pollen grains. Sexual Plant Reproduction 11: 65–80.

Pinheiro MC, Teixeira Ormond W, Alves de Lima H, and Rodrigues Correia MC (1995) Biologia da reproducao de *Norantea brasiliensis* Choisy (Marcgraviaceae). Revista Brasileira de Biologia 55(Suppl 1): 79–88.

Platt KA, Huang AH, and Thompson WW (1998) Ultrastructural study of lipid accumulation in tapetal cells of *Brassica napus* L. cv. Westar during microsporogenesis. International Journal of Plant Sciences 159: 724–737.

Ressayre A, Dreyer L, Triki-Teurtroy S, Forchioni A, and Nadot S (2005) Post-meiotic cytokinesis and pollen aperture pattern ontogeny: comparison of development in four species differing in aperture pattern. American Journal of Botany 92: 576–583.

Reznickova SA and Dickinson HG (1982) Ultrastructural aspects of storage lipid mobilization in the tapetum of *Lilium hybrida* var. enchantment. Planta 155: 400–408.

Richter S (1929) Über den Öffnungsmechanismus der Antheren bei einigen Vertretern der Angiospermen. Planta 8: 154–184.

Rose MJ and Barthlott W (1995) Pollen-connecting threads in *Heliconia* (Heliconiaceae). Plant Systematics and Evolution 195: 61–65.

Sazima I, Buzato S, and Sazima M (1993) The bizarre inflorescence of *Norantea brasiliensis* (Marcgraviaceae): visits of hovering and perching birds. Botanica Acta 106: 507–513.

Schill R and Wolter M (1986) On the presence of elastoviscin in all subfamilies of the Orchidaceae and the homology to pollenkitt. Nordic Journal of Botany 6: 321–324.

Shukla AK (1984) A clarification on the use of the term viscin thread in Orchidaceae. Grana 23: 127.

Skvarla JJ, Raven PH, and Praglowski J (1975) The evolution of pollen tetrads in Onagraceae. American Journal of Botany 62: 6–35.

Skvarla JJ, Raven PH, Chissoe WF, and Sharp M (1978) An ultrastructural study of viscin threads in Onagraceae. Pollen Spores 20: 5–144.

Troll W (1928) Über *Spathicarpa sagittifolia* Schott. Flora 123: 286–316.

Ubera Jimenez J, Hidalgo Fernandez P, Schlag MG, and Hesse M (1996) Pollen and tapetum development in male fertile *Rosmarinus officinalis* L. (Lamiaceae). Grana 34: 305–316.

Vijayaraghavan MR and Shukla AK (1980) Viscin threads in *Zeuxine strateumatica* (Orchidaceae). Grana 19: 173–175.

Vogel S (1959) Organographie der Blüten kapländischer Ophrydeen mit Bemerkungen zum Koaptations-Problem. Teil I: *Disinae* und *Satyrinae*. Akademie der Wissenschaften und Literatur, Abhandlungen der mathematisch-naturwissenschaftlichen Klasse, Jahrgang 1959, Nr. 6, Verlag der Akademie der Wissenschaften und der Literatur in Mainz, in Kommission bei Franz Steiner. Verlag GmbH, Wiesbaden.

Waha M (1984) Zur Ultrastruktur und Funktion pollenverbindender Fäden bei Ericaceae und anderen Angiospermenfamilien. Plant Systematics and Evolution 147: 189–203.

Wang Y, Zhang D, Renner SS, and Chen Z (2004) A new selfpollination mechanism. Nature 431: 30–40.

Weber M (1992) The formation of pollenkitt in *Apium nodiflorum* (Apiaceae). Annals of Botany 70: 573–577.

Willemstein SC (1987) An evolutionary basis for pollination ecology. Leiden University Press, Leiden.

Wolter M, Seuffert C, and Schill R (1988) The ontogeny of pollinia and elastoviscin in the anther of *Doritis pulcherrima* (Orchidaceae). Nordic Journal of Botany 8: 77–88.

Zetter R and Hesse M (1996) The morphology of pollen tetrads and viscin threads in some Tertiary, *Rhododendron*-like Ericaceae. Grana 35: 285–294.

Zetter R, Weber M, Hesse M, and Pingen M (2002) Pollen, pollenkitt and orbicules in *Craigia bronnii* flower buds (Tilioideae, Malvaceae) from the Miocene of Hambach, Germany. International Journal of Plant Sciences 163: 1067–1071.

2 Deadly Glue – Adhesive Traps of Carnivorous Plants

Wolfram Adlassnig, Thomas Lendl, Marianne Peroutka and Ingeborg Lang

Contents

Abstract

Carnivorous plants trap and utilize animals in order to improve their supply with mineral nutrients. One strategy for prey capture is the use of adhesive traps, i.e., leaves that produce sticky substances. Sticky shoots are widespread in the plant kingdom and serve to protect the plant, especially flowers and seeds. In some taxa, mechanisms have been developed to absorb nutrients from the decaying carcasses of animals killed by the glue. In carnivorous plants *sensu stricto*, additional digestive enzymes are secreted into the glue to accelerate degradation of prey organisms.

The glues are secreted by glands that are remarkably uniform throughout all taxa producing adhesive traps. They follow the general scheme of plant glandular organs: the glands consist of a stalk, a neck equipped with a suberin layer that separates the gland from the rest of the plant, and the glandular cells producing sticky secretions. This glue always forms droplets at the tip of the glandular hairs. In most genera, these glands produce only glue whereas enzymes for prey digestion are secreted by a second type of gland. Two types of glue can be distinguished, polysaccharide mucilage in Droseraceae, Lentibulariaceae and their relatives, and terpenoid resins in Roridulaceae. On the ultrastructural level, mucilage is produced by the Golgi apparatus. Resins can be expected to be produced by the endoplasmic reticulum and by leucoplasts.

Adhesive traps are suitable not only for the capture of small animals but also for the collection of organic particles like pollen grains. The glue may contain toxic compounds but the prey usually dies from suffocation by clogging of its tracheae. In *Pinguicula* and *Drosera*, the performance of the traps is improved by a slow movement, i.e., the folding of the leaf around the prey animal upon stimulation. In some species of *Nepenthes*, a pitcher with smooth walls is filled with a sticky digestive fluid.

Some organisms, however, have developed strategies to survive on the deadly traps. Several species of Hemiptera are able to walk on the sticky traps and nourish on the prey; their faeces are absorbed by the plant. In *Roridula*, this relationship is highly specialized and essential for both the plant and the insect. Mutualistic fungi and bacteria are common in many adhesive traps where they degrade and dissolve the plant's prey. The traps of *Drosera*, on the other hand, are virtually sterile.

In spite of the extensive literature on adhesive traps, numerous questions still remain. Only a small percentage of "sticky" plants have actually been tested for carnivory. The properties and composition of their glues are widely unknown. In advanced adhesive traps, the mechanisms regulating secretion and absorption are poorly understood. Thereby, some glues may be applicable for human as they are non-toxic, quite stable under environmental conditions, and partly exhibit mildly antibiotic properties. Some carnivorous plants with adhesive traps have been used by humans for the capture of insects as well as for food processing.

2.1 Introduction

2.1.1 Carnivorous Plants

Carnivorous plants trap and absorb animals to supplement their mineral nutrition. *Per definitionem*, the complete process of prey utilization – the carnivorous syndrome (Juniper et al., 1989) – comprises four steps: **(1) attraction** of animals by means of optical signals, scents or nectar, **(2) retention** by specialized leaves or, rarely, axes – the traps, **(3) degradation** of the prey by digestive enzymes, and **(4) uptake** of soluble compounds. Plants lacking one of these features are called protocarnivorous and are regarded as ancestors of carnivorous plants *sensu stricto*. Most protocarnivorous plants lack enzyme production; prey digestion is performed by mutualistic animals, bacteria or fungi that may form complex communities inhabiting the otherwise deadly traps (e.g., Kitching, 2000; Fauland et al., 2001).

The benefit of carnivory is an improved supply with minerals, especially phosphorus and nitrogen (Ellison, 2006). Recent research gave evidence for the absorption of trace elements (Adlassnig et al., 2009) and of organic carbon (Sirova et al., 2010). The utilization of animals enables carnivorous plants to colonize nutrient-poor habitats like peat bogs or tropical table mountains. Detailed models for the ecology of carnivorous plants were published by Givnish (1989) and Ellison (2006). So far, about 600 species of vascular plants (Barthlott et al., 2004) and a few mosses (Barthlott et al., 2000) are known to use this strategy. Besides higher plants (*sensu stricto*), some unicellular algae (Raven, 1997) and fungi (Dowe, 1987) also trap animals.

Five different strategies have been developed for prey retention: **(1) adhesive traps** produce sticky secretions; the prey is glued to the trap surface (Fig. 2.1); **(2) pitcher traps** consist of cone shaped leaves with smooth walls; animals fallen into the pitcher are not able to climb out and therefore drown in a pool of digestive fluid; **(3) snap traps** consist of two moveable lobes; the prey is enclosed upon touch; **(4) hollow suction traps** produce a low hydrostatic pressure; after mechanical stimulation, a small volume of water is sucked in together with the prey; and **(5) tubular eel traps** with inward pointing hairs allow for easy movement towards a digestive chamber at the end of the trap but obstruct the way out in the opposite direction. Some species of carnivorous plants combine two of these strategies (Rice, 2007). For a more detailed survey on the trapping mechanisms of carnivorous plants, compare Barthlott et al. (2004) and Peroutka et al. (2008).

2.1.2 Evolution and Diversity of Adhesive Traps

Many non-carnivorous plants produce sticky secretions to defend themselves against animals; glue production is often concentrated at the reproductive organs. Animals trapped by the glue die after some time and are degraded by bacteria and fungi. At least in some species, the epidermis of the plant is permeable and absorption of inorganic nutrients from the carcasses is possible. In such case, the plant would gain a small benefit from the trapped animals. Under nutrient poor conditions, an evolutionary selection for improved prey capture and nutrient uptake would take place. Though this model of adhesive trap evolution was already published by Darwin (1875), little research has been done on plants that are on the turn towards protocarnivory. In general, little specific adaptations for the capture of animals can be detected in those species. Most of them inhabit eutrophic or mesotrophic soils, so it is not clear if prey-derived nutrients provide a significant benefit. Plants at the base of the evolution of carnivorous plants are distributed all over the subclass Rosopsida but are

Table 2.1: Diversity of insect trapping plants with adhesive traps. Dubious cases are marked with a "?"

Carnivory	Species	Family	Remarks	References
Trapping insects, but not carnivorous or dubious (selected taxa)	*Plumbago scandens*	Plumbaginaceae	Sticky inflorescence. Probably closely related to the common ancestor of Nepenthales	Beal (1876), Rachmilevitz and Joel (1976), Schlauer (1997)
	Silene spp.	Caryophyllaceae		Beal (1876), Chase et al. (2009)
	Lychnis spp.	Caryophyllaceae		Beal (1876), Chase et al. (2009)
	Salvia glutinosa	Lamiaceae	Sticky inflorescence. Capture of large insects, but no uptake	Pohl (2009)
	Rhododendron sp.	Ericaceae	Sticky buds and young shoots. Related to *Roridula*	Beal (1876)
	Erica tetralix	Ericaceae		Darwin (1875)
	Aesculus hippocastanum	Hippocastanceae	Sticky flower buds	Darwin (1875)
	Cleome droserifolia	Cleomaceae		Plachno et al. (2009)
	Rubus vitis-idaea	Rosaceae		Juniper et al. (1989)
	Hyoscyamus desertorum	Solanaceae		Plachno et al. (2009)
	Solanum tuberosum	Solanaceae	Isolated accounts even for enzyme production	Beal (1876), Chase et al. (2009)
	Physalis spp.	Solanaceae		Beal (1876)
	Nicotiana tabacum	Solanaceae		Juniper et al. (1989)
	Proboscidea parviflora	Martyniaceae		Plachno et al. (2009)
	Martynia sp.	Martyniaceae		Beal (1876)
	Genlisea (21 spp.)	Lentibulariaceae	Sticky inflorescences combined with subsoil eel traps	Lendl (2007)
Protocarnivorous	*Roridula* (2 spp.)	Roridulaceae		Midgley and Stock (1998)
	Potentilla arguta	Rosaceae		Spomer (1999)
	Rubus phoeniculasius (?)	Rosaceae		Krbez et al. (2001), Pohl (2009)
	Geranium viscosissimum	Geraniaceae		Spomer (1999)
	Saxifraga (≥3 spp.)	Saxifragaceae	Sticky inflorescence	Darwin (1875)
	Stylidium spp.	Stylidaceae	Sticky inflorescence	Darnowski (2002), Darnowski (2003), Darnowski et al. (2006)
	Ibicella lutea (?)			Plachno et al. (2009)
Carnivorous *sensu stricto*	*Triphyophyllum peltatum*			Green et al. (1979)
	Drosophyllum lusitanicum	Drosophyllaceae		Schnepf (1963a)
	Drosera (200 spp.)	Droseraceae		
	Nepenthes (≥3 of ≥80 spp.)	Nepenthaceae	Pitcher traps in some species combined with adhesive fluid	Devecka (2007), Rice (2007)
	Byblis (7 spp.)	Byblidaceae	Contradictory data on enzyme production	Hartmeyer (1998), Wallace and McGhee (1999)
	Pinguicula (85)	Lentibulariaceae		Heslop-Harrison (2004)

lacking in Magnoliopsida and Liliopsida. They include widespread genera like *Rubus* (Rosaceae) (Krbez et al., 2001), *Saxifraga* (Saxifragaceae) (Darwin, 1875), *Geranium* (Geraniaceae) or *Potentilla* (Rosaceae) (Spomer, 1999). Chase et al. (2009) review the diversity of protocarnivorous plants.

Highly elaborated adhesive traps – protocarnivorous as well as carnivorous *sensu stricto* – are restricted to three orders of angiosperms (Albert et al., 1992). Table 2.1 gives an overview on adhesive traps and presents references concerning specific taxa.

- **Ericales:** the whole order exhibits a strong tendency towards oligotrophic habitats and alternative nutrition, e.g., mycotrophy or parasitism. In some species of *Rhododendron* (Ericaceae), large insects have been found trapped on sticky shoots. One genus, *Roridula* (Roridulaceae) has developed highly sticky leaves trapping numerous insects but has no own enzyme production. More than 70% of the plants' nitrogen comes from prey (Midgley and Stock, 1998). *Roridula's* sister taxon, Sarraceniaceae, forms pitcher traps (Albert et al., 1992).
- **Nepenthales** (also known as Droserales) are thought to be derived from *Plumbago*-like ancestors that were already equipped with sticky glands (Schlauer, 1997). Two genera, *Triphyophyllum* (Dioncophyllaceae) and *Drosophyllum* (Drosophyllaceae), are equipped with large and highly complex glands. In the more advanced *Drosera* (Droseraceae), the glands are morphologically simplified but acquired the power of movement (see Chapter 2.3.1, p. 22). In *Dionea* and *Aldrovanda* (Droseraceae), the trap leaves have been developed further into snap traps without glue production (Williams, 1976). In *Nepenthes* (Nepenthaceae), pitcher traps have evolved. In some species, the fluid that fills the pitcher is highly viscoelastic and sticky and therefore contributes to prey retention.
- **Lamiales:** Glands with toxic or sticky protective secretions are widespread in this order (Müller et al., 2004). In *Byblis* (Byblideceae), highly specialized adhesive traps are found but digestion is dubious. The same is true for Martyniaceae. The entire family Lentibulariaceae is carnivorous *sensu stricto*. The most primitive genus, *Pinguicula*, has adhesive traps that perform limited movements. In the further advanced genera *Genlisea* and *Utricularia*, adhesive traps were transformed to eel and suction traps, respectively. In *Genlisea*, sticky hairs are still present in the inflorescence.

2.2 Glues and Their Production

2.2.1 Morphology and Anatomy of Glue-Producing Glands

Glands play a key role in all elements of the carnivorous syndrome: they produce **(1) volatiles and nectar** for attraction, **(2) mucilage or pitcher fluid** for prey retention, and **(3) digestive enzymes** for prey degradation. Unlike animal glands, glands of carnivorous plants do not only secrete metabolites but also **(4) absorb substances** from the surroundings. This is due to a specific feature of plant shoots, i.e., their coverage by a continuous hydrophobic cuticle. Secretion of glue or other compounds is only possible *via* pores in the cuticle which also provide access for external substances (Jeffree, 2006). Furthermore, glue producing glands in plants are not submerged into the tissue but localized on the tips of hairs and therefore elevated above the surface of the leaf or shoot. For a general discussion of glands in carnivorous plants, compare Fenner (1904), Juniper et al. (1989), and Adlassnig et al. (2005).

Though carnivorous plants with adhesive traps are polyphyletic, all glands that produce glues are similar and based on the same scheme. *Byblis liniflora* (Byblidaceae) is a typical example (Fig. 2.2): The leaves of *B. liniflora* are covered with small hairs that measure about 1 mm. *B. liniflora* has two kinds of glandular hairs: sessile and stalked ones. The sessile glands consist of eight to ten cells arranged in a circle. This glandular "head" is responsible for secretion and uptake. Only in these cells, the cuticle contains pores. The center of the head is formed by a neck cell, which is in direct and close contact with base cells submerged into the epidermis. The base cells provide a connection between the gland and the vascular system of the leaf. The cell wall of the neck cell is equipped with an endodermic suberin incrustation. The neck cell hence regulates the transport of substances from the leaf towards the gland cells and *vice versa*. The heads of the stalked glands resemble closely the sessile glands and are formed by the same number of cells, also arranged in a circle with a neck cell in its center. In addition, stalked glands are supported by one long, single cell (the stalk) connecting the gland with the base cells in the epidermis (which is also in contact with base cells). Different types of glue are produced by the stalked and sessile glands: the secretion of the stalked glands is more viscous and probably stickier whereas the fluid of the sessile glands is more liquid and serves as a solvent for digestive enzymes. The glues are secreted at the tip of the

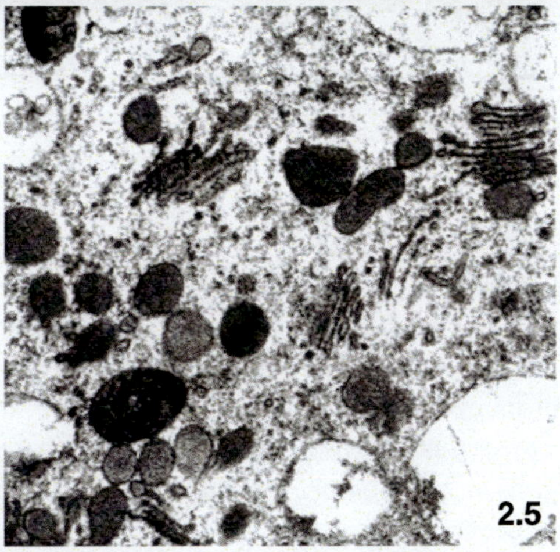

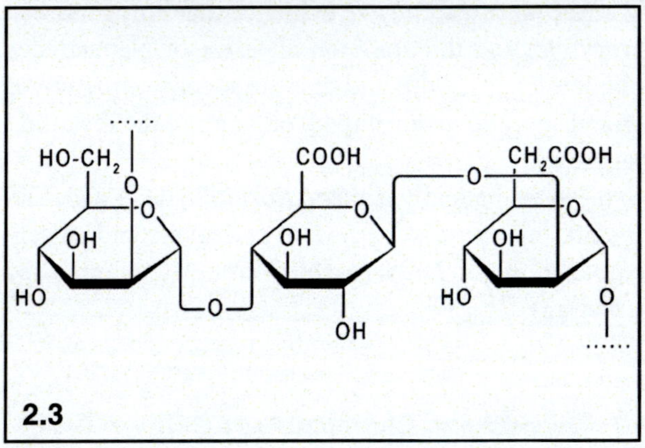

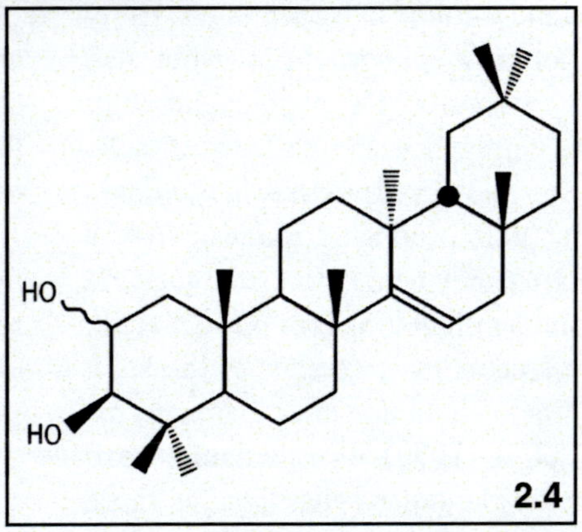

Fig. 2.1 Adhesive trap of *Drosera capensis f. minor*. The leaf is covered by stalked glands, each bearing a droplet of glue

Fig. 2.2 Glands of *Byblis liniflora*. The stalked gland produces sticky mucilage. The sessile gland produces less viscous mucilage and digestive enzymes and absorbs nutrients (UV micrograph)

Fig. 2.3 Chemical structure of the backbone of the polysaccharide mucilage of *Drosera capensis*. Adapted from Gowda et al. (1983) and reproduced with permission

Fig. 2.4 Chemical structure of the backbone of the resin 2,3β-Taraxeradiol from *Roridula gorgonias*. Adapted from Simoneit et al. (2008) and reproduced with permission

Fig. 2.5 TEM micrographs of dictyosomes in the glandular cells of *Drosera capensis* during glue production, huge Golgi vesicles filled with mucilage can be distinguished

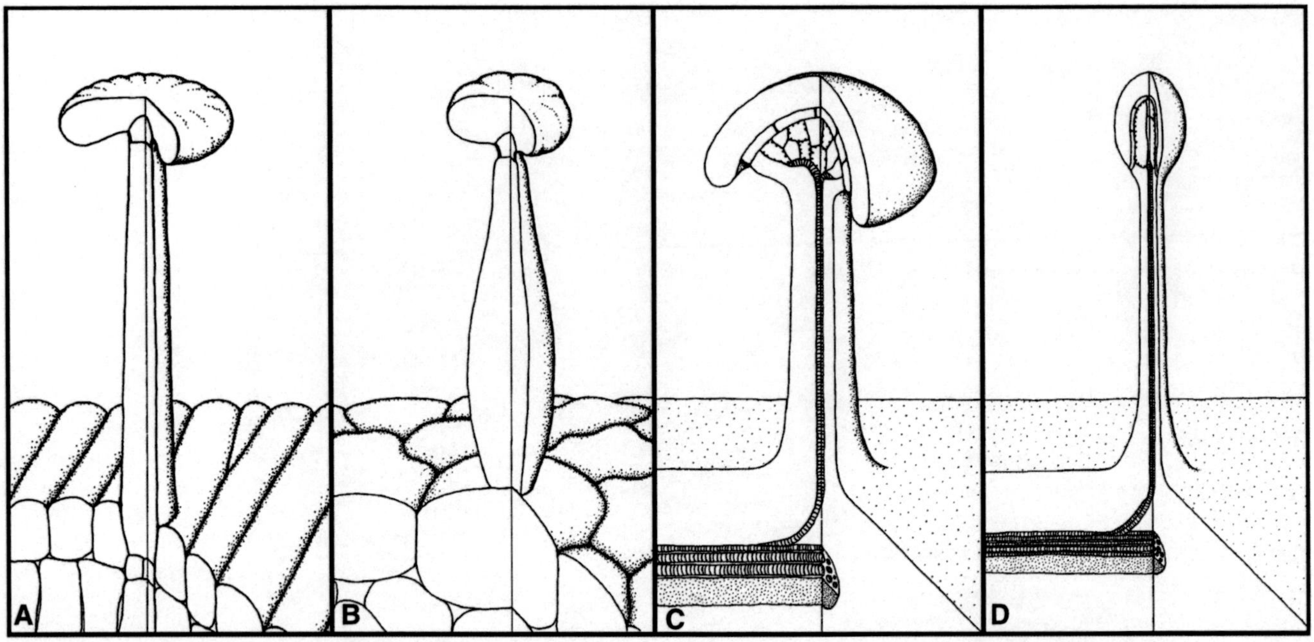

Fig. 2.6 Schemes of the glue producing glands of (A) *Byblis*, (B) *Pinguicula*, (C) *Drosophyllum*, and (D) *Drosera*

hairs thus glittering in the sun to attract insects (Fenner, 1904; Heslop-Harrison, 1976; Pauluzzi, 1995).

Figure 2.6 shows the remarkable similarity of glue-producing glands from unrelated taxa. In *Pinguicula* or *Ibicella*, the anatomy of the glue-producing glands is virtually the same as in *Byblis*, only the number of cells forming the head or the stalk differs (Juniper et al., 1989).

A specific type of gland is found in Nepenthales. Here, the stalk consists of not only epidermal tissue but also vascular and parenchymal cells. The whole glandular structure is therefore an emergence, also called a tentacle. *Triphyophyllum peltatum* and *Drosophyllum lusitanicum* have extremely similar glue-producing glands. With a size of several millimetres, they belong to the largest and most complex glands in the plant kingdom (Green et al., 1979). The vascular elements in the center of the stalk consist of numerous tracheids and a thick parenchyma (Fenner, 1904; Green et al., 1979). The glandular head exhibits several layers of glandular cells. Besides these tentacles, multicellular sessile glands are responsible for the production of digestive enzymes. Absorption is performed by both types of glands simultaneously (unpublished observation of the authors). In *Drosera*, the glands are secondarily simplified. The emergence has only one central tracheid, one layer of parenchyma and two layers of glandular cells. All functions are fulfilled by the stalked emergences, i.e., the production of glue and digestive enzymes as well as the absorption of nutrients. The sessile glands are highly reduced and without apparent function. This development is completed in the most derived *Dionaea* and *Aldrovanda*, where the emergence is completely reduced and only the glandular head is left (Lloyd, 1942). Similar glands producing mucilage are found in various Polygonaceae (Schnepf, 1968). Recent research gave evidence that Polygonaceae are very close to the common ancestor of Nepenthales (Meimberg et al., 2000) and that the specific structure of the glands may have developed long before the invention of carnivory.

In the systematically isolated *Roridula*, the glands and their stalks also consist of several cell layers and closely resemble those of *Drosera* (M. Peroutka, unpublished observation).

2.2.2 Physical and Chemical Properties of Glues

The glue secreted by the gland usually forms a clearly distinct droplet at the tip of each glandular hair. It never covers the epidermis; therefore, respiration is not disturbed. Though macroscopic visualization of the glue is easy, high magnifications are difficult to realize, e.g., to study interactions between the glue and solid bodies. A useful technique is Cryo Scanning Electron Microscopy used by Gorb et al. (2007) and Peroutka et al. (2008). Gorb et al. (2007) measured contact angles between the glue and solid bodies to estimate the adhesive strength of the glue.

Two types of glue are produced by carnivorous plants, i.e., polysaccharide mucilage and lipophilic resins. Formulas of two characteristic compounds are given in Figs. 2.3 and 2.4. Glues based on polysaccharides are found in *Triphyophyllum*, *Drosera*, *Drosophyllum*, and *Pinguicula* (Vintejoux and Shoar-Ghafari, 2000). Though no detailed analyses are available, the sticky substances produced by *Nepenthes*, *Byblis*, *Ibicella*, and their relatives are probably similar. Resins are found in *Roridula* (Simoneit et al., 2008).

Several studies deal with the chemical composition of the glue of *Drosera*: the mucilage is a viscoelastic, completely homogenous, about 4% aqueous solution of a single polysaccharide with a molecular weight over $2 \cdot 10^6$ Da (Rost and Schauer, 1977). The shear viscosity is about 10^2 Pa·s (Erni et al., 2008). Stickiness is lost after denaturation by acidification, alkalinization, freezing or heating (Rost and Schauer, 1977). After hydrolysis, the sugars L-arabinose, D-xylose, D-galactose, D-mannose, and D-glucuronic acid are found in molar ratio of 3.6:1.0:4.9:8.4:8.2 (Gowda et al., 1983). The backbone of the polysaccharide consists of a repeating dimer of glucuronic acid and mannose (Fig. 2.3); the other sugars are present in end groups and side chains (Gowda et al., 1982, 1983). The polysaccharide exhibits a structural similarity to adhesives of bacteria, fungi, and algae but is more homogenous (Haag, 2006). Sulfate ester bonds have also been described from the glues of diatoms (Chiovitti et al., 2006).

Within the genus *Drosera*, the differences concerning the composition of the mucilage are probably negligible. The glues of *D. capensis* and *D. binnata* differ only in the proportions of sugar residues in the side chains (Aspinall and Puvanesarajah, 1984). Besides polysaccharides, the mucilage of *Drosera* contains inorganic cations (22 mM Ca^{2+}, 19 mM Mg^{2+}, 0.9 mM K^+, and 0.2 mM Na^+) and up to 1.2% sulfur, present as an ester sulfate. Surprisingly, no proteins or any other nitrogenous compounds are present in the mucilage before stimulation by prey (Rost and Schauer, 1977); digestive enzymes are secreted only afterwards.

In *Drosophyllum lusitanicum*, the mucilage shows a similar composition. The monomers are arabinose, galactose, xylose, rhamnose, glucuronic acid, and ascorbic acid. The mucilage therefore shows an acid reaction (Schnepf, 1963a) and has a strong odor of honey (Meyer and Dewèvre, 1894). Its production depends on functioning respiration (Schnepf, 1963b) and can be increased by feeding (Schnepf, 1963a), and it can be secreted in such quantities that it runs down the leaf surface in droplets (Darwin, 1875). Furthermore, a yellow fluorescent substance is observed in the mucilage of *Drosophyllum* (Schnepf, 1963a). Concerning its physical properties, the glue is not drawn out into slender viscous threads as in *Drosera* but it is easily pulled off the gland as a whole (Darwin, 1875). In this xeromorphic species, the mucilage is highly hygroscopic and seems to collect water from fog as additional water supply for the plant (Adamec, 2009). The water of the mucilage does not evaporate even at relative humidities lower than 40% (Adlassnig et al., 2006). In the hydrophilic *Drosera*, on the other hand, the glands become dry at humidities lower than 70% (Volkova and Shipunov, 2009).

Little is known about the composition of the sticky, viscoelastic pitcher fluid in some species of *Nepenthes*. Besides its surface tension, it exhibits a relatively low shear viscosity of $1.5 \pm 4 \cdot 10^{-2}$ Pa·s combined with an extremely high extensional viscosity about 10^4 times larger than the shear viscosity. Dilution by water, which is common in a humid climate, does not affect the viscoelasticity of the fluid. These properties indicate the presence of linear polymeric molecules. Although no chemical analysis is available, the close relation between *Nepenthes*, *Drosophyllum*, and *Drosera* suggests the presence of a polysaccharide (Gaume and Forterre, 2007).

In *Roridula*, the glue consists of a mixture of resins (Voigt and Gorb, 2008). A detailed chemical analysis was published by Simoneit et al. (2008): In both *Roridula* species, triterpenoids with the formula $C_{30}H_{50}O_2$ and a molecular weight of 442 Da count for most of the glue. In *R. dentata*, two compounds (dihydroxyolean-12-ene and dihydroxyurs-12-ene) were identified; in *R. gorgonias*, additionally taraxeradiol (olean-18-en-2,3-diol) was found. Also small amounts of other triterpenols were detected in both species. All major compounds of the glue would be crystalline solids after purification but in the mixture, crystallization is prevented resulting in a highly viscous and sticky fluid. Additionally, small amounts of flavones are found in *R. gorgonias* whereas the glue of *R. dentata* contains flavonols (Wollenweber, 2007).

2.2.3 Cytological Aspects of Glue Production

In all traps using polysaccharide mucilage, the glue is produced by the Golgi apparatus of the gland cells

(Vintejoux and Shoar-Ghafari, 2000). A complex gland anatomy, however, is no prerequisite for the production of mucilage, as pointed out by Juniper et al. (1977); e.g., polysaccharide mucilage is also produced in root caps by morphologically undifferentiated cells. As in carnivorous plants, mucilage production is performed by a hypertrophic Golgi apparatus (Mollenhauer et al., 1961).

The most detailed studies concerning glue production were carried out in *Drosera* and *Drosophyllum* using electron microscopy:

In *Drosera* glandular cells, Schnepf (1961) described large vesicles deriving from the Golgi apparatus. They contain polysaccharides and upon secretion, they form the glistening droplets of mucilage. These vesicles stain with different intensity; therefore the mucilage may be processed stepwise within the Golgi apparatus. The mucilage exits the glandular cells through pores in the cuticula and appears here and further on the surface as reticulate material (Williams and Pickard, 1974). For interpretation of the structural and staining characteristics of the mucilage, it must be considered, however, that the fixation, dehydration, staining, and embedding procedures required for electron microscopy may cause severe structural changes to the mucilage. The secretion of mucilage is facilitated by numerous ingrowths of the outer walls of the glandular cells which increase the contact area between the plasma membrane and the cell wall; similar structures are widely distributed in adhesive traps (Pate and Gunning, 1972).

The development of glandular cells from juvenile to mature state was described by Outenreath and Dauwalder (1982, 1986) using ultrastructural studies and radioautography. In juvenile outer glands mainly from the apex, Golgi stacks are not extensive in number or size and consist of three to eight cisternae per Golgi stack. With the onset of secretion, large numbers of vesicles are derived from the Golgi apparatus (Fig. 2.5). They are clearly associated with the Golgi cisternae and contain fine fibrillar material similar in appearance to the outer excreted mucilage. This material is secreted through the plasma membrane. In mature glands, large vesicles are no longer in contact with the Golgi cisternae and only a few small vesicles are attached to their borders. Often, the stacks are larger with 9–20 cisternae per stack which can be either curled or in a stair-stepped form, especially in the outer glandular cells. Whereas the mucilage appears fine and loosely fibrillar in the vesicles, it gains a more consistent texture outside the convoluted plasma membrane. Different densities of the mucilage were described depending on their origin from either inner or outer glandular cells (Outenreath and Dauwalder, 1982, 1986). The production of glue and other secretions obviously require much energy. In *Drosera prolifera*, the dark respiration rate of the tentacles exceeds that of the photosynthetic leaf lamina more than seven times (Adamec, 2010).

In *Drosophyllum*, similar to *Drosera*, Golgi bodies are responsible for the production of mucilage (Schnepf, 1963c). When they grow, their content becomes partly darkly, partly lightly stained in electron micrographs and forms fine flakes. Their size and number corresponds to the amount of secreted mucilage. The temperature optimum for mucilage production is at 32°C (Schnepf, 1961). In darkness and without watering, secretion is reduced and stopped.

Muravnik (1988) describes the formation of mucilage in *Pinguicula vulgaris*: Glue production starts at leaf maturation and proceeds continuously until senescence. The mucilage is produced by the Golgi apparatus, which is significantly enlarged. After synthesis, the glue is stored inside the cell in vacuoles and between the plasma membrane and the cell wall, before it is released to the gland surface *via* incontinuities of the cuticle. This intracellular storage of mucilage is highly uncommon in plants but was also described for the related but non-carnivorous *Mimulus tilingii* (Schnepf and Busch, 1976).

No cell biological studies have been carried out so far to localize the production of sticky resins in *Roridula*. It might be similar to the situation in the sticky but non-carnivorous *Salvia* and other Lamiaceae where terpenoids are produced as well (Kampranis et al., 2007). The cytoplasm of glandular cells in these plants is characterized by abundant, smooth endoplasmic reticulum, and leucoplasts (Kolalite, 1994).

2.3 Interactions of Adhesive Traps and Animals

2.3.1 Prey Capture

The process of prey utilization starts with the attraction of animals towards the traps. Specific mechanisms include optical signals like ultraviolet patterns in *Drosera binata*, *D. capensis*, *Drosophyllum lusitanicum*, *Pinguicula gypsicola*, *P. ionantha,* or *P. zecheri* (Joel et al., 1985; Gloßner, 1992). Furthermore, the glistening droplets of glue seem to have a strong attractive effect on insects (Voigt and Gorb, 2008). The production of nectar or volatiles is rare in adhesive traps, with the exception of *Dro-*

sophyllum producing a honey-like scent (Schnepf, 1963a). In *Pinguicula*, the glue effuses a delicate fungus-like odor (Lloyd, 1942) that might add to the attractiveness of the traps for potential prey, especially fungus gnats.

Plants using polysaccharide mucilage are restricted to the capture of very small insects. In *Drosera rotundifolia*, the majority of the prey consists of small Diptera, Lepidoptera, Nematocera, Collembola, Acarina, Aphidoidea, and Cocoidea (Darwin, 1875; Thum, 1986). In other species of *Drosera*, the situation is similar: The species composition of the prey of *D. intermedia* differs significantly from that of *D. rotundifolia* but small forms like Collembola or Diptera form the vast majority. In spite of the comparatively large trap of *D. intermedia*, bigger insects like Odonata or Saltatoria are trapped only exceptionally (Thum, 1986). In *D. anglica*, more than 90% of the prey consists of tiny Ceratopogonidae and Chironomidae (Murza et al., 2006; Hagan et al., 2008). In *D. filiformis*, the upper limit for prey retention is a body size of about 10 mm (Gibson, 1999). In *Drosophyllum lusitanicum*, small gnats and lacewings were found (Adlassnig et al., 2006). In *Pinguicula longifolia*, the prey consists almost exclusively of Diptera with a size of 1–4 mm (Antor and Garcia, 1994). In some habitats, 97% of the prey of *P. vulgaris* is Diptera of the genus *Cnephia* (Adler and Malmquist, 2004). In *P. lutea*, the maximum prey size is about 5 mm (Gibson, 1999).

Besides the capture of animals, the traps may be suitable for collecting small parts of other plants. Harder and Zemlin (1968) found that feeding with pollen has almost the same effect as feeding with animals in *Pinguicula*. *P. alpina* may collect and digest dead leaves on its traps (Darwin, 1875; Klein, 1887). Juniper et al. (1989) suggest that rainforest species of *Drosera* may use their traps to utilize nutrient-rich canopy leaching.

Because traps using polysaccharides have only a limited capacity to retain animals, several species combine glue with other mechanisms. *Drosera* and *Pinguicula* use so-called active adhesive traps. Only in the first phase of the trapping process, the prey is exclusively retained by the glue. After a few seconds to hours, the leaf starts to fold and rolls around the animal (Barthlott et al., 2004). In *Drosera*, the glandular tentacles are moveable and bend towards the prey (Darwin, 1875). The mechanism of movement was clarified in *Drosera*: Mechanical and chemical stimulation of the prey stimulate action potentials in the glandular cells which are transmitted towards the leaf epidermis similar to animal neurons (Williams and Spanswick, 1972). This electric signal initiates the exudation of the growth hormone auxin which causes the local elongation of cells and leads to the bending of leaves and tentacles (Bopp, 1985).

In some species of the pitcher plant *Nepenthes*, the fluid is highly sticky and viscous. Animals are either retained by touching the surface of the fluid, or they are not able to leave it after falling into the pitcher (Rice, 2007). Di Giusto et al. (2008) showed in *Nepenthes rafflesiana* that a high viscosity of the fluid enables the trap to retain a greater diversity of animals. Devecka (2007) studied a variety of *Nepenthes* species and found that in *N. talangensis* × *ventricosa*, 70% of the trapped ants are retained exclusively by the fluid, in *N. inermis* 50%, but only 10% in *N. gracilis*. In all species of *Nepenthes*, however, the pitcher fluid is combined with the retentive effect of the steep and smooth pitcher wall.

Traps using sticky resins instead of mucilage are more effective to affix larger animals. The sticky but little specialized and probably non-carnivorous *Salvia glutinosa* successfully retains big insects like bees or earwigs (Pohl, 2009). *Roridula* is considered to have the most effective adhesive traps. Large insects like wasps can be retained (unpublished observation of the authors), though the mean prey length is only 3.6 mm (Ellis and Midgley, 1996). The effectiveness of the trap is enhanced by a special arrangement of the glandular hairs (Voigt et al., 2009): long, medium sized and small trichomes can be distinguished. The longest hairs are most flexible but exert the smallest adhesive force. The first contact between leaf and insect is usually provided by the long hairs. If the prey animal tries to get free of these hairs, it comes into contact with the short ones which serve for the final retention.

Many carnivorous plants form rosettes of adhesive traps with the inflorescence in the center. The stalk of the inflorescence may also be equipped with sticky glands. According to Kerner von Marilaun (1876), the ultimate function of the traps was the protection of flowers and fruits. Though this hypothesis was not correct, the protection of the inflorescence *via* glue is an important feature in many non-carnivorous plants, e.g., *Salvia glutinosa* or *Aesculus hippocastanum* (Table 2.1). In protocarnivorous and carnivorous plants, the function of trapping is usually taken over by the leaves. Still, sticky hairs can be found in the inflorescence of genera like *Byblis*, *Stylidium*, or *Pinguicula*. Hanslin and Karlsson (1996) found that the absorptive capacities of glands in the inflorescence are more limited than in the leaf in *Pinguicula*. The eel trap *Genlisea* descended from adhesive traps similar to *Pinguicula* (Barthlott et al., 2004). The traps of *Genlisea* produce no glue but in some species hairs within the inflorescence produce sticky resins (Lendl, 2007).

Fig. 2.7 The Hemipteran *Dicyphus pallidus* on the highly sticky inflorescence of *Salvia glutinosa*

2.3.2 Life on Adhesive Traps

Trap leaves, glands and glues are produced to retain and kill the prey. The abundance of dead and decaying animals turns the traps into attractive habitats for different organisms. Carnivorous and protocarnivorous plants employ two different strategies to face this challenge: some species use antibiotic compounds within the glue to repel or kill invasive organisms. Others form mutualistic associations with trap inhabitants and use their digestive capacities in the process of prey degradation.

In the extremely sticky but non-carnivorous *Salvia glutinosa*, various species of Hemiptera inhabit the traps, e.g., *Dicyphus pallidus*, *Macrotylus quadrilineatus,* or *Eysarcoris venstissiumus* (Fig. 2.7). The animals step only on the epidermis of the shoots and avoid direct contact with the sticky glands (Pohl, 2009). The association between *Salvia* and the Hemipterans is very loose; the insects do not depend on the plant and no benefit for the plant can be recognized. In the protocarnivorous *Byblis*, a more specialized relationship can be observed (Hartmeyer, 1998): the Hemiptera *Setocornis bybliphilus* seems to occur exclusively on this plant and nourishment on the prey is highly probable. Because *Byblis* does not produce effective digestive enzymes, the faeces of the Hemipteran may be a more accessible source of carbon than the intact carcasses of the prey.

In *Roridula*, virtually all plants are inhabited by the Hemipteran *Pameridea roridulae* (Ellis and Midgley, 1996) which is found exclusively on this plant (Dolling and Palmer, 1991). Voigt and Gorb (2008) studied the protection used by *Pameridea* in detail: The animal is not retained by the resinous glue, even if force is used to bring it in contact with the glands. To relieve a leg from the glue, *Pameridea* needs an energy demand of 0.07 ± 0.026 J whereas a leg of a fly is retained by a force of 0.18 ± 0.093 J. This protection is due to a thick layer of grease covering the epidermis of *Pameridea*. A documentary film by Carrow et al. (1997) shows how *Pameridea* is able to move on *Roridula* and how it sucks on the plant's prey, including big flies. The faeces of the Hemipteran are absorbed *via* pores by *Roridula's* epidermis, not through the glands as in all other plants with adhesive traps (Anderson, 2005). Since *Roridula* does not produce digestive enzymes, the presence of *Pameridea* is essential for prey utilization. However, the mutualistic relationship between *Roridula* and its symbionts is maintained only if sufficient prey is available. Otherwise, the *Pameridea* nourishes on the sap of the plant and turns towards parasitism (Anderson and Midgley, 2007). A second inhabitant of the trap disturbs this mutualistic relationship as well. The spider *Synaema marlothi* overcomes the sticky glands by forming a cobweb all over the plant and nourishes on *P. roridulae*, thus significantly reducing the benefit for the plant (Anderson and Midgley, 2002).

Little is known about mutualistic relationships between protocarnivorous plants and bacteria although aqueous mucilage can be expected to provide a suitable habitat for microorganisms. *Potentilla arguta*, *Rubus phoeniculasius* or *Geranium viscosissimum* trap and kill animals and have been shown to incorporate nitrogen *via* the leaf but lack digestive enzymes (Spomer, 1999; Krbez et al., 2001). It is suspected that prey degradation is performed by bacteria and fungi inhabiting the mucilage (Fauland et al., 2001). *Drosera* employs a completely different strategy: though the traps may be covered by dead animals, the trapping mucilage is virtually sterile. Bacteria inoculated to the mucilage die within a few hours (Pranjic, 2004). There is also some evidence for antimicrobial activity in the mucilage of *Pinguicula* (Chase et al., 2009).

No animals are known to use the traps of *Drosera* as a permanent habitat. This effect is probably due to naphtochinons like droseron that are secreted together with the mucilage. These secondary metabolites are directed against both microbes and insects, and therefore serve as a universal protection for the plant (Didry et al., 1998; Tokunaga et al., 2004). Only few insects are known to overcome both the sticky mucilage and the chemical defence. In *D. rotundifolia*, up to 70% of the prey is removed and consumed by ants (*Formica picea*, *Leptothorax acer-*

vorum and *Myrmica scabrinodis*; Thum, 1989); on *Drosera capillaris*, the caterpillar *Trichoptilus parvulus* shows the same behavior (Eisner and Shepard, 1965). A similar type of kleptoparasitism is found in *Pinguicula vallisnerifolia* where the slug *Deroceras hilbrandi* removes up to 90% of the prey (Zamora and Gomez, 1996).

2.4 Future Aspects and Practical Applications

Carnivorous plants are fascinating organisms that are cultivated and propagated by many enthusiasts and professional breeders. In scientific research, they are valuable objects in areas as diverse as taxonomy (Meimberg et al., 2006), plant physiology (Peroutka et al., 2008), ecology (Srivastava et al., 2004) or cell biology (Adlassnig et al., 2005). Still, there are numerous open questions, especially concerning the diversity and function of adhesive traps.

- Most species with highly specialized and eye-catching traps have been described. However, many plants with sticky leaves but without morphologic adaptations may have escaped our attention (Chase et al., 2009). Studies scanning a great variety of sticky plants for carnivorous features, as carried out by Pohl (2009) or Plachno et al. (2009), are rare.
- The glands of carnivorous plants are unique in the plant kingdom since their activity is regulated by external stimuli (Jones and Robinson, 1989). Little information is available on the nature of these stimuli, their perception and the transduction of the signals.
- Some glues of adhesive traps have been characterized. However, this is not true for the viscoelastic fluid in some species of *Nepenthes*. Though multifunctionality is common for biological adhesives (Smith and Callow, 2006b), *Nepenthes* exhibits some specific features: the fluid does not only serve as a glue but contains also digestive enzymes, reactive oxygen species, detergents, acids, narcotics, etc. (reviewed by Adlassnig, 2007); with the exception of the enzymes, none of these compounds has been studied in detail up to now.
- Hemipterans seem to be pre-adapted to colonize sticky plants; they resist both sticky mucilage and resins. The underlying mechanism was clarified in one case of extreme specialization (Voigt and Gorb, 2008) but little is known about its evolutionary development and its ethologic implications.

Practical applications of biological glues are a hot topic of recent research (Smith and Callow, 2006a). The mucilage of *Drosera* or *Drosophyllum* might exhibit interesting features for pharmaceutic use: it is non-toxic, remains stable under varying environmental conditions and even exhibits antibiotic properties (Pranjic, 2004). Until today, species with adhesive traps are virtually the only carnivorous plants that found practical applications: *Drosophyllum lusitanicum* was used to keep houses free of insects in Portugal (Darwin, 1875). More recently, *Pinguicula* and *Drosera* were recommended to reduce fungus gnats (Mycetophilidae) in terrariums. The extraordinary large *Drosera dichotoma* var. *giant* is even suitable for the use in greenhouses (D'Amato, 1998). Furthermore, antimicrobial metabolites of various Droseraceae – though not the glue itself – are widely used in pharmacy (Krolicka et al., 2008). The mucilage of *Drosera* is used for food processing in Northern Europe: milk proteins are precipitated and partly digested by the addition of *Drosera* leaves to create a drink known as "Ropy milk" (=*Tættemælk* in Sweden, *Tettemelk* or *Tjukkmjølk* in Norway, *Viili* in Finland; Thomas and McQuillin, 1953; Chase et al., 2009). The use of *Pinguicula* results in a different consistency (Furuset, 2008).

Acknowledgments

Thanks are due to Prof. Dr. I. K. Lichtscheidl (University of Vienna) and Prof. emer. Dr. P. Hepler (University of Massachusetts) for providing TEM images, to Prof. Dr. J. Derksen (Radboud University Nijmegen) and to M. Edlinger (Bundesgärten Schönbrunn). This study was supported by grant H-02319/2007 of the Hochschuljubiläumsstiftung der Stadt Wien for M. Peroutka and W. Adlassnig.

References

Adamec L (2009) Ecophysiological investigation on *Drosophyllum lusitanicum*: Why doesn't the plant dry out? Carnivorous Plant Newsletter 38: 71–74.

Adamec L (2010) Dark respiration of leaves and traps of terrestrial carnivorous plants: are there greater energetic costs in traps? Central European Journal of Biology 5: 121–1214.

Adlassnig W (2007) Ökophysiologie karnivorer Kesselfallenpflanzen. PhD thesis, Universität Wien: 541 pp.

Adlassnig W, Peroutka M, Eder G, Pois W, and Lichtscheidl IK (2006) Ecophysiological observations on *Drosophyllum lusitanicum*. Ecological Research 21: 255–262.

Adlassnig W, Peroutka M, Lang I, and Lichtscheidl IK (2005) Glands of carnivorous plants as a model system in cell biological research. Acta Botanica Gallica 152: 111–124.

Adlassnig W, Steinhauser G, Peroutka M, Sterba JH, Lichtscheidl IK, and Bichler M (2009) Expanding the menu for carnivorous plants: uptake of potassium, iron and manganese by carnivorous pitcher plants. Applied Radiation and Isotopes 67: 2117–2122.

Adler PH and Malmquist B (2004) Predation on black flies (Diptera: Simuliidae) by the carnivorous plant *Pinguicula vulgaris* (Lentibulariaceae) in northern Sweden. Entomologica Fennica 15: 124–128.

Albert VA, Williams SE, and Chase MW (1992) Carnivorous plants: phylogeny and structural evolution. Science 257: 1491–1495.

Anderson B (2005) Adaptions to foliar absorption of faeces: a pathway in plant carnivory. Annals of Botany 95: 757–761.

Anderson B and Midgley JJ (2002) It takes two to tango but three is a tangle: mutualists and cheaters on the carnivorous plant *Roridula*. Oecologia 132: 369–372.

Anderson B and Midgley JJ (2007) Density-dependent outcomes in a digestive mutualism between carnivorous *Roridula* plants and their associated hemipterans. Oecologia 152: 115–120.

Antor RJ and Garcia MB (1994) Prey capture by a carnivorous plant with hanging adhesive traps: *Pinguicula longifolia*. American Midland Naturalist 131: 128–135.

Aspinall GO and Puvanesarajah V (1984) Selective cleavage of b-D-Glucopyranosiduronic acid linkages in methylated polysaccharide acids from *Drosera* species. Carbohydrate Research 131: 53–60.

Barthlott W, Fischer E, Frahm JP, and Seine R (2000) First experimental evidence for zoophagy in the Hepatic *Colura*. Plant Biology 2: 93–97.

Barthlott W, Porembski S, Seine R, and Theisen I (2004) Karnivoren. Biologie und Kultur fleischfressender Pflanzen. Eugen Ulmer, Stuttgart

Beal WJ (1876) Carnivorous plants. The American Naturalist 10: 588–591.

Bopp M (1985) Leaf blade movement of *Drosera* and auxin distribution. Naturwissenschaften 72: 434.

Carrow T, Hirschel K, Winkelmann H, and Radke R (1997) Fleischfressende Pflanzen: tödliche Fallen. Audiovisual Material ZDF, Germany: length 45 min.

Chase MW, Christenhusz MJM, Sanders D, and Fay MF (2009) Murderous plants: Victorian gothic, Darwin and modern insights into vegetable carnivory. Botanical Journal of the Linnaean Society 161: 329–356.

Chiovitti A, Dugdale TM, and Wetherbee R (2006) Diatom adhesives: Molecular and mechanical properties. In: Smith AM and Callow JA (eds) Biological Adhesives. Springer, Berlin: pp 79–103.

D'Amato P (1998) The savage garden. Cultivating Carnivorius Plants, 1st Ed. Ten Speed Press, Berkely.

Darnowski DW (2002) Triggerplants, 1st Ed. Rosenberg Publishing, Kenthurst.

Darnowski DW (2003) Triggerplants (*Stylidium*; Stylidiaceae): a new floral and horticultural crop with preliminary analysis of hardiness. Acta Horticulturae 624: 93–101.

Darnowski DW, Carroll DM, Plachno B, Kabanoff E, and Cinnamon E (2006) Evidence of protocarnivory in triggerplants (*Stylidium* spp.; Stylidiaceae). Plant Biology 8: 1–8.

Darwin C (1875) Insectivorous Plants, 1st Ed. John Murray, London.

Devecka A (2007) Diversität der *Nepenthes*-Falle unter besonderer Berücksichtigung der Oberflächenstrukturen und ihrer Rolle beim Beutefang. Diploma thesis, Technical University Zvolen: 123 pp.

Di Giusto B, Grosbois VV, Fargeas E, Marshall DJ, and Gaume L (2008) Contribution of pitcher fragance and fluid viscosity to high prey diversity in a *Nepenthes* carnivorous plant from Bornea. Journal of Biosciences 33: 121–136.

Didry N, Dubreuil L, Trotin F, and Pinkas M (1998) Antimicrobial activity of aerial parts of *Drosera peltata* Smith on oral bacteria. Journal of Ethnopharmacology 60: 91–96.

Dolling WR and Palmer JM (1991) *Pameridea* (Hemiptera, Miridae) – predaceous bugs specific to the highly viscid plant genus *Roridula*. Systematic Entomology 16: 319–328.

Dowe A (1987) Räuberische Pilze, 2nd Ed. A. Ziemsen, Wittenberg.

Eisner T and Shepard J (1965) Caterpillar feeding on a sundew plant. Science 150: 1608–1609.

Ellis AG and Midgley JJ (1996) A new plant-animal mutualism involving a plant with sticky leaves and a resident hemipteran insect. Oecologia 106: 478–481.

Ellison AM (2006) Nutrient limitation and stoichiometry of carnivorous plants. Plant Biology 8: 740–747.

Erni P, Varagnat M, and McKinley GH (2008) Little shop of horrors: rheology of the mucilage of *Drosera* sp., a carnivorous plant. In: Co A, Leal LG, Colby RH, and Giacomin AJ (eds) The XVth International Conference on Rheology, Monterey: pp 579–581.

Fauland K, Krbez P, and Heinrich G (2001) Indirekte Karnivorie von *Rubus phoenicolasius* Maxim. durch symbiontische Pilze im Sekret der Drüsenhaare? 14. Tagung des Österreichischen Arbeitskreises für Pflanzenphysiologie, Forstliche Bundesversuchsanstalt, Waldforschungszentrum, Neuberg an der Mürz: 112 pp.

Fenner CA (1904) Beiträge zur Kenntnis der Anatomie, Entwicklungsgeschichte und Biologie der Laubblätter und Drüsen einiger Insektivoren. Flora 93: 335–434.

Furuset K (2008) Tettegrasets rolle i tettemelk. Blyttia 66: 55–62.

Gaume L and Forterre Y (2007) A viscoelastic deadly fluid in carnivorous pitcher plants. PLoS ONE 2007: 1–7.

Gibson TC (1999) Differential escape from insects from carnivorous plant traps. American Midland Naturalist 125: 55–62.

Givnish TJ (1989) Ecology and evolution of carnivorous plants. Plant-animal interactions. McGraw-Hill, New York: pp 243–290.

Gloßner F (1992) Ultraviolet patterns in the traps and flowers of some carnivorous plants. Botanische Jahrbücher für Systematik 113: 577–587.

Gorb SN, Voigt D, and Gorb EV (2007) Visualisation of small fluid droplets on biological and artificial surfaces using the Cryo-SEM approach. In: Mendez-Vilas A and Diasz J (eds) Modern Research and Educational Topics in Microscopy, 2nd Ed. Formatex, Badajoz: pp 812–819.

Gowda DC, Reuter G, and Schauer R (1982) Structural features of an acidic polysaccharide from the mucin of *Drosera binata*. Phytochemistry 21: 2297–2300.

Gowda DC, Reuter G, and Schauer R (1983) Structural studies of an acidic polysaccharide from the mucin secreted by *Drosera capensis*. Carbohydrate Research 113: 113–124.

Green S, Green TL, and Heslop-Harrison Y (1979) Seasonal heterophylly and leaf gland features in *Triphyophyllum* (Dioncophyllaceae) a new carnivorous plant genus. Botanical Journal of the Linnaean Society 78: 99–116.

Haag AP (2006) Mechanical properties of bacterial exopolymeric adhesives and their commercial development. In: Smith AM and Callow JA (eds) Biological Adhesives. Springer, Berlin: pp 1–19.

Hagan DV, Grogan W, Murza GL, and Davis AR (2008) Biting midges (Diptera: Ceratopogonidae) from the English sundew, *Drosera anglica* Hudson (Droseraceae), at two fens in Saskatchewan, Canada. Proceedings of the Entomological Society of Washington, Vol. 110: pp 397–401.

Hanslin HM and Karlsson PS (1996) Nitrogen uptake from prey and substrate as affected by prey capture level and plant reproductive status in four carnivorous plant species. Oecologia 106: 370–375.

Harder R and Zemlin I (1968) Blütenbildung von *Pinguicula lusitanica* in vitro durch Fütterung mit Pollen. Planta 78: 72–78.

Hartmeyer S (1998) Carnivory in *Byblis* revisited II: the phenomenon of symbiosis on insect trapping plants. Carnivorous Plant Newsletter 27: 110–113.

Heslop-Harrison Y (1976) Enzyme secretion and digest uptake in carnivorous plants. In: Sunderland N (ed) Perspectives in Experimental Biology. S.E.B. Symposium, Vol. 2. Proceedings of the 50th Anniversary Meeting, Pergamon Press, Oxford, Cambridge: pp 463–476.

Heslop-Harrison Y (2004) Biological flora of the British Isles: *Pinguicula* L. Journal of Ecology 92: 1071–1118.

Jeffree CE (2006) The fine structure of the plant cuticle. In: Riederer M and Müller C (eds) Biology of the plant cuticle. Blackwell Publishing, Oxford: pp 11–125.

Joel DM, Juniper BE, and Dafni A (1985) Ultraviolet patterns in the traps of carnivorous plants. New Phytologist 101: 585–593.

Jones RL and Robinson DG (1989) Protein secretion in plant. New Phytologist 111: 567–597.

Juniper BE, Gilchrist AJ, and Robins RJ (1977) Some features of secretory systems in plants. Histochemical Journal 9: 659–680.

Juniper BE, Robins RJ, and Joel DM (1989) The Carnivorous Plants, 1st Ed. Academic Press Limited, London.

Kampranis SC, Ioannidis D, Purvis A, Mahrez W, Ninga E, Katerelos NA, Anssour S, Dunwell JM, Degenhardt J, Makris AM, Goodenough PW, and Johnson CB (2007) Rational conversion of substrate and product specificity in a *Salvia* monoterpene synthase: structural insights into the evolution of terpene synthase function. Plant Cell 19: 1994–2005.

Kerner von Marilaun A (1876) Die Schutzmittel der Blüthen gegen unberufene Gäste. Verhandlungen der k.u.k. zoologisch-botanischen Gesellschaft (Festschrift) 26: 190–261.

Kitching RL (2000) Food webs and container habitats. The natural history and ecology of phytotelmata. Cambridge University Press, Cambridge.

Klein I (1887) *Pinguicula alpina* als insektenfressende Pflanze und in anatomischer Beziehung. Beiträge zur Biologie der Pflanzen 3: 163–186.

Kolalite MR (1994) Dynamics of ultrastructure in peltate glands in *Nepeta cataria* and *Dracocephalum moldavica* (Lamiaceae) in connection with terpene biosynthesis. Botanicheskii Zhurnal 79: 17–27.

Krbez P, Fauland K, and Heinrich G (2001) Untersuchungen zur möglichen Karnivorie von *Rubus phoenicolasius*. 14. Tagung des Österreichischen Arbeitskreises für Pflanzenphysiologie, Forstliche Bundesversuchsanstalt, Waldforschungszentrum, Neuberg an der Mürz: 112 pp.

Krolicka A, Szpitter A, Gilgenast E, Romanik G, Kaminski M, and Lojkowska E (2008) Stimulation of antibacterial naphthochinones and flavonoid accumulation in carnivorous plants grown in vitro by addition of elicitors. Enzyme and Microbial Technology 42: 216–221.

Lendl T (2007) Aspekte der Karnivorie der Gattung *Genlisea*. Diploma Thesis, University of Vienna: 146 pp.

Lloyd FE (1942) The Carnivorous Plants, 9th Ed. Ronald Press, New York.

Meimberg H, Dittrich P, Bringmann G, Schlauer J, and Heubl G (2000) Molecular phylogeny of Caryophyllidae *s.l.* based on *MatK* sequences with special emphasis on carnivorous taxa. Plant Biology 2: 218–228.

Meimberg H, Thalhammer S, Brachmann A, and Heubl G (2006) Comparative analysis of a translocated copy of the *trnK* intron in carnivorous family Nepenthaceae. Molecular Phylogenetics and Evolution 39: 478–490.

Meyer A and Dewèvre A (1894) Über *Drosphyllum lusitanicum*. Botanisches Zentralblatt 60: 33–41.

Midgley JJ and Stock WD (1998) Natural abundance of $\delta^{15}N$ confirms insectivorous habit of *Roridula gorgonias*, despite it having no proteolytic enzymes. Annals of Botany 82: 387–388.

Mollenhauer HH, Whaley WG, and Leech JH (1961) A function of the Golgi apparatus in outer rootcap cells. Journal of Ultrastructure Research 5: 193–200.

Müller K, Borsch T, Legendre L, Porembski S, Theisen I, and Barthlott W (2004) Evolution of carnivory in Lentibulariaceae and Lamiales. Plant Biology 6: 477–490.

Muravnik LE (1988) The slime gland ultrastructure in *Pinguicula vulgaris*, Lentibulariaceae, in the course of their development and activity. Botanicheskii Zhurnal 73: 523–1535.

Murza GL, Heaver JR, and Davis AR (2006) Minor pollinator-prey conflict in the carnivorous plant, *Drosera anglica*. Plant Ecology 184: 43–52.

Outenreath R and Dauwalder M (1982) Ultrastructural and radioautographic studies of the digestive gland cells of *Drosera capensis*: I. Development and mucilage secretion. Journal of Ultrastructure Research 80: 71–88.

Outenreath R and Dauwalder M (1986) Ultrastructural and radioautographic studies of the digestive gland cells of *Drosera capensis*: II. Changes induced by stimulation. Journal of Ultrastructure Research 95: 164–174.

Pate JS and Gunning BES (1972) Transfer cells. Annual Review of Plant Physiology 23: 173–196.

Pauluzzi I (1995) Ultraviolett-Mikroskopie von Pflanzenzellen. Diploma thesis, University of Vienna: 121 pp.

Peroutka M, Adlassnig W, Lendl T, Pranic K, and Lichtscheidl IK (2008) Functional biology of carnivorous plants. In: Teixeira da Silva JA (ed) Floriculture, Ornamental and Plant Biotechnology. Advances and Topical Issues, 5th Ed. Global Science Books, Isleworth: pp 266–287.

Plachno BJ, Adamec L, and Huet H (2009) Mineral nutrient uptake from prey and glandular phosphatase activity as a dual test of carnivory in semi-desert plants with glandular leaves suspected of carnivory. Annals of Botany 104: 649–654.

Pohl SA (2009) Untersuchungen zur möglichen Protokarnivorie von *Lathraea squamaria*, *Salvia glutinosa* und *Rubus phoenicolasius*. Diploma thesis, University of Vienna: 101 pp.

Pranjic K (2004) Zur Ökologie karnivorer Pflanzen: Die Rolle von Mikroorganismen beim Abbau von Tieren durch fleischfressende Pflanzen. Diploma thesis, University of Vienna: 145 pp.

Rachmilevitz T and Joel DM (1976) Ultrastructure of the calyx glands of *Plumbago capensis* Thumb. in relation to the process of secretion. Israel Journal of Botany 25: 127–139.

Raven JA (1997) Phagotrophy in phototrophs. Limnology and Oceanography 42: 198–205.
Rice B (2007) Carnivorous plants with hybrid trapping strategies. Carnivorous Plan 36: 23–27.
Rost K and Schauer R (1977) Physical and chemical properties of the mucin sectreted by *Drosera capensis*. Phytochemistry 16: 365–368.
Schlauer J (1997) "New" data relating to the evolution and phylogeny of some carnivorous plant families. Carnivorous Plant Newsletter 26: 34–38.
Schnepf E (1961) Quantitative Zusammenhänge zwischen der Sekretion des Fangschleimes und den Golgi-Strukturen bei *Drosophyllum lusitanicum*. Zeitschrift für Naturwissenschaften 16: 605–610.
Schnepf E (1963a) Zur Cytologie und Physiologie pflanzlicher Drüsen: 1. Teil. Über den Fangschleim der Insectivoren. Flora 153: 1–22.
Schnepf E (1963b) Zur Cytologie und Physiologie pflanzlicher Drüsen: 2. Teil. Über die Wirkung von Sauerstoffentzug und von Atmungsinhibitoren auf die Sekretion des Fangschleimes von *Drosophyllum* und auf die Feinstruktur der Drüsenzellen. Flora 153: 23–48.
Schnepf E (1963c) Zur Cytologie und Physiologie pflanzlicher Drüsen. 3. Teil. Cytologische Veränderungen in den Drüsen von *Drosophyllum* während der Verdauung. Planta 59: 351–379.
Schnepf E (1968) Zur Feinstruktur der schleimsezernierenden Drüsenhaare auf der Ochrea von *Rumex* und *Rheum*. Planta 79: 22–34.
Schnepf E and Busch J (1976) Morphology and kinetics of slime secretion in glands of *Mimulus tilingii*. Zeitschrift für Pflanzenphysiologie 79: 62–71.
Simoneit BRT, Medeiros PM, and Wollenweber E (2008) Triterpenoids as major components of the insect-trapping glue of *Roridula* species. Zeitschrift für Naturforschung 63: 625.
Sirova D, Borovec J, Santruckova H, Santrucek J, Vrba J, and Adamec L (2010) Utricularia carnivory revisited: plants supply photosynthetic carbon to traps. Journal of Experimental Botany 61: 99–103.
Smith AM and Callow JA (2006a) Biological Adhesives. Springer-Verlag, Berlin.
Smith AM and Callow JA (2006b) Preface. In: Smith AM and Callow JA (eds) Biological Adhesives. Springer-Verlag, Berlin: pp V–VII.
Spomer GG (1999) Evidence of protocarnivorous capabilities in *Geranium viscosissimum* and *Potentilla arguta* and other sticky plants. International Journal of Plant Sciences 160: 98–101.
Srivastava DS, Kolasa J, Bengtsson J, Gonzalez A, Lawler SP, Miller TE, Munguia P, Romanuk T, Schneider DC, and Trzcinski MK (2004) Are natural microcosms useful model systems for ecology? Trends in Ecology and Evolution 19: 379–384.
Thomas SB and McQuillin J (1953) Ropy milk organism isolated from insectivorous plants. Dairy Industry 18: 40–42.
Thum M (1986) Segegation of habitat and prey in two sympatric carnivorous plant species, *Drosera rotundifolia* and *Drosera intermedia*. Oecologia 70: 601–605.
Thum M (1989) The significance of opportunistic predators for the sympatric carnivorous plant species *Drosera intermedia* and *Drosera rotundifolia*. Oecologia 81: 397–400.
Tokunaga T, Takada N, and Ueda M (2004) Mechanism of antifeedant activity of plumbagin, a compound concerning the chemical defense in carnivorous plant. Tetrahedron Letters 45: 7115–7119.
Vintejoux C and Shoar-Ghafari A (2000) Mucigenic cells in carnivorous plants. Acta Botanica Gallica 147: 5–20.
Voigt D, Gorb E, and Gorb S (2009) Hierarchical organisation of the trap in the protocarnivorous plant *Roridula gorgonias* (Roridulaceae). Journal of Experimental Botany 212: 3184–3191.
Voigt D and Gorb S (2008) An insect trap as habitat: cohesion-failure mechanism prevents adhesion of *Pameridea roridulae* bugs to the sticky surface of the plant *Roridula gorgonias*. Journal of Experimental Biology 211: 2647–2657.
Volkova PA and Shipunov AB (2009) The natural behaviour of *Drosera*: Sundews do not catch insects on purpose. Carnivorous Plant Newsletter 38: 114–120.
Wallace J and McGhee K (1999) Testing for Carnivory in *Ibicella lutea*. Carnivorous Plant Newsletter 28: 49–50.
Williams SE (1976) Comparative sensory physiology of the Droseraceae – the evolution of a plant sensory system. Proceedings of the American Philosophical Society, Vol. 120: pp 187–204.
Williams SE and Pickard BG (1974) Connections and barriers between cells of *Drosera* tentacles in relation to their electrophysiology. Planta 116: 1–16.
Williams SE and Spanswick RM (1972) Intracellular recordings of the action potentials which mediate the thigmnastic movements of *Drosera*. Plant Physiology 49: 64.
Wollenweber E (2007) Flavonoids occuring in the sticky resin on *Roridula dentata* and *Roridula gorgonias* (Roridulaceae). Carnivorous Plant Newsletter 36: 77–80.
Zamora R and Gomez JM (1996) Carnivorous plant-slug interaction: a trip from herbivory to kleptoparasitism. Journal of Animal Ecology 65: 154–160.

3 Bonding Tactics in Ctenophores – Morphology and Function of the Colloblast System

Janek von Byern, Claudia E. Mills and Patrick Flammang

Contents

"The Jelly-fishes are among the most wonderful of all the animals of the sea. Their jelly-like bodies, curious forms and structure, their beautiful colors, of claret, rose, and pink, their varied and almost magical movements, as varied and graceful as those of the birds and insects of the air, their phosphorescence by night, causing them to be called the "Lamps of the Sea" and their curious changes in passing from the young to the adult state, have interested all intelligent visitors to the sea-side, and have caused these animals to be carefully studied by some of the most eminent of naturalists."

Tenney (1875), p. 456

3.1 Introduction

Ctenophores are a group of animals found in all the world's seas, from coastal areas to the deep sea and from the tropics to the poles (Hyman, 1940). They are sometimes called "comb jellies" because they have a jelly-like appearance and distinctive rows of comb plates (ctenes) that are used for locomotion. Most ctenophores are transparent or translucent, and range from millimeters up to two meters in length, although most are in the few centimeter range (Ruppert et al., 2004).

The members of the phylum Ctenophora differ from those in the phylum Cnidaria in well-defined morphological characteristics such as the occurrence of swimming comb rows, the presence of an ectomesoderm, and the presence of special adhesive cells (termed "colloblasts", "Klebzellen", "collocytes", or "lasso cells") instead of venomous cells ("nematocytes" or "cnidoblasts") (Pang and Martindale, 2008). Some of the best-known ctenophores are the small and round "sea gooseberry" genus *Pleurobrachia*, the lobate "sea walnut" genus *Mnemiopsis*, and the "Venus' girdle" genus *Cestum*.

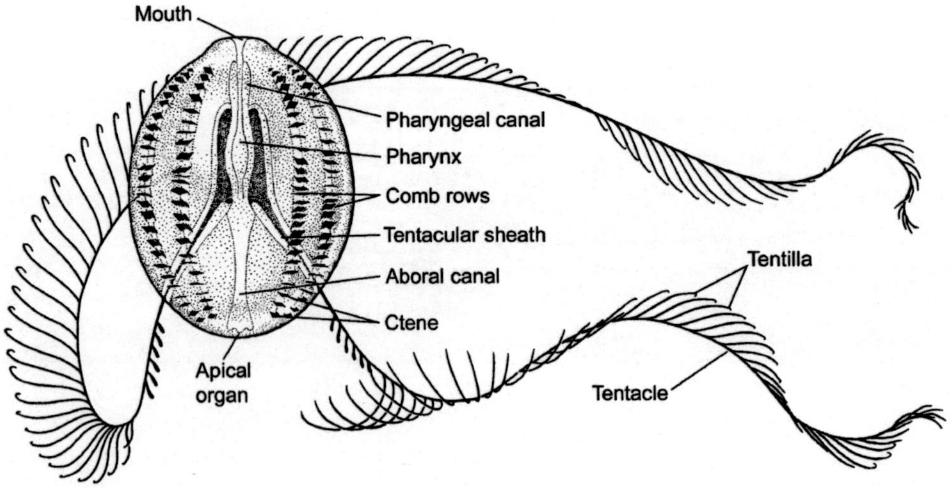

Fig. 3.1 General organization of the cydippid comb jelly *Pleurobrachia*. Not shown are the eight meridional canals underlying the eight comb rows, which connect to the pharynx and pharyngeal canal. Adapted from Ruppert et al. (2004) and reproduced with permission

Almost all ctenophores are capable of bioluminescence, producing flashes of light in photocytes beneath the comb plates (Hyman, 1940; Haddock, 2007). In addition, the refraction of light from the comb plates gives ctenophores their characteristic iridescent appearance. Nearly all ctenophores are hermaphroditic, possessing both eggs and sperm in separate gonads beneath the ctene rows.

All ctenophores are carnivores, feeding on many other planktonic organisms, while beroids prey nearly exclusively on other ctenophores or occasionally on salps (Harbison et al., 1978; Tamm and Tamm, 1991). The comb jellies, except those in the family Beroidae (Carré and Carré, 1989), have two, opposed, very strong and fast ductile tentacles with which they catch and deliver marine zooplankton to the digestive tract (Fig. 3.1). In most families, the tentacles have side branches, or tentilla, that contribute to a much larger catch-surface, but a few cydippid ctenophores capture prey with a pair of simple tentacles (Mills and Miller, 1984; Haddock, 2007). In the resting state, the tentacles are shortened to a ball and retracted in two pouches beside the pharynx. For prey capture the tentacles stretch and fan out into the water, providing a net area of up to 400 cm^2 in *Pleurobrachia* (Greve, 1974). A film sequence published together with the observations of Greve (1974) provides a short sequence demonstrating the shape, length, and delicateness of the tentacles in the sea gooseberry *Pleurobrachia pileus*. The drifting catching net forms a trap for plankton. During capture, the comb jelly wraps its adhesive tentacles around the prey and then pulls the tentacles back. A detailed description of the capture behavior and food intake is provided in Sect. 3.6 "Capture Phenomenon" (p. 37) below.

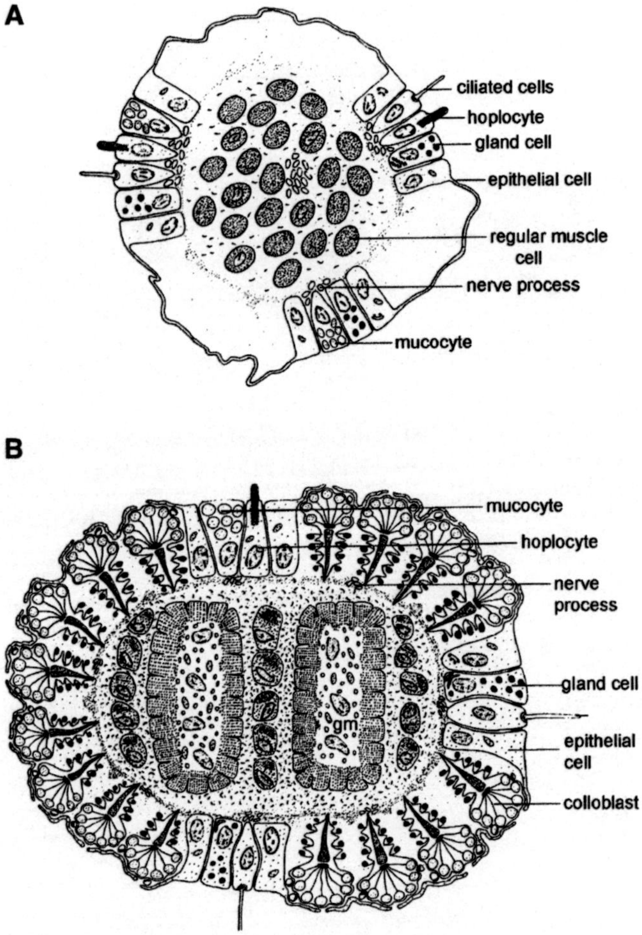

Fig. 3.2 Transverse sections through (A) a tentacle and (B) a tentillum of *Coeloplana bannworthi*. Adapted from Eeckhaut et al. (1997) and reproduced with permission

3.2 General Tentacle Morphology

Each individual tentacle consists of a main branch, usually with numerous, small side branches, the tentilla. The tentilla appear either as thin, extensible filaments, as in *Pleurobrachia* (Bargmann et al., 1972), or rarely as stouter, tightly-coiled appendages as in *Euplokamis* (Mills, 1987; Mackie et al., 1988).

Both tentacle and tentilla have the same double-layer organization (Fig. 3.2): (1) a dense fibrillar mesoglea enclosing the tightly-packed, longitudinally-oriented muscle fibers, and tentacular nerves and (2) a peripheral cortex consisting of epithelial cells, gland cells, sensory cells, and colloblasts (Eeckhaut et al., 1997). Depending on the species, the latter are either restricted to the tentilla (Carré and Carré, 1993; Eeckhaut et al., 1997) or present on both the tentilla and tentacles (Bargmann et al., 1972; Franc, 1978). Generally, colloblasts are more numerous on the tentilla than on the tentacles (Fig. 3.3).

The epithelial cells form a continuous layer (Fig. 3.2) between the sensory elements, the gland cells, and the heads of mature colloblasts, to which they are connected by a zonula adherens (Eeckhaut et al., 1997). The gland cells include mucous and granular cells, which are easily identifiable by their secretory content.

The sensory system comprises ciliated cells and hoplocytes. In ciliated cells, a single $9 \times 2+2$ non-motile cilium associated with a characteristic spherical striated root rises from the cell apex. The hoplocytes bear a single or a few stout pegs filled with a dense fibrillar core (Hernandez-Nicaise, 1974; Eeckhaut et al., 1997). Both sensory cells are grouped into clusters and are always as-

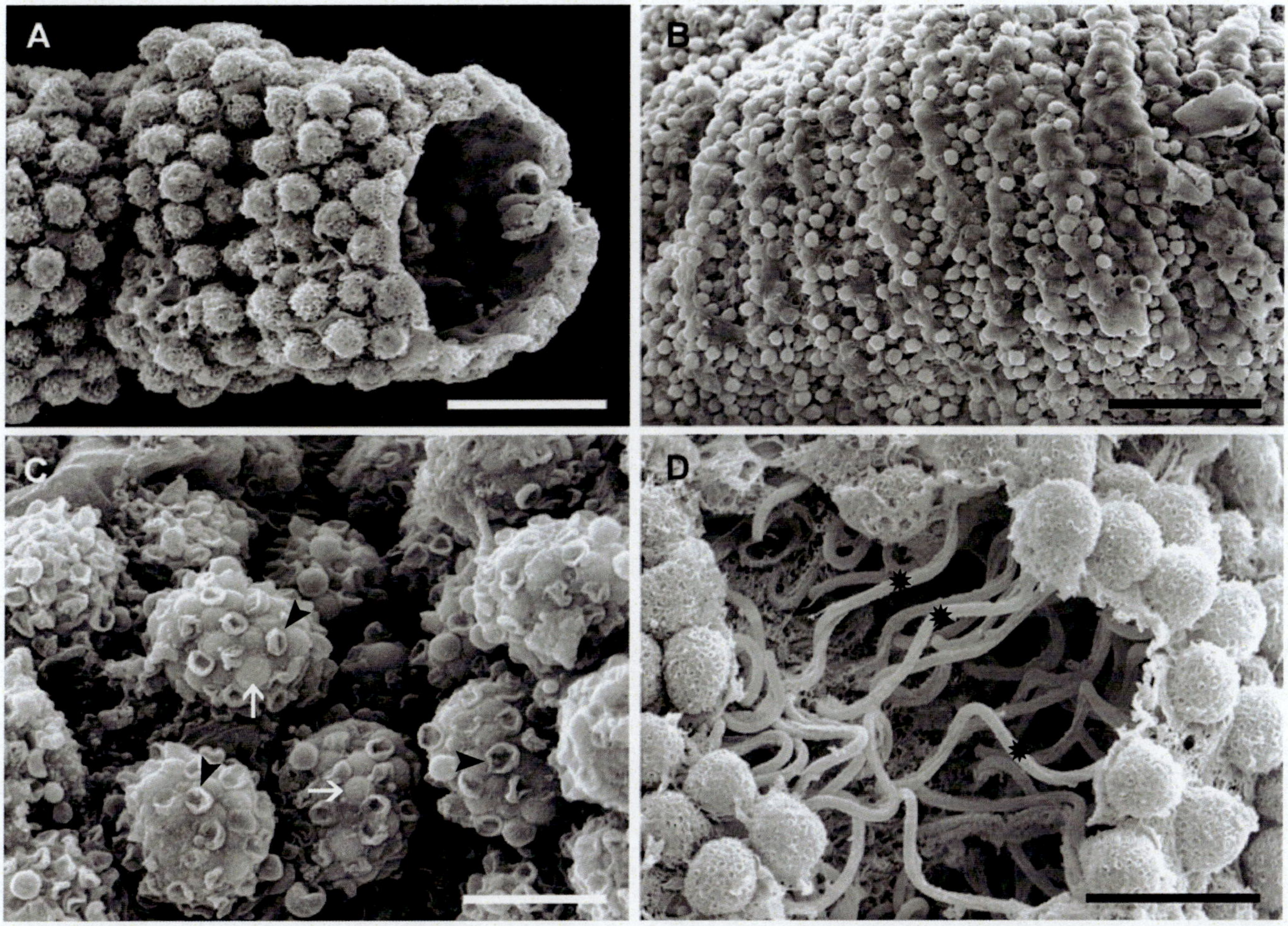

Fig. 3.3 Scanning electron microscope images of *Pleurobrachia pileus* show that the number of colloblasts is low (A) on the tentacle, but higher (B) on the tentillum. (C) At higher magnification, sickle-shaped external granules (*black arrowhead*) as well as the bulges created by the internal granules (*white arrow*) become visible. Image D shows a torn tentillum in which the collospheres and the coiled spiral filaments (*black asterisks*) are visible. Images A, B, and C by the first author, image D provided by Emeline Wattier from the Université de Mons, Belgium. Scale bar in (A) = 20 μm; (B) = 50 μm; (C) = 5 μm; (D) = 10 μm

sociated with the gland cells. These sensory clusters form an extensive "monitoring" system over the surface of the widespread tentacular apparatus (Eeckhaut et al., 1997).

3.3 Colloblast Organization

Still, little is known about the development, function, and bonding mechanisms of the colloblasts. Early light microscope studies (Chun, 1880; Hertwig, 1880; Schneider, 1908; Komai, 1922; Weill, 1935b; Hyman, 1940) were complemented by electron microscopy investigations performed by Hovasse and de Puytorac (1962, 1963). More recently, Bargmann et al. (1972), Benwitz (1978), Franc (1978), Mackie et al. (1988), Carré and Carré (1993), and Eeckhaut et al. (1997) published detailed ultrastructural studies on the adhesive structures in ctenophores.

In general, colloblasts have a hemispherical head (collosphere) and a conical stalk (collopod) (Fig. 3.4). The stalk is surrounded by a helical thread (Figs. 3.4 and 3.5), which is coiled in the form of a helix and ends in the mesoglea by a root (Weill, 1935a). The colloblast head is decorated by external granules, attached outside the plasma membrane in a regular manner. Those granules are either empty or include only a small, electron-dense border area. Within the collosphere a further granular type of medium electron density occurs.

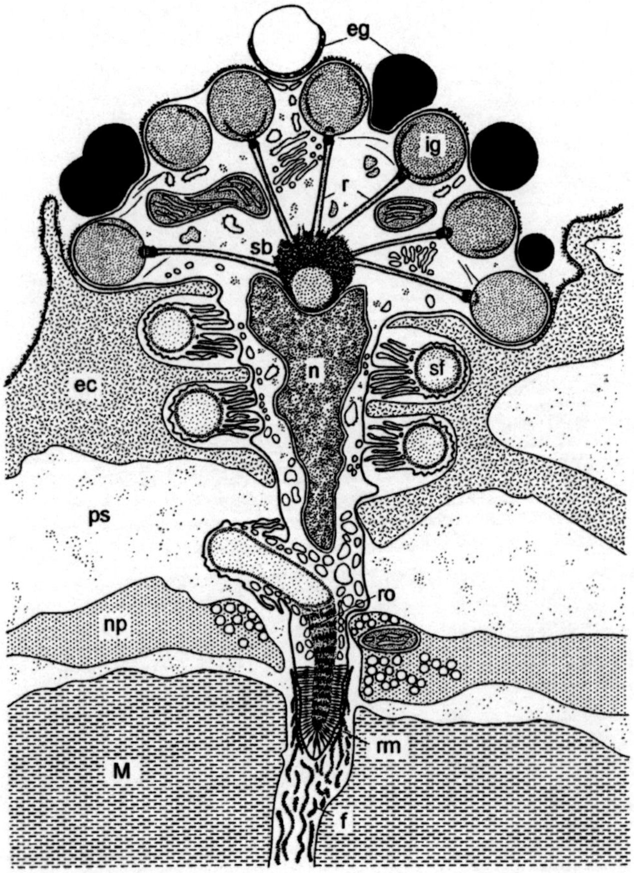

Fig. 3.4 Schematic drawing of the colloblast and adjacent epithelial cells within a tentillum of *Pleurobrachia*. Adapted from Benwitz (1978) and reproduced with permission

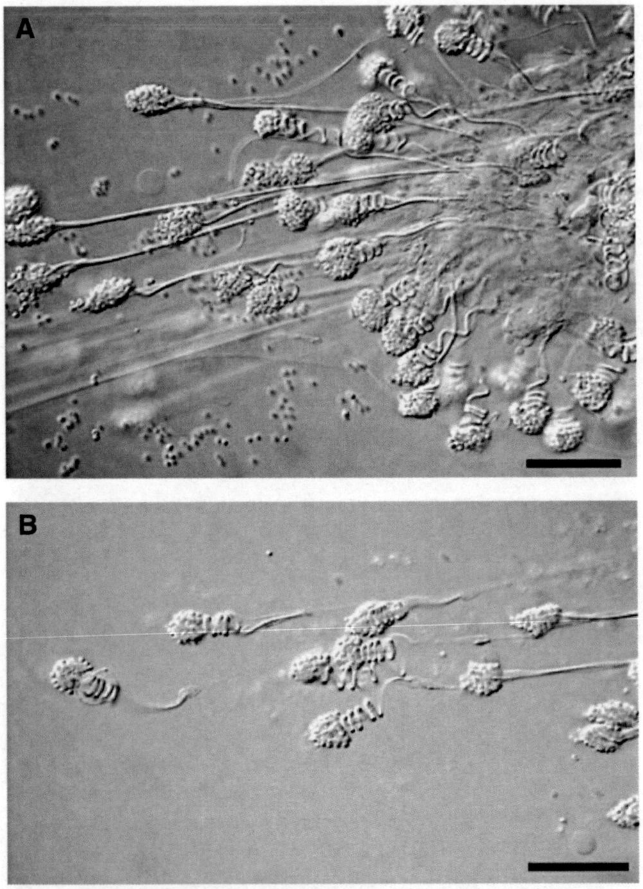

Fig. 3.5 Light microscope images of colloblasts on the tentacles of the benthic ctenophore *Vallicula multiformis*. (A) Lightly-smashed preparation of colloblasts at the tip of a tentillum; (B) six colloblasts isolated at the tip of a tentillum. Images of *Vallicula multiformis* reproduced with permission of Alvaro Migotto from the Center of Marine Biology of the University of São Paulo, in São Sebastião, Brazil, their copyrights © remain with Alvaro Migotto. Scale bar in both images = 20 μm

3.3.1 Head (Collosphere) and Spheroidal Body

The head of the colloblast has a bulb- to kidney-like shape (∅ 5 μm in *Pleurobrachia*) (Bargmann et al., 1972). Its cytoplasm consists of loose, fine-grained ma-

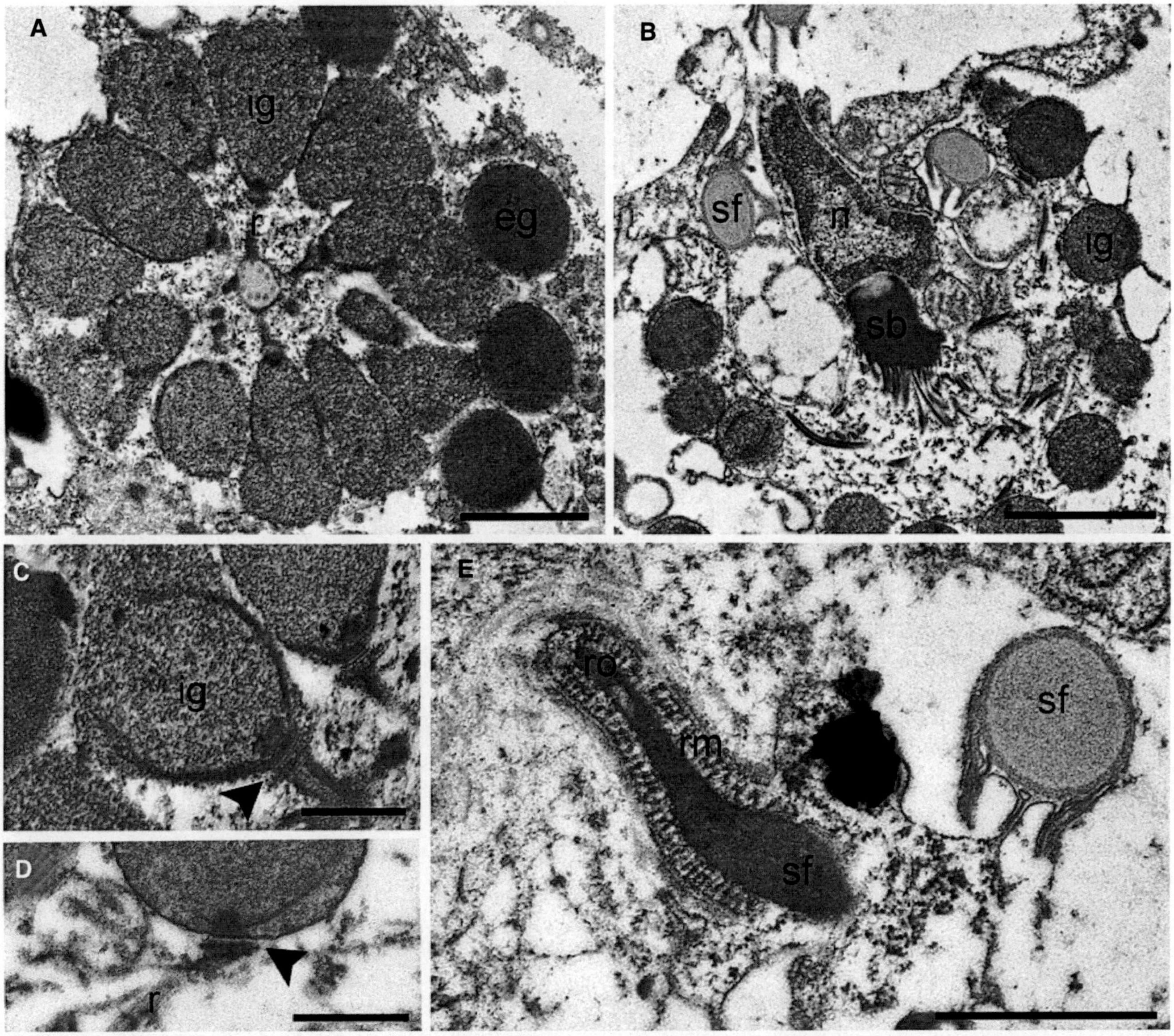

Fig. 3.6 Transmission electron microscope images of developing (stage 2; A, C) and mature (B, D, E) colloblasts in *Pleurobrachia pileus*. (A) Comparison of the inner granules in the developing colloblast and the outer granule in the cap cell. (B) The central spheroidal body and nucleus. (C, D) Higher magnification of the small plate (*arrowhead*) and stamp-like pad connecting the internal granules and the radial fibers. (E) Overview of the spiral filament and the root. All images provided by Emeline Wattier from the Université de Mons, Belgium. Scale bar in A, E = 1 μm, B = 2 μm, C = 400 nm, and D = 500 nm

terial and contains typical cell organelles such as mitochondria, endoplasmic reticulum, Golgi network, and microtubules. Very specific structures are two granule types (see below) and the spheroidal body (Fig. 3.6A and B). The latter forms the end of the spiral filament and is located close to the stalk. From its center, fibers composed of about 20 filamentous sub-units, the so-called radii, stream toward the head periphery and end at a stamp-like fibrillar pad adjacent to the internal granules. A small plate, consisting of small rods, connects this pad to one granule.

To date, the function of the spheroidal body remains unresolved. Carré and Carré (1993) suggest that it serves as a "nucleation center" for the spiral filament or "acts to organize the structure of the collosphere". It is also conceivable that the radii orient the internal granules and lift them up against the collosphere membrane and likewise create the external bulges visible on its outer surface (Fig. 3.3C).

3.3.2 Stalk (Collopod) and Spiral Filament

3.3.2.1 Stalk

The collopod consists mostly of a straight, elongated nucleus surrounded by the typical nuclear membrane and containing evenly dispersed chromatin. No nucleoli have been detected in the mature colloblasts. Apart from the nucleus, cell organelles such as mitochondria, endoplasmic reticulum, few vesicles, and microtubules occur in the collopod.

3.3.2.2 Spiral Filament

The helical thread, also termed as filament (Benwitz, 1978; Carré and Carré, 1993), spiral filament (Benwitz, 1978; Mackie et al., 1988), or shaft (Bargmann et al., 1972), has a tubular appearance (∅ 0.25–1.25 μm) (Franc, 1978) and is filled with dense, homogeneous material (Fig. 3.6E). The filament runs from the spheroidal body along the collopod axis by oblique right-handed spiral turns (in some cases also left-handed) and ends as a root. The number of these coils and their interspaces are not constant but vary between and within species (e.g., 1–2 in *Lampea* (formerly named as *Lampetia*) and *Cestum* (formerly named as *Cestus*); 5 in *Vallicula* (see Fig. 3.5); 6–7 in *Leucothea* (formerly named as *Eucharis*); 9 in *Pleurobrachia*; and up to 11 and 14 in *Euplokamis* and in *Minictena*, respectively (Weill, 1935b; Bargmann et al., 1972; Franc, 1978; Mackie et al., 1988; Carré and Carré, 1993).

In the retracted state, the spiral filament always appears in the typical coiled position, never inclined or broken. In contrast, in the extended state, the spiral filament is several times longer, but it does not rotate during stretching and remains loosely coiled (see Fig. 3.5).

3.3.2.3 Root

The spiral filament ends in a cone-shaped, electron-dense root structure (Fig 3.6E). Its surface is bristled with numerous, thin filaments (calculated number: 700–900 fibrils) (Benwitz, 1978), providing a bottle–brush appearance. The filaments are enclosed by a membrane from which fibers (0.03–0.05 μm in diameter) proceed into the mesoglea.

Furthermore, in the basal region of the spiral filament, just above the root structure, a synaptic junction occurs, connecting a nerve cell to the colloblast. In the synapses, small electron-lucent vesicles may be present, oriented toward the colloblast (Franc, 1978). The innervation of colloblasts reported by Franc was not, however, seen in *Euplokamis* by Mackie et al. (1988).

3.3.3 Secretion Granules

The colloblasts have two types of granules, which have been given various names in the literature (see below). To simplify matters here, we differentiate the types based on their location as "external" and "internal" granules.

Both granule types are synthesized only during early colloblast development, whereas the respective organelles participating in their formation are degenerated when the colloblast is mature. The animals are therefore unable to replace or re-synthesize "used" granules, but must develop new colloblasts. Concerning their origin, orientation, and content, both granule types differ strongly.

3.3.3.1 Internal Granules

There are different terms for this granule type: "eosinophilic granules" (Schneider, 1902; Franc, 1978; Mackie et al., 1988; Eeckhaut et al., 1997), "P-Körperchen" (Benwitz, 1978) or "secretory globules" (Bargmann et al., 1972).

The internal granules are formed by the fusion of Golgi vesicles in the collosphere during late colloblast development (see Sect. 3.4 "Colloblast Development" p. 35). Fully-developed granules measure about 0.8 μm in diameter and are filled with a fine-grained content. In mature colloblasts, the granules form a regular layer at the head periphery. Each granule is bordered by the small plate and the stamp-like pad, terminating the radiating fibers (Fig. 3.6C and D). Since the granules are clearly pushed toward the plasma membrane, they form a bulge on the collosphere surface, but remain separated from the membrane by a small, electron-dense cleft. The outer part of the bulges is covered with very fine filamentous projections, which are absent in other areas of the colloblast membrane.

3.3.3.2 External Granules

The external granules are situated in cavities delimited by the bulges on the surface of the colloblast heads. As in the internal granules, the external granules have been variously named based on their features: "Klebkörnchen" (Bargmann et al., 1972), "osmiophilic droplets" (Benwitz, 1978), "refractive vesicles" (Franc, 1978), "refringent" or "refractive granules" (Mackie et al., 1988; Eeckhaut et al., 1997), or "secretion granules" (Storch and Lehnert-Moritz, 1974).

The external granules are located in hollows between the bulges of the outer collosphere membrane and, as with the internal granules, are not directly connected to the colloblast membrane but separated by a narrow, electron-dense layer. They have a variable content: from obviously empty to filled with coarse and/or homogeneous electron-dense material. All these types can occur associated with the same cell.

Earlier studies revealed that these granules do not derive from the colloblast but rather from another cell type, the "cap cells" (see Sect. 3.4) (Benwitz, 1978; Mackie et al., 1988).

3.4 Colloblast Development

As in cnidarian nematocysts, the colloblasts of ctenophores can be used only once. The spiral filament cannot be re-compressed and the granules (internal and external) need to be synthesized *de novo* during new colloblast development. In cnidarians, about 25% of the nematocysts are lost and can be replaced within 48 hours (Moore, 2001). No observations about colloblast replacement times are available for comb jellies.

Detailed descriptions of colloblast origin and development are given by Storch and Lehnert-Moritz (1974) and Mackie et al. (1988). These authors divided colloblast development into three successive, seamlessly merging stages. The first description, however, omitted the development of the external granules in the "cap cells"; this glandular structure was later described by Benwitz (1978) and Mackie et al. (1988) (named "accessory cells"):

3.4.1 Stage 1

The colloblast originally evolves from a choanocyte-like cell (Fig. 3.7A) and becomes strongly modified during

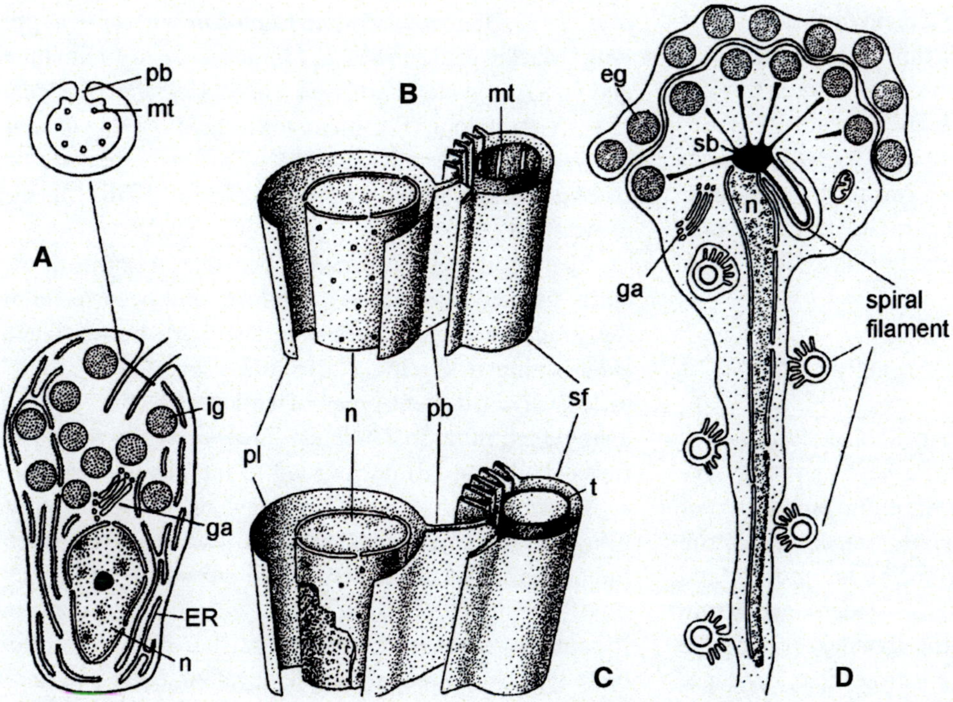

Fig. 3.7 Schematic illustration of colloblast development in *Pleurobrachia pileus*. (A) The colloblasts originally derive from choanocyte-like cells which, in addition to a cilium, have all the typical cell organelles. (B, C) With further development, the cilium grows spirally around the basal part of the cell and is connected with the cell body by means of a thin strand of cytoplasm. An electron-dense cylinder, the tube, arises outside the microtubules which then degenerate. Parallel to this, additional membrane projections (cristae) arise from the membrane of the former cilium which has become the spiral filament. (D) In this stage, the developing colloblast corresponds almost to its final appearance, with a head and stalk. The internal granules occupy the whole apical area of the head, while the cap cells are reduced, leaving the external granules on the colloblast head. The main characteristic of this stage is the development of the spheroidal body from the cilium rootlet. Its radii initially develop on each internal granule and later converge at the spheroidal body. Adapted from Storch and Lehnert-Moritz (1974) and reproduced with permission

development. Initially, its nucleus is centrally located and surrounded by a large cytoplasm, which contains typical cell organelles such as endoplasmic reticulum, Golgi network, and mitochondria along with vesicles of middle electron density. The cap cells arise during this early colloblast development and cover the choanocyte totally. The cap cells are filled with small vesicles that derive from the endoplasmic reticulum or Golgi network. Fusion of the small vesicles yields larger granules, which are filled with the same contents as in the mature stage.

3.4.2 Stage 2

With ongoing differentiation, the colloblast expands and tapers basally (Fig. 3.7B and C). The nucleus is translocated to the basal area and becomes elongated. The upper part of the cell includes large amounts of rough endoplasmic reticulum. The internal granules are no longer uniformly distributed, but migrate apically (Fig. 3.6A, C). In some individual cells, the mitochondria start to degenerate. The cilium starts to grow spirally around the basal part of the cell, adjoined lengthwise by a thin plasma strip. With increasing differentiation the connection between this developing spiral filament and cell body is reduced to a narrow bridge. Parallel to this, membrane furrows (cristae) on the spiral filament are oriented toward the collopod. Inside, the spiral filament is supported by microtubules embedded in an electron-lucent matrix (Fig. 3.6A).

3.4.3 Stage 3

In this stage (Fig. 3.7D), the pre-colloblast corresponds almost to its final appearance, with a head and stalk.

The internal granules occupy the entire apical area of the head and are filled with material of medium electron density. Outside, the external granules are present and may be likewise filled with a material of medium electron density. In most cases, however, the granules appear to be visually empty or include only a narrow, electron-dense border area. In later development, the number of these granules becomes reduced.

The main characteristic of this stage is the development of the spheroidal body in the center of the head; it is derived from the rootlet of the spiral filament. Its radii initially develop at the internal granules and then extend toward the spheroidal body.

With increasing size of the spiral filament, the microtubules shift into the periphery. An electron-dense cylinder arises between the microtubules and the thread membrane and forms a central tube. As soon as the filament is completely developed, the microtubules disappear. The filament turns spirally around the elongated collopod, but the connecting plasma bridges remain during the entire process. In the proximal tip of the filament, the root is derived with all its features.

As soon as the colloblast becomes functionally active, cap cells degenerate but leave the external granules on top of the collosphere. The mature external granules are surrounded by two membranes – an inner, strongly osmiophilic vesicle membrane and the outer cap cell membrane.

3.5 Colloblast Polymorphisms

Just as animal size varies, the size of the colloblast also differs with species. The smallest ones are found in *Pleurobrachia* (10 μm long, maximum 4 μm wide) (Bargmann et al., 1972; Benwitz, 1978); in *Euplokamis* they are 12 μm long (Mackie et al., 1988), in *Minictena* up to 14 μm long and 4–10 μm wide (Carré and Carré, 1993), in *Vallicula* about 8–9 μm long (Emson and Whitfield, 1991) whereas in *Leucothea* (formely named as *Eucharis*) they are up to 25 μm long and 8–10 μm wide (Franc, 1978; Mackie et al., 1988).

Apart from the size and number of spiral turns of the spiral filament (see above), the form and occurrence of the granules in the colloblasts also vary between and even within a species. Carré and Carré (1993) found at least five different types of colloblasts (Fig. 3.8) on a single tentillum in *Minictena luteola*. While three cell types (I, II, and III) correspond to the general structure as described above, types IV and V differ significantly. They are larger and lack external granules. Also, their internal granules are larger than in types I–III and sometimes completely filled with an electron-dense content. It remains unclear whether these five different colloblast types represent different developmental stages or are differently shaped mature types. The function of polymorphic colloblast types remains debatable. In the tentacles of cnidarians, however, different categories of nematocysts occur and fulfill different functions such as prey capture, locomotion, and defense (Kass-Simon and Scappaticci, 2002). A similar situation is possible for the ctenophores.

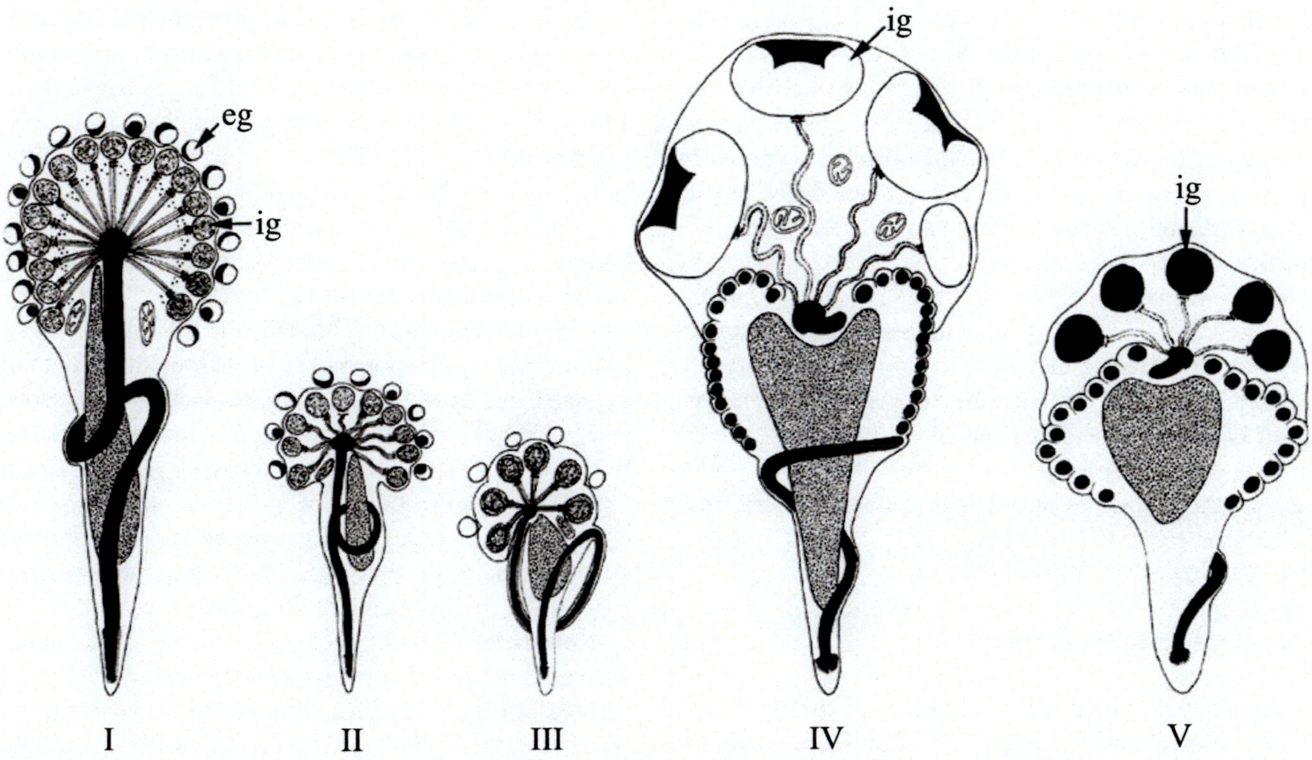

Fig. 3.8 Colloblast polymorphism in *Minictena luteola.* A total of five types (I to V) of colloblast are present in one tentillum. They differ in size, structure, and partly in the absence of external granules. Adapted from Carré and Carré (1993) and reproduced with permission

3.6 Capture Phenomenon

3.6.1 Capture Behavior

Comb jellies are widely distributed marine predators that feed on small crustaceans, invertebrate larvae, and other pelagic organisms. Although some species of ctenophores swim while searching for prey (e.g., *Cestum*, *Eurhamphaea*, *Leucothea*, and *Beroe*) (Harbison et al., 1978; Hamner et al., 1987; Matsumoto and Harbison, 1993; Haddock, 2007), most display a sit-and-wait strategy that has been well described by Moss and Tamm (1993), Tamm and Moss (1985), and Mills (1987) among others.

Through rhythmic contraction of the tentacles, *Pleurobrachia* determines if the frictional resistance of the catching nets in the water is increased by adhering particles or potential prey. This behavior enables even those organisms that do not move very much to be recognized as prey. At the same time, this elastic jerking movement serves to tire the prey, much as in angling.

Together with the stretchable spiral filaments and the tentacle's stem, ctenophores have developed a highly elastic system which can compensate for the escape behaviors of prey organisms and maintain bonding strength despite strong movements of the victims (Greve, 1975).

Although marine zooplankton is captured passively by contact with the tentacle/tentilla net, ctenophores seem to be able to trigger the bonding mechanisms actively. Unsuitable food items, or contact of the tentilla with "unusual objects" (e.g., forceps), stimulates no bonding effect (pers. observation by the first author). Also remarkable is that the tentacles always spread out easily from the tentacle sheath without being tangled or stuck together. In contrast, the relaxed, isolated, and even chemically fixed tentilla always clump together irreversibly (pers. observation by the first author).

Suitable prey, caught in the net trap, stimulates the tentacle. It then retracts from its tip to its base, provoking the retraction of the tentilla one by one. By rotating around their transverse axis through increased beating frequency of the cilia in the comb rows, *Pleurobrachia* wrap their tentacle around the prey and strengthen the contact. The tentacle is curled-up over the mouth and taken into the mouth by ciliary action. This brings the adhering prey into the pharynx (Greve, 1975).

After the prey becomes detached in the pharynx, the tentacles contract and thereby draw themselves out of the mouth. The animal supports this process by accelerating the cilia-beating in the comb rows.

As one tentacle is rolled in, the other one sometimes remains spread out as a catching net. These different tentacle formations increase the comb jelly's capture efficiency. After a rapid drawing-in, the tentacles are soon spread out again for further prey catching. By swimming in a helix while relaxing its tentacles, called "Veronica display" by Mackie and Boag (1963) for its similarity to the movement of a bullfighter's cape, *Pleurobrachia* is able to extend and spread out its tentacles in an evenly-spaced horizontal net (Greve, 1975). Other genera set their tentacles in different characteristic patterns (Harbison et al., 1978; Haddock, 2007).

3.6.2 Capture Mechanism

Despite the detailed characterization of the colloblast system, its capture mechanism has only rarely been investigated and remains unsolved (Franc, 1978); nevertheless, similarities of the colloblast structures with those of nematocysts may indicate a similar function.

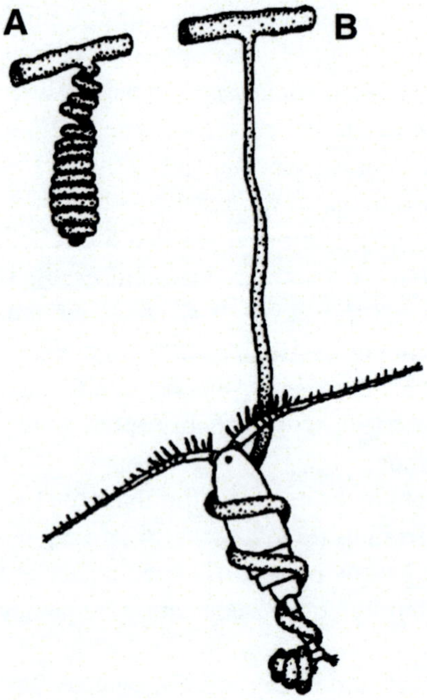

Fig. 3.9 Schematic drawing of (A) a coiled resting tentillum of *Euplokamis*, anchored to the tentacle; and of (B) a discharged tentillum, adhering to the prey by its colloblasts. Adapted from Mackie et al. (1988) and reproduced with permission

The prominent exposure of the sensory cells with their protruded cilia or pegs, adjoining the colloblasts, indicates their role as primary sensors and effectors (Eeckhaut et al., 1997). The detection of suitable prey (perhaps by chemical, mechanical and/or electrical stimuli as proposed for cnidarians – see Kass-Simon and Scappaticci, 2002) provokes an elevation of all colloblasts located on one tentillum. In such stimulated colloblasts, the spiral filament is slightly extended but remains coiled (Franc, 1978).

In *Euplokamis* instead, not only the colloblast but also its complete tentillum is moveable. The animals are able to uncoil and recoil the tentillum, performing slow spontaneous movements that resemble a wriggling worm. In the case of small prey (e.g., copepods about 1–2 mm long), a single tentillum becomes discharged within 40–60 ms (Fig. 3.9), but larger copepods (4–5 mm) induce a discharge of several tentilla, all in contact with the prey (Mills, 1987; Mackie et al., 1988).

Nonetheless, how colloblasts ultimately become discharged remains unknown for all comb jellies. This could involve active projection of the colloblast head toward the prey by the coiled spiral filament. In another probable scenario, the prey is captured passively during contact with the colloblast adhesive. In this case, the filament is extended after prey capture and serves to absorb escape movements of the prey or to reduce tensile loads during tentacle movement (Franc, 1978). The complex anchoring of the root within the mesoglea, does indicate that comb jellies can hold strong loads with their spiral filaments. The nematocysts of cnidarians are also anchored in the tentacle mesoglea, by a fibrillar basket of microtubules and microfilaments (Cormier and Hessinger, 1980).

Although studies on the discharge of the colloblast are lacking, observations by Franc (1978) indicate that the internal granules release the "gluing substance" that bonds the prey. According to Franc, most internal granules burst during prey contact and release their secretory content on the prey cuticle. The external granules, however, remain mostly intact and still located on the collosphere as well as distributed singularly on the prey surface (Franc, 1978).

In our opinion, secretory components in the internal granules are favorable candidates for glue production. Thus, species with a colloblast system having only internal granules (e.g., those of types IV and V in *Minictena luteola*) (Carré and Carré, 1993) would also be able to capture prey in this manner. The contribution of the external granules to the capture process is still unresolved. The vesicles may strengthen the bonding effect by an additional adhesive, provide a paralyzing/toxic substance to immobilize the prey, or have a completely different purpose.

So far it remains unresolved how the animals release the prey within the pharynx and dispose of the "used" colloblasts.

3.6.3 Sensory Cells

The function of sensory cells does not seem to be limited to prey capture but also provides information for defense and escape reactions. If *Pleurobrachia* is caught, for example by the tentacles of the scyphomedusa *Cyanea*, it secretes a layer of slime from the gland cells within the tentacles. This enables the ctenophore to lift off the predator's tentacles and swim away (Greve, 1975). Moreover, at the slightest touch of its tentacles/tentilla against a predator (e.g., *Beroe gracilis*), *Pleurobrachia* recognizes its enemy immediately and shows a fast escape reaction. Such behavior is also induced by strong mechanical (water currents) or thermal stimuli (fluctuation of 2–4°C) against the tentacles (Esser et al., 2004). This leads to an interruption of swimming and causes the tentacles to retract into their pouch.

3.6.4 Glue Composition

Although the colloblast system has been characterized morphologically in detail, our knowledge about the secretory composition and its bonding principle remains fragmentary. So far only behavioral observations confirm that the colloblast glue induces a target-oriented fast and strong adhesion. Interestingly, this glue could be reversible and enable prey to be released into the mouth without any visible macroscopic change to the tentillum or tentillum loss. Whether prey release is caused by a duo-gland system within the colloblast system, an enzymatic reaction within the mouth and/or by an expelling mechanism of the complete colloblast is unknown. It is conceivable that the attached colloblasts are released from the tentacles and swallowed with the prey.

Histochemical analyses are rare and are unsuitable to differentiate between the internal and external granules. Unpublished histochemical results by the first author indicate that the granules (internal and/or external) react positively to stains specific for neutral sugars (PAS) and for neutral proteins (Biebrich scarlet) at pH 6.0, weakly at pH 8.5, but not at higher levels (pH 9.5 and 10.5). Tests for acidic macromolecules (AB pH 1.0 and 2.5) showed no positive reaction for the granules.

The ultrastructural descriptions by Bargmann et al. (1972) and other researchers suggest that the osmiophilic external granules contain lipoproteins.

Nevertheless, it remains unclear how the animals are able to recognize the "right" prey spectra, stick to and then later release them in the mouth. Further investigations of active adhering and "used" tentilla are necessary to provide further details about the bonding mechanisms in ctenophores.

Abbreviations

ec	epithelial cell
eg	external granules
er	endoplasmic reticulum
f	fiber of the mesoglea
ga	golgi apparatus
gm	giant muscle cell
ig	internal granules
m	muscle
mt	microtubules
n	nucleus
np	nerve plexus
pb	plasma bridge
pl	plasma membrane
ps	perimuscular space
r	radius
ro	root
rm	root membrane
sb	spheroidal body
sf	spiral filament
t	tube

Acknowledgments

We thank Emeline Wattier from the Université de Mons, Belgique, for providing the ultrastructural images for this contribution. The first author also thanks the staff of the Biological Anstalt Helgoland, Germany, for their help and support in collecting comb jellies for the histological/histochemical analysis.

We are also very grateful to Alvaro Migotto from the Center of Marine Biology of the University of São Paulo, in São Sebastião, Brazil for providing Fig. 3.5.

Patrick Flammang is Senior Research Associate of the Fund for Scientific Research of Belgium (F.R.S.–FNRS).

References

Bargmann W, Jacob K, and Rast A (1972) Über Tentakel und Colloblasten der Ctenophore *Pleurobrachia pileus*. Zeitschrift für Zellforschung und mikroskopische Anatomie 123(1): 121–152.

Benwitz G (1978) Elektronenmikroskopische Untersuchungen der Colloblasten-Entwicklung bei der Ctenophora *Pleurobrachia pileus* (Tentaculifera, Cydippea). Zoomorphologie 89: 257–278.

Carré C and Carré D (1989) *Haeckelia bimaculata* sp. nov., une nouvelle espéce méditerranéenne de cténophore (Cydippida Haeckeliidae) pourvue de cnidocystes et de pseudocolloblastes. Comptes Rendus de l'Académie des Sciences, Paris 308 (Serie III): 321–327.

Carré D and Carré C (1993) Five types of colloblast in a cydippid ctenophore, *Minictena luteola* Carré and Carré: an ultrastructural study and cytological interpretation. Philosophical Transactions of the Royal Society: Series B, Biological Sciences 341: 437–448.

Chun C (1880) Die Ctenophoren des Golfes von Neapel und der angrenzenden Meeres-Abschnitte. Zoologische Station zu Neapel. Verlag von Wilhelm Engelmann, Leipzig.

Cormier SM and Hessinger DA (1980) Cellular basis for tentacle adherence in the Portuguese man-of-war (*Physalia physalis*). Tissue Cell 12(4): 713–721.

Eeckhaut I, Flammang P, Lo Bue C, and Jangoux M (1997) Functional morphology of the tentacles and tentilla of *Coeloplana bannworthi* (Ctenophora, Platyctenida), an ectosymbiont of *Diadema setosum* (Echinodermata, Echinoida). Zoomorphology 117: 165–174.

Emson RH and Whitfield PJ (1991) Behavioural and ultrastructural studies on the sedentary platyctenean ctenophore *Vallicula multiformis*. Hydrobiologia 216/217: 27–33.

Esser M, Greve W, and Boersma M (2004) Effects of temperature and the presence of benthic predators on the vertical distribution of the ctenophore *Pleurobrachia pileus*. Marine Biology 145: 595–601.

Franc JM (1978) Organization and function of ctenophore colloblasts: an ultrastructural study. Biological Bulletin 155: 527–541.

Greve W (1974) Organisation der Rippenqualle *Pleurobrachia pileus* (Ctenophora). Institut für wissenschaftlichen Film C1186/1972: pp 3–13.

Greve W (1975) Verhaltensweisen der Rippenquallen *Pleurobrachia pileus* (Ctenophora). Institut für wissenschaftlichen Film C1181/1975: pp 3–10.

Haddock SHD (2007) Comparative feeding behavior of planktonic ctenophores. Integrative and Comparative Biology 47(6): 847–853.

Hamner WM, Strand SW, Matsumoto GI, and Hamner PP (1987) Ethological observations of foraging behavior of the ctenophore *Leucothea* sp. in the open sea. Limnology and Oceanography 32: 645–652.

Harbison GR, Madin LP, and Swanberg NR (1978) On the natural history and distribution of oceanic ctenophores. Deep-Sea Research 25: 233–256.

Hernandez-Nicaise ML (1974) Ultrastructural evidence for a sensory-motor neuron in ctenophora. Tissue and Cell 6(1): 43–47.

Hertwig R (1880) Über den Bau der Ctenophoren. Verlag von Gustav Fischer, Jena.

Hovasse R and de Puytorac P (1962) Contributions á la connaissance du colloblaste, grace á la microscopie électronique. Comptes Rendus de l'Académie des Sciences, Paris 255: 3223–3225.

Hovasse R and de Puytorac P (1963) Le colloblaste des ctenophores: Ultrastructure, signification. In: Moore JA (eds) Proceedings of the XVI International Congress of Zoology 20–27 August 1963 1 XVI International Congress of Zoology, Washington: 27 pp.

Hyman LH (1940) The Radiate Phyla – Phylum Ctenophora. The Invertebrates: Protozoa through Ctenophora. McGraw-Hill Book Company, New York: pp 662–696.

Kass-Simon G and Scappaticci AA (2002) The behavioral and developmental physiology of nematocysts. Canadian Journal of Zoology 80: 1772–1794.

Komai T (1922) Studies on two aberrant ctenophores, Coeloplana and Gastrodes. Kyoto Imperial University, Kyoto.

Mackie GO and Boag DA (1963) Fishing, feeding and digestion in siphonophores. Pubblicazioni della Stazione Zoologica di Napoli 33: 178–196.

Mackie GO, Mills CE, and Singla CL (1988) Structure and function of the prehensile tentilla of *Euplokamis* (Ctenophora, Cydippida). Zoomorphology 107: 319–337.

Matsumoto GI and Harbison GR (1993) *In situ* observations of foraging, feeding, and escape behavior in three orders of oceanic ctenophores: Lobata, Cestida, and Beroida. Marine Biology 117: 279–287.

Mills CE (1987) Revised classification of the genus *Euplokamis* Chun, 1880 (Ctenophora: Cydippida: Euplokamidae n. fam.) with a description of a new species *Euplokamis dunlapae*. Canadian Journal of Zoology 65: 2661–2668.

Mills CE and Miller RL (1984) Ingestion of a medusa (*Aegina citrea*) by the nematocyst-containing ctenophore *Haeckelia rubra* (formerly *Euchlora rubra*): phylogenetic implications. Marine Biology 78: 215–221.

Moore J (2001) Cnidaria. An Introduction to the Invertebrates, 1st Ed. Cambridge University Press, Cambridge: pp 30–46.

Moss AG and Tamm SL (1993) Patterns of electrical activity in comb plates of feeding *Pleurobrachia* (Ctenophora). Philosophical Transactions of the Royal Society: Series B, Biological Sciences 339: 1–16.

Pang K and Martindale MQ (2008) Ctenophores. Current Biology 18(24): R1119–R1120.

Ruppert EE, Fox RS, and Barnes RD (2004) Ctenophora. Invertebrate Zoology – A Functional Evolutionary Approach, 7th Ed. Thomson Brooks/Cole, Belmont: pp 181–195.

Schneider KC (1902) Lehrbuch der vergleichenden Histologie der Tiere. Verlag von Gustav Fischer, Jena.

Schneider KC (1908) Histologisches Praktikum der Tiere für Studenten und Forscher. Verlag von Gustav Fischer, Jena.

Storch V and Lehnert-Moritz K (1974) Zur Entwicklung der Kolloblasten von *Pleurobrachia pileus* (Ctenophora). Marine Biology 28: 215–219.

Tamm SL and Moss AG (1985) Unilateral ciliary reversal and motor responses during prey capture by the ctenophore *Pleurobrachia*. Journal of Experimental Biology 114(1): 443–461.

Tamm SL and Tamm S (1991) Reversible epithelial adhesion closes the mouth of *Beroe*, a carnivorous marine jelly. Biological Bulletin 181: 463–473.

Tenney S (1875) Elements of Zoology – A Text Book. Scribner, Armstrong & Co Publishers, New York.

Weill MR (1935a) Le fonctionnement des colloblastes. Comptes Rendus de l'Académie des Sciences, Paris 201: 850–853.

Weill MR (1935b) Structure, origine et interpretation cytologique des colloblastes de *Lampetia pancerina* Chun (Ctenophores). Comptes Rendus de l'Académie des Sciences, Paris 17: 1628–1630.

4 Gastropod Secretory Glands and Adhesive Gels

Andrew M. Smith

Contents

4.1 Introduction

Gastropod molluscs are known for slime, yet the complexity and variety of their slimes is not always appreciated. These snails and slugs secrete visco-elastic mucous gels with functions that include feeding, protection, reproduction, locomotion, lubrication, defense, and adhesion (Denny, 1983). While the functional demands of such disparate tasks obviously vary widely, there has been little work on the biochemical variations and different secretory structures that give rise to these functional differences. The general structure and mechanics of Molluscan mucus have been reviewed (Denny, 1983; Smith, 2002), and the biochemical structure of some adhesive gels has been analyzed (Smith, 2006). Nevertheless, we still have a long way to go in characterizing the diversity of these gels and linking differences in structure to differences in their functional properties.

This is too bad, as gels are unusual materials with many important practical applications. We have much to learn about the diversity and potential of gels from molluscs; they are sophisticated gel architects, manufacturing custom materials with properties tailored to different functions. Detailed investigation of these gels could lead to significant advances in the development of novel materials.

While these gels are all named mucus, that term does not denote a common biochemical structure. In fact, among gastropods and probably most other animals, the term mucus is used for any viscous secretion from an epithelium (Davies and Hawkins, 1998). Such secretions can be created by any polymers that form giant complexes. Mammalian mucus is based on glycoproteins from the mucin family (Perez-Vilar and Hill, 1999; Silberberg and Meyer, 1982). These heavily glycosylated proteins form huge complexes that entangle at low concentrations to create a loose, slippery gel. Gastropod mucus can form gels by tangling of similarly huge polymers, though the polymers may be quite different from mucin with different amino acid compositions and different carbohydrates. Gastropod mucus often consists of protein-polysaccharide complexes with heavily charged carbohydrates that may or may not be firmly linked to proteins (Denny, 1983), as opposed to the small, numerous oligosaccharides that are covalently bound to mucin. Denny (1983) emphasizes the range of different protein-polysaccharide complexes that may be involved in mucus formation. The polysaccharides in gastropod mucus may consist of repeating units (Shashoua and Kwart, 1959), as in a glycosaminoglycan. In mucus, such polysaccharides are typically referred to as mucopolysaccharides. Mucopolysaccharides can be neutral or acidic, with varying degrees of charge. Because of the high charge density of some acidic polysaccharides, this difference is likely to play a role in the gel's properties. There are other important compositional differences among different forms of mucus. The adhesive mucus of the limpet *Lottia limatula* consists mostly of relatively short, cross-linked proteins with relatively little carbohydrate (Smith et al., 1999). The defensive secretion of the slug *Arion subfuscus* is also dominated by small proteins, and it also has a substantial metal component that plays an essential mechanical role (Werneke et al., 2007). The structural differences in both these gels coincide with physical properties that differ markedly from common lubricating gels, showing much greater elasticity and adhesiveness.

A diverse set of secretory glands corresponds to the diversity of gel types. Rarely do molluscan epithelia consist of merely one or two secretory cell types (Simkiss and Wilbur, 1977). There are often a variety of different glands secreting different materials. An analysis of these secretory structures and the components they produce should help our understanding of the structure and function of these gels. At present, little work has been done linking structure to function in these adhesive systems. While many gastropod epithelial glands have been identified, evidence supporting specific functions often comes only from the location and number of the glands. Simply put, the most common glands in a region secreting an adhesive might be the adhesive glands. In some cases, a gland is localized and common enough that its secretion can be uniquely identified as having a specific function. Often, though, the glands are scattered amongst each other over broad regions. In some cases a comparative approach can be used to identify adhesive glands, looking for glands that are more common in animals that form stronger attachments. Overall though, few histological studies in gastropods unequivocally identify a secretory gland as having a particular function. Instead, assessment of function has been based on circumstantial evidence or untested assumptions.

This chapter reviews the histology and histochemistry of glands that are involved, or potentially involved, in gastropod adhesion. Because of the number and variety of different glands associated with gastropod epithelia, and the frequent lack of evidence clearly demonstrating an adhesive function, this review will evaluate all the mucus-secreting glands that may contribute to adhesion. Through this comparative analysis, some common features may emerge that merit more detailed study.

4.2 Background

Notable adaptations for adhesion are widely spread among gastropods. The keyhole limpets and abalone of the Vetigastropoda and the true limpets of the Patellogastropoda (using the classification scheme of Bouchet and Rocroi (2005)) are best known for strong adhesion. Most of the work in this area has been on the true limpets. Further work on the keyhole limpets would be informative. Within the Caenogastropoda, the adhesive systems of periwinkles and slipper shells have been studied. The Heterobranchia contains terrestrial slugs and snails that produce adhesive secretions, as well as siphonarian limpets. Some of these appear to be defensive secretions. Finally, it should also be noted that the remaining major gastropod groups, the Cocculiniformia and Neritomorpha also include species that probably have adhesive systems. The Cocculiniformia includes deep water limpets while the Neritomorpha includes intertidal snails that may use an adhesive gel to adhere to rocks (pers. obs.).

In analyzing these adhesive systems, histochemical methods involving stains for specific components are often used. Some caution should be used in analyzing histochemical data, though. At best, they can only broadly characterize the adhesive secretion. The commonly used dyes often bind by fairly non-specific mechanisms. Many histochemical studies address this, using carefully controlled variations in staining conditions to distinguish different types of polysaccharides for instance. A common stain is alcian blue, which is used to detect mucopolysaccharides. At different pH levels it distinguishes between acid and neutral mucopolysaccharides. It is worth noting, though, that while the stain binds strongly to such molecules, its mode of action is merely based on electrostatic attractions to negatively charged polymers (Scott et al., 1964). Mucopolysaccharides are among the most negatively charged polymers in a tissue, but the dye will also interact with strongly negative proteins such as phosphoproteins (Scott et al., 1964). Such proteins may be present in glues; in fact, they are the main component of the cement of some tube worms (Stewart et al., 2004). Similarly, the commonly used dye mercuric bromophenol blue typically stains all proteins, but may do so by binding to a variety of groups, such as amino groups, sulfhydryl groups, aromatic groups and free carboxyls (Mazia et al., 1953). Furthermore, staining can be weakened if the main target groups are blocked or already interacting with something else (Mazia et al., 1953).

4.3 Limpets and Limpet-Like Molluscs

4.3.1 True Limpets

Limpets such as those in the genera *Lottia* and *Patella* adhere strongly to most surfaces using a mucus-based secretion. Limpet adhesion has attracted interest for many years due to its high attachment strength (Smith, 1991). Limpets can attach with a force per unit area of several kilograms per square centimeter (Smith, 2002). This is enough to render them almost immovable in the face of predators and crashing waves. They achieve this adhesion with a dilute gel. Limpets are also interesting because they can alternate between suction and gluing (Smith, 1991, 1992). This suggests the use of different gels. The gel used for gluing has been studied in *Lottia*, and it contains a protein that is absent from the mucus used during locomotion and suction (Smith et al., 1999). The correlation between the presence of this protein and an adhesive function suggests that this 118 kDa protein plays an important role in adhesion. Further research on similar proteins in other gastropods suggests that this is a gel-stiffening protein that may cross-link the gel (Pawlicki et al., 2004).

Another interesting aspect of limpet adhesion is that while most other mucous gels depend on the tangling of gigantic protein-carbohydrate complexes, Lottiid and Patellid limpets, and probably other limpets as well, crawl upon a gel that consists predominantly of proteins. These range in size from 20 to 200 kDa (Grenon and Walker, 1980; Smith et al., 1999). In several different limpet species that have been studied, the adhesive gels consist of roughly 3% protein (ranging from 1.8 to 3.9%) and 1% carbohydrate (ranging from 0.3 to 1.8%) (Smith, 2002). While the carbohydrate values are likely to be low because amino sugars are often not detected by such assays (Smith and Morin, 2002), this still represents a relatively protein-dominated secretion compared to most forms of mucus.

Grenon and Walker (1978) describe the glands and their secretory products in two limpet species, *Patella vulgata* and *Acmaea tessulata*. In *P. vulgata*, they identify nine different glands, with six of these secreting onto the sole (Fig. 4.1). Similar glands are found on the sole

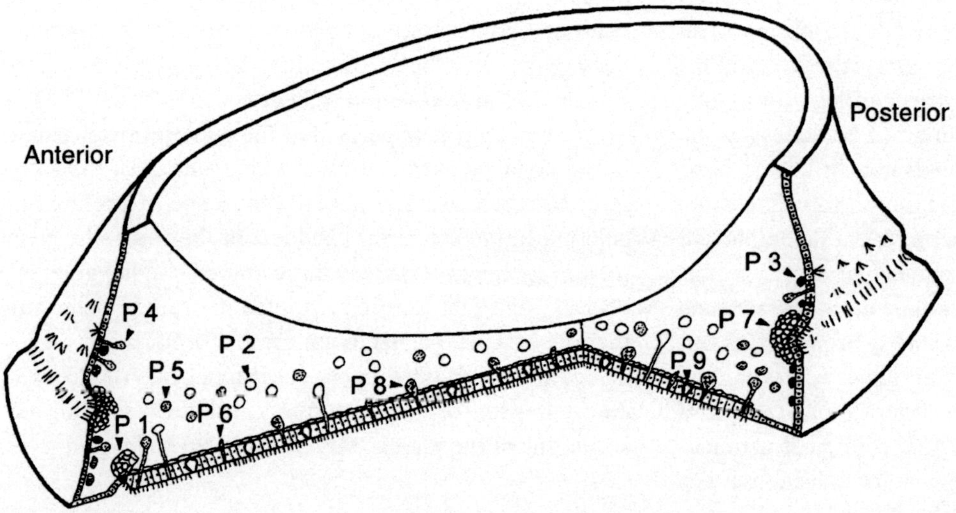

Fig. 4.1 Different secretory glands identified on the foot of the limpet *P. vulgata*. The secretory products of the different glands are as follows: *P1* proteins; *P2* acidic sulfated and non-sulfated and neutral mucopolysaccharides; *P3* proteins; *P4* mucoproteins; *P5* acidic sulfated and non-sulfated mucopolysaccharides; *P6* proteins; *P7* acidic sulfated and non-sulfated mucopolysaccharides; *P8* acidic sulfated and non-sulfated mucopolysaccharides; *P9* weakly acidic sulfated and non-sulfated mucopolysaccharides. Adapted from Grenon and Walker (1978) and reproduced with permission

of *A. tessulata*. The large number of distinct glands highlights the potential diversity of gels that may be secreted. There are four glands that are common over the entire surface of the sole, and these are identified as P2, P5, P8, and P9. Note that in most studies of gastropod secretory cells, the glands are identified by the first letter of the genus name, and an arbitrarily assigned number. These four glands secrete acid mucopolysaccharides, with a mixture of sulfated and non-sulfated sugars and some neutral mucopolysaccharides. It is striking that the histochemistry did not detect any proteins, knowing that the adhesive secretion in this species was found to contain twice as much protein as carbohydrate (Grenon and Walker, 1980). The stains used for protein were mercuric bromphenol blue, and a diazotization coupling to detect tyrosine.

The gland identified as P9 was described as a typical epithelial mucocyte, similar to the secretory cells found throughout the gastropoda on all their epithelia, including those that have poor adhesive ability (Grenon and Walker, 1978). It is a goblet cell and secretes weakly acidic mucopolysaccharides. The secretions of such glands generally serve a lubricating function. Davies and Hawkins (1998) suggest that these cells are actually adhesive in function, but there seems to be no evidence to support that other than their high density on the pedal sole.

Grenon and Walker (1978) hypothesize that the other three common glands of the sole (P2, P5, and P8) function in adhesion. They base this on the location of the glands and the fact that they secrete acidic mucopolysaccharides. Hunt (1973) suggested that acidic mucopolysaccharides might be more viscous. While a number of other researchers cite this observation as evidence to support an adhesive function, many common lubricating secretions also consist of acidic mucopolysaccharides. Thus, they are not inherently adhesive. In order to be adhesive, the viscosity of large polysaccharides is not sufficient; some cross-linking typically occurs (Smith, 2002). It is true, however, that acidic side groups provide a favorable site for cross-links. In particular, a number of acidic groups form strong interactions with alkaline earth metals such as calcium, and transition metals such as iron. This is important, given the finding that metals play a central role in cross-linking the glue of the terrestrial slug *A. subfuscus* (Werneke et al., 2007). Thus, while acidity of a mucopolysaccharide alone is not sufficient for adhesion, it is a plausible component of an adhesive mixture.

Glands P2, P5, and P8 are all subepithelial (Fig. 4.1) (Grenon and Walker, 1978), which is common for gastropod adhesive glands. The club-shaped P8 cells extend into the subepithelial space with a neck that is roughly 60 μm long, while flask-shaped P2 and P5 cells occur much deeper in the subepithelial space, with a long neck leading between epithelial cells to the outside. The detailed structure of the secretory vesicles of these glands was not visible in the wax sections used (7 μm). Interestingly, the P5 glands are more common around the edge of the foot. This would be the best place for an adhesive secretion as adhesive failure often begins at the edge (Smith, 2006), and the glue would help prevent that. Notably, Grenon and Walker (1978) did not detect any calcium using the von Kossa and Alizarin red stains in any of the glands from the limpets studied. Nevertheless, Grenon and Walker (1980) report that magnesium and calcium make up almost two-thirds of the inorganic content of the secreted gel. This discrepancy is important to clarify because calcium can have a large impact on gel mechanics (Smith, 2002), and may play a substantial role in terrestrial gastropod adhesives (see Sect. 4.5).

In addition to the common mucopolysaccharide-secreting glands of the sole, Grenon and Walker (1978) identify two glands that secrete protein on the sole. The P6 glands are relatively thin cells that penetrate through the epithelium into the subepithelial space much like the P8 glands. The P6 glands occur in very small numbers. The P1 glands are clusters of cells secreting into the marginal groove. This is a groove separating the ciliated epithelium of most of the sole from the thin strip of non-ciliated epithelia at the periphery of the sole. The P1 glands were noticed to be more active during locomotion following extended adhesion, and were interpreted to provide a mucus to crawl upon. The fact that the marginal groove only extends around the anterior third of the foot of *P. vulgata* argues against an adhesive function, though it extends all the way around in *A. tessulata* (Grenon and Walker, 1978).

In summary, it appears that the sub-epithelial glands of the pedal sole are the most likely source of the adhesive. Nevertheless, the apparent absence of protein and calcium in the secretory products of these glands, given their presence in relatively large amounts in the glue, is a central point that must be resolved in order to determine the role of these glands in adhesion. Furthermore, the focus of the work on limpets and limpet-like molluscs to date has been on histochemistry; higher resolution descriptions of the glands' structure would be helpful.

4.3.2 Abalone

Abalone provide an interesting comparison to limpets as they belong to a different clade than the true limpets and are often much larger, but they have a limpet-like body plan. It is not clear whether they adhere using glue, suction or some other mechanism. Thus, information on their epithelial glands is purely descriptive at this point. Lee et al. (1999) compare five species of abalone and describe a combination of epithelial and subepithelial mucocytes. Epithelial mucocytes occur primarily in the peripheral region of the sole. These contain predominantly neutral mucopolysaccharides in some species and acidic mucopolysaccharides in others. As with the patellid limpets, there are also abundant subepithelial glands, though these contain more neutral than acidic mucopolysaccharides. The authors did not use any protein stains, so it is unknown whether the secretions contain significant amounts of protein.

4.3.3 Slipper Shells

Slipper shells of the genus *Crepidula* also provide an interesting comparison to limpets. These are limpet-like gastropods belonging to the Caenogastropoda. They are known to attach strongly, leading a mostly sessile lifestyle. Early in life, however, they are relatively mobile. This allows a comparison between the secretory structures of younger, mobile individuals to those of older individuals that have firmly attached (Chaparro et al., 1998). Chaparro et al. (1998) identify four common types of mucus-secreting gland. Three are epithelial mucocytes, while a fourth is subepithelial and is more abundant in sessile animals. While this difference was statistically significant, the authors do not indicate the magnitude of the difference. Nevertheless, this correlation provides intriguing evidence for an adhesive function. Again, it is the subepithelial cells that are implicated in adhesion, while the epithelial mucocytes appear to play a larger role in locomotion. The authors do not provide any histochemical data for the glands.

4.4 Periwinkle Snails

Periwinkle snails such as the marsh periwinkle *Littorina irrorata* (also known as *Littoraria*) also form strong attachments using an adhesive gel (Smith and Morin, 2002). They glue the lip of their shell to intertidal surfaces in order to maintain their position above the tideline. Their adhesive mucus contains roughly equal amounts of protein and carbohydrate. In contrast, the mucus used in locomotion contains roughly the same amount of carbohydrate but much less protein. The carbohydrates appear to be in the form of mega-dalton sized polysaccharides or complexes of polysaccharides (Pawlicki et al., 2004). The protein in the glue consists almost exclusively of 41 and 36 kDa proteins that are unique to the glue. These were isolated and shown to have potent gel-stiffening ability (Pawlicki et al., 2004). Thus, they are likely to play a central role in cross-linking the glue.

The location of the adhesive glands in periwinkles has not been firmly established. Periwinkles produce the glue by "licking" the edge of their shell with the sole of their foot, starting with the anterior edge of the foot (Bingham, 1972). Based on this, the anterior pedal gland and anterior region of the sole may both contribute to the glue. Sirbhate and Cook (1987) describe five specific gland types within these regions. All five of the gland types are sub-epithelial. Two of the glands, L1 and L2, are associated into a structure called the anterior pedal gland. This forms an arc along the foot's leading edge. The L1 cells are the most common, filling a large part of the anterior pedal gland. They secrete a mucoprotein, and the secretory material is reticular. The L2 cells are smaller, with slender necks, and they group at the anterior end of the gland. They secrete a granular material that stains positively for neutral mucoprotein, with strong protein staining and weaker staining for neutral mucopolysaccharides. The presence of protein in both cases is based on mercuric bromophenol blue staining. Protein in the L1 glands was also verified by tyrosine staining using diazotization coupling. On the sole of the foot are three subepithelial glands, named L3, L4, and L5. L3 secretes sulfated mucopolysaccharides, while L4 secretes carboxylated mucopolysaccharides. L4 occurs throughout the sole but more commonly at the anterior end of the foot. Finally, L5 is relatively rare and secretes protein and neutral mucopolysaccharides. As with limpets, the work focuses on histochemical staining, and does not provide high resolution of the secretory structures.

Overall, the sole contributes mostly mucopolysaccharides, as was true for limpets, while the anterior pedal gland contributes both mucopolysaccharides and protein. Given the importance of protein in cross-linking the glue, and the location of the anterior pedal gland where glue formation begins, the anterior pedal gland is a likely candidate for a primary adhesive gland. The glands of the pedal sole may contribute, and based on the description in Shirbhata and Cook (1987), are similar to the subepithelial pedal glands that have been implicated in limpet adhesion. The authors assume that acid mucopolysaccharides are adhesive, again based on Hunt's note that these would form more viscous gels than neutral mucopolysaccharides (1973). As with limpets, little calcium was detected with the von Kossa and Alizarin red stains (Shirbhate and Cook, 1987).

4.5 Land Snails

Some land snails adhere to surfaces with an epiphragm (Campion, 1961). This is a layer of mucus that dries into a thin, translucent sheet connecting all around the lip of the shell to the substratum. It simultaneously attaches the animal to the substratum and seals the edge to prevent moisture loss during estivation. The epiphragm contains primarily mucoprotein with some calcium, particularly in a slightly porous area that allows ventilation (Barnhart, 1983). Roughly half of the organic material in the epiphragm of *Helix aspersa* consists of three proteins with masses of 82, 97, and 175 kDa (Pawlicki et al., 2004). When purified, these proteins exert a similar gel-stiffening effect to periwinkle glue proteins. The remainder is a megadalton-sized component containing protein and carbohydrate.

Campion (1961) describes the glandular structures responsible for secreting the epiphragm in *H. aspersa*. For the snail as a whole, a total of eight types of gland cell occur in different regions of the skin. The glands that secrete the epiphragm appear to be those on the mantle collar, because this is the region that produces the epiphragm. There are three common gland cells in this region (Fig. 4.2), with a fourth that is more rare. They are large, single-celled sub-epithelial glands, typically ranging from 500 to 800 μm long.

One of the most common glands of the mantle collar is the mucus gland type A (Campion, 1961). This secretes an acidic, probably sulfated, mucopolysaccharide. Inside the gland, the secretion has a reticular appearance. The secretion of the less-common type B gland is histochemically similar but has a granular appearance. Campion (1961) attributes a lubricating function to these acid mucopolysaccharides, commenting that secretions that are histochemically similar to the acid mucopolysaccharides of the type A gland are found throughout the animal kingdom serving lubricating functions. Furthermore, a lubricating function would make sense for the ventral mantle epithelium. A similar type of gland cell

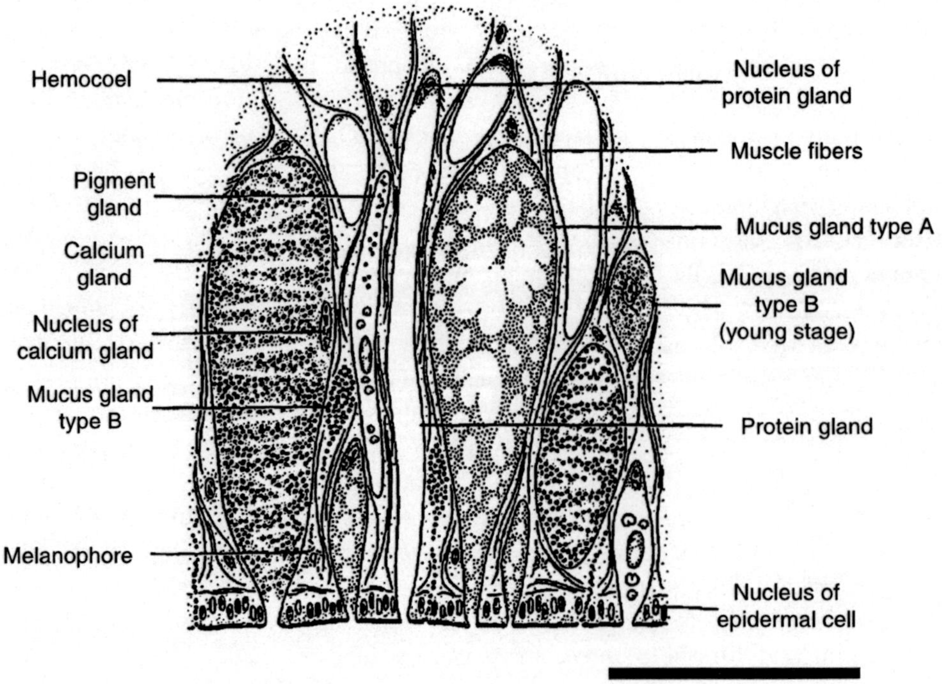

Fig. 4.2 The glands from the epiphragm-secreting region of the mantle collar of *H. aspersa*. Mucus-gland type A contains a reticular material that stains positively for acid mucopolysaccharides. Mucus-gland type B is similar, but less abundant and with a more granular secretion. The protein gland secretes a homogeneous material staining positively for protein. It may be homologous to the channel cells of terrestrial slugs. The calcium gland secretes calcium granules in a matrix of basic protein. Adapted from Campion (1961) and reproduced with permission. Scale bar = 200 μm

is also the primary gland on the pedal sole, which is not known to secrete any form of adhesive mucus.

The second prominent cell type in the ventral mantle collar is the calcium gland (Campion, 1961). This is physically similar to the mucus type A and B glands, but contains rod-shaped or spherical granules in a matrix of basic protein. Several methods confirm the presence of significant amounts of calcium in this material. Campion (1961) suggests that this calcium may play a role in cross-linking, causing the mucus to become more viscous. The reported presence of basic protein is intriguing, as this would interact with the negatively charged mucopolysaccharides. Furthermore, these cells appear to be much more common in the epiphragm-secreting region. They are not described in the pedal sole, though they are shown as a minor component in the diagrammatic sketch accompanying the text. Campion (1961) also noted that

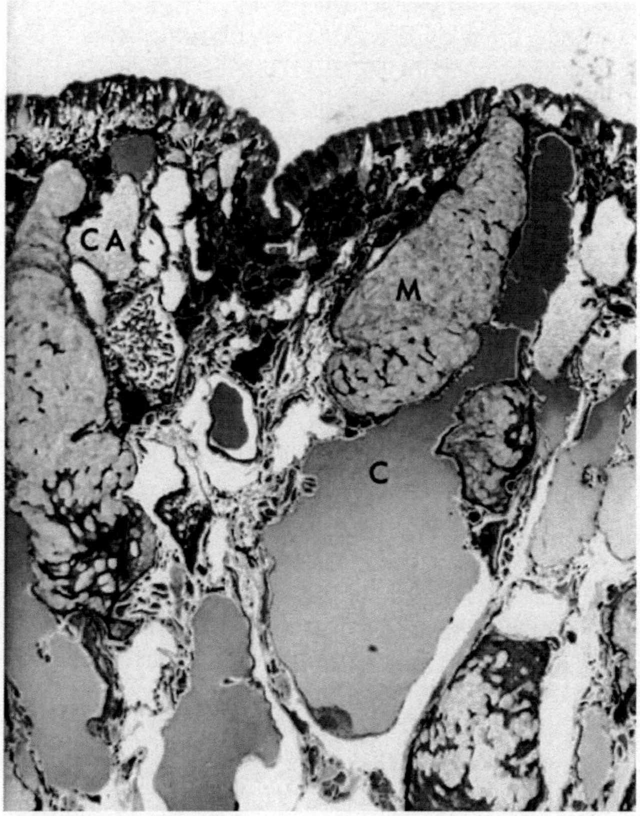

Fig. 4.3 The glands from the dorsal surface of the slug *A. columbianus*. The overlying epithelium is simple cuboidal. The primary subepithelial glands are as follows: the channel cells (*C*) secrete protein and transmit a large fluid volume, the mucous cells (*M*) secrete acidic mucopolysaccharides, and the calcium cells (*CA*) secrete calcium and protein. Thin section (1 μm) stained with toluidine blue, ×300. Adapted from Luchtel et al. (1984) and reproduced with permission

the calcium content was much lower in the pedal mucus secreted by the sole.

The final gland type common in the epiphragm-secreting region is the protein gland (Campion, 1961). These have homogeneous, lightly granular secretions that contain mostly protein. Their secretion is described as watery, and large quantities of this secretion are released to flush away irritants. This gland may be homologous to the channel cells in the slug body wall (Fig. 4.3).

4.6 Terrestrial Slugs

The adhesive glands of terrestrial slugs in the genera *Ariolimax* and *Arion* are perhaps the best studied in the gastropoda, and the connection between these glands and adhesion seems to be the firmest. These slugs secrete copious amounts of a thick mucus from their dorsal surface. The total amount of the secretion can reach 5.5% of the slug's body weight (Martin and Deyrup-Olsen, 1986). This is likely a defensive secretion (Mair and Port, 2002), as it is highly sticky and elastic, may deter predation and is secreted in response to irritation. It starts out as a viscous slime, but within 15–120 sec sets into a sticky, resilient mass. Deyrup-Olsen et al. (1983) note that this is quite distinct from the thin, slippery viscous mucus used in locomotion. In addition to the macroscopic differences, these authors note that the sticky secretion forms diffuse networks of strands when stressed and when in the presence of calcium. The secretion of the slug *Arion subfuscus* was studied in depth (Pawlicki et al., 2004). As with limpets, periwinkles and land snails, the adhesive mucus contains specific proteins that distinguish it from the pedal mucus used in locomotion. Both forms of slug mucus contain giant, carbohydrate-containing protein complexes and smaller proteins, but about a quarter of the organic material in the glue consists of a 15 kDa protein. This protein also triggers gel-stiffening (Pawlicki et al., 2004).

The adhesive-secreting dorsal epithelial of the slug *Ariolimax columbianus* has been studied the most thoroughly. As with the land snail epiphragm-secreting cells, there are three primary glands in the dorsal surface. These were identified as mucus cells, calcium cells and channel cells (Luchtel et al., 1984) (Fig. 4.3). Luchtel et al. (1984) suggest that the channel cells are homologous to the protein glands identified by Campion (1961). Martin and Deyrup-Olsen (1986) state that the calcium and mucus glands are responsible for secreting the adhesive. They found that inhibiting the channel cells does not seem to interfere with production of the sticky mucus. The mucus cells contain vesicles with a reticulated appearance while the calcium cells have calcium-containing granules ("calcospheres") (Luchtel et al., 1984).

The secretory cells of the slug *Incilaria fruhstorferi* appear roughly similar, but with only two types of secretory cells common on the dorsal surface (Yamaguchi et al., 2000). One is a channel cell and the other a tubular mucocyte. The mucocyte secretes acidic mucopolysaccharides. A similar round mucocyte is more common on the ventral surface (the pedal sole) and stains weakly for protein but not acid mucopolysaccarides. The absence of calcium cells is notable. It is possible that they were not distinguished from other mucocytes by their staining methods, as no calcium stains were used, but the calcium glands would likely still have been distinct. They would also likely have been detected by the protein stain if they were present. Alternatively, the slug may not have that gland because they do not produce a strongly adhesive mucus. The authors do not attribute any adhesive properties to the dorsal mucus, and the description in the paper suggests that it serves primarily to lubricate and protect the epithelium. Arcadi (1967) also found only two types of mucus-secreting cell in the slug *Lehmania poirieri*. These corresponded morphologically to the channel cells and mucus cells described previously, with no calcium cells. This slug has also not been noted for the production of a sticky defensive secretion, though this has not been studied systematically. Cook and Shirbhate (1983) found three primary secretory glands in the dorsal epithelium of *Limax pseudoflavus*; two of these secrete protein and one secretes acid mucopolysaccharides. One of the protein secreting cells appears similar to channel cells and the mucopolysaccharide-secreting cell seems typical of the other slugs, but the second protein-secreting cell is different in being deeper in the sub-epithelial space. Most notably, no calcium staining was detected. Cook and Shirbhate (1983) report that the only functions of the slug's mucus are locomotion, cleansing and communication. Unlike *Ariolimax* and *Arion*, these slugs reportedly only secrete a "very watery" mucus from their dorsal surface in response to irritation or mechanical stimulation.

Given that there is a wide variety of terrestrial slugs, some of which produce markedly sticky forms of mucus while others do not, there is potential for an interesting comparative study. By comparing the glands present in the epithelia of slugs that produce strong glue with the glands present in the epithelia of slugs that only produce lubricating slime, it may be possible to identify the essential adhesive glands with more confidence. The preliminary analysis here suggests a correlation between adhesion and the presence of the calcium glands. This would be consistent with the importance of proteins and the finding that chelation of metals including calcium disrupts the glue of Arionid slugs (Werneke et al., 2007; Smith et al., 2009).

The mode of mucus secretion is note-worthy, as it may hold information relevant to understanding how the glue sets. Slugs secrete the mucus in microscopic packets that can be collected and kept in stable form in the laboratory (Deyrup-Olsen et al., 1983). These packets rupture to form a uniform visco-elastic secretion. Similar packets are also present in land snail mucus. Packets of mucus have also been seen in the secretion of epithelial mucocytes from *P. vulgata* (Davies and Hawkins, 1998). Before packet rupture, the secretion from slugs flows easily, but it thickens into a strong adhesive upon rupture (Luchtel et al., 1991). Rupture depends on calcium and can be triggered by ATP or shear stress (Luchtel et al., 1991). Given that the mucus probably plays a defensive role, it is interesting that shear easily ruptures the packets. Thus, anything that rubbed the dorsal surface of the slug would trigger the secretion and formation of a sticky mucus that would then adhere to their own skin. Without shear, the mucus would just flow off the slug (pers. obs.). More interestingly, the process of rupturing the packets to release their contents raises the potential of a two-component adhesive where packets from two different cell types can be mixed together then ruptured to allow reaction and cross-linking between the different materials.

Finally, the channel cells may play an indirect role in adhesion. These cells appear adapted to move large volumes of fluid across the epithelium. Thus, they may control the concentration and hence to some extent the mechanics of the secretion (Luchtel et al., 1984; Martin and Deyrup-Olsen, 1986). More likely, they provide a way to cleanse irritants or debris off the dorsal surface (Luchtel et al., 1984). The channel glands also contain basic protein, though, being strongly eosinophilic. They may also contain other substances as many large molecules that were injected into the body cavity show up in the secretion as well – even those as large as ink particles and hemoglobin (Luchtel et al., 1984).

The histological evidence combined with the biochemical analysis of the glue suggest a possible model for adhesion. The mucus glands secrete megadalton-size mucoproteins or mucopolysaccharides, while the calcium cells may secrete smaller proteins. This makes the calcium glands a likely source of the gel-stiffening proteins identified by Pawlicki et al. (2004). The importance of calcium may partially account for the impact of EDTA in disrupting the gel and blocking gel stiffening (Werneke et al., 2007; Smith et al., 2009). When the mucus packets rupture, these components would be mixed, which could allow cross-linking by metals and the gel-stiffening proteins.

An interesting comparison to the temperate slugs *Arion* and *Ariolimax* is the tropical slug *Veronicella*. This slug also produces a sticky mucus on its dorsal surface in response to irritation, but the slug belongs to a different group within the pulmonates, and is relatively distinct morphologically. Of the 11 gland types on the skin, two or three secrete on the dorsal surface (Cook, 1987). Unlike other slugs, the gland cells secrete into a common duct, which then empties onto the surface. The most common cell (V9) secretes neutral and weakly acidic mucopolysaccharides. The V11 cells are similar in location and histochemistry, and Cook (1987) suggests that they may just be V9 cells in a different stage of secretion. The V10 cells secrete protein.

Cook (1987) compares the dorsal secretion to that of *Limax pseudoflavus*. *Limax* dorsal mucus consists of heavily sulfated mucopolysaccharides, while the bulk of *Veronicella* mucus is weakly acidic, non-sulfated mucopolysaccharides. Nevertheless, observationally, *Veronicella* mucus is at least as viscous, if not more. This provides further evidence against the common assumption that acidic mucopolysaccharides should be more adhesive by their nature, and suggests that the protein secretion of the V10 glands is important.

A final adhesive system that has been studied are the defensive glands of limpets in the genus *Siphonaria* (Pinchuck and Hodgson, 2009). Siphonarians are pulmonates that have re-invaded the marine habitat and converged on a limpet-like form. They secrete a white, sticky defensive secretion in response to mechanical stimulation. Some chemical analysis has been performed identifying defensive compounds. It is worth noting that there is also anecdotal evidence for defensive chemicals in the mucus of Arionid slugs.

The defensive secretion of *Siphonaria* comes from clearly defined multi-cellular glands that are unusual among gastropods (Pinchuck and Hodgson, 2009). The glands are oval capsules that connect to the outside through distinct pores. Each capsule contains a number of cells emptying into a common duct (Fig. 4.4). There are two histochemically distinct cell types within the gland, which Pinchuck and Hodgson (2009) identify as Type I and II. Both contain mucopolysaccharides, with Type I secreting neutral and sulfated mucopolysaccharides and Type II secreting strongly acidic, sulfated mucopolysaccharides. Neither secretes protein, based on mercuric bromophenol blue staining, and there is little rough endoplasmic reticulum. The secretion of Type I cells is more electron lucent and reticular, while that of the Type II cells is more electron dense, vesicular and granular. The authors show several micrographs where the glands are in the process of secreting, and interestingly the secretions of the two cells are distinct and do not appear to mix within the duct, though the authors state that that occurs. The mucus does not appear to be secreted in packets.

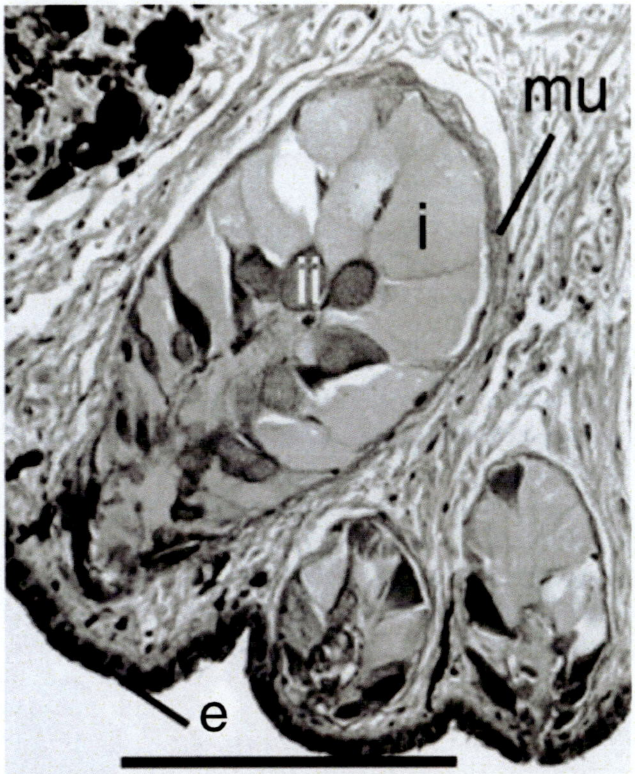

Fig. 4.4 The multicellular secretory gland of *S. capensis*. This section was stained with Mallory's trichrome and shows Type I and II secretory cells (*i* and *ii*). Type I cells secrete neutral and sulfated mucopolysaccharides and Type II cells secrete strongly acidic, sulfated mucopolysaccharides. The epithelium (*e*) and muscle (*mu*) are also indicated. Scale bar is 100 μm. Adapted from Pinchuck and Hodgson (2009) and reproduced with permission

4.7 Summary

A wide variety of glandular cells contribute to gastropod mucus. While there are hypotheses for the function of each of these, it is often uncertain exactly which cells contribute to the adhesive mucus, and what roles the different components play. Unicellular subepithelial glands typically appear important, though in three unrelated genera (*Littorina*, *Veronicella*, and *Siphonaria*) the gland cells unite around a common duct to form a multicellular gland. In many animals, acidic mucopolysaccharides are common in the putative adhesive cells, but this is not universally true. Furthermore, acidic mucopolysaccharides are unlikely to be adhesive on their own, as they are often present in lubricating secretions as well. Finally, another gland cell implicated in adhesion in terrestrial slugs secretes a granular material that stains positively for protein and calcium.

Once the adhesive glands have been definitively identified, further histological work needs to be performed. Much of the work on these glands has been in the context of a general survey of all the glands of a given species, typically using 5–10 μm wax sections which do not provide as much detail as thinner sections. A more thorough description of the glands that produce adhesive secretions would be helpful in comparing the adhesive systems.

References

Arcadi JA (1967) The two types of mucous gland cells in the integument of the slug, *Lehmania poirieri* (Mabille): a study in metachromasy. Transactions of the American Microscopical Society 86: 506–509.

Barnhart MC (1983) Gas permeability of the epiphragm of a terrestrial snail, *Otala lactea*. Physiological Zoology 56: 436–444.

Bingham FO (1972) The mucus holdfast of *Littorina irrorata* and its relationship to relative humidity and salinity. The Veliger 15: 48–50.

Bouchet P and Rocroi JP (2005) Classification and nomenclator of gastropod families. Malacologia 47(1–2): 1–397.

Campion M (1961) The structure and function of the cutaneous glands in *Helix aspersa*. Quarterly Journal of Microscopical Science 102: 195–216.

Chaparro OR, Bahamondes-Rojas I, Vergara AM, and Rivera AA (1998) Histological characteristics of the foot and locomotory activity of *Crepidula dilatata* Lamarck (Gastropoda: Calyptraeidae) in relation to sex changes. Journal of Experimental Marine Biology and Ecology 223: 77–91.

Cook A (1987) Functional aspects of the mucus-producing glands of the systellommatophoran slug, *Veronicella floridana*. Journal of Zoology, London 211: 291–305.

Cook A and Shirbhate R (1983) The mucus producing glands and the distribution of the cilia of the pulmonate slug *Limax pseudoflavus*. Journal of Zoology, London 201: 97–116.

Davies MS and Hawkins SJ (1998) Mucus from marine molluscs. Advances in Marine Biology 34: 1–71.

Denny M (1983) Molecular biomechanics of molluscan mucous secretions. In: Hochachka PW (ed) The Mollusca – Metabolic Biochemistry and Molecular Biomechanics, 1st Ed. Academic Press, New York: pp 431–465.

Deyrup-Olsen I, Luchtel DL, and Martin AW (1983) Components of mucus of terrestrial slugs (Gastropoda). American Journal of Physiology 245: R448–R452.

Grenon JF and Walker G (1978) The histology and histochemistry of the pedal glandular system of two limpets, *Patella vulgata* and *Acmaea tessulata* (Gastropoda: Prosobranchia). Journal of the Marine Biological Association of the United Kingdom 58: 803–816.

Grenon JF and Walker G (1980) Biochemical and rheological properties of the pedal mucus of the limpet, *Patella vulgata* L. Comparative Biochemistry and Physiology Series B Biochemistry & Molecular Biology 66: 451–458.

Hunt S (1973) Fine structure of the secretory epithelium in the hypobranchial gland of the prosobranch gastropod mollusc *Buccinum undatum* L. Journal of the Marine Biological Association of the United Kingdom 53: 59–71.

Lee C, Moon DY, Jee YJ, and Choi BT (1999) Histochemistry of mucosubstances on the pedal sole of five abalone species. Korean Journal of Biological Sciences 3: 253–258.

Luchtel DL, Martin AW, and Deyrup-Olsen I (1984) The channel cell of the terrestrial slug *Ariolimax columbianus* (Stylommatophora, Arionidae). Cell Tissue Research 235(1): 143–151.

Luchtel DL, Deyrup-Olsen I, and Martin AW (1991) Ultrastructure and lysis of mucin-containing granules in epidermal secretions of the terrestrial slug *Ariolimax columbianus* (Mollusca: Gastropoda: Pulmonata). Cell Tissue Research 266: 375–383.

Mair J and Port GR (2002) The influence of mucus production by the slug, *Deroceras reticulatum*, on predation by *Pterostichus madidus* and *Nebria brevicollis* (Coleoptera: Carabidae). Biocontrol Science and Technology 12: 325–335.

Martin AW and Deyrup-Olsen I (1986) Function of the epithelial channel cells of the body wall of the terrestrial slug *Ariolimax columbianus*. Journal of Experimental Biology 121: 301–314.

Mazia D, Brewer PA, and Alfert M (1953) The cytochemical staining and measurement of protein with mercuric bromphenol blue. Biological Bulletin 104: 57–67.

Pawlicki JM, Pease LB, Pierce CM, Startz TP, Zhang Y, and Smith AM (2004) The effect of molluscan glue proteins on gel mechanics. Journal of Experimental Biology 207: 1127–1135.

Perez-Vilar J and Hill RL (1999) The structure and assembly of secreted mucins. Journal of Biological Chemistry 274: 31751–31754.

Pinchuck SC and Hodgson AN (2009) Comparative structure of the lateral pedal defensive glands of three species of *Siphonaria* (Gastropoda: Basommatophora). Journal of Molluscan Studies 75: 371–380.

Scott JE, Quintarelli G, and Dellovo MC (1964) The chemical and histochemical properties of alcian blue: I. The mechanism of alcian blue staining. Histochemie 4: 73–85.

Shashoua VE and Kwart H (1959) The structure and constitution of mucus substances. II. The chemical constitution of *Busycon* mucus. Journal of the American Chemical Society 81: 2899–2905.

Shirbhate R and Cook A (1987) Pedal and opercular secretory glands of *Pomatias*, *Bithynia* and *Littorina*. Journal of Molluscan Studies 53: 79–96.

Silberberg A and Meyer FA (1982) Structure and function of mucus. In: Chantler EN, Elder JB, and Elstein M (eds) Mucus in Health and Disease – II, Vol. 144. Plenum Press, New York: pp 53–74.

Simkiss K and Wilbur KM (1977) The molluscan epidermis and its secretions. Symposia of the Zoological Society of London 39: 35–76.

Smith AM (1991) The role of suction in the adhesion of limpets. Journal of Experimental Biology 161: 151–169.

Smith AM (1992) Alternations between attachment mechanisms by limpets in the field. Journal of Experimental Marine Biology and Ecology 160: 205–220.

Smith AM (2002) The structure and function of adhesive gels from invertebrates. Integrative and Comparative Biology 42: 1164–1171.

Smith AM (2006) The biochemistry and mechanics of gastropod adhesive gels. In: Smith AM and Callow JA (eds) Biological Adhesives. Springer-Verlag, Heidelberg: pp 167–182.

Smith AM and Morin MC (2002) Biochemical differences between trail mucus and adhesive mucus from marsh periwinkle snails. Biological Bulletin 203: 338–346.

Smith AM, Quick TJ, and St. Peter RL (1999) Differences in the composition of adhesive and non-adhesive mucus from the limpet *Lottia limatula*. Biological Bulletin 196: 34–44.

Smith AM, Robinson TM, Salt MD, Hamilton KS, Silvia BE, and Blasiak R (2009) Robust cross-links in molluscan adhesive gels: testing for contribution from hydrophobic and electrostatic interactions. Comparative Biochemistry and Physiology Series B Biochemistry & Molecular Biology 152: 110–117.

Stewart RJ, Weaver JC, Morse DE, and Waite JH (2004) The tube cement of *Phragmatopoma californica*: a solid foam. Journal of Experimental Biology 207: 4727–4734.

Werneke SW, Swann CL, Farquharson LA, Hamilton KS, and Smith AM (2007) The role of metals in molluscan adhesive gels. Journal of Experimental Biology 210: 2137–2145.

Yamaguchi K, Seo N, and Furuta E (2000) Histochemical and ultrastructural analyses of the epithelial cells of the body surface skin from the terrestrial slug, *Incilaria fruhstorferi*. Zoological Science 17: 1137–1146.

5 Characterization of the Adhesive Systems in Cephalopods

Norbert Cyran, Lisa Klinger, Robyn Scott, Charles Griffiths, Thomas Schwaha, Vanessa Zheden, Leon Ploszczanski and Janek von Byern

Contents

"Amongst the many things which are in the ocean, and concealed from our eyes, or only presented to our view for a few minutes, is the Kraken. This creature is the largest and most surprising of all the animal creation, and consequently well deserves such an account as the nature of the thing, according to the Creator's wise ordinances, will admit of. Such I shall give at present, and perhaps much greater light on this subject may be reserved for posterity [...]."

Pontoppidan (1755)

5.1 Introduction

Cephalopods are highly evolved invertebrates; since ancient times, they have been admired for their intelligence, their ability to change color within milliseconds and their flexible arms, equipped with suckers or hooks. The suckers are versatile, mainly used to attach mechanically (by a reduced-pressure systems with a low pressure up to 0.01 MPa) to hard or soft surfaces (Smith, 1996; Kier and Smith, 2002; Pennisi, 2002); its usage and force strength varies, from a "soft sensing" of unknown objects to a fast and forceful holding of resisting prey. The suckers also have a sensory function and are equipped with a large repertoire of numerous mechano- and chemoreceptors (Nixon and Dilly, 1977).

In addition to this well-known mechanical bonding system, four genera of cephalopods belonging to four different families (*Euprymna*, Sepiolidae; *Idiosepius*, Idiosepiidae; *Nautilus*, Nautilidae and *Sepia*, Sepiidae) produce glue for temporary attachment (von Byern and Klepal, 2006). Although in all four cases the glue is provided by epithelial gland structures, the localization and function vary according to the species.

Euprymna uses adhesives to cover itself with a coat of sand or mud for camouflage. When threatened, the animals release the cover instantaneously to deflect predators (Singley, 1982; Shears, 1988).

By contrast, in *Idiosepius* only a small area on the dorsal mantle side is concerned with adhesion. The animals attach themselves to leaves of sea grass or algae for camouflage and/or prey capture (Sasaki, 1921; Moynihan, 1983; Suwanmala et al., 2006b).

In *Nautilus* the adhesive structures are present only on the digital tentacles. They are used to hold prey and to attach to the substratum or to other individuals during mating.

Adhesion in *Sepia* is mainly induced mechanically by a defined dermal structure on the ventral mantle. Additionally, chemical substances secreted from this adhesive area might be used to increase the strength of attachment.

So far chemical adhesive systems in molluscs are mostly associated with mussels (see Chapter 18, p. 273) and partly with gastropods (see Chapter 4, p. 41), however, cephalopods are still far out of the focus. Nevertheless these four species demonstrate a larger and much more specialized rang of glue applications (camouflage, prey capture, for egg spawning, mating) than the simple attachment to the substratum. Nevertheless, our knowledge is still marginal and we are far away from rudimentary understanding their bonding mechanisms as given for the byssus threads of mussels.

So far, for some of the here presented cephalopod adhesive systems (*Euprymna*, *Nautilus*, *Sepia*) an duo-gland mechanism was proposed as given for echinoderms (Flammang, 1996; Haesaerts et al., 2005; Santos et al., 2005) and other temporarily binding animals (Adams and Tyler, 1980; Tyler and Rieger, 1980; Gelder and Tyler, 1986).

This type of bonding describes an antagonistic system carried out by an adhesive and a release secretion which requires at least two distinct cell types (Hermans, 1983). However, several aspects of the following re-evaluation argue against a duo-gland mechanism in the four cephalopods bonding systems.

Remarks

The designations of the respective secretory epithelial cell types depend sometimes less on functional aspects or shape than on the nomenclature used by the authors. Since the adhesive system in *Sepia* and partly in *Nautilus* is described in this book for the first time, we avoided naming the glandular cells, choosing rather to number them consecutively. Table 5.1 in the chapter "Conclusion" is designed to shed some light on the overall confusion and provide a summary of each secretory cell type and its histochemical reactivity.

5.2 *Euprymna*
(*Lisa Klinger, Janek von Byern and Norbert Cyran*)

5.2.1 Introduction

The association between *Euprymna scolopes* and the luminous bacterium *Vibrio fischeri* has been studied for over 15 years as a model for the establishment, development, and maintenance of horizontally transmitted symbioses (McFall-Ngai and Ruby, 1991, 1998; McFall-Ngai, 2002; Nyholm and McFall-Ngai, 2004). The squid maintains the bacteria extracellularly in a ventral tissue complex, the light organ, and feeds a sugar and amino acid solution to its symbiont. *Vibrio fischeri* produces light that hides the squid's silhouette when viewed from above, which means camouflage and protection from predators (Singley, 1982 McFall-Ngai and Montgomery, 1990; Claes and Dunlap, 2000).

Nonetheless, the adhesive substances of these animals are used in different ways, and studies on these different mechanisms allow a comparison between the various genera. Morphological studies show that *E. scolopes* possesses multiple adhesive glands in the dorsal epidermis by which it affixes sand to its body surface (Singley, 1982). Earlier observations indicated a duo-gland adhesive system (Hermans, 1983) to be responsible for adhesion and de-adhesion, but the present study suggests a chemically induced bonding and a release by muscles.

Extensive studies, however, have only been conducted in *Euprymna scolopes*. These concern gene activity (Small and McFall-Ngai, 1999; Foster et al., 2000; Nyholm et al., 2002), structure and function of the hatching gland (Hoyle organ) (Arnold et al., 1972; Arnold, 1990), as well as structure and histochemistry of the adhesive region (Singley, 1982, 1983). Information about an adhesive behavior in other *Euprymna* species still needs to be verified.

5.2.2 Systematics

Within the genus *Euprymna*, most of the species are well defined according to their morphological characteristics. The genus *Euprymna* was first defined by Steenstrup in 1887 within the family Sepiolidae in the order Sepioidea (Voss, 1977). As it currently stands, twelve nominal species have been characterized:

- *E. morsei* (Verrill, 1881) from Japan
- *E. albatrossae* Voss, 1963 from the Philippines
- *E. berryi* Sasaki, 1929 from Japan, China, and the Gulf of Tonkin
- *E. hoylei* Adam, 1986 from the Sulu Archipelago, Philippines
- *E. hyllebergi* Nateewathana, 1997 from the Andaman Sea
- *E. penares* (Gray, 1849) from Singapore
- *E. phenax* Voss, 1963 from the Philippines
- *E. scolopes* Berry, 1913 from the Hawaiian Islands
- *E. stenodactyla* (Grant, 1833) from the Western Indo-Pacific (Mascarene Islands to Queensland and Polynesia)
- *E. tasmanica* (Pfeffer, 1884) from the Bass Strait, Australia
- *E. bursa* (Pfeffer, 1884) from Hong Kong
- *E. similis* (Sasaki, 1913) from Japan

Most species are distinguished only by the number and position of enlarged suckers in mature males. No diagnostic characters are available to identify females or immature male specimens. At least two additional unresolved taxa exist in Australian waters. Preliminary DNA analysis has demonstrated a distinction into locally occurring species, but no morphological characters have yet been found to distinguish these taxa from each other (Norman and Lu, 1997).

5.2.3 Ecology

Euprymna scolopes is a very small species, endemic to the shallow waters of the Hawaiian Islands. It has a mantle length of approximately 3.5 cm and a weight of up to 2.7 g. Juveniles appear to behave identically to adults: they remain buried in the sand during the day and hunt for prey at night. Light intensity is apparently decisive for successful feeding: lower light enhances feeding, whereas bright light seems to retard it (Hanlon et al., 1997).

Females deposit serial clutches of eggs on the underside of coral ledges or other hard substrates, but do not tend the eggs, as is characteristic for some cephalopod species. Instead, they cover the eggs with a patina of sand, and the embryos develop independent of parental care. The embryonic period depends on the temperature and ranges from 18 to 26 days (Arnold et al., 1972). Like other cephalopods, *E. scolopes* does not have a true larval stage; the juvenile animal hatches as a miniature adult. This quick development is inevitable because of their short lifespan (3–10 months).

Since *E. scolopes* spends much of its life buried in the sand, it developed a special technique to adhere sand grains to its dorsal mantle and head, using the second arm pair to rake sand over, to form a "sand coat" (Singley, 1982; Shears, 1988). This sand coat remains attached to the animal when flushed from the substrate during daylight hours, but not at night. Presumably the coat acts as camouflage over the matching substrate during daylight hours. The animals are capable of instantaneously shedding this coat (Singley, 1982).

5.2.4 Gland Morphology

The adhesive structures of *E. scolopes* are located in the dorsal epidermis (Fig. 5.6A). The epithelial secretory system produces a mucous coat and gives the squid the ability to fix sand and other bottom materials to the dorsal surface of its body while moving (Singley, 1982). Four types of gland cells occur in this dorsal adhesive region, whereby only two of them are present ventrally. All cell types span the full thickness of the epithelium and release secretory material to the surface (Fig. 5.1).

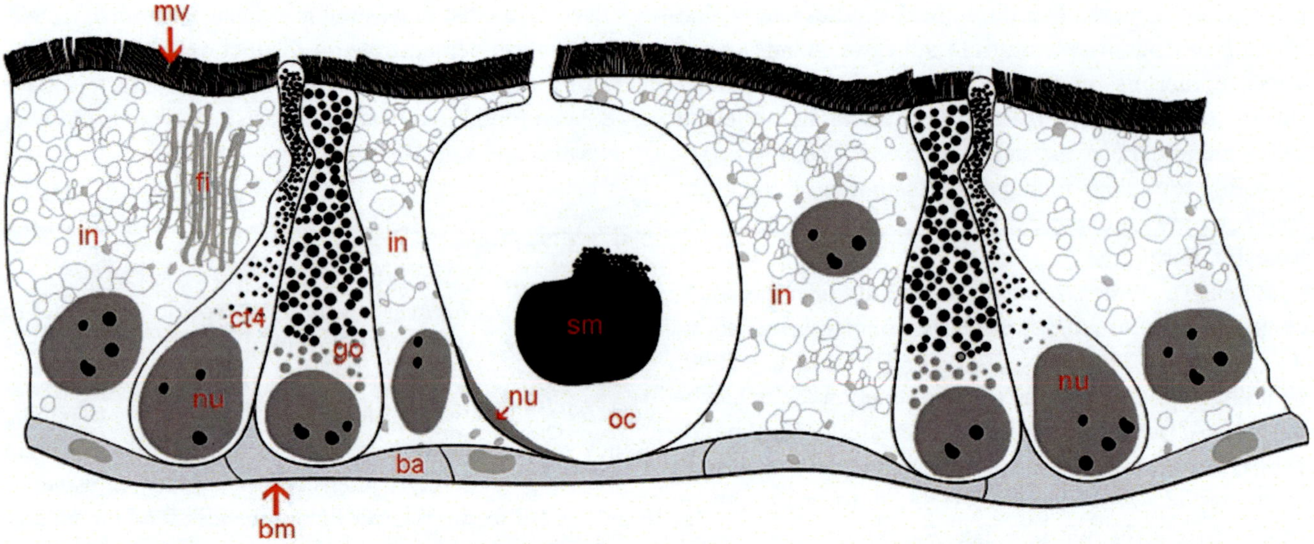

Fig. 5.1 Schematic drawing of the adhesive epithelium in *Euprymna scolopes*. Cross section of dorsal epithelium showing the different cell types (ovate, goblet, interstitial and cell type 4) within the adhesive epithelium

5.2.4.1 Earlier Studies

Singley (1982) was the first and, until now, sole author to investigate ultrastructurally and histochemically the epidermal constitution of *E. scolopes*. He describes the epidermis as a pseudostratified columnar epithelium, composed of three morphologically distinct cell types and delimited by a thick basement membrane. The basal region of the epithelium is often undulated and the surface layer is comprised in part of a pile of microvilli.

The first gland cell type Singley (1982) describes is the polymorphic **interstitial cell**. These cells are the most abundant type in the dorsal epidermis and less frequently observed in the ventral epidermis. On the apical surface is a layer of microvilli (Fig. 5.2B). The rounded nucleus is located near the epithelial surface. The cytoplasm is typically filled with thick (10–15 nm) and thin (5–8 nm) filaments, which are especially prominent in the basal region. The cells possess small vesicles containing material of varying electron density. No conclusive evidence of exocytosis was observed by Singley (1982).

The **ovate cell** occurs in all regions of the epidermis except for the distal surfaces of the fins and arms. This cell type is the largest of all and is ovate to sac-like. It contains a fine granular secretory material of uniform

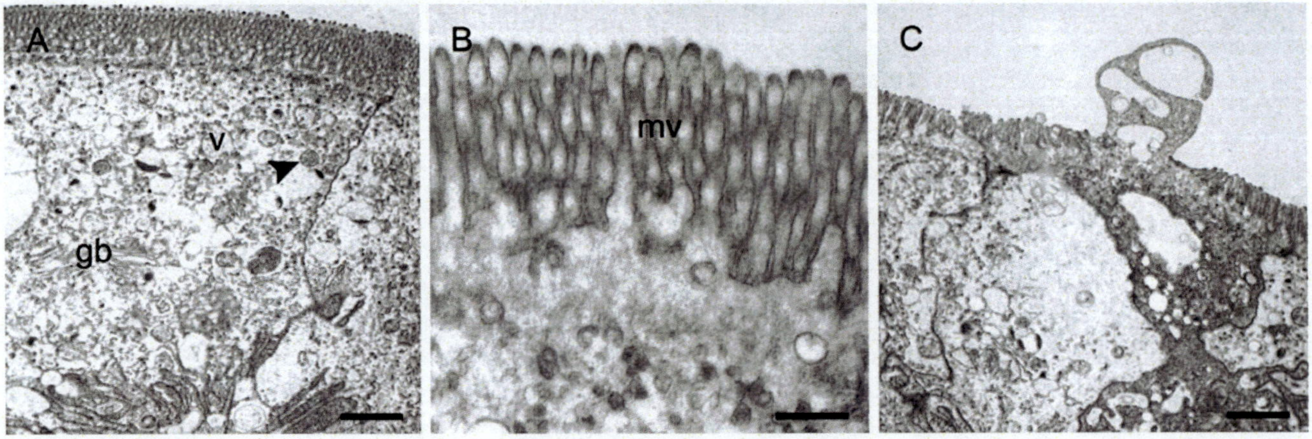

Fig. 5.2 Interstitial cell. (A) Terminal web (arrowhead) through which vesicles transit to and from the apical membrane; (B) The surface is characterized by microvilli with glycocalix in between; (C) Exocytosis to form glycocalix. Scale bars in (A) = 2 μm; (B) = 0.5 μm; (C) = 2 μm

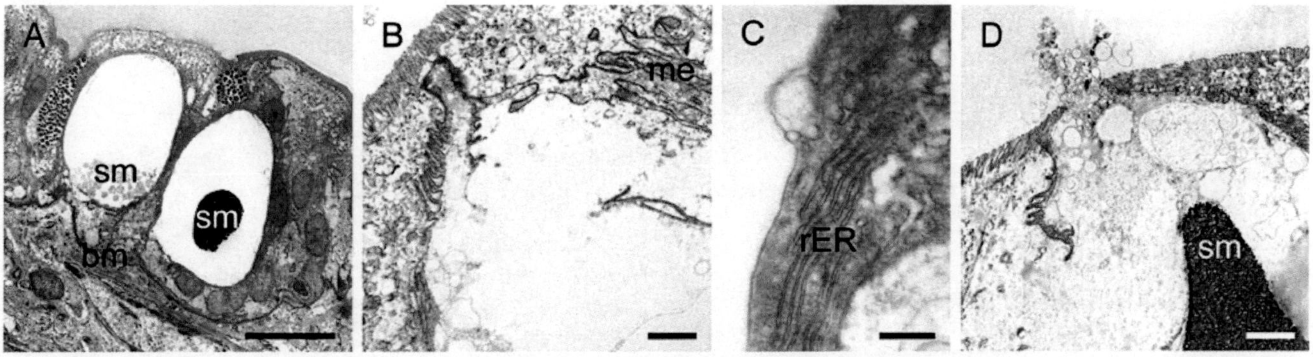

Fig. 5.3 Ovate cell. (A) Fine granular secretory material in different density stages; (B) Membranes produce deep folds beginning near the apical end of the cells and progressing toward the basal end; (C) The cytoplasm laterally contains rER; (D) Secretion process. Scale bars in (A) = 20 μm; (B) = 2 μm; (C) = 0.5 μm; (D) = 2 μm

appearance, and flattens the nucleus against the basal surface (Fig. 5.3A). Various amounts of rough endoplasmatic reticulum occur within the cytoplasm surrounding the nucleus (Fig. 5.3C), but only few Golgi bodies are present. Within the peripheral cytoplasm Singley found a reticulum of filaments (5–8 nm) that becomes rearranged and tightly packed during secretory activity.

Additionally, these filaments appear to produce deep folds in the plasma membrane, beginning near the apical end of the cells and progressing toward the basal end (Fig. 5.3B).

The third cell type described is the **goblet cell**, which occurs only in the dorsal epidermis and is especially abundant in the head and dorsal mantle. Goblet cells are rounded at their base, taper inward toward their apical end and have rounded cross sectional profiles (Fig. 5.4A). The cells are infolded laterally toward their apices, displaying a branching profile in longitudinal section. The surface area of the goblet cells is much smaller than that of the interstitial cells, and the microvilli are shorter (Fig. 5.4B).

A large rough endoplasmatic reticulum (rER)-Golgi complex is present near the bases of the goblet cells. The Golgi apparatus is generally more massive than the rER.

Especially in actively secreting cells numerous microtubules are oriented along the longitudinal axes of goblet cells and frequently appear in close proximity to the secretory granules. These secretory granules are characterized as highly electron-dense, spherical and with a considerable variation in size. The density is uniform and each granule is surrounded by a single unit membrane (Fig. 5.4C). They are distributed from the region of the elliptical nucleus near the basal end up to the apical surface. During active secretion, the granules appear to break up into smaller units as they approach the apical surface.

5.2.4.2 Recent Re-characterization

Recent ultrastructural and histochemical investigations of the epithelium of *Euprymna scolopes* provide new details of the adhesive region, and additional information can be added to Singley's (1982) observations. The thickness of the mantle epidermis in an average-sized animal ranges dorsally from 34 to 60 μm, at the ventral side from 33 to 43 μm and at the fins from 13 to 25 μm. The basal lamina is about 0.3 μm thick. Beneath the epithelium are variously orientated muscle layers. Contrary to Singley (1982) observations, four different cell types can be distinguished (Fig. 5.1).

Interstitial cells occur in the entire mantle epithelium. Typically, interstitial cells are positioned among the gland cells and are very variable in shape and size. These cells separate ovate from goblet cells in the adhesive region and are dominant in the normal mantle tissue. At the apical surface of the cell is a dense layer of microvilli (1.3–1.5 μm long) (Fig. 5.2B) (Singley, 1982: 0.5–0.6 μm). Nuclei are found either at the basal or at the apical end of the cell and are usually rounded to ovate. The cells contain numerous, evenly distributed organelles, especially mitochondria, but also rER and Golgi bodies (Fig. 5.2A). Along the cell walls membranes can be found frequently, often interspersed with mitochondria.

Singley (1982) did not demonstrate any secretion by these cells, but our observations show evidence of (rare) exocytosis (Fig. 5.2C). This function might be essential to form the glycocalix on the apical surface of the cells between and on top of the microvilli.

Ovate cells are found not only in the adhesive region of *E. scolopes*, but also in the regular mantle epithelium at the ventral side of the animal and on the base of the

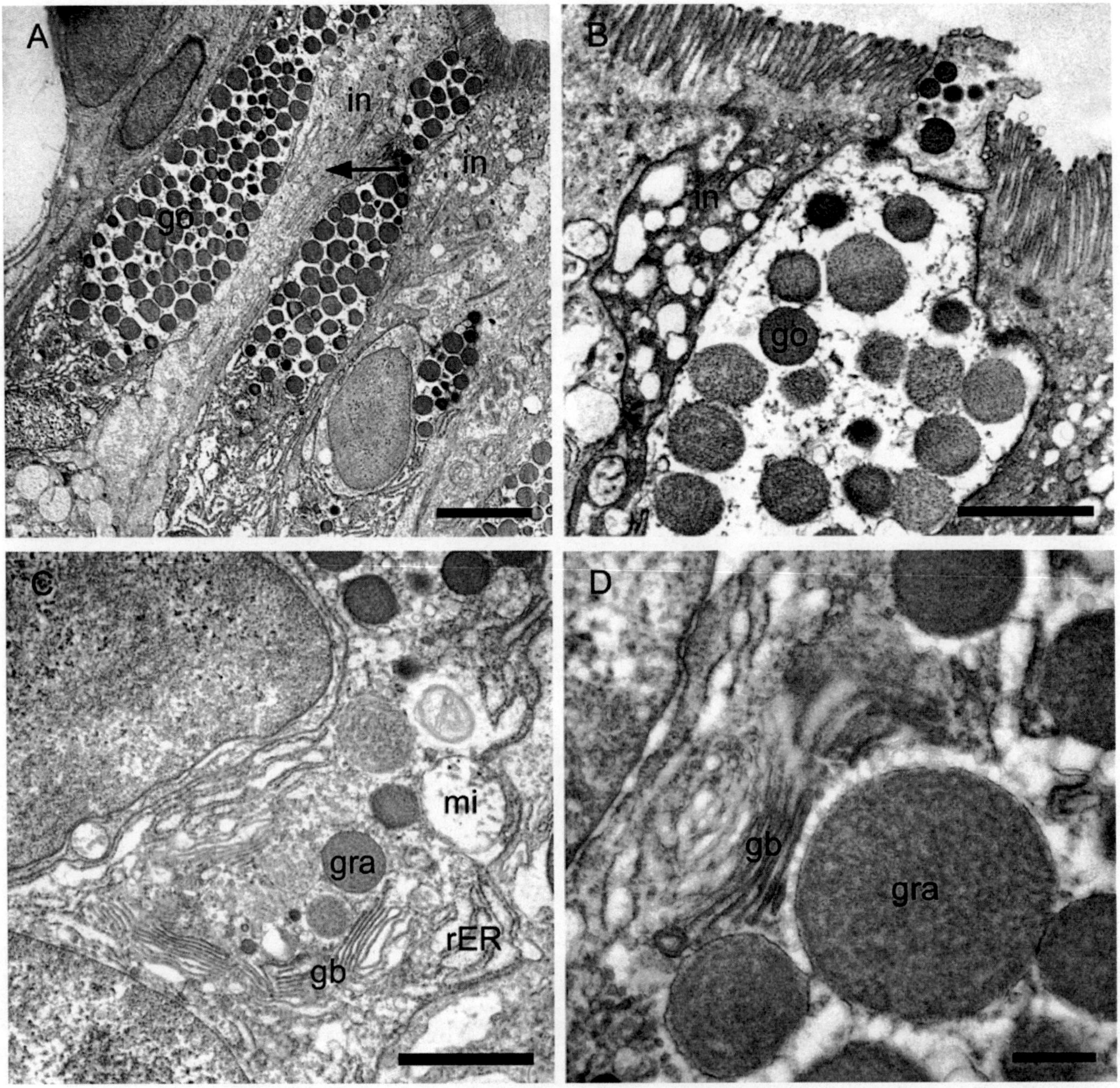

Fig. 5.4 The goblet cells (A) often appears grouped, but always interspersed among interstitial cells. Note intracellular filaments in the interstitial cells (arrow); (B) Secretion process; (C) Golgi apparatus, rER, and mitochondrium. Vesicles released from the Golgi apparatus appear electron lucent, while the granules are more electron dense. (D) High activity of secretory material synthesis. Scale bars in (A) = 5 μm; (B) = 2 μm; (C) = 2 μm; (D) = 0.5 μm

fins. Contrary to Singley (1982), who described a lack of a microvilli layer at the surface of the cells, we detected 0.8–2 μm long microvilli.

We observed ovate cells in different phases of activity (Fig. 5.3A). Some cells have very dense secretory material, which accumulates centrally in the cell. Others contain loose secretory material evenly distributed over the entire cell area. Finally, completely empty cells are also present. The latter seem to represent a stage at which secretory material is completely released. Further investigation is needed to determine whether ovate cells "refill" themselves.

Ultrastructural observations of the secretory process indicate a direct release of (secretion) material through an opening in the cell's surface (Fig. 5.3D).

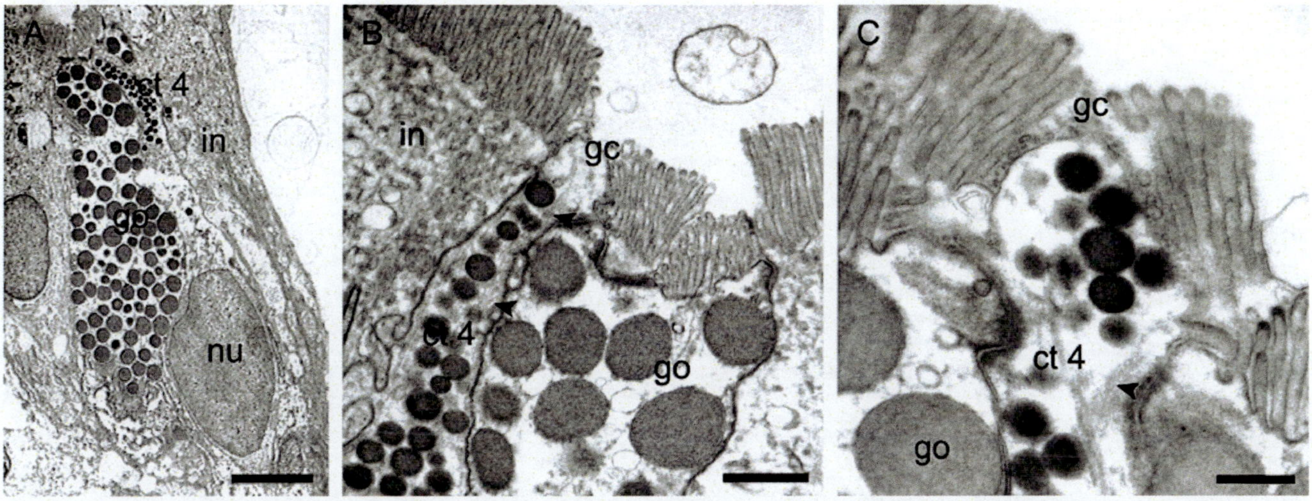

Fig. 5.5 Cell type 4 (A) occurs in immediate vicinity of a goblet cell. Its elongated nucleus is located at the basal end; (B) The granules are transported by filaments (arrowhead) to the cell surface. The surface area covered with glycocalix but no microvilli as given for the goblet cells; (C) Secretion process. Only few granules are transported by the filaments (arrowhead) to the surface and secreted. Scale bars in (A) = 5 µm; (B) = 0.5 µm; (C) = 1 µm

Finally, the third known cell type, the **goblet cell**, occurs in the dorsal part of the mantle and at the fin base. Their microvilli are approximately 0.7 µm long and thus shorter than the microvilli of the interstitial cells (Fig. 5.4B). Organelles, especially rER and Golgi apparatus and few mitochondria are present in the area around the nucleus (Fig. 5.4C). Vesicles, released from the Golgi apparatus, are electron lucent (Fig. 5.4D), but become more electron-dense the closer they are to the apical end of the cell. They are always characterized by a uniform appearance, showing a homogeneous density and shape (Fig. 5.4C). The vesicles have a diameter of 0.8–1 µm (Singley, 1982: 0.5–0.7 µm) and are spherical and membrane-bound.

Goblet cells differ from the other cell types in their secretion process. Instead of releasing their secretory material through an opening in their surface, as occurs in the interstitial and ovate cells, they secrete the most apical part of the cell content, which includes some cytoplasm and secretory vesicles. The cells probably reproduce the released part immediately.

Additionally, an atypical appearance of the goblet cell was detected. This involves an augmented condensation of vesicles and therefore an increased presence of secretory material. One possible explanation is that a dysfunction of the secretion process causes an accumulation of the secretory vesicles; another is a simple overproduction of such material.

Beside the aforementioned cell types described by Singley (1982), a **further cell type** (cell type 4) was identified. Similar to the goblet cells this cell type contains spherical, membrane-bound, and very electron-dense granules. These granules, however, are much smaller (0.2–0.4 µm) and evenly distributed within the cell. They are transported to the surface by filaments inside the cell.

The cell type 4 always occurs in the immediate vicinity of a goblet cell (Fig. 5.5A) and was found only in the adhesive region of *Euprymna scolopes*. It differs from the other epithelium cells by the absence of the microvilli layer (Fig. 5.5C). The cell is elongated with a rounded base and is more slender than the goblet cell. Organelles (mitochondria, rER, and Golgi apparatus) are located at the basal end of the cell near the elongated nucleus (Fig. 5.5B). The cell shows, as opposed to the goblet cell, a release of secretory material through a small opening on the surface (Fig. 5.5C).

5.2.4.3 Histochemistry

Histochemical observations of Singley (1982) showed that **goblet cells** contain neutral sugars (positive PAS reaction) (Fig. 5.6F) and show only a weak reaction for basic (Biebrich Scarlet at pH 6.0–10.5) (Fig. 5.6C, D) and acidic protein stains (Alcian Blue at pH 1.0 and 2.5) (Fig. 5.6E).

The secretory material of the **ovate cells** appeared to consist of highly sulfated proteins; the granules are unreactive for sugars (PAS negative) (Fig. 5.6F) but remain strongly reactive to basic proteins (Biebrich Scarlet test) (Singley, 1982) (Fig. 5.6C, D). Glandular material tested

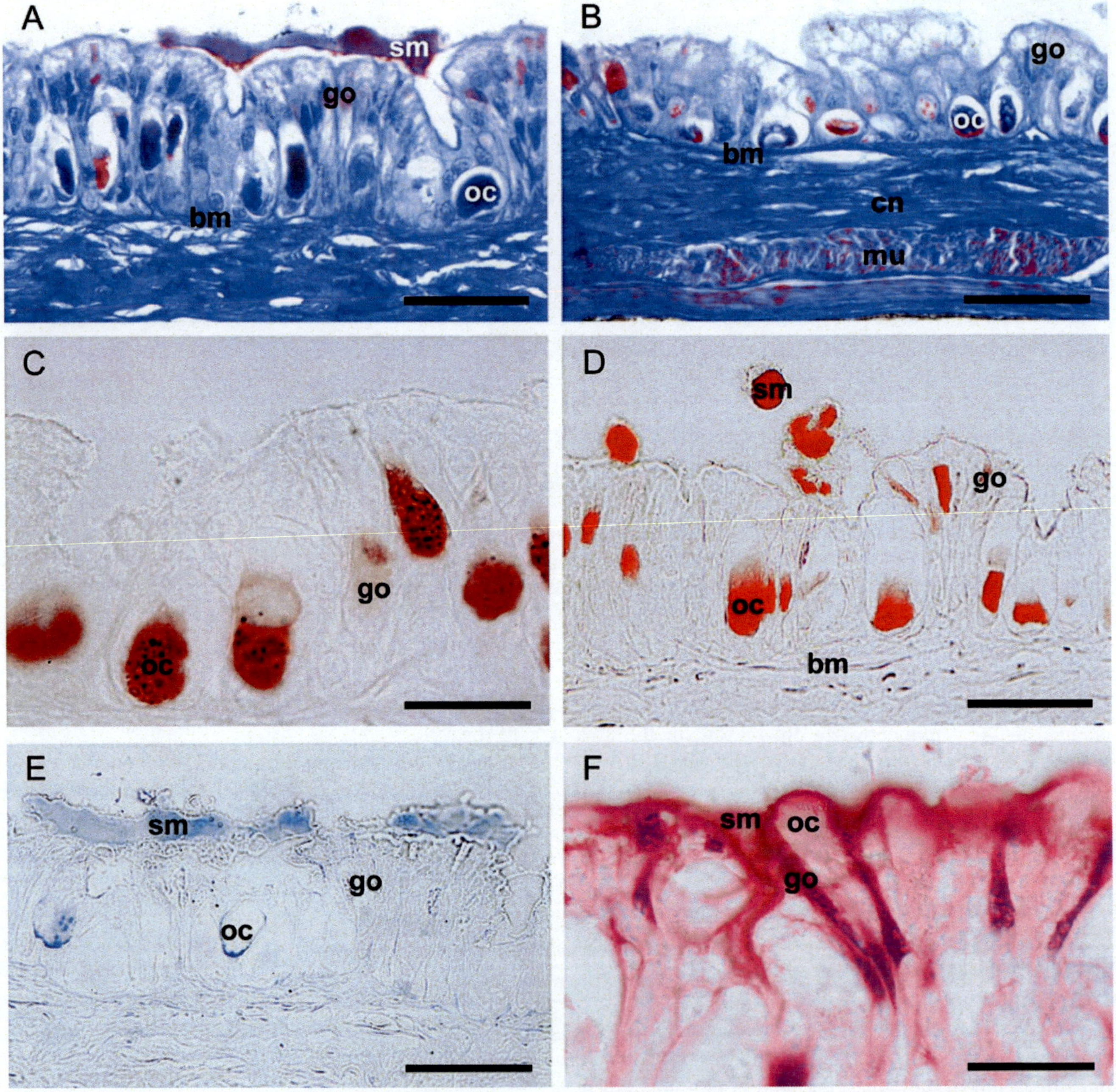

Fig. 5.6 Histochemistry. (A) The dorsal adhesive epithelium is higher than (B) the ventral epithelium. Both contain goblet, ovate and interstilial cells. Beneath is a high dermal layer consiting of connective tissue and muscle fibers (stain AZAN); The ovate cells (C) on the ventral and (D) dorsal side bear basic proteins around pH 6.0 (in C) up to pH 8.5 (in D, E). The ovate cells are also light reactive to acid proteins (Alcian Blue pH 2.5) while the secretory material of the goblet cells consists of neutral sugars (PAS reaction). Scale bars in (A, B) and (E) = 50 µm; (C) and (F) = 20 µm; (D) = 100 µm

in the process of active secretion, termed "evacuating ovate cells" by Singley (1982), however, indicates a pH level shift from basophilia to acidophilia. Singley (1982) suggested that this protein modification could be induced by the surrounding sea water, in which masked acidic proteins become unmasked and effect the visible pH shift.

The present re-characterization, however, showed some slight histochemical differences to the descriptions in Singley (1982) (Fig. 5.6A, B). While the results given for the goblet and ovate cells could be confirmed, the pH shift in the secreted material of the ovate cells could not. Staining of cells in different secretion states, e.g., with the presence of sea water during fixation, always stained strongly positive for basic (Fig. 5.6C, D) and only very weakly for acidic proteins (Fig. 5.6E).

The secretory content of the **interstitial cells** and newly described **cell type 4** remained too small to show any clear positive reactivity to the staining tested so far.

5.2.5 Bonding Mechanism

Temporary adhesion is often effected by a duo-gland system, which includes two types of secretory cells, i.e., cells responsible for adhesion and cells releasing a de-adhesive secretion (Hermans, 1983). Singley (1982) assumed that neutral mucopolysaccharides, secreted from the goblet cells, were responsible for adhesion, whereas de-adhesion was caused by acidic mucoproteins released from the ovate cells.

Our histochemical re-characterization showed that the secretory material of ovate cells does not react strongly to tests for acidic groups. We were unable to confirm the hypothesis of Singley (1982) because the pH-value of the glandular material within as well as outside the cells remained steady.

Nevertheless, a recent ultrastructural and histochemical re-characterization of the epithelium of *Euprymna scolopes* argues against a duo-gland adhesive system, as suggested earlier (Singley, 1982). Ovate cells presumably do not play a major role in the adhesion/de-adhesion process. These cells were not only present in the adhesive region, but also found in the regular mantle tissue. They probably solely produce mucus to cover the squid's body surface and do not participate further in the adhesion/de-adhesion process. Interstitial cells potentially help produce the glycocalix, which is found at the microvilli layer on the surface of the epithelium.

The cell type 4 seems to be too small to produce sufficient secretory material for adhesion or release alone. Since this cell type always closely adjoins the goblet cells, a duo-component system involving both cell types seems to be more likely, as proposed for *Idiosepius* and *Sepia* (see Sects. 5.3 and 5.5).

Several differently oriented dermal and mantle muscle layers were found beneath the epithelium of *Euprymna scolopes*. This makes a mechanically related release of sand particles more probable than chemically induced de-adhesion. Interestingly, the presence of the dermal muscle layer as well as a muscular-related release was never mentioned before.

As already considered for *Sepia* and *Nautilus*, a mechanical de-attachment mechanism seems to be likely also for *Euprymna*. The animals react considerably faster and more precisely than a chemically induced release mechanism would allow.

5.3 *Idiosepius*
(Norbert Cyran and Janek von Byern)

The genus *Idiosepius* (Steenstrup, 1881) comprises some of the smallest species among cephalopods, with a mantle length <6 mm in females and 3 mm in male representatives of *I. pygmaeus* (von Byern and Klepal, 2010). The animals have an elongated mantle that is slightly pointed posteriorly; their fins are small and kidney-shaped (Fig. 5.7). The main characteristic of this genus is an oval attachment area with a rough surface located dorsally on the posterior half of the mantle (Nesis, 1982).

Several aspects of these animals have been studied, including their development, physiology, and behavior (Nabhitabhata, 1998; Kasugai, 2000; Shigeno and Yamamoto, 2002; Yamamoto et al., 2003). Nonetheless, the systematics of this genus as well as the phylogenetic relationship of *Idiosepius* with the other cephalopod taxa

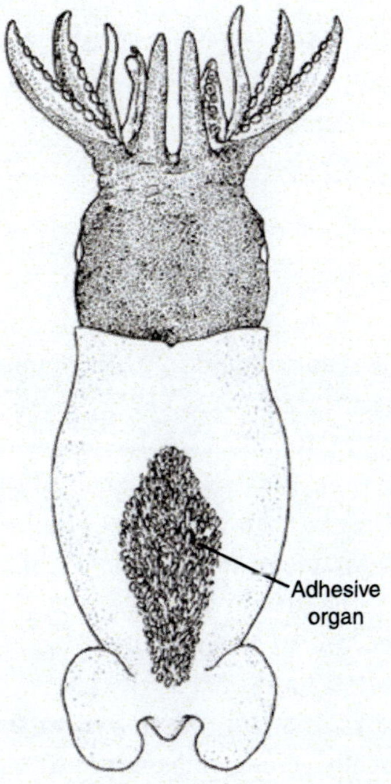

Fig. 5.7 Dorsal side of *Idiosepius pygmaeus* with the adhesive organ. Adapted from Voss (1963) and reproduced with permission

is far from being settled. One reason is that only sparse information is available about the *Idiosepius* life cycle, geographical distribution or ecological adaptations.

5.3.1 Systematics

The phylogenetic relationship of *Idiosepius* with the other cephalopod taxa remains unclear, some researchers place the genus into the group Sepiidae (Lindgren et al., 2004; Strugnell et al., 2005; Akasaki et al., 2006), while others argue for Teuthidae (Bonnaud et al., 1996, 1997, 2005; Carlini et al., 2000; Takumiya et al., 2005).

The conspicuous morphological character in representatives of this monogeneric family (Idiosepiidae Appellöf, 1898) is the adhesive organ (also known as adhesive gland). To date, the genus comprise eight species (Jereb and Roper, 2005):

- *I. biserialis* Voss, 1962
- *I. macrocheir* Voss, 1962
- *I. minimus* (D'Orbigny and de Ferussac, 1848)
- *I. notoides* Berry, 1921
- *I. paradoxus* Ortmann, 1888
- *I. picteti* (Joubin, 1894)
- *I. pygmaeus* Steenstrup, 1881 (= *I. pygmaeus hebereri* Grimpe, 1931)
- *I. thailandicus* Chotiyaputta et al., 1991

However, latest analyses raise questions about the validity of *I. macrocheir* and *I. thailandicus* (von Byern et al., 2010; von Byern and Klepal, 2010).

5.3.2 Ecology

Idiosepius has been reported from the tropical Indo-Pacific, Japan, southern Australia including Tasmania and in southern East African waters (Voss, 1962; Nesis, 1982), but some species were found in cooler Russian and Japanese waters (Nesis et al., 2002; Sato et al., 2009). Within the genus, *I. biserialis* has the widest geographical distribution, ranging from Japan to the Indo-Pacific (Thailand and Indonesia) (Voss, 1963; von Byern et al., 2005); it has also been recorded in Mozambique (incorrectly annotated by Voss in 1962 as South Africa).

Most species inhabit sea grass areas; only *I. pygmaeus* and *I. thailandicus* apparently prefer mangrove rivers (Hylleberg and Nateewathana, 1991; Lewis and Choat, 1993a; Nabhitabhata, 1994), and *I. notoides* was found beneath floated seaweeds (Tracey et al., 2003) as well as between aggregations of tunicates (von Byern, unpubl. data).

Fig. 5.8 *Idiosepius* attaches to mangrove roots or sea grass waiting to capture prey, for camouflage or to escape against predators. Females furthermore adhere for spawning (Image by J. von Byern)

Idiosepius camouflage during the day by sticking to the underside of sea grass leaves or algae (Moynihan, 1983) (Fig. 5.8). Hiding there, the animals wait to capture prey swimming by. Females, moreover, adhere for spawning (Natsukari, 1970; Lewis and Choat, 1993b; Kasugai, 2000).

5.3.3 Gland Morphology

Structurally, the adhesive organ of *Idiosepius* can be easily discerned from the regular mantle skin (Sasaki, 1921). While the mantle epithelium is 30–40 μm thick and has a smooth surface, the adhesive organ has deep furrows and is thicker (60–100 μm depending on the species) (von Byern et al., 2008).

The dermal layer beneath the adhesive and regular mantle epithelium contains the typical musculature layer

(5 µm thick) and small nerve fiber networks, followed by connective tissue including chromatic elements (chromatophores, iridophores, and reflector cells).

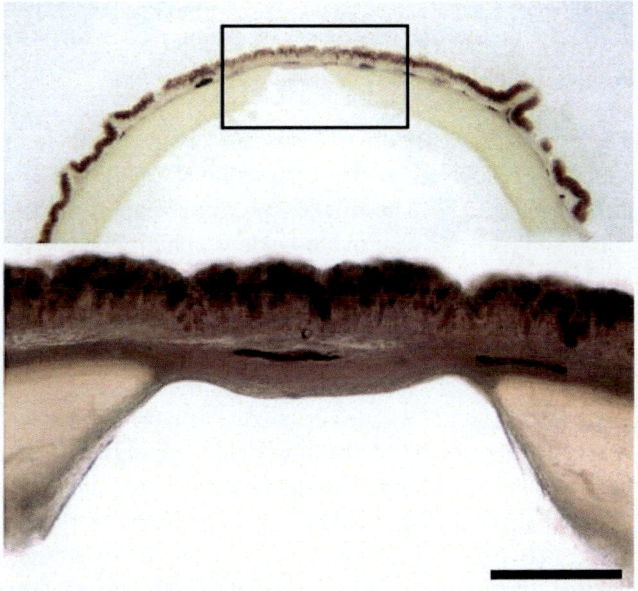

Fig. 5.9 Beneath the center of the adhesive organ (box in the upper and its lower image magnification) the typical mantle musculature is absent. One suggestion is that this muscular absence increases the flexibility of the adhesive organ resulting in a more adaptable and precise attachment to different formed surfaces. Scale bar in the lower image 250 µm

In contrast to the normal mantle, the area below the adhesive system lacks of the typical mantle musculature (Fig. 5.9).

5.3.3.1 The Adhesive Organ

Ultrastructural studies show that the adhesive organ (Fig. 5.10) consists of two distinct cell layers: (1) the proximal layer (about 5 µm high) consists exclusively of basal cells; (2) the distal cell layer contains two glandular (columnar and granular cells) and two non-secretory cell types (interstitial and small ciliated fusiform cells) (Cyran et al., 2008, 2010). The cell types in the adhesive epithelium can be distinguished morphologically and by their secretory components:

The **basal cells** adjoin the basal membrane and form a continuous cell layer (about 5 µm high). The cells are flat ovoid to cone-shaped and mostly have a globular nucleus, lacking nucleoli. The cytoplasm is largely full of vesicles (∅ 0.5 µm), which bud by endocytosis from the membrane of the cell base. Endoplasmatic reticulum and mitochondria are rare, but cytoskeletal elements and polyosomes are abundant.

The **columnar cells** contain small vesicles of 1 µm in diameter, densely filled with electron-dense material. This material consists mainly of neutral carbohydrates and to

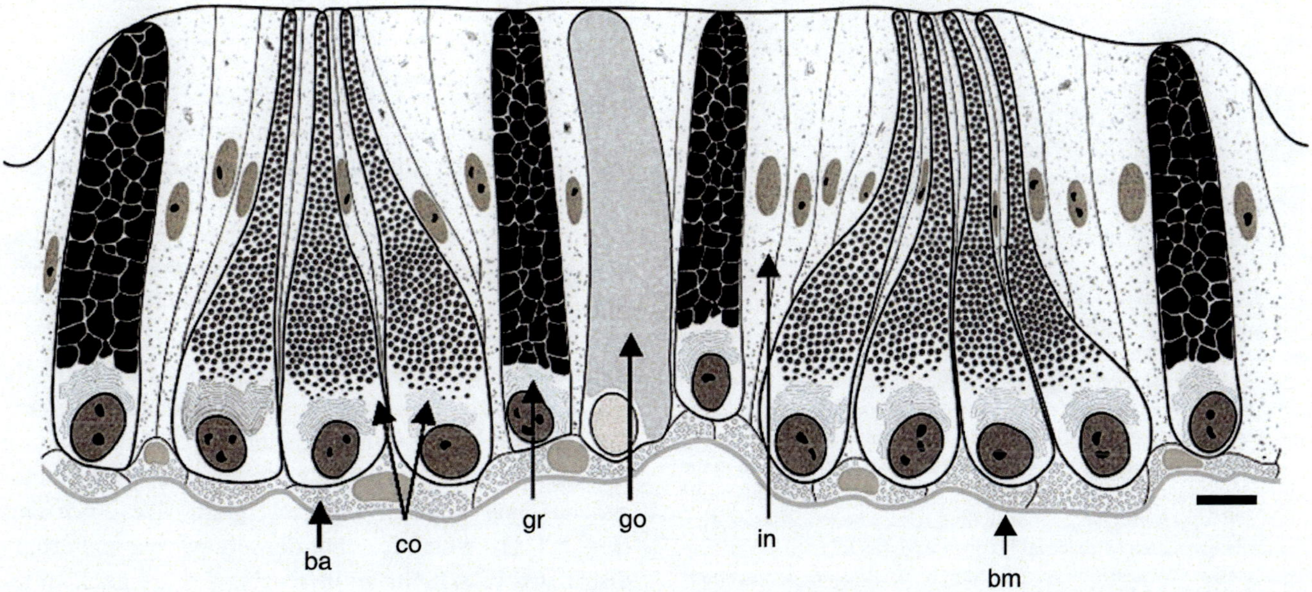

Fig. 5.10 Schematic drawing of the adhesive epithelium in *Idiosepius*. While the columnar cells and granular cells are restricted to the adhesive area only, the goblet cells, interstitial cells and basal cells are also given for the regular mantle epithelium with its basal membrane. The fusiform cells are missing in this earlier drawing. Scale bar = 20 mm. Image from Cyran et al. (2005) and reproduced with permission

a lesser extent of basic proteins (von Byern et al., 2008). The cells are pear shaped, tapered toward the surface and end in a hump. Columnar cells always occur in aggregates of 5–15 cells, which are regularly distributed in the adhesive organ (Fig. 5.11A, B).

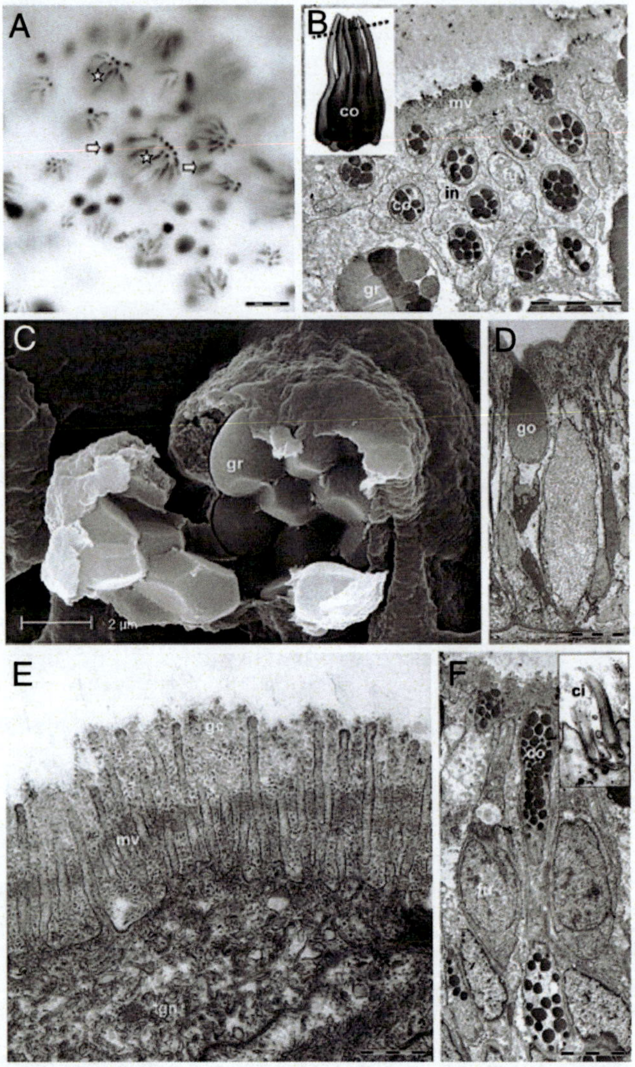

Fig. 5.11 (A) Vibratome section showing the distribution of columnar cell aggregates (stars) and granular cells (arrows) in the epithelium; (B) A horizontal (see inlet) section of columnar and fusiform cells illustrates the encasement of columnar cells within interstitial cells; (C) SEM micrograph of a granular cell with its larger polygonal granules, packed tightly together; (D) Goblet cells have an uncontinous release mode, the secretory material is accumulated in the cell lumen and discharged entirely. The low density of contents in the right cell represents an earlier charching state, whereas in the mature left cells the secretory material is condensed; (E) Interstitial cells are characterized by dense microvilli with glycocalix and a distinct trans golgi network; (F) Two fusiform cells encircled by columnar cells. The cells bear cilia on the surface (inlet). Scale bar in (A) = 20 μm, (B) = 5 μm, (D) = 10 μm, (E) = 500 nm, (F) = 5 μm. Image C adapted from von Byern et al. (2008) and reproduced with permission

The **granular cells** are cylindrical (10 μm in diameter on average) and densely packed with granules (2–5 μm) (Fig. 5.11C) containing basic proteins and carbohydrates (von Byern et al., 2008). In the loosely packed stage these granules appear globular, but when tightly packed they are polygonal. Frequently, granules fuse partly and incorporate membrane residues. The granular cells are also regularly distributed.

The pear-shaped **fusiform cells** are about 15 μm long and always located in the surface layer of the epithelium (Fig. 5.11F). The thicker, proximal part is almost completely filled by an ovate nucleus containing several nucleoli. One or more cilia (250 nm in cross section and 2 μm long; $9 \times 2 + 2$ arrangement) extrude from the apical pole of the cells. The cells occur exclusively between the columnar cell aggregations.

Interstitial cells build the unspecific epithelium parts between the glandular cells and provide the micovilli-covered surface (Fig. 5.11E). The cytoplasm is rich in Golgi network near the surface, associated with producing the glycocalix. Modified interstitial cells encasing the columnar cells have a significantly smaller cell volume and contain more cell organelles. Their densely arranged microvilli layer on the apical surface of the cells is longer (1.5 μm instead of 0.5–1 μm long in the regular epithelium) and forms a microvillous collar, which surrounds the tips of the columnar cells.

5.3.3.2 The Regular Mantle Epithelium

The regular mantle epithelium is likewise two-layered, composed of a basal cell layer and a distal layer containing non-glandular (interstitial) and secretory (goblet, saccular, and type 1) cell types. Goblet and saccular cells are found in lower abundance in the adhesive epithelium as well, although they do not participate in the bonding process. In contrast, neither columnar, granular nor fusiform cells were detected in the regular mantle epithelium.

The **interstitial** and **basal cells** correspond structurally and in their arrangement to those in the adhesive organ.

Goblet cells are cylindrical or barrel-shaped with a diameter of 10–20 μm. The cells are filled with homogeneous fine granular material (grain size 20–50 nm) (Fig. 5.11D), which is either densely packed and evenly distributed or exhibits an increasing density gradient toward the apical pole. In contrast to the other glandular cells, the goblet cells release their contents completely. The secretory material of the goblet cells consists of a large amount of basic proteins (positive reaction to Biebrich Scarlet) and a smaller amount of neutral mucopolysaccharides (PAS) (von Byern et al., 2008).

The **saccular cells** mostly extend from the epithelium surface to half the height of the epithelium, infrequently contacting the basal cells. They are ovoid to saccular and have a flat to sickle-shaped nucleus at their base. Some cells contain somewhat fine-grained material, but most cells appear to be empty. It is unclear so far whether this is an artifact caused by fixation or whether the material is released before fixation. The histochemical analyses revealed no specific reactivity of the secretory content, and the cells therefore appear as empty spaces between the other cells.

Cell type 1, named mucus cells by von Byern et al. (2008), contains spherical to oval granules (∅ 5 μm) of dense material. In contrast to the glandular cells in the adhesive organ, the granules are not membrane-bound and therefore frequently form a uniform mass.

The granules react strongly positive to carbohydrates (PAS and D-mannose lectin GNA) and basic proteins (Biebrich Scarlet pH level 6.0–10.5 and Fast Green). Acidic proteins (Alcian Blue) and lipids are lacking in the secretory material, as is the case for the other gland cells (von Byern et al., 2008).

5.3.4 Development of the Adhesive Organ

Earlier studies provide an overview of the embryonic development in *Idiosepius*, indicating that *I. paradoxus* hatch after 18 days (at 20°C) (Natsukari, 1970; Yamamoto, 1988) and *I. pygmaeus* as well as *I. biserialis* already 6 days (at 28–30°C) after oviposition (Suwanmala et al., 2006a). Cultivation of hatchlings in aquaria is still limited to 4–6 days because of unsuitable food sources. Nevertheless, rare observations are available about the postembryonic development of this genus (von Byern et al., 2006).

Recent ultrastructural studies on the embryogenesis in *Idiosepius pygmaeus* indicate that the adhesive organ is not established during early development (stage 24) as suggested earlier (Natsukari, 1970; Yamamoto, 1988). The noticeable structure refers rather to another glandular system, the Hoyle organ:

The Hoyle organ was first described in 1889 by Hoyle in *Sepia*, but its function was unclear. Later studies by Wintrebert (1928), Yung Ko Ching (1930), and von Orelli (1959) suggested a digestive function during hatching. The Hoyle organ is located on the posterior part of the dorsal mantle surface and solely active in the late embryonic developmental phase (von Orelli, 1959; Fioroni, 1962; Arnold and Singley, 1989).

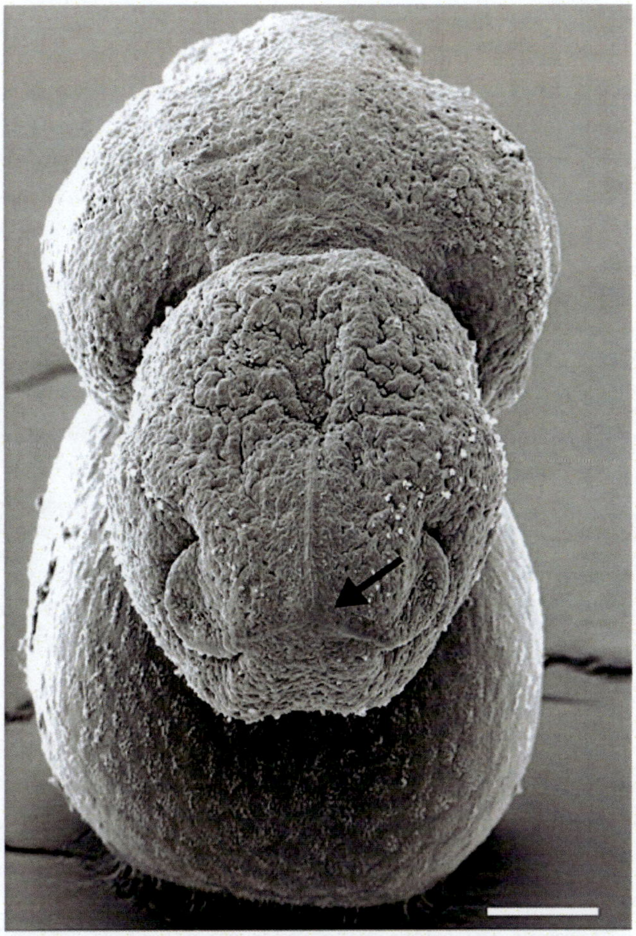

Fig. 5.12 In the late embryonic development (stage 25) *Idiosepius* develops a digestive gland system (arrow), the so called Hoyle organ. Following hatch, the production of secretory material decreases and the Hoyle organ degrades within 1–2 days. Scale bar = 100 μm

In *Idiosepius* the Hoyle organ has an anchor-like shape (Fig. 5.12) and comprises one cell type (so far unnamed), filled with small granules. So far, no information is available on the composition of the digestive enzymes. Following hatching, the production of secretory material decreases and the Hoyle organ degrades within days.

Latest unpublished results by N. Cyran indicate that the adhesive organ in *Idiosepius* starts to develop about one week after hatching. Initially, unspecific glandular cells are randomly distributed within the adhesive area. These cells appear near the basal membrane, initially without contact to the epithelial surface, and contain a roundish nucleus surrounded by endoplasmatic reticulum.

In eight-day-old juveniles the cells extend toward the epithelial surface and start to synthesize secretory granules. Depending on the granula size and morphology, a first differentiation into columnar and granular cells is possible. The study shows that both cell types develop simultaneously. At day 10 the columnar cells form the typical aggregations also found in the adhesive organ of adults. First adhesive behavior was observed 14 days after hatching (von Byern et al., 2006).

5.3.5 Process of Secretion and Bonding Mechanisms

A predominant mechanical adhesion or release by a reduced-pressure system as in the cephalopod suckers (Kier and Smith, 2002; Pennisi, 2002) or in *Sepia* (see chapter below) can be excluded for *Idiosepius*. The absence of the typical mantle musculature below the adhesive organ as shown in Fig. 5.9, however, increases the flexibility of the adhesive organ and thus allow a more precise contact to the substrate surface (Cyran et al., 2010).

Nevertheless, adhesion is primarily realized by the secretory substances from the columnar and granular cells. Both glandular cells show a high synthesis activity and are tightly packed with secretory material. The cells generally release only parts of their material during secretion; this material fuses on contact with the outer medium into a uniform mass (von Byern et al., 2008).

Based on histochemical evidence the secretory material in both secretory cells consists of protein–carbohydrate complexes; however, the ratio of proteins to carbohydrates varies considerably among the cells (von Byern et al., 2008): The columnar cells primarily contain sugars (such as fucose) and relatively small amounts of basic proteins (pH 6.0–10.5). The granular cells, in contrast, have a balanced ratio of sugar (primarily mannose with a small proportion of glucose) and basic proteins. Acidic proteins were not detected within either the glandular cells or their secretions.

Smear, scratched from the adhesive organ of *I. pygmaeus*, was analyzed by SDS-Page, SELDI-and MALDI-TOF-MS (von Byern and Grunwald, 2008). Up to 6 proteins around 14, 17, 25, 39, 50, and 75 kD were obtained. Although the detected proteins have not yet been assigned to the respective glandular cell type, they differ compared to those in other glue-producing animals:

In the adhesive systems of gastropods (see Chapter 4, p. 41) and echinoderms (see Chapter 7, p. 99), which also possess cross-linked protein–carbohydrate glues, the molecular weight of the glue proteins ranges from 20 to 200 kD (Smith and Morin, 2002; DeMoor et al., 2003). The permanent cement of barnacles is composed of proteins with a molecular weight of 16, 19, 20, 52, 68, and 100 kD (Kamino, 2006).

The current results indicate that adhesion and release in *Idiosepius* is not based on a duo-gland mechanism. *Idiosepius* probably attaches to surfaces by a gel, as is known for gastropods (Smith, 1991, 2002; Smith et al., 1999). Adhesion produced solely by proteins, as is the case for sessile organisms such as mussels (see Chapter 18, p. 273), barnacles (see Chapter 9, p. 153), and polychaetes (see Chapter 10, p. 169), can be excluded for *Idiosepius*.

Nevertheless, other release mechanisms such as body movements, contraction of the dermal muscles or dilution of the glue by sea water or enzymes also are conceivable. Unfortunately, due to its small size and the fact that it is easily startled no detailed observations of attachment and release have been made to date.

5.4 *Nautilus*

(*Janek von Byern, Thomas Schwaha, Leon Ploszczanski and Norbert Cyran*)

Among the cephalopods, nautiloids are the oldest group. They originated about 500 million years ago and are famed as "living fossils" (Teichert and Matsumoto, 1987). The name *Nautilus*, originally applied to the *Argonauta* (also named "paper nautilus"), was already described by Aristotle's in his Historia Animalium, page 622[b] (Thompson, 1910).

In general, *Nautilus* is similar to the other cephalopod groups, with a prominent head and tentacles; however, it is the only living cephalopod whose developed a brown and white externalized shell (Fig. 5.13). This shell is coiled, about 20 cm in diameter and 9 cm in breadth, and divided by regularly spaced septa into chambers (Stenzel, 1964). The shell has different functions: on one hand the animals withdraw completely into their shell, closing the opening with a cartilaginous hood, when threatened. On the other hand, it enables the animals to maintain buoyancy during vertical movement in the habitat (Roper and Young, 1975; Greenwald and Ward, 1987).

5.4.1 Systematics

The Nautiloidea Agassiz, 1847 consist of the single family Nautilidae De Blainville, 1825 comprising two genera: *Nautilus* Linne, 1758 and *Allonautilus* Ward and Saunders, 1997.

Fig. 5.13 *Nautilus pompilius* with its digital tentacles (arrow). Image reproduced with permission of Uwe Möckel Germany, its copyright © remains with the photograph

During the past centuries, 11 species and 7 variants of living *Nautilus* have been proposed, mostly based on descriptions of drifted shells (Sowerby, 1848). More detailed evaluations, including characterizations of their morphology (Tanabe et al., 1983) as well as morphometric and geographical data on the population level (Tanabe et al., 1985), indicate that most formerly described species and variants are merely "intrapopulation variations" of two species.

Today, only seven species remain valid (Jereb and Roper, 2005):

- *Nautilus pompilius* Linne, 1758 (*N. p. pompilius* Linne, 1758 & *N. p. suluensis* Habe and Okutani, 1988)
- *N. belauensis* Saunders, 1981
- *N. macromphalus* (Sowerby, 1848)
- *N. stenomphalus* Sowerby, 1848
- *N. repertus* Iredale, 1944
- *Allonautilus scrobiculatus* Lightfood, 1786
- *A. perforatus* (Conrad, 1849).

These species are separated by morphological attributes such as shell coloration, size, shell form, ornaments, radula, and hood texture (Nesis, 1982). However, geographic isolation, sexual dimorphism, life history as well as other attributes tend to increase the morphological and genetic variability in *Nautilus* and lead to a broad range of inter- and intraspecific variations (Saunders and Ward, 1987; Swan and Saunders, 1987; Woodruff et al., 1987).

5.4.2 Ecology

Nautilus occurs only in tropical waters of the Eastern Indian Ocean and Western Pacific Ocean down to Australia, ranging from approximately 30°S to 30°N and approximately 90–185°W (Saunders, 1987). Nevertheless, drifted specimens as well as floating shells are recorded occasionally off Japan (Zhengzhi, 2010), along the Indian coasts and down Eastern Africa (House, 1973) to the Western Cape in South Africa (House, 1987).

Until now, our knowledge on the ecology of *Nautilus* has been largely based on trapping results, remote telemetry experiments (Ward et al., 1984) and observations of captured animals held in shallow water or aquarium systems (Carlson, 1987).

Nautilus is a warm-water (≈10–25°C) fore-reef slope inhabitant, active both nocturnally and diurnally. During the day the animals can descend to 600 m, with a preferred water depth ranging from 150 to 300 m; at night they rise to shallower waters (100 m) to forage (Saunders and Landman, 1987; Ward, 1988). The upper limit seems to be determined mostly by the water temperature; conditions above 25°C for several days are lethal for the animals (Saunders and Spinosa, 1979).

The animals are bottom-dwelling scavengers and opportunistic predators, feeding mainly on dead or dying prey such as shrimp, small fish, and crustaceans. The food is picked by the digital tentacles (Packard et al., 1980). The animals further use the tentacles to attach to vertical/horizontal substrata or other individuals for mating. During swimming, the tentacles are retracted into their sheaths (Fig. 5.14).

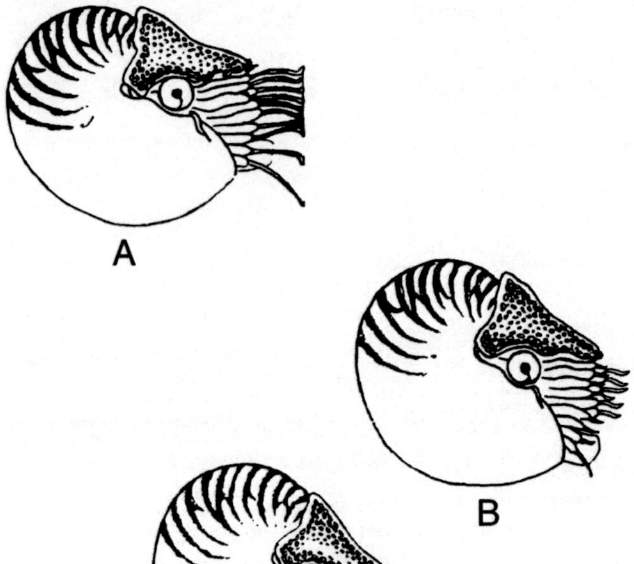

Fig. 5.14 *Nautilus* lacks of suckers on its tentacles as given for the other cephalopod genera. Instead, this "living fossil" uses glue to bond to the substratum (A and C), to hold on prey or to adhere to other individuals during copulation. During swimming the animals retract most of the digital tentacles (B) and only leave its tips hanging outside. Adapted from Kakinuma et al. (1995) and reproduced with permission

With its coiled shell, the animals adjust their buoyancy by gas and liquid exchange in the chambers (Ward et al., 1977; Ward and Greenwald, 1981; Greenwald and Ward, 1987). The shell itself consists mainly of calcium carbonate ($CaCO_3$) (Crick and Mann, 1987) and is strong enough to withstand high hydrostatic pressure without internal equalization of chamber pressure (Shapiro and Saunders, 1987). This aspect has been the subject of numerous studies in the material sciences (Mayer, 2005; Barthelat, 2007; Munch et al., 2008; Luz and Mano, 2009).

This pearly shell with its mathematically defined spiral and subtle coloration has fascinated humans for millennia. On the down side, this fascinating structure is currently contributing to this group's extinction. The extensive overcatch of the animals by inhabitants of the Philippines and other islands for the button industry, shell handcrafts, and collector's items is responsible for an ongoing decline in *Nautilus*. At least one population in the Philippines is already known to be wiped (Ward, 2008).

5.4.3 Tentacles

The gross anatomy and morphology of the tentacles has been previously described (Owen, 1843; Haswell, 1895; Griffin, 1898; Willey, 1898a, b; Bidder, 1962). The detailed histological and ultrastructural characterization of the structures and their functional properties is based mostly on studies by Barber and Wright (1969), Fukuda (1988), Muntz and Wentworth (1995), and Ruth et al. (2002).

Nautilus has typically up to 90 tentacles which, after Owen (1832), are classified in three groups: (1) a variable number of labial tentacles surrounding the buccal region, which participate mainly in reproductive behavior; (2) two thick pre- and postocular tentacles, one of each situated in front of and behind each eye, thought to be sensory; (3) 19 pairs of digital tentacles surrounding the labial tentacles, equipped with an adhesive tactile surface (Fig. 5.13).

Each of the ocular and digital tentacles can be completely retracted into a muscular sheath during resting or swimming; the labial tentacles, in contrast, are less retractile (Dean, 1901). As opposed to other cephalopod groups the tentacles in *Nautilus* lack suckers or hooks; instead, *Nautilus* produce glue in a specialized glandular structure on the oral side of the digital tentacles.

5.4.4 Gland Morphology

The digital tentacles are slender, up to 10 cm long and 0.4 cm thick at their base, and slightly tapered, with a blunt, rounded tip. Transverse, tongue-like ridges surround the tentacle trunk like sickle-shaped lamellae (Fig. 5.15); these are more prominent on the oral *versus* aboral side (Stenzel, 1964).

In cross sections, an axial nerve cord is visible in the center of the tentacles (Kier, 1987). One large, thick-walled

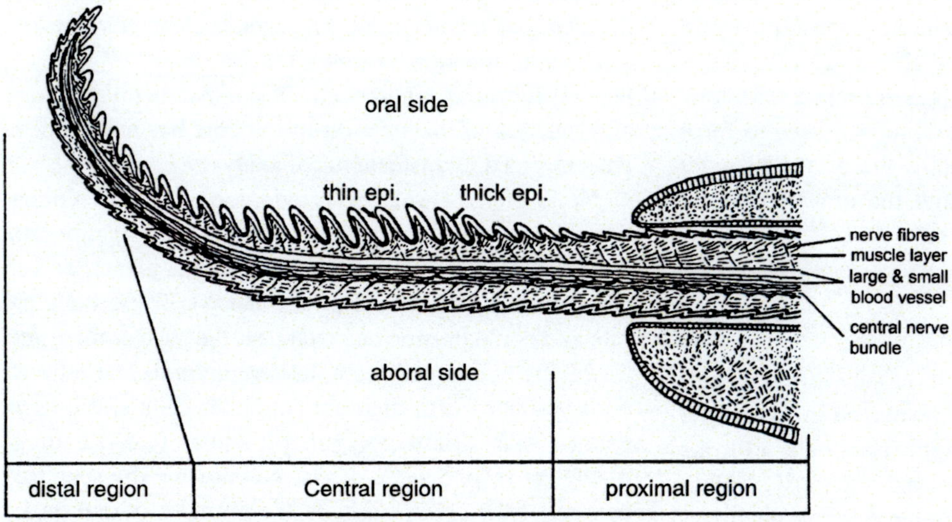

Fig. 5.15 Structure of a longitudinal digital tentacle with the muscular cirrus (subdivided into three regions). Many transverse ridges are present on the dorsal (oral) surface of the cirrus. The epithelium of the ridges is thin in the distal surface of each ridge, but thick on the proximal region. Adapted from Fukuda (1988) and reproduced with permission. There are instances where we have been unable to trace or contact the copyright holder. If notified the publisher will be pleased to rectify any errors or omissions at the earliest opportunity

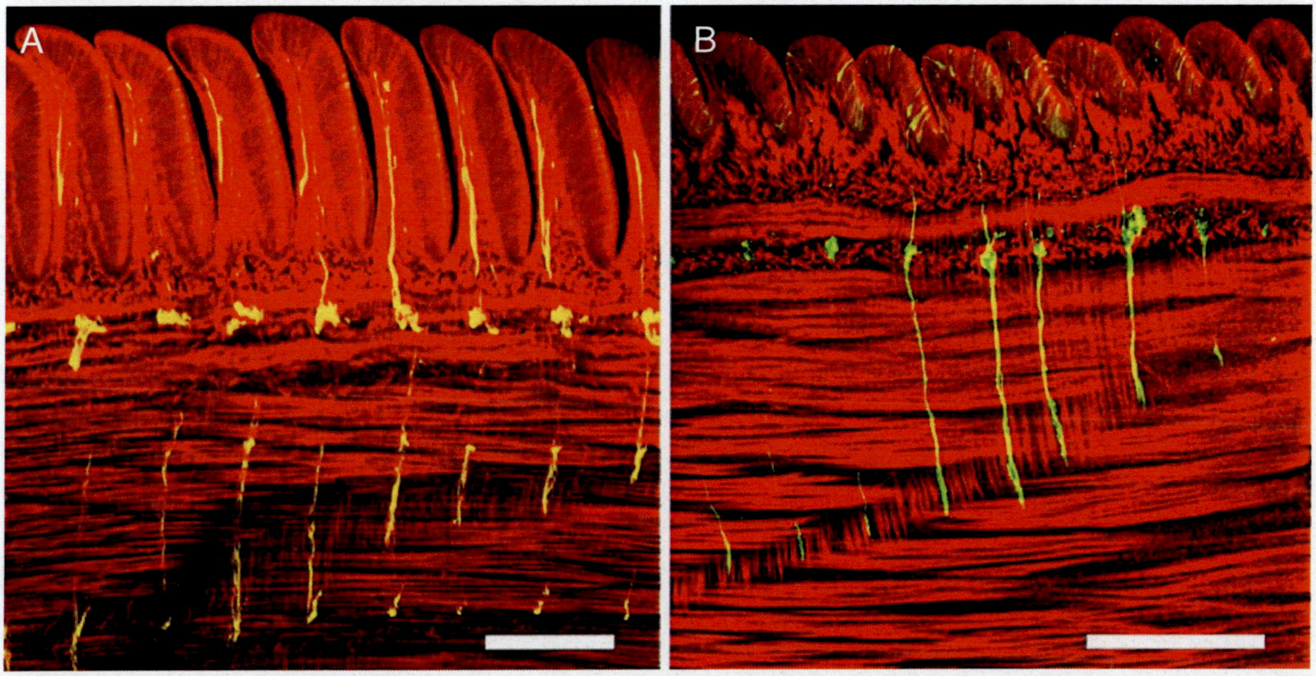

Fig. 5.16 Characterization of the muscle layers (red) and nerve fibers (yellow in A and green in B) of the oral side (A) and aboral side (B) of the *Nautilus* digital tentacle. Below each ridge a sublamellar nerve plexus is formed, interconnected to each other as well as with the central nerve cord. The nerve fibers traverse the epithelial ridge and end as sensory ciliated cell on the tentacle surface. Scale bar in (A, B) = 150 μm

artery and one small, thinner-walled vein run close-by the oral surface. Light- and immunofluorescence microscopic analyses show that a large network of radial muscle fibers surrounds the central nerve bundle. The longitudinal muscles are divided into bundles and form the main body of the tentacle; several layers of oblique muscle fibers and connective tissue enclose the muscle layers. Tentacle retraction is effected by the longitudinal muscles, while elongation is performed by the radial musculature.

From the central nerve cord, nerves branch off below each oral ridge and merge with circular nerves to form a sublamellar nerve plexus. Finally, the nerves proceed together with muscular fibers into the oral and aboral ridges (Fig. 5.16).

5.4.4.1 Oral Side

The ridges on the oral side (Fig. 5.15) are low and shallow in the distal region but well developed in the central region, about 500–700 μm in height. They have a serrate appearance and are slightly concaved anteriorly. In longitudinal section the epithelium of the ridges is heterogeneous, thick (approx. 40 μm wide at its base, up to 80 μm at the ridge top) on the proximal side but thin (about 8 μm wide at the base, increasing to 30 μm at the top) on the distal surface (Figs. 5.17A and 5.18A).

In contrast to previous investigations, the thick epithelium possesses two glandular cell types that differ in their morphology and secretory composition; the thin epithelium, however, bears only one glandular cell type. Additionally, different types of ciliated cells are widely distributed within both epithelia (Ruth et al., 2002). In the distal region of the thick epithelium, numerous taste-bud-like chemoreceptors are present, which are rare in the thin epithelium and on the aboral side.

Proximal to the basal membrane (about 0.3–0.6 μm thick) are rare clusters of small granules, presumably pigments. Below the basal lamina and within the center of the oral ridges are bundles of nerves, muscles and collagen fibers.

5.4.4.1.1 Thick Epithelium

So far, the literature mentions only one glandular cell type in the thick epithelium, named **columnar cells** (Fukuda, 1988). These abundant cells are long, slender (about 2.5 μm wide), and scattered throughout the epithelium (Fig. 5.17B). The cells contain small, round to polygonal granules (about 1 μm in diameter, with an electron-dense central core) (Fig. 5.17G).

The granules react moderately to PAS staining (Fukuda, 1988 and unpubl. results of the first author) and show a weak positive reaction to protein stain (Millon reaction), suggesting the presence of mucopolysaccharides (Fukuda, 1980, 1988; Muntz and Wentworth, 1995). Basic or acidic proteins (Biebrich Scarlet and Alcian Blue) as well as lipid components seem to be absent. Since the granules are so small, however, a clear positive histochemical reaction as given for the other cell types is lacking for the columnar cells. Also, the chemical nature and function of the electron-lucent core has not yet been characterized by histochemical analyses.

None of the earlier descriptions of the thick epithelium (Kier, 1987; Fukuda, 1988; Muntz and Wentworth, 1995) mentioned or described a second glandular cell type (named **cell type 1** in this study). These cells are rare and could only be detected near the base of the ridge, traversing likewise the complete epithelium. The cells are wider (about 7 μm) than the columnar cells and loosely packed with almost polygonal granules (1.5–1.8 μm in diameter) (Fig. 5.17C). As is the case for the secretory content of the columnar cells, the granules of cell type 1 also have an electron-lucent central core.

The granules react strongly to neutral sugars (PAS) (Fig. 5.17C) but lack basic (Biebrich Scarlet) and acidic proteins (Alcian Blue) as well as lipids (Sudan Black B).

5.4.4.1.2 Thin Epithelium

The glandular cells in the thin epithelium are named mucus (Muntz and Wentworth, 1995), granular (Fukuda, 1988) or **goblet cells** (Kier, 1987); they are rare and singularly distributed within the epithelium from the ridge base to its top. The goblet cells are round to oval (Fig. 5.17D, E) and loosely filled with round granules (Ø 1.2–1.7 μm).

Unpublished results by the first author indicate that the granules consist solely of neutral mucopolysaccharides (Fig. 5.17F). Earlier analyses, however, also determined a weak positive reaction to protein stain (Millon reaction) in the secretory material (Fukuda, 1988; Muntz and Wentworth, 1995). The present re-characterization could not verify this.

5.4.4.2 Aboral Surface

Its ridges are low and flat (140–160 μm high), and possess an epithelium without regional variation. A detailed characterization of the aboral epithelium, as given for the oral side, was lacking so far in the literature. Pigment clusters

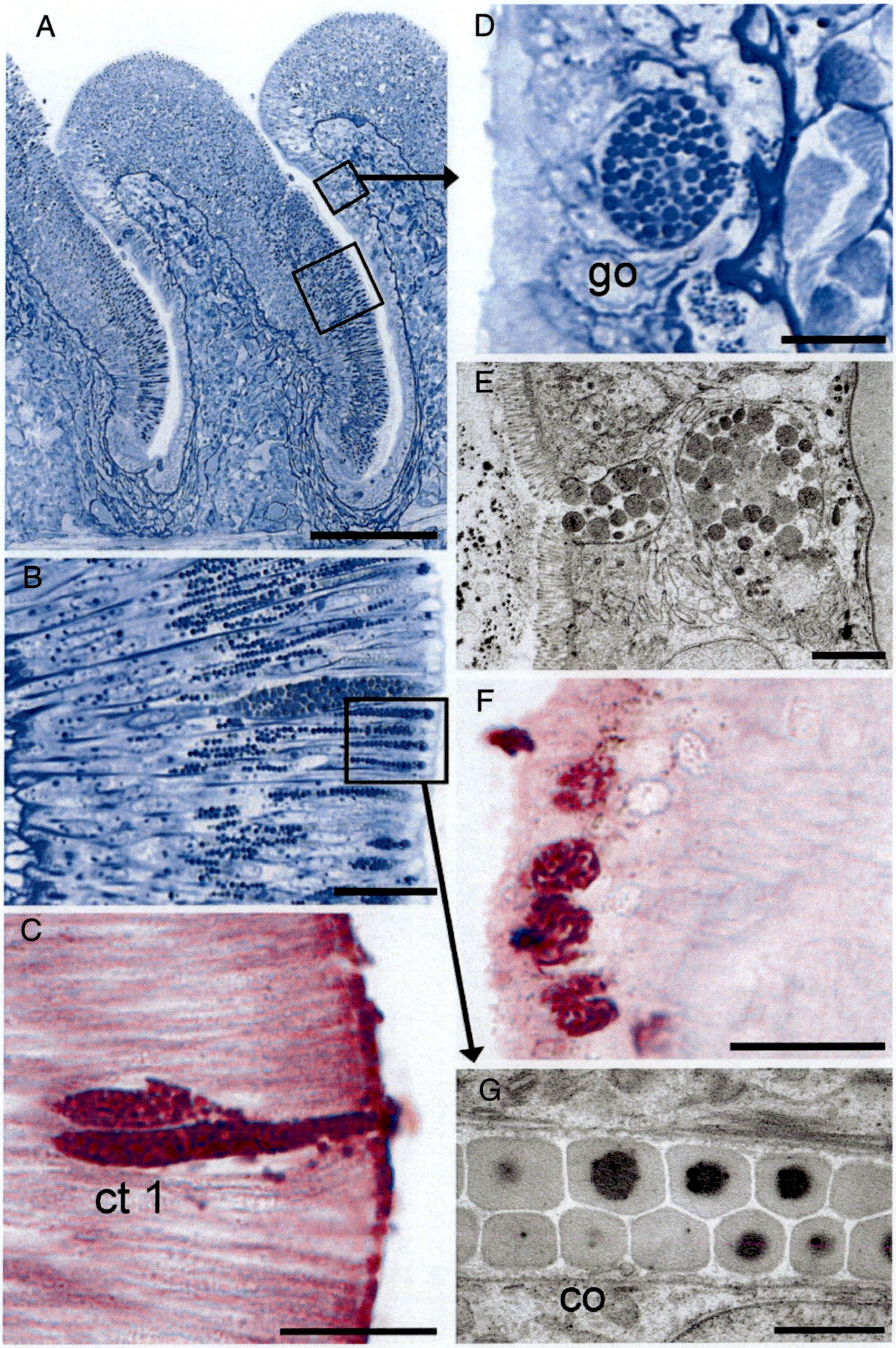

Fig. 5.17 (A) Ridge on the oral side of the digital tentacle (B, C and G = thick epithelium, D–F = thin epithelium); (B) The thick epithelium bears two glandular cells: The columnar cells with small granules (higher magnification of the granules in G) and cell type 1 (C) with large, PAS-positive granules. (D–F) The thin epithelium possesses one glandular cell type (named goblet cells), containing large PAS-positive granules. Scale bar in (A) = 150 μm, (B, C and F) = 20 μm; (D) = 10 μm; (E) = 4 μm; (G) = 1 μm

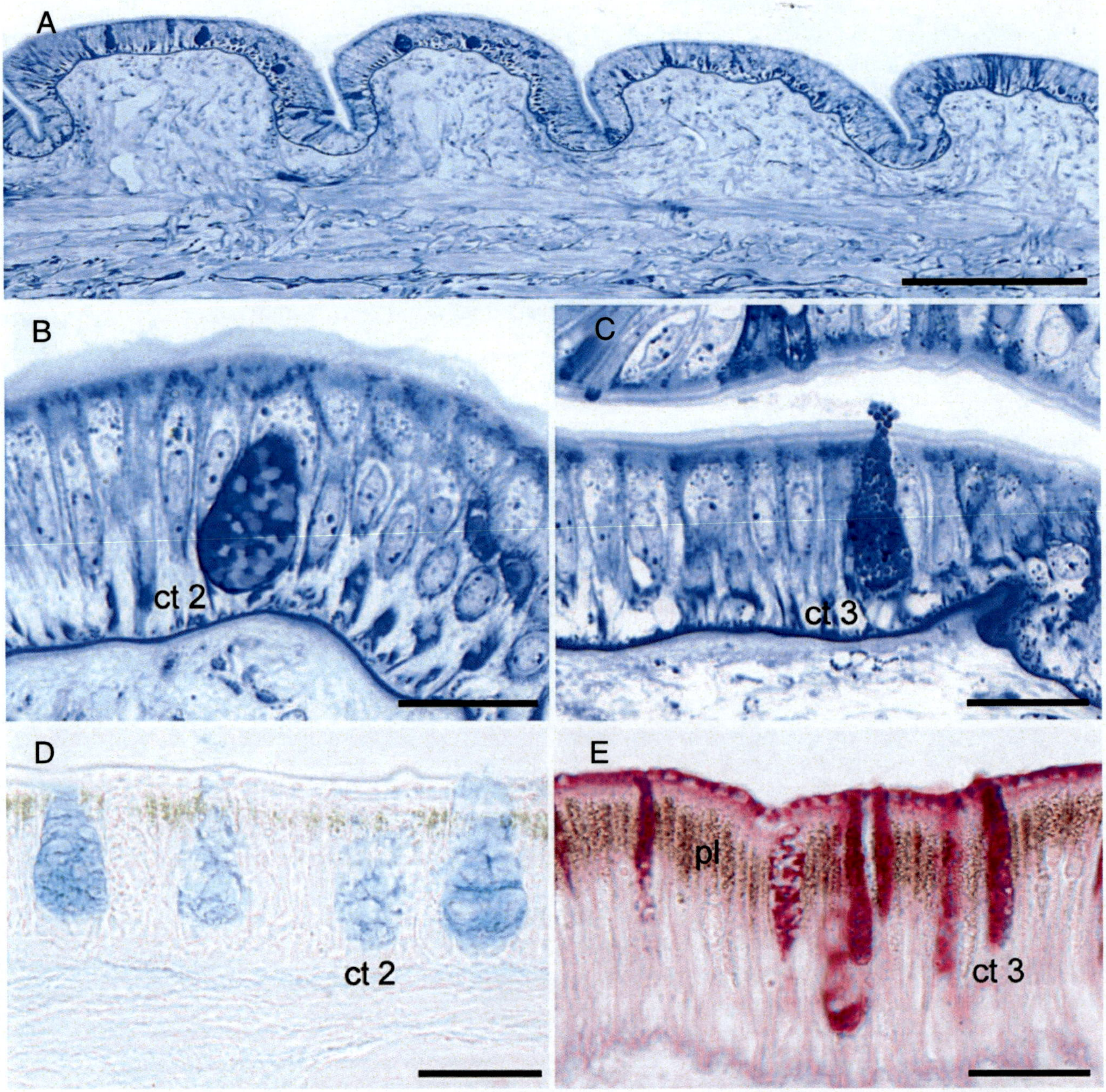

Fig. 5.18 (A) Ridge on the aboral side of the digital tentacle with two glandular cell types. (B) and (D) Cell type 2 possesses large polygonal granules, which are different in density and consists of acid proteins (D = Alcian Blue reaction at pH 2.0); (C) and (E) The granules of cell type 3 are slightly smaller in size and contain neutral acidic sugars (E = PAS positive reaction). Scale bar in (A) = 150 µm; (B–E) = 20 µm

form a prominent thin layer (about 10 µm thick) beneath the surface of the epithelium (Fig. 5.18A). An earlier description denied the presence of muscles within the ridge (Fukuda, 1988); the present re-examination, however, shows a high abundance of muscle fibers and nerves. Additionally, ciliated cells are also widely distributed within the epithelium, as is also the case for the oral side.

Although two distinct cell types are present in the aboral epithelium (Fig. 5.18B, C), both were never mentioned and named in the earlier literature. We therefore avoid naming them but rather number them consecutively:

Cell type 2 is pear-shaped and tightly packed with large polygonal granules (∅ 2 µm) of different density

(Fig. 5.18B). The secretory material reacts strongly for neutral sugars and moderately for acid proteins (Fig. 5.18D) but lacks basic proteins. A chemical differentiation of the different dense granules was negative.

Cell type 3 corresponds in its shape and chemical nature (only positive for neutral sugars) to cell type 1 from the thick epithelium. However, its granules are slightly smaller (∅ 1.3–1.5 µm) and more loosely packed (Fig. 5.18C, E). Furthermore, the granules lack the prominent electron-dense core as observed for type 1. Whether both cell types are the same and could perhaps be summarized under one term needs further verification.

5.4.5 Mechanism of Bonding

Nautilus has a "powerful grip" (Willey, 1902) when attaching with its tentacles. Willey (1902) described its adhesive power as so strong that a tentacle, attached to an object, remains fixed even when the tentacles break off.

Many morphological studies have been conducted on *Nautilus*, yet its bonding mechanism remains unknown and speculative. Early researchers suggested an exclusively mechanically related mechanism similar to the laminated sucker of the fish *Remora* (Owen, 1843); later morphological studies favored a chemical bonding effect, induced by the columnar cells alone (Fukuda, 1988) or by a combination of the columnar and goblet cells (Muntz and Wentworth, 1995).

As for *Euprymna* and *Idiosepius*, an antagonistic duo-gland system can be ruled out in *Nautilus* as well. Since the cell types in the thick epithelium and their secretions all possess a similar composition (neutral sugars), a release by chemical modification is uncertain; *Nautilus* detach by mechanical mechanisms (muscle contraction and retraction).

The present data indicate that the glue in *Nautilus* consists mainly of neutral sugars, perhaps with a small protein ratio.

The glue seems to be more a viscous carbohydrate-like gel as known for *Idiosepius* and gastropods (see Chapter 4, p. 41) than of proteomic nature as in mussels (see Chapter 18, p. 273) and barnacles (see Chapter 9, p. 153).

Until now, we do not know whether one, two, or all three glandular cells jointly generate the glue. All three cell types provide advantages (high secretion amount in goblet cells and cell type 1; abundance and location of the columnar cells), but other facts (location of goblet cells and cell type 1; low secretion amount in the columnar cells) argue against a sole bonding responsibility of any one type.

Currently we cannot support the idea of a three-parted glue in *Nautilus* because too little is known about the three cell types and the glue itself. Further analyses are planned to provide detailed knowledge about the bonding behavior, the cellular secretion process and its possible modification to a functional glue in *Nautilus.*

5.5 *Sepia*

(Janek von Byern, Robyn Scott, Charles Griffiths, Vanessa Zheden and Norbert Cyran)

The order Sepiida, commonly known as cuttlefish, is – apart from the Octopoda and Teuthida – the third largest group of cephalopods. Cuttlefishes are dorso-ventrally compressed and oval (Tompsett, 1939). Encircling the mouth are eight arms with suckers; the animals also possess two elongated tentacles, which can be rapidly extended and are used to catch prey. Sepiida have an internal hard structure (the cuttlebone), which is porous and composed of calcium carbonate; it serves for the animals' buoyancy (Sherrard, 2000).

Cuttlefishes have an unusual biogeographic pattern: The animals are totally absent along the American coasts, but occur in East and South Asia, Western Europe, the Mediterranean, as well as all along the coasts of Africa and Australia. Young et al. (1998) suggests that by the time the family evolved, the Northern Atlantic had become too cold and deep for these warm-water species to cross.

Sepiida are primarily demersal inhabitants of shallow-water zones and occupy all habitats, from rocky, sandy and muddy bottoms to grassflats, seaweed beds, coral reefs, etc. Although excellent swimmers, they generally are bottom-dwelling and spend most of their time lying on the substratum. Sepiida typically lack adhesive structures to camouflage themselves in contrast to *Euprymna* and *Idiosepius* or to hold onto prey (as in *Nautilus*).

At the same time, there are at least four *Sepia* species that have developed distinct dermal structures on the ventral mantle surface for active attachment to the substratum. Although this bonding is primarily mechanical and based on a reduced-pressure system, chemical components seem to enhance the attachment. This allows the animal to immobilize itself by bonding on a hard substrate and resisting shearing forces of water movement.

5.5.1 Description of the Glue-producing Sepiida Species

In South Africa more than 18 *Sepia* species – separable by their mantle and tentacular morphology – are known. Nonetheless, our knowledge about their ecology and behavior are still marginal and based mostly on records of the South African Museum and on ecological surveys of the University of Cape Town.

Of these 18 species, only four (*S. papillata*, *S. pulchra*, *S. tuberculata*, and *S. typica*) possess special glandular structures on the ventral mantle side. These four occur predominately along the South African coast, from Western Cape to the Natal province; their absence in the Mozambique-Malagasy province does not seem to be due to the water temperature but depends rather on the different coastal conditions there.

5.5.1.1 *Sepia papillata* (Quoy and Gaimard, 1832)

Sepia papillata has a broadly oval mantle, with a mean mantle length of 12 cm. The skin is tuberculate dorsally and laterally on head, arms and mantle but smooth ventrally in the non-adhesive skin. Most of the ventral mantle consists of two large adhesive wrinkled patches (Fig. 19A); two smaller patches also occur along the ventral surfaces of the fourth arm pair (Roeleveld, 1972).

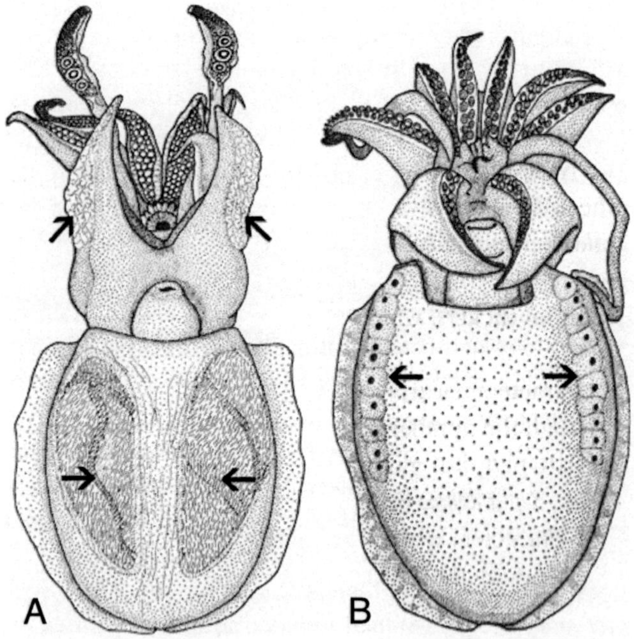

Fig. 5.19 Ventral mantle of *Sepia papillata* (A) and *Sepia typica* (B) with the adhesive areas (arrow). In *S. tuberculata* and *S. papillata* the area consists of two large oval patches with a wrinkled surface (A) while *S. typica* and *S. pulchra* bear two flesh ridges near the fin base, each ridge with numerous pores. In *S. papillata* and *S. tuberculata* apart from the mantle also the ventral arms bears adhesive pads. Adapted from Roeleveld (1972) and reproduced with permission

The color of the dorsal mantle is dark reddish-brown to purple but pale ventrally, with scattered chromatophores. The animals were collected at depths from 26 to 120 m at different sites around the Cape of Good Hope (Roeleveld, 1972).

From its outer morphology, *S. papillata* is closely related to *Sepia tuberculata* (see below), although both species differ in sucker morphology on the tentacular club and in cuttlebone shape.

5.5.1.2 *Sepia pulchra* (Roeleveld and Liltved, 1985)

The animals are small with mantle lengths of 1.5–2.2 cm in males and 1.9–2.4 cm in females. The entire dorsal side of mantle, head, and arms is covered with tubercles of variable size, orientation, and complexity. The ventral surface forms a sole, made up of a pair of fleshy keels on the mantle and the swollen under-surfaces of the ventral arms.

As in *S. papillata*, *S. pulchra* is also reddish-brown dorsally on the mantle, head, and arms but buff on the complete ventral side; a large oval purplish patch is present middorsally on the mantle.

On a series of scuba dives of the Cape Peninsula, this species was collected from 15 to 35 m, adhering to vertical rock faces in a head-down position (Roeleveld and Liltved, 1985).

5.5.1.3 *Sepia tuberculata* (de Lamarck, 1798)

The animals have a short and broad mantle and are about 5–8 cm long. As in *S. papillata* and *S. pulchra* also the dorsal mantle skin in *Sepia tuberculata* is covered with small (∅ 1–5 mm) oval to roundish tubercles (Fig. 5.20). The ventral mantle as well as the fourth arm pair bears two oval wrinkled adhesive areas as given for *Sepia papillata* (Roeleveld, 1972).

The dorsal mantle side is reddish-brown to black; the ventral side always has a pale shade of color. Scattered chromatophores are visible in the regular skin of the ventral mantle side but are lacking completely in the adhesive area.

Sepia tuberculata lives in wave- and windswept rocky intertidal regions (between 20 and 60 cm deep), strongly exposed to turbulent water movement (Roeleveld, 1972).

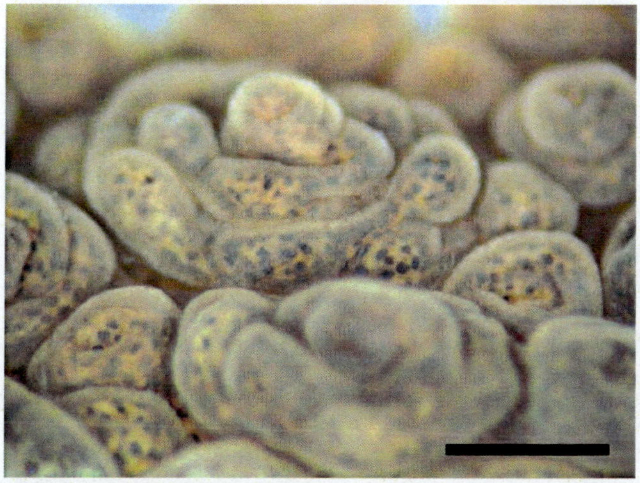

Fig. 5.20 The black-spotted dorsal surface of the arms, head, mantle, and fins consists of small (∅ 1–5 mm) oval to roundish tubercles, which arise spirally and exhibit the typical chromatic elements. The ventral surface is pale white and lacks tubules. Scale bar = 5 mm

This species was collected by the first author mainly under small, flat, and minimally fouled stones.

5.5.1.4 *Sepia typica* (Steenstrup, 1875a, b)

The animals are small, with a mean mantle length of 2.5 cm. The mantle is broadly oval, almost as wide as long, and covered with numerous dorsal papillae. Instead of two large patches, *Sepia typica* bears two small fleshy ridges near the fin bases on the ventral mantle side (Fig. 5.19B). Each ridge has several pores anteriorly, most commonly 10–12 on each side, but the number varies between 5 and 15. The ridges become less distinctive posteriorly (Roeleveld, 1972).

The animals are dark reddish-purple dorsally on the head, arms, and mantle, with a darker diamond-shaped region mid-dorsally over the shell. Ventrally the mantle is pale, except for the darker pores.

Sepia typica was found during low tides, attached to jetties, and other hard surfaces. Nevertheless, distribution records indicate that the animals occur also in greater depth, from 2 m in Cape Town down to 156 m around the Durban coast (Roeleveld, 1972).

The above characterization shows that the shape and form of the adhesive areas do not correlate with the different habitat type and depth ranges. It also remains unclear why two different adhesive types are formed and why this special anatomical character occurs only in South African *Sepia* species.

5.5.2 Gland Morphology

In *S. papillata* and *S. tuberculata* most of the ventral mantle consists of the adhesive structure (Roeleveld, 1972; von Byern and Klepal, 2006). It is divided into two equal areas. *Sepia typica* and *S. pulchra*, in contrast, bear only two small fleshy ridges near the fin base, each ridge with numerous pores anteriorly (Roeleveld, 1972).

So far, only the adhesive system of *Sepia tuberculata* has been investigated in detail:

The adhesive areas on the ventral mantle are mostly furrowed, divided by a medial, smooth band into two large oval wrinkled pads. In contrast, the adhesive pads on the arms lack such a medial band.

The glycocalyx of the entire mantle epithelium contains neutral sugar units and weakly acidic proteins. In cross section, the adhesive epithelium can be easily distinguished from the regular mantle epithelium (30–40 μm thick) by its greater thickness (about 100 μm) and its different cell composition. Nerve fibers are present in the proximal part of both the regular and adhesive mantle epithelium as well as directly beneath the basement membrane. No connection toward and with the glandular cells was found.

Beneath the epithelia is a high dermal layer (up to 500 μm), consisting of connective tissue as well as longitudinal and circular muscle fibers. In the regular mantle the dermal layer also includes the typical chromatic elements (leuco-, irido-, and chromatophores); the ventral mantle contains only chromatophores. The longitudinal and circular muscle fibers end in the thick main mantle musculature.

5.5.2.1 The Adhesive Area

There are no differences in cell composition between the adhesive pads on the arms and mantle. Both possess three glandular cell types (cell type 1, type 2, and goblet cells) as well as the non-secretory interstitial cells (Fig. 5.21).

While cell type 1 occurs also sporadically in the regular mantle epithelium, it is very abundant in the adhesive area and exhibits a different chemical composition. In contrast, the goblet cells are quite sparse in the adhesive area but more prominent in the regular mantle epithelium.

The **cell type 1** is slightly pear-shaped (Fig. 5.22B) around its oval, basally located, nucleus and tapers towards the apical pole. This cell type contains different types of secretory granules. These include large round, membrane-bound but loosely packed granules (∅ 1–1.5 μm) of different density (Fig. 5.22B). In some cases the granules loosen and appear as fine and bright material (Fig. 5.22A). The granules are oriented along

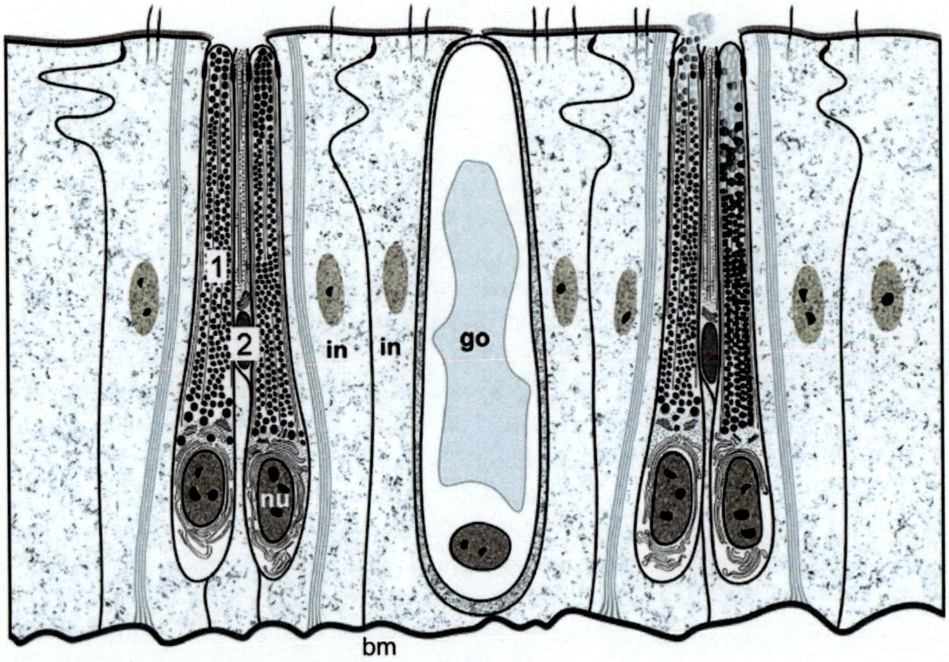

Fig. 5.21 Schematic drawing of the adhesive epithelium in *Sepia tuberculata.* While cell type 1 (*1*) and type 2 (*2*) are restricted to the adhesive area only, the goblet cells and interstitial cells are also given for the regular mantle epithelium. The nucleus, rER and Golgi apparatus are located basally. The microvilli covered interstitial cells form the main part of the epithelium surface. They bear several rooted cilia and longitudinal filaments closed to cell type 1

strands of tubules running beside the lateral cell membrane (as in cell type 2 in Fig. 5.22F).

The secretory product of cell type 1 contains only neutral sugars (Fig. 5.22D, E) and basic proteins. No histochemical difference between the different granula forms could be determined.

Cell type 2 is narrower than cell type 1 and always contains bright granules 50–250 nm in diameter, which are likewise lined along tubules (Fig. 5.22F). This type is either joined with type 1 to form a doublet (Fig. 5.22A, C) or mostly encased by them, forming a triplet aggregation.

Unpublished histochemical tests indicate that this cell type reacts weakly only to acidic proteins at pH 2.5. As this cell type and its secretory granules are very small, a clear histochemical characterization is complicated.

5.5.2.2 The Regular Mantle Epithelium

The epithelium contains mainly of the non-secretory **interstitial cells** with their nuclei (Fig. 5.22G), interrupted mainly by the goblet cells and to a lesser extent by cell type 1. In the interstitial cells the nucleus is oblong and located in the middle region. Cilia (250 nm in cross section and 2 μm long) are randomly distributed on the cellular surface; filaments, sometimes aggregated into longitudinal bundles, traverse the cell longitudinal.

The cell content does not react with any histochemical tests applied.

Cell type 3 is shorter and narrower than cell type 1 from the adhesive epithelium. The cells possess only one form of glandular material: small round granules about 1 μm in diameter (Fig. 5.23A). The secretory material of cell type 3 was strongly reactive only for basic proteins (Fig. 5.23B).

The **goblet cells** are round oval to pear shaped and contain a central, balloon-like cavity. This cavity is empty or partly contains fine-grained or vesicular material (Fig. 5.23C). Their nuclei are flat and located at the base of this cavity. The goblet cells occur in the regular mantle epithelium and are prominent around the suckers of the fourth arm pair.

Histochemical characterization indicates that both the glandular material and secretions of goblet cells mainly consist of acidic proteins; their cell membrane is, moreover, strong reactive for neutral sugars (Fig. 5.23D).

The goblet cells in *Sepia tuberculata* closely resemble the goblet and saccular-shaped cells present in the regular epithelium in *Idiosepius* (see previous chapter). A detailed characterization and alignment to one of those two

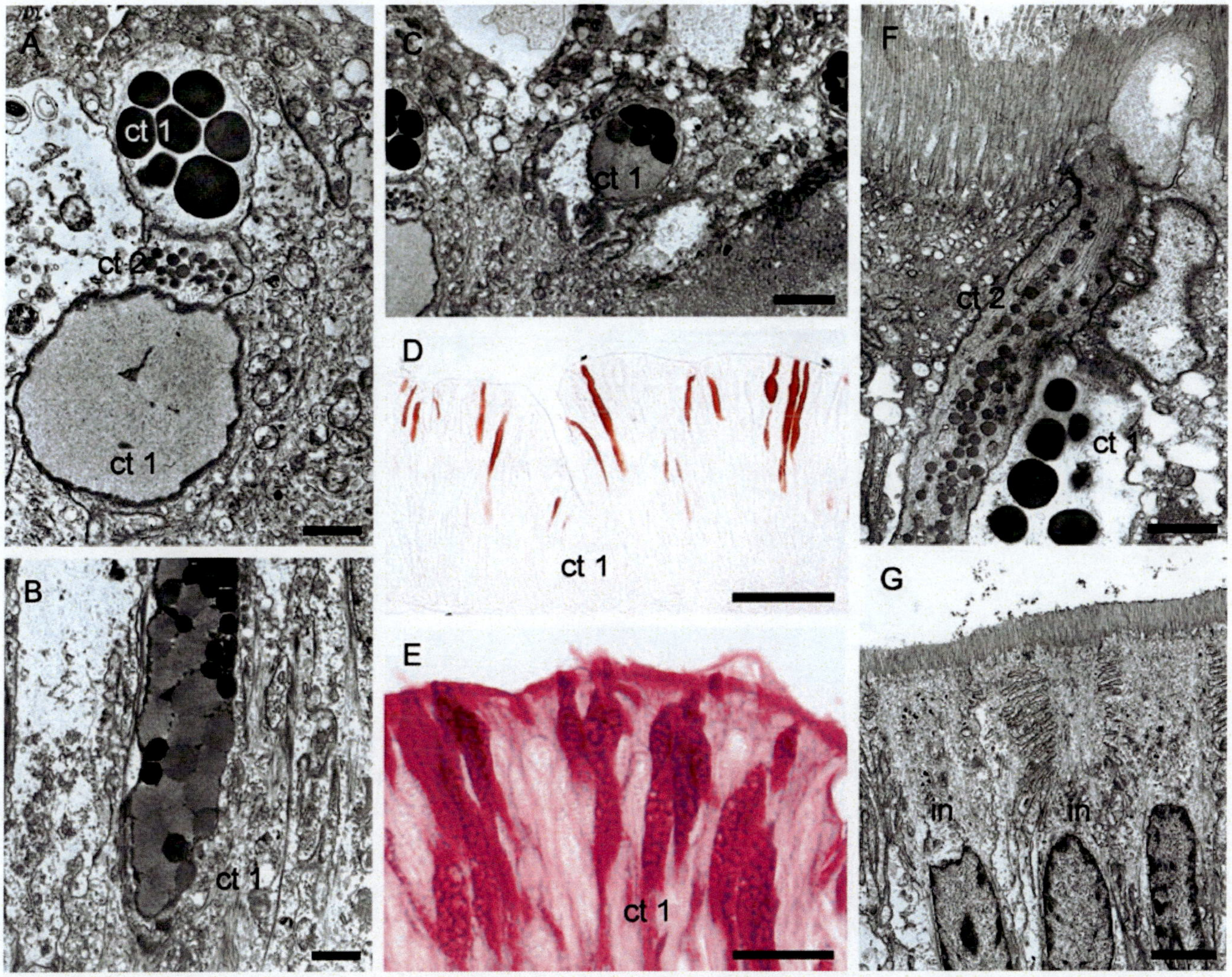

Fig. 5.22 Overview of the two glandular cell types in the adhesive epithelium in *Sepia tuberculata*. Both cell types are joined to a doublet (C) or triplet aggregation (A). (A) Cell type 1 (*ct 1*) is cylindrical and contains secretory granules, which appear in different forms e.g. as electron-dense granules (B, C); bright (B) or as fine and bright material (A, C). The secretory product of cell type 1 contains neutral sugars (D) and basic proteins (E). (F) Cell type 2 (*ct 2*) is narrower than cell type 1 and contains bright granules, which are lined along tubules. (G) The interstitial cells lack secretory granules but contain rather loosely distributed filaments, sometimes aggregated to longitudinally bundles. Scale bar in (A) = 1 µm; (B, C) = 2 µm; (D) = 50 µm; (E) = 20 µm; (F) = 3 µm; (G) = 5 µm

cells, however, is still absent for the goblet cell of *Sepia tuberculata*.

5.5.3 Mechanism of Bonding

Observations by the second author as well as von Boletzky and Roeleveld (2000) showed that *Sepia tuberculata* involves a defined movement of the entire body. The animals use the adhesive pads on the arms first to get in contact with the substratum and then tighten the bonding with the adhesive structure on the ventral mantle side.

Apart from this known mechanical bonding, the occurrence of specific glandular cells in the adhesive area argues for a chemical participation in bonding, leastwise acting as a sealant in order to amplify the force of adhesion by preventing water inflow. Probably the release procedure is likewise achieved primarily mechanically by assignment of the well-developed dermal and mantle musculature.

The chemical properties of the glandular contents are not resolved so far. Since two gland cell types are apparently participated, both a cooperating and an antagonistic effect are conceivable.

Cell type 2 may act as an accelerator and makes the cell type 1 secretions sticky. The observation of always

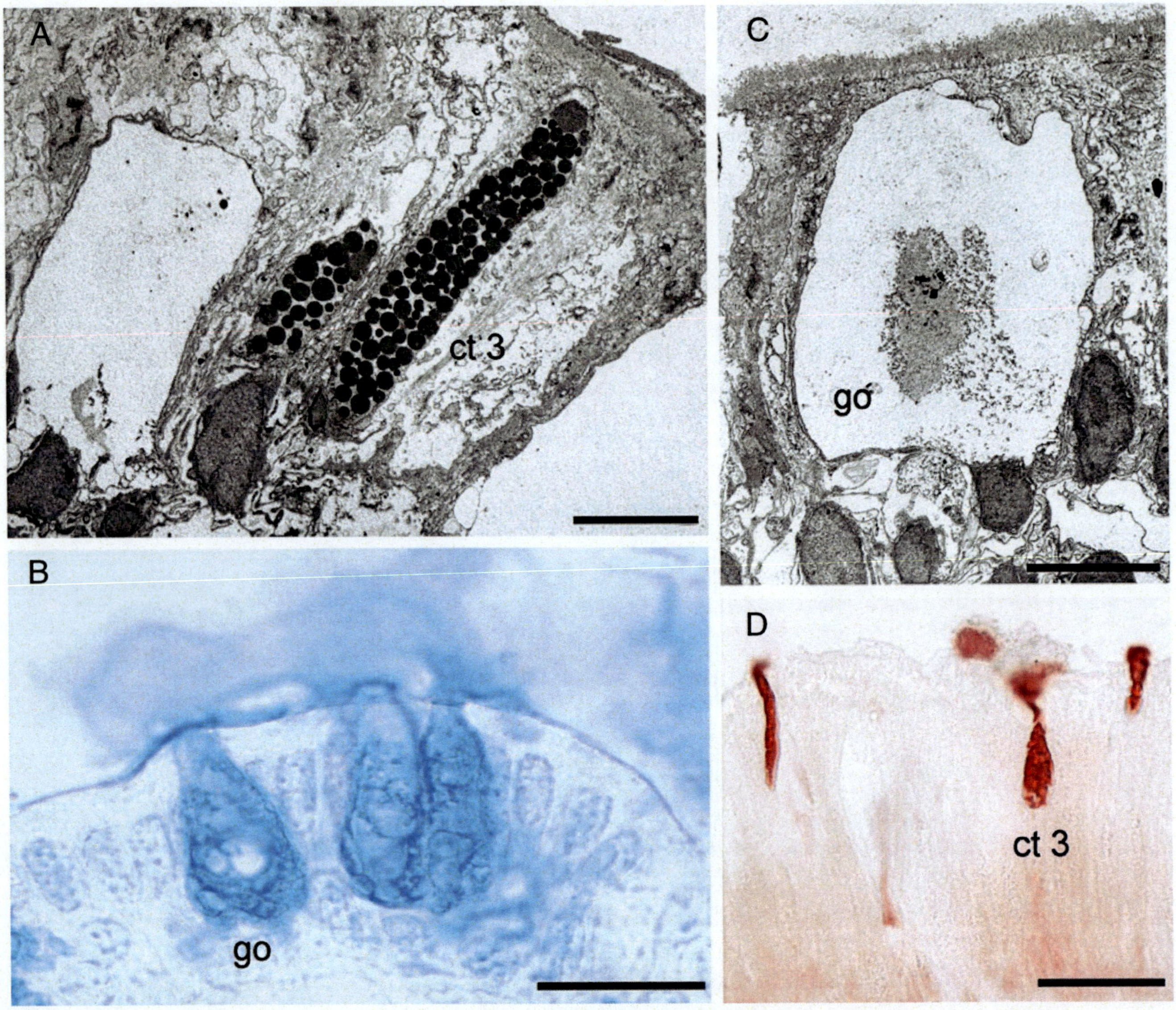

Fig. 5.23 Overview of the glandular cell types in the regular mantle epithelium. (A) Cell type 3 is long slender and possesses small round granules about 1 μm in diameter. (B) Its secretory material is strongly reactive for basic proteins (Biebrich Scarlet at pH 6.0 and weakly pH 8.5). (C) The goblet cells are round oval to pear shaped and contain a central balloon-like cavity. (D) Histochemical characterization indicates that its glandular material as well as the secretions is mainly acid proteins (Alcian Blue at pH 2.5). Scale bar in (A and C) = 10 μm; (B and D) = 20 μm

simultaneous release of material in both cell types rather speaks for this variant.

As in the case of the other glue-producing cephalopods (see previous chapters), the presence of an antagonistic duo-gland system is uncertain in *S. tuberculata* as well. Although cell type 2 contains weak acidic protein components, which are thought to effect a modification of the glue resulting in de-adhesion (Hermans, 1983), several facts argue against this hypothesis.

Anyway a muscular-effected release would be faster and effective in a such an agile animal. Moreover, cell type 2 with its marginal amount of material, seems poorly suited to effect a glue dissembling.

Conclusion

Although the different adhesive systems vary between the cephalopod genera regarding their location, application, and function, they possess certain similarities in their morphology and secretory composition.

Generally, at least two specific glandular cells are exclusively present in the adhesive epithelia (Table 5.1):

Table 5.1: Summary of the glandular cell types within the epithelia of all four cephalopod genera. Cells which occur exclusively in the adhesive epithelium (AO) are marked in darker green. Cell types occurring both in the adhesive as well as regular epithelium are classified as "global". Secretory content was characterized histochemically for sugars (PAS reaction), proteins (Alcian Blue B at pH 1.0 and 2.5 as well as Biebrich Scarlet at pH 6.0–10.5), and lipids (Sudan Black B test)

	Euprymna scolopes				*Idiosepius spec.*						
Name	Type 4	Goblet	Ovate	Interstitial	Columnar	Granular	Fusiform	Goblet	Saccular	Basal	Interstitial
Nucleus	elongated	elliptical	flat	spheroidal	spheroidal	spheroidal	oval	spheroidal	flat to sickle shaped	spheroidal	oval
Location	basal	basal	basal	basal and apical	basal	basal	basal	basal	basal	undetermined	middle
Other major cell compartments	mitochondria, rER, Golgi	Golgi, rER, mitochon-dria	rER, Golgi	mitochondria, rER, Golgi	rER, Golgi	rER, Golgi	polysomes, cilia	rER	rER, filament	vesicles, cytoskeleton	Golgi network, cytoskeleton
Location	AO	AO	global	global	AO	AO	AO	global	global	global	global
Shape	elongate pear-shaped	elongate pear-shaped, infolded laterally	ovate to sac-like	polymorphic	elongate pear-shaped	cylindrical	pear-shaped	cylindrical	sac-to balloon-shaped	polymorphic	polymorphic
Gimmick	–	microtubules along long axes	peripheral filaments	thick and thin filaments	–	–	cilia, syn-apse	–	lateral filaments	vesicles with e--dense granules	longitudinal bundles
Secretory content	granules ø 0.2–0.4 mm	granules ø 0.5–1.3 mm	fine granules	glycocalix	granules ø 1 mm	granules ø 2–5 mm	–	fine-grained material	fine-grained material	–	glycocalix
Histochemistry	undetermined	++ neutral sugars	proteins ++ (pH 6.0) + (pH 8.5)	undetermined	++ neutral sugar + protein (pH 6.0–10.5)	++ neutral sugar ++ protein (pH 6.0–10.5)	–	+ neutral sugar ++ protein (pH 6.0–10.5)	–	–	–
Arrangement	always in close vicinity to goblet	singular	singular	singular	aggregates of 8–15 cells	singular	singular	singular	singular	singular	singular
Secretory release	partial release of granules	partial release of granules	partial release of granules	exocytosis	partial release of granules	partial release of granules	–	release of entire secretory content	?	–	exocytosis

(Continued)

Table 5.1: (Continued)

	Nautilus pompilius					*Sepia tuberculata*				
Name	Columnar	Type 1	Goblet	Type 2	Type 3	Type 1	Type 2	Goblet	Type 3	Interstitial
Nucleus	oval	oval	oval to flat	flat	flat	oval to flat	?	oval to flat	oval to flat	oblong
Location	basal	basal	basal	basal	basal	basal	middle	basal	basal	middle
Organelles	rER, Golgi	rER, Golgi	?	?	?	rER, Golgi	Golgi	rER, Golgi	rER, Golgi	Golgi network, cytoskeleton
Location	AO (thick epithelium)	AO (thick epithelium)	AO (thin epithelium)	Aboral side	Aboral side	AO	AO	global, prominent around suckers	global	global
Shape	long slender	cylindrical	round to oval	elongate pear-shaped	cylindrical	elongate pear-shaped	elongate pear-shaped	oval to pear-shaped	elongate pear-shaped	high prismatic
Gimmick	translucent granular core	translucent granular core	–	–	–	tubuli between granules	tubuli between granules	–	–	longitudinal bundles
Secretory content	granules ø 1 mm	granules ø 1.5–1.8 mm	granules ø 1.2–1.7 mm	granules ø 2 µm different density	granules ø 1.3–1.5 µm	granules ø 1–1.5 mm different types	granules ø 50–250 nm	partly filled with fine-grained content	granules ø 1 µm	glycocalix
Histochemistry	+– neutral sugars	++ neutral sugars	++ neutral sugars	++ neutral sugars +– proteins (pH 1.0 and 2.5)	++ neutral sugars	++ neutral sugar ++ proteins (pH 6.0–10.5)	+– proteins (pH 2.5)	++ proteins (pH 2.5) cell membrane ++ neutral sugars	++ proteins (pH 6.0 + 8.5)	–
Arrangement	singular	singular	singular	singular	singular	appear as doublet/triplet aggregation		singular	singular	singular
Secretory release	partial release of granules	partial release of granules	partial release of granules	partial release of granules	partial release of granules	simultaneously, partial release of granules		?	?	exocytosis

The first cell type (goblet in *Euprymna*, granular in *Idiosepius*, type 1 in *Sepia*, and type 1 in *Nautilus*) mostly contains large, membrane-bound secretory granules (0.5–2 μm in diameter) consisting of neutral sugars. In some cases this secretory material also contains a large amount of basic proteins.

The second cell type (type 4 in *Euprymna*, columnar in *Idiosepius*, type 2 in *Sepia*, columnar in *Nautilus*) usually contains smaller granules (0.05-1 μm) and, likewise, neutral sugars (except cell type 2 in *Sepia tuberculata*).

Most glandular cells in the adhesive organ have in common that they lack exclusive basic or acidic protein compositions opposite to those in the regular epithelium. Instead its adhesive secretory material is always a mix between carbohydrates and proteins, even in different ratio.

The present data suggest, in all four cephalopod genera, a cooperation of both secretory cells in terms of a two-component system. The presence of an antagonistic gland system (duo-gland mechanisms as proposed by Hermans, 1983) is uncertain due to the location and secretory composition of the glandular cell types.

In three cases (*Euprymna*, *Nautilus*, and *Sepia*) a muscular de-adhesion seems to be the primary mechanism, where as in *Idiosepius* the absence of an explicit mantle musculature beneath the adhesive area speaks against such a release mechanisms.

The here investigated Coleoidea possess within the regular epithelium a common cell type (ovate in *Euprymna* and goblet in *Idiosepius* and *Sepia*), which resembles to each other and are filled with fine-grained material. In the literature this cell type is often named generally as "mucus cell" (Budelmann et al., 1997). Although this cell type often also occur within the adhesive systems, we most likely exclude a contribution in the bonding/release mechanism but rather indicate a more general function, e.g., mucus synthesis, defence components, etc. Further investigations are currently in progress to verify the occurrence of this ubiquitous-like mucous cell type in different cephalopod groups and epithelia areas.

Opposite to the well analyzed and established systems, e.g., those of *Mytilus* or *Balanus* the chemical bonding mechanisms in cephalopods are recent and barely investigated. One main reason is that cephalopods are too agile and skittish, therefore complicates *in vivo* adhesion measurements or non-invasive glue samplings.

However, the present state of knowledge show that, although differently located and applied, the adhesive systems of all four cephalopod species exhibit certain "cellular similarities" as the number of glandular cells, their shape and composition as well as others. This effective temporary fastening system with its "on demand" release mechanism differentiates them clearly from the well-known permanent bonding system of mussels and barnacles.

We are still far away from understanding these adhesive systems and bonding mechanism. But we work on it, in particular since the glue components of cephalopods offer in future new opportunities for adhesion in wet environments, for example as surgical glues.

Abbreviations

ao	adhesive organ
ba	basal cells
bm	basal membrane
ce	chromatic elements
ci	cilia
cn	connective tissue
co	columnar cell
ct 1	cell type 1
ct 2	cell type 2
ct 3	cell type 3
ct 4	cell type 4
fi	filaments
fu	fusiform cell
gb	golgi body
gc	glycocalix
go	goblet cell
gr	granular cells
gra	granulum
in	interstitial cells
me	membrane
mi	mitochondria
mu	muscle
mv	microvilli
nu	nucleus
oc	ovate cell
pl	pigment layer
rER	rough endoplasmatic reticulum
sm	secretory material
tgn	trans golgi network
v	vesicle

Acknowledgments

We thank Andrea Micossi from the University of Vienna Austria for providing some of the ultrastructural images for the *Sepia* contribution. Special thanks go to the Austrian Science Fund (FWF, Project No. P 17193 – B12 & P 21135 – B12) for funding the presented investigations. The editorial assistance of Sylvia Nürnberger for figure arrangement, Dr. Michael Stachowitsch, University of Vienna, Dept of Marine Biology, Austria and Dr. Shuichi Shigeno from the University of Chicago, USA for critically reading the manuscript is greatly appreciated.

References

Adam W (1986) Contribution a la connaissance du genre *Euprymna* Steenstrup, 1887 (Mollusca, Cephalopoda). Bulletin de L'Institut Royal des Sciences Naturelles de Belgique 56: 131–136.

Adams PJM and Tyler S (1980) Hopping locomotion in a nematode: functional anatomy of the caudal gland of *Theristus caudasaliens* sp. n. Journal of Morphology 164: 265–285.

Agassiz L (1847) An Introduction to the Study of Natural History in a Series of Lectures Delivered in the Hall of the College of Physicians and Surgeons, New York. Greeley & McElrath, New York.

Akasaki T, Nikaido M, Tsuchiya K, Segawa S, Hasegawa M, and Okada N (2006) Extensive mitochondrial gene arrangements in coleoid Cephalopoda and their phylogenetic implications. Molecular Phylogenetics and Evolution 38(3): 648–658.

Appellöf A (1898) Cephalopoden von Ternate. Abhandlungen der Senckenbergischen Naturforschenden Gesellschaft 24(4): 570–637.

Arnold JM (1990) Embryonic development of the squid. In: Gilbert DL, Adelmann WJ, and Arnold JM (eds) Squid as Experimental Animals. Plenum Press, New York: pp 77–92.

Arnold JM and Singley CT (1989) Ultrastructural changes in the cells of the Hoyle organ during hatching of the squid *Loligo peali*. Journal of Cephalopod Biology 1(1): 1–14.

Arnold JM, Singley CT, and Williams-Arnold LD (1972) Embryonic development and post-hatching survival of the sepiolid squid *Euprymna scolopes* under laboratory conditions. The Veliger 14(4): 361–364.

Barber VC and Wright DE (1969) The fine structure of the sensoe organs of the cephalopod mollusc *Nautilus*. Zeitschrift für Zellforschung 102: 293–312.

Barthelat F (2007) Biomimetics for next generation materials. Philosophical Transactions of the Royal Society: Series A, Mathematical, Physical and Engineering Sciences 365(1861): 2907–2919.

Berry SS (1913) Some new Hawaiian cephalopods. Proceedings of the United States National Museum 45(1996): 563–566.

Berry SS (1921) Cephalopods of the genera *Sepiolidea*, *Sepiadarium*, and *Idiosepius*. The Philippine Journal of Science 47(1): 39–55.

Bidder AM (1962) Use of the tentacles, swimming and buoyancy control in the pearly *Nautilus*. Nature 196(4853): 451–454.

Bonnaud L, Boucher-Rodoni R, and Monnerot M (1996) Relationship of some coleoid cephalopods established by 3′end of the 16S rDNA and cytochrome oxidase III gene sequences comparison. American Malacological Bulletin 12(1/2): 87–90.

Bonnaud L, Boucher-Rodoni R, and Monnerot M (1997) Phylogeny of Cephalopods inferred from mitochondrial DNA Sequences. Molecular Phylogenetics and Evolution 7(1): 44–54.

Bonnaud L, Pichon D, and Boucher-Rodoni R (2005) Molecular approach of decabrachia phylogeny: is *Idiosepius* definitely not a sepiolid. Phuket Marine Biological Center Research Bulletin 66: 203–212.

Budelmann BU, Schipp R, and von Boletzky S (1997) Cephalopoda. In: Harrison FW and Kohn AJ (eds) Microscopic Anatomy of Invertebrates, Vol. 6A Mollusca II, 1997. Wiley-Liss, New York: pp 119–414.

Carlini DB, Reece KS, and Graves JE (2000) Actin gene family evolution and the phylogeny of coleoid cephalopods (Mollusca: Cephalopoda). Molecular Biology and Evolution 17(9): 1353–1370.

Carlson BA (1987) Collection and aquarium maintenance of *Nautilus*. In: Saunders WB and Landman NH (eds) *Nautilus*. The Biology and Paleobiology of a Living Fossil. Plenum Press, New York: pp 563–578.

Chotiyaputta C, Okutani T, and Chaitiamvong S (1991) A new pygmy cuttlefish from the Gulf of Thailand *Idiosepius thailandicus* n. sp. (Cephalopoda: Idiosepiidae). Venus, the Japanese Journal of Malacology 50(3): 165–174.

Claes MF and Dunlap PV (2000) Aposymbiotic culture of the sepiolid squid *Euprymna scolopes*: role of the symbiotic bacterium *Vibrio fischeri* in host animal growth, development, and light organ morphogenesis. Journal of Experimental Zoology 286(3): 280–296.

Conrad TA (1849) Art. XVI. – Notes on shells, with description of new genera and species. Journal of the Academy of Natural Science of Philadelphia, Series 2 1(3): 210–214.

Crick RE and Mann KO (1987) Biomineralization and systematic implications. In: Saunders WB and Landman NH (eds) *Nautilus*. The Biology and Paleobiology of a Living Fossil. Plenum Press, New York: pp 115–134.

Cyran N, von Byern J, and Klepal W (2005) Ultrastructure of the adhesive organ of *Idiosepius* (Mollusca, Cephalopoda). Microscopy Conference 6. Dreiländertagung 28.08.–02.09.2005 Paul Scherrer Institut, Davos, Switzerland: p 164.

Cyran N, Klepal W, and von Byern J (2008) Ultrastructural characterization of the adhesive organ of Idiosepiidae Voss, 1962 (Mollusca, Cephalopoda). In: Faber A, Weiss R, and Fuchs D (eds) 3rd International Symposium "Coleoid cephalopods through time" Musée National d'Histoire Naturelle Luxembourg, Luxembourg: pp 97–98.

Cyran N, Klepal W, and von Byern J (2010) Ultrastructural characterization of the adhesive organ of *Idiosepius biserialis* (Voss, 1962) and *Idiosepius pygmaeus* (Steenstrup, 1881) (Mollusca, Cephalopoda). Journal of the Marine Biological Association of the United Kingdom: submitted.

De Blainville MHD (1825) Céphalophores. Cephalophora. Manual de Malacologie et de Conchyliologie. Paris: pp 364–387.

de Lamarck MC (1798) Extrait d'un Mémoire sur le genre de la séche, du calmar et du poulpe, vulgairement nommés, polypes de mer. Bulletin de la Societe Philomathique de Paris 2(17): 129–131.

Dean B (1901) Notes on living *Nautilus*. The American Naturalist XXXV: 819–837.

DeMoor S, Waite JH, Jangoux M, and Flammang P (2003) Characterization of the adhesive from cuvierian tubules of the sea cucumber *Holothuria forskali* (Echinodermata, Holothuroidea). Marine Biotechnology 5(1): 45–57.

D'Orbigny ACV and de Ferussac A (1848) Histoire naturelle generale et particuliere des cephalopodes acetabuliferes, vivantes et fossiles. Libraire de L'Academie Nationale De Medecine, Paris.

Fioroni P (1962) Die embryonale Entwicklung der Hautdrüsen und des Trichterorganes von *Octopus vulgaris* Lam. Acta Anatomica 50: 264–295.

Flammang P (1996) Adhesion in echinoderms. In: Jangoux M and Lawrence JM (eds) Echinoderm Studies. A.A. Balkema, Rotterdam: pp 1–60.

Foster JS, Apicella MA, and McFall-Ngai MJ (2000) *Vibrio fischeri* Lipopolysaccharide induces developmental Apoptosis, but not complete Morphogenesis, of the *Euprymna scolopes* symbiotic light organ. Developmental Biology 226: 242–254.

Fukuda Y (1980) Observations by SEM. In: Hamada T, Obata I, and Okutani T (eds) *Nautilus macromphalus* in captivity – Japanese Expert Consultation on Living *Nautilus*. Tokai University Press, Tokyo: pp 23–33.

Fukuda Y (1988) Histology of the long digital tentacles. In: Saunders WB and Landman NH (eds) *Nautilus*. The Biology and Paleobiology of a Living Fossil. Plenum Publishing Corporation, New York: pp 249–256.

Gelder SR and Tyler S (1986) Anatomical and cytochemical studies on the adhesive organs of the ectosymbiont *Histriobdella homari* (Annelida: Polychaeta). Transactions of the American Microscopical Society 105: 348–356.

Grant RE (1833) On a new species of *Sepiola* (*Sepiola stenodactyla*) from the Mauritius, presented by C. Telfair, Esq. Proceedings of the Zoological Society of London 1: 42–43.

Gray JE (1849) Cephalopoda antepedia. In: Gray JE (ed) Catalogue of the Mollusca in the Collection of the British Museum. London: pp 1–164.

Greenwald L and Ward PD (1987) Buoyancy in *Nautilus*. In: Saunders WB and Landman NH (eds) *Nautilus*. The Biology and Paleobiology of a Living Fossil. Plenum Press, New York: pp 547–560.

Griffin LE (1898) Notes on the anatomy of *Nautilus pompilius*. Zoölogical Bulletin I: 147–161.

Grimpe G (1931) Teuthologische Mitteilungen XIII. Über die Cephalopoden der Sunda-Expedition Rensch. Zoologischer Anzeiger 95(5/8): 149–174.

Habe T and Okutani T (1988) A new subspecies of living *Nautilus* (Cephalopoda; Nautiloidea) from the Sulu Sea. Venus, the Japanese Journal of Malacology 47(2): 91–94.

Haesaerts D, Jangoux M, and Flammang P (2005) The attachment complex of brachiolaria lavae of the sea star *Asterias rubens* (Echinodermata): an ultrastructural and immunocytochemical study. Zoomorphology 124: 67–78.

Hanlon RT, Claes MF, Ashcraft SE, and Dunlap PV (1997) Laboratory culture of the sepiolid squid *Euprymna scolopes*: a model system for bacteria-animal symbiosis. Biological Bulletin 192: 364–374.

Haswell WA (1895) Note on certain points in the arrangement and structure of the tentaculiferous lobes in *Nautilus pompilius*. Proceeding of the Linnean Society of New South Wales 10 (Second series): 544–548.

Hermans CO (1983) The duo-gland adhesive system. Oceanography and Marine Biology: An Annual Review 21: 283–339.

House MR (1973) An analysis of Devonian Goniatite distributions. In: Hughes NF (ed) Organisms and Continents Through Time: A Symposium Volume of 23 Papers. Special papers in Palaeontology No. 12 Ed. The Palaeontological Association, London: pp 305–317.

House MR (1987) Geographic distribution of *Nautilus* shells. In: Saunders WB and Landman NH (eds) *Nautilus*. The Biology and Paleobiology of a Living Fossil. Plenum Press, New York: pp 53–64.

Hoyle WE (1889) VI. On a tract of modifies epithelium in the embryo of *Sepia*. Proceedings of the Royal Physical Society of Edinburgh X: 58–60.

Hylleberg J and Nateewathana A (1991) Redescription of *Idiosepius pygmaeus* Steenstrup, 1881 (Cephalopoda: Idiosepiidae), with mention of additional morphological characters. Phuket Marine Biological Center Research Bulletin 55: 33–42.

Iredale T (1944) Australian pearly *Nautilus*. The Australian Zoologist 10(3): 294–298.

Jereb P and Roper CFE (2005) Cephalopods of the world – An annotated and illustrated catalogue of cephalopod species known to date. No. 4, Vol. 1: Chambered nautiluses and sepioids (Nautilidae, Sepiidae, Sepiolidae, Sepiadariidae, Idiosepiidae and Spirulidae), Food and Agriculture Organization of the United Nations, Rome.

Joubin L (1894) Céphalopodes d'Amboine. Revue Suisse de Zoologie et Annales du Musèe d'Historie Naturelle de Genéve 2: 23–64.

Kakinuma K, Hisanaga K, Tsukahara J, and Tabata M (1995) The predatory activity of captured *Nautilus belauensis*. In: Kakinuma Y (ed) Studies of *Nautilus belauensis* in Palau, 27th Ed. Kagoshima University, Kagoshima: pp 83–90.

Kamino K (2006) Barnacles underwater attachment. In: Smith AM and Callow JA (eds) Biological Adhesives. Springer-Verlag, Heidelberg: pp 145–166.

Kasugai T (2000) Reproductive behavior of the pygmy cuttlefish *Idiosepius paradoxus* in an Aquarium. Venus, the Japanese Journal of Malacology 59(1): 37–44.

Kier WM (1987) The functional morphology of the tentacle musculature of *Nautilus pompilius*. In: Saunders WB and Landman NH (eds) *Nautilus*. The Biology and Paleobiology of a Living Fossil. Plenum Publishing Corporation, New York: pp 257–269.

Kier WM and Smith AM (2002) The structure and adhesive mechanism of *Octopus* suckers. Integrative and Comparative Biology 42: 1146–1153.

Lewis AR and Choat JH (1993) Spawning mode and reproductive output of the tropical cephalopod *Idiosepius pygmaeus*. Canadian Journal of Fisheries and Aquatic Sciences 50: 20–28.

Lightfood J (1786) A catalogue of the Portland Museum lately the property of the. Duchess Dowager of Portland deceased; which will be sold by auction by Mr Skinner & Co. on Monday the 24th of April, 1786 and the thirty-seven following days at 12 o'clock Sundays & the 5th June (the day his Majesty's birthday is kept) excepted, at her late Dwelling-House, in Privy-Garden, Whitehall; by order of the acting executrix. Skinner & Co., London.

Lindgren AR, Giribet G, and Nishiguchi MK (2004) A combined approach to the phylogeny of Cephalopoda (Mollusca). Cladistics 20: 454–486.

Linne C (1758) Systema naturae per regna tria naturae, secundum classes, ordines, genera, species cum characteribus, differentiis, synonymis, locis. Tomus I. Ed., 267, Laurentii Salvii, Upsala.

Luz GM and Mano JF (2009) Biomimetic design of materials and biomaterials inspired by the structure of nacre. Philosophical Transactions of the Royal Society: Series A, Mathematical, Physical and Engineering Sciences 367(1893): 1587–1605.

Mayer G (2005) Rigid biological systems as models for synthetic composites. Science 310(5751): 1144–1147.

McFall-Ngai MJ (2002) Unseen forces: the influence of bacteria on animal development. Developmental Biology 242: 1–14.

McFall-Ngai MJ and Montgomery MK (1990) The anatomy and morphology of the adult bacterial light organ of *Euprymna scolopes* berry (Cephalopoda: Sepiolidae). Biological Bulletin 179: 332–339.

McFall-Ngai MJ and Ruby EG (1991) Symbiont recognition and subsequent morphogenesis as early events in an animal-bacterial mutualism. Science 254(5037): 1491–1494.

McFall-Ngai MJ and Ruby EG (1998) Sepiolids and vibrios: when first they meet. Bioscience 48(4): 257–265.

Moynihan M (1983) Notes on the behavior of *Idiosepius pygmaeus* (Cephalopoda: Idiosepiidae). Behavior 85: 42–57.

Munch E, Launey ME, Alsem DH, Saiz E, Tomsia AP, and Ritchie RO (2008) Tough, bio-inspired hybrid materials. Science 322(5907): 1516–1520.

Muntz WRA and Wentworth SL (1995) Structure of the adhesive surface of the digital tentacles of *Nautilus pompilius*. Journal of the Marine Biological Association of the United Kingdom 75: 747–750.

Nabhitabhata J (1994) Rearing of Thai pygmy cuttlefish, *Idiosepius thailandicus* Chot., Okut. & Chait., I.: some biological aspects. Technical Paper, Rayong Coastal Aquaculture Station, Coastal Aquaculture Division, Department of Fisheries 13/1994, 12.

Nabhitabhata J (1998) Distinctive behaviour of Thai pygmy squid, *Idiosepius thailandicus* Chotiyaputta, Okutani & Chaitiamvong, 1991. Phuket Marine Biological Center Special Publication 18(1): 25–40.

Nateewathana A (1997) The sepiolidae (Cephalopoda) of the Andaman sea, Thailand, with description of *Euprymna hyllebergi* sp. nov. Phuket Marine Biological Center Special Publication 17(2): 465–481.

Natsukari Y (1970) Egg-Laying behavior, embryonic development and hatched larva of the pygmy cuttlefish, *Idiosepius pygmaeus paradoxus* Ortmann. Bulletin of the Faculty of Fisheries, Nagasaki University 30: 15–29.

Nesis K (1982) Cephalopods of the World. Translated from Russian by B.S. Levitov (ed), V.A.A.P. Copyright Agency of the UdSSR for Light and Food Industry Publishing House, Moscow, 1987 T.F.H. Publications, Inc. Ltd., for English Translation.

Nesis K, Katugin ON, and Ratnikov AV (2002) Pygmy cuttlefish *Idiosepius paradoxus* (Ortmann, 1888) (Cephalopoda) – First record of Idiosepiidae in Russian seas. Ruthenica 12(1): 81–84.

Nixon M and Dilly PN (1977) Sucker surfaces and prey capture. Symposia of the Zoological Society of London 38: 447–511.

Norman MD and Lu CC (1997) Redescription of the southern dumpling squid *Euprymna tasmanica* and a revision of the genus *Euprymna* (Cephalopoda: Sepiolidae). Journal of the Marine Biological Association of the United Kingdom 77: 1109–1137.

Nyholm SV, Deplancke B, Gaskins HR, Apicella MA, and McFall-Ngai MJ (2002) Roles of *Vibrio fischeri* and nonsymbiotic bacteria in the dynamics of mucus secretion during symbiont colonization of the *Euprymna scolopes* light organ. Applied and Environmental Microbiology 68(10): 5113–5122.

Nyholm SV and McFall-Ngai MJ (2004) The winnowing: establishing the squid-vibrio symbiosis. Nature Reviews Microbiology 2(8): 632–642.

Ortmann A (1888) Japanische cephalopoden. Zoologische Jahrbücher 3: 639–670.

Owen R (1843) XXXVII. On the structure and homology of the cephalic tentacles in the pearly *Nautilus*. The Annals and Magazine of Natural History Including Zoology, Botany, and Geology XII: 305–311.

Packard A, Bone Q, and Hignette M (1980) Breathing and swimming movements in a captive *Nautilus*. Journal of the Marine Biological Association of the United Kingdom 60: 313–327.

Pennisi E (2002) Biology reveals new ways to hold on tight. Science 296: 250–251.

Pfeffer G (1884) Die Cephalopoden des Hamburger Naturhistorischen Museums. Abhandlungen aus dem Gebiet der Naturwissenschaft, herausgegeben vom Naturwissenschaftlichen Verein Hamburg 8(1): 1–30.

Pontoppidan E (1755) Kraken, or Korven, the Largest Creature in the World. The Natural History of Norway: Containing, a Particular and Accurate Account of the Temperature of the Air, the Different Soils, Waters, Vegetables, Metals, Minerals, Stones, Beasts, Birds, and Fishes; Together with the Dispositions, Customs, and Manner of Living of the Inhabitants: Interspersed with Physiological Notes from Eminent Writers, and Transactions of Academies. A. Linde, London: pp 210–218.

Quoy JMC and Gaimard JP (1832) Animaux Mollusques – Céphalés. Voyage de Décuvertes de L'Astrolabe Exécuté par order du Roi, pendant les années 1826–1827–1828–1829, sours le Commandement de M.J. Dumont D'Urville. Tome II. Zoologie Ed., Tastu, Paris: pp 61–90.

Roeleveld MAC (1972) A review of the Sepiidae (Cephalopoda) of Southern Africa. Annals of the South African Museum 59(10): 193–313.

Roeleveld MAC and Liltved WR (1985) A new species of *Sepia* (Cephalopoda, Sepiidae) from South Africa. Annals of the South African Museum 96(1): 1–18.

Roper CFE and Young RE (1975) Vertical distribution of pelagic cephalopods. Smithsonian Contributions to Zoology 209: 1.

Ruth P, Schmidtberg H, Westermann B, and Schipp R (2002) The sensory epithelium of the tentacles and the rhinophore of *Nautilus pompilius* L. (Cephalopoda, Nautiloidea). Journal of Morphology 251(3): 239–255.

Santos R, Gorb SN, Jamar V, and Flammang P (2005) Adhesion of echinoderm tube feet to rough surfaces. Journal of Experimental Biology 208: 2555–2567.

Sasaki M (1913) Decapod cephalopods found in Japan (Sepiolidae). Zoolological Magazine Tokyo 25: 247–252.

Sasaki M (1921) On an adhering habit of a pygmy cuttlefish, *Idiosepius pygmaeus* Steenstrup. Annotationes Zoologicae Japonenses 10(21): 209–213.

Sasaki M (1929) A monograph of the dibranchiate cephalopods of the Japanese and adjacent waters. Journal of the Faculty of Agriculture, Hokkaido Imperial University, Supplementary 20: 1–357.

Sato N, Awata S, and Munehara H (2009) Seasonal occurrence and sexual maturation of Japanese pygmy squid (*Idiosepius paradoxus*) at the northern limits of their distribution. ICES Journal of Marine Science 66: 811–815.

Saunders WB (1981) A new species of *Nautilus* from Palau. The Veliger 24(1): 1–7.

Saunders WB (1987) The species of *Nautilus*. In: Saunders WB and Landman NH (eds) *Nautilus*. The Biology and Paleobiology of a Living Fossil. Plenum Press, New York: pp 35–52.

Saunders WB and Spinosa C (1979) *Nautilus* movement and distribution in Palau, Western Caroline Islands. Science 204(4398): 1199–1201.

Saunders WB and Landman NH (1987) *Nautilus*. The Biology and Paleobiology of a Living Fossil. Plenum Press, New York.

Saunders WB and Ward PD (1987) Ecology, distribution, and population characteristics of *Nautilus*. In: Saunders WB and Landman NH (eds) *Nautilus*. The Biology and Paleobiology of a Living Fossil. Plenum Press, New York: pp 137–162.

Shapiro EA and Saunders WB (1987) *Nautilus* shell hydrostatics. In: Saunders WB and Landman NH (eds) *Nautilus*. The Biology and Paleobiology of a Living Fossil. Plenum Press, New York: pp 527–545.

Shears J (1988) The use of a sand-coat in relation to feeding and diel activity in the sepiolid squid *Euprymna scolopes*. Malacologia 29(1): 121–133.

Sherrard KM (2000) Cuttlebone morphology limits habitat depth in eleven species of *Sepia* (Cephalopoda: Sepiidae). Biological Bulletin 198(3): 404–414.

Shigeno S and Yamamoto M (2002) Organization of the nervous system in the pygmy cuttlefish, *Idiosepius paradoxus* Ortmann (Idiosepiidae, Cephalopoda). Journal of Morphology 254(1): 65–80.

Singley CT (1982) Histochemistry and fine structure of the ectodermal epithelium of the sepiolid squid *Euprymna scolopes*. Malacologia 23(1): 177–192.

Singley CT (1983) *Euprymna scolopes*. In: Boyle PR (ed) Cephalopod Life Cycles I Species Accounts. Academic Press, New York: pp 69–74.

Small A and McFall-Ngai MJ (1999) Halide peroxidase in tissues that interact with bacetria in the host squid *Euprymna scolopes*. Journal of Cell Biology 72: 445–457.

Smith AM (1991) The role of suction in the adhesion of limpets. Journal of Experimental Biology 161: 151–169.

Smith AM (1996) Cephalopod sucker design and the physical limits to negative pressure. Journal of Experimental Biology 199: 949–958.

Smith AM (2002) The structure and function of adhesive gels from invertebrates. Integrative and Comparative Biology 42: 1164–1171.

Smith AM and Morin MC (2002) Biochemical differences between trail mucus and adhesive mucus from marsh periwinkle snails. Biological Bulletin 203: 338–346.

Smith AM, Quick TJ, and St. Peter RL (1999) Differences in the composition of adhesive and non-adhesive mucus from the limpet *Lottia limatula*. Biological Bulletin 196: 34–44.

Sowerby GB (1848) Monograph of the genus *Nautilus*. Thesaurus Conchyliorum, or Monographs of Genera of Shells, Vol. II Ed. London: pp 463–465.

Steenstrup J (1875a) *Hemisepius*, en ny slaegt af *Sepia*-blaekspruternes familie, med bemaerkninger om *Sepia*-formerne i almindelighed. Det Kongelige Danske Videnskabernes Selskabs skrifter, Femte Raekke, Naturvidenskabelig og mathematisk 10(VII): 465–482.

Steenstrup J (1875b) Sur l'Hemisepius un genre nouveau de la famille. Det Kongelige Danske Videnskabernes Selskabs skrifter, Femte Raekke, Naturvidenskabelig og mathematisk 10: I–VI.

Steenstrup J (1881) *Sepiadarium* and *Idiosepius* two new genera of the family of Sepia. With remarks on the two related forms *Sepioloidea* d'Orb. and *Spirula* Lmk. Det Kongelige Danske Videnskabernes Selskabs skrifter Raekke 6(Bd. 1): 211–242.

Steenstrup J (1887) Notae Teuthologicae, 6. International Bibliography, Information, Documentation, pp 47–66.

Stenzel HB (1964) Living nautilus. In: Moore RC (ed) Treatise on Invertebrate Paleontology. The University of Kansas Press & The Geological Society of America, Kansas: pp K59–K93.

Strugnell J, Norman MD, Jackson J, Drummond AJ, and Cooper A (2005) Molecular phylogeny of coleoid cephalopods (Mollusca: Cephalopoda) using a multigene approach; the effect of data partitioning on resolving phylogenies in a Bayesian framework. Molecular Phylogenetics and Evolution 37(2): 426–441.

Suwanmala J, Szaffich CA, von Byern J, and Nabhitabhata J (2006a) Ultrastructural insights in the embryonic development of *Idiosepius biserialis* (Mollusca, Cephalopoda). In: Moltschaniwskyj NA, Pecl GT, Semmens J, and Jackson GD (eds) Cephalopod International Advisory Council Symposium 2006, Hobart, Tasmania: p 106.

Suwanmala J, von Byern J, and Nabhitabhata J (2006b) Observation of *Idiosepius pygmaeus* (Cephalopoda, Idiosepiidae) at Klong Bangrong, Phuket Island, Thailand. Phuket Marine Biological Center Research Bulletin 67: 49–51.

Swan ARH and Saunders WB (1987) Morphological variation in *Nautilus* from Papua New Guinea. In: Saunders WB and Landman NH (eds) *Nautilus*. The Biology and Paleobiology of a Living Fossil. Plenum Press, New York: pp 85–103.

Takumiya M, Kobayashi M, Tsuneki K, and Furuya H (2005) Phylogenetic relationships among major species of Japanese coleoid cephalopods (Mollusca: Cephalopoda) using three mitochondrial DNA sequences. Zoological Science 22: 147–155.

Tanabe K, Hayasaka S, Saisho T, Shinomiya A, and Aoki K (1983) Morphological variation of *Nautilus pompilius* from the Philippines and Fiji Islands. In: Hayasaka S (ed) Studies on *Nautilus pompilius* and its Associated Fauna from Tanon Strait, the Philippines. 1st Ed. Kagoshima University, Kagoshima: pp 9–21.

Tanabe K, Hayasaka S, and Tsukahara J (1985) Morphological analysis of *Nautilus pompilius*. In: Hayasaka S (ed) Marine Ecological Studies on the Habitat of *Nautilus Pompilius* in the Environs of Viti Levu, Fiji. 4th Ed. Kagoshima University, Kagoshima: pp 38–49.

Teichert C and Matsumoto T (1987) The ancestry of the genus *Nautilus*. In: Saunders WB and Landman NH (eds) *Nautilus*. The Biology and Paleobiology of a Living Fossil. Plenum Press, New York: pp 25–32.

Thompson DW (1910) Historia animalium. In: Ross D (ed) The works of Aristotle. Book IV, At the Clarendon Press, Oxford: pp 486a–633a.

Tompsett DH (1939) *Sepia*, XXXIInd Ed. The University Press, Liverpool.

Tracey SR, Steer MA, and Pecl GT (2003) Life history traits of the temperature mini-maximalist *Idiosepius notoides* (Cephalopoda:Sepioidea). Journal of the Marine Biological Association of the United Kingdom 83: 1297–1300.

Tyler S and Rieger GE (1980) Adhesive organs of the Gastrotricha I. Duo-Gland organs. Zoomorphologie 95: 1–15.

Verrill AE (1881) The Cephalopods of the North-Eastern Coast of America Part II. The smaller Cephalopods, including the "Squids" and the Octopi, with other allied Forms. Transactions of the Connecticut Academy of Arts and Sciences VI: 259–446.

von Boletzky S and Roeleveld MAC (2000) "Ventral adhesion" to hard substrates: a thigmotactic response in sepiid cuttlefish (Mollusca, Cephalopoda). Vie Milieu 50(1): 59–64.

von Byern J and Klepal W (2006) Adhesive mechanisms in cephalopods: a review. Biofouling 22(5): 329–338.

von Byern J and Grunwald I (2008) Glue components in *Idiosepius* (Mollusca, Cephalopoda). In: Gelpey JC, Hamers RJ, Muralt P, and Orme CA (eds) From Biological Materials to Biomimetic

Material Synthesis, Material Research Society Spring Meeting 2008 March 24–28, San Francisco, USA: p 645.
von Byern J and Klepal W (2010) Re-evaluation of taxonomic characters of *Idiosepius* (Cephalopoda, Mollusca). Malacologica 52(1): 43–65.
von Byern J, Nürnberger S, and Shigeno S (2005) Distribution pattern of a minimalist – New records for *Idiosepius biserialis* (Idiosepiidae, Cephalopoda). In: Kostak M and Marek J (eds) 2nd International Symposium "Coleoid cephalopods through time" 26.–28.09.2005, Institute of Geology and Palaeontology, Faculty of Science, Charles University of Prague, Prague, Czech Republic: pp 38–43.
von Byern J, Shigeno S, Klepal W, and Kasugai T (2006) Postembryonic development of the adhesive organ in *Idiosepius* under artificial conditions. In: Moltschaniwskyj NA, Pecl GT, Semmens J, and Jackson GD (eds) Cephalopod International Advisory Council Symposium 2006, Hobart, Tasmania: p 54.
von Byern J, Rudoll L, Cyran N, and Klepal W (2008) Histochemical characterization of the adhesive organ of three *Idiosepius* spp. species. Biotechnic & Histochemistry 83(1): 29–46.
von Byern J, Söller R, and Steiner G (2010) Phylogenetic characterization of the genus *Idiosepius* (Cephalopoda; Idiosepiidae). in preparation.
von Orelli M (1959) Über das Schlüpfen von *Octopus vulgaris*, *Sepia officinalis* und *Loligo vulgaris*. Revue Suisse de Zoologie 66(18): 330–343.
Voss GL (1962) South African cephalopods. Transaction of the Royal Society of South Africa 36(4): 245–272.
Voss GL (1963) Cephalopods of the Philippine Islands. United States National Museum Bulletin 234: 1–180.
Voss GL (1977) Present status and new trends in cephalopod systematics. Symposia of the Zoological Society of London 38: 49–60.
Ward PD (1988) In Search of *Nautilus*. Simon & Schuster Trade Division, New York.
Ward PD (2008) *Nautilus*: Chamber of secrets. New Scientist Magazine 2650: 40–43.
Ward PD and Greenwald L (1981) Chamber refilling in *Nautilus*. Journal of the Marine Biological Association of the United Kingdom 62: 469–475.
Ward PD and Saunders WB (1997) *Allonautilus*: a new genus of living nautiloid cephalopod and its bearing on phylogeny of the nautilida. Journal of Paleontology 71(6): 1054–1064.
Ward PD, Stone R, Westermann G, and Martin A (1977) Note on animal weight, cameral fluids, swimming speed, and color polymorphism of the cephalopod *Nautilus pompilius* in the Fiji Islands. Paleobiology 3: 377–388.
Ward PD, Carlson BA, Weekly M, and Brumbaugh B (1984) Remote telemetry of daily vertical and horizontal movement of *Nautilus* in Palau. Nature 309: 248–250.
Willey A (1898a) The adhesive tentacles of *Nautilus*, with some notes on its pericardium and spermatophores. Quarterly Journal of Microscopical Science 40: 207–209.
Willey A (1898b) The pre-ocular and post-ocular tentacles and osphradia of *Nautilus*. Quarterly Journal of Microscopical Science 40: 197–201.
Willey A (1902) Contribution to the natural history of the pearly *Nautilus*. In: Willey A (ed) Zoological results based on material from New Britain, New Guinea, loyalty islands and elsewhere collected during the years 1895, 1896 and 1897 Part VI. University Press, Cambridge: pp 691–830.
Wintrebert P (1928) L'Eclosion par digestion de la coque chez les poissons, les amphibiens et les céphalopodes dibranchiaux décapodes. Comptes rendus de l'Association des Anatomistes Nancy 1928: 496–503.
Woodruff DS, Carpenter MP, Saunders WB, and Ward PD (1987) Genetic variation and phylogeny in *Nautilus*. In: Saunders WB and Landman NH (eds) *Nautilus*. The Biology and Paleobiology of a Living Fossil. Plenum Press, New York: pp 65–83.
Yamamoto M (1988) Normal embryonic stages of the pygmy cuttlefish, *Idiosepius pygmaeus paradoxus* Ortmann. Zoological Science 5: 989–998.
Yamamoto M, Shimazaki Y, and Shigeno S (2003) Atlas of the embryonic brain in the pygmy squid, *Idiosepius paradoxus*. Zoological Science 20(2): 163–179.
Young RE, Vecchione M, and Donovan DT (1998) The evolution of coleoid cephalopods and their present biodiversity and ecology. In: Payne AIL, Lipinski MR, Clarke MR, and Roeleveld MAC (eds) Cephalopod Biodiversity, Ecology and Evolution, 20th Ed. South African Journal of Marine Science: pp 393–420.
Yung Ko Ching M (1930) Contribution á l'etude cytologique de l'ovogenese, du developpment et de quelques organes chez les cephalopodes. Annales de L'Institut Oceanographique 7(Fasc. 8): 300–364.
Zhengzhi D (2010) On the geographical distribution of recent *Nautilus*. Marine Scienes 4: 52–56.

6 Unravelling the Sticky Threads of Sea Cucumbers – A Comparative Study on Cuvierian Tubule Morphology and Histochemistry

Pierre T. Becker and Patrick Flammang

Contents

6.1 Introduction

Cuvierian tubules are peculiar organs found in several species of sea cucumbers, all belonging to the family Holothuriidae (order Aspidochirotida). Two main types of tubules can be differentiated on the basis of their gross external morphology, lobulated and smooth (Fig. 6.1; Table 6.1) (Lawrence, 2001). Lobulated tubules occur exclusively in the genus *Actinopyga*; they are never expelled and are not sticky (VandenSpiegel and Jangoux, 1993). On the other hand, smooth tubules are present in the genera *Bohadschia*, *Holothuria,* and *Pearsonothuria*, in which they generally appear as sticky white threads that function as a defence mechanism (Hamel and Mercier, 2000; Flammang, 2006). Indeed, once ejected, they can release a glue allowing instantaneous adhesion on any object, and can therefore entangle a predator in a matter of seconds (Zahn et al., 1973; VandenSpiegel and Jangoux, 1987). Smooth Cuvierian tubules will be the main focus of the present Chapter.

Smooth Cuvierian tubules occur in great numbers (Table 6.1) in the posterior part of the body cavity of the holothuroid. Proximally they are attached to the basal part of the left respiratory tree and their distal, blind ends float freely in the coelomic fluid (Fig. 6.2A). The mechanism leading to their discharge is as follows: when the animal is disturbed, it directs its posterior end toward the stimulating source and undergoes a general body contraction. Consequently, the wall of the cloaca breaks and the free ends of a few tens of tubules are expelled through the tear and the cloacal orifice (Fig. 6.2B). The water of the respiratory tree is then forcefully injected into the lumen of the tubules causing their elongation, up to 20 times their initial length (Fig. 6.2B). Upon contact with

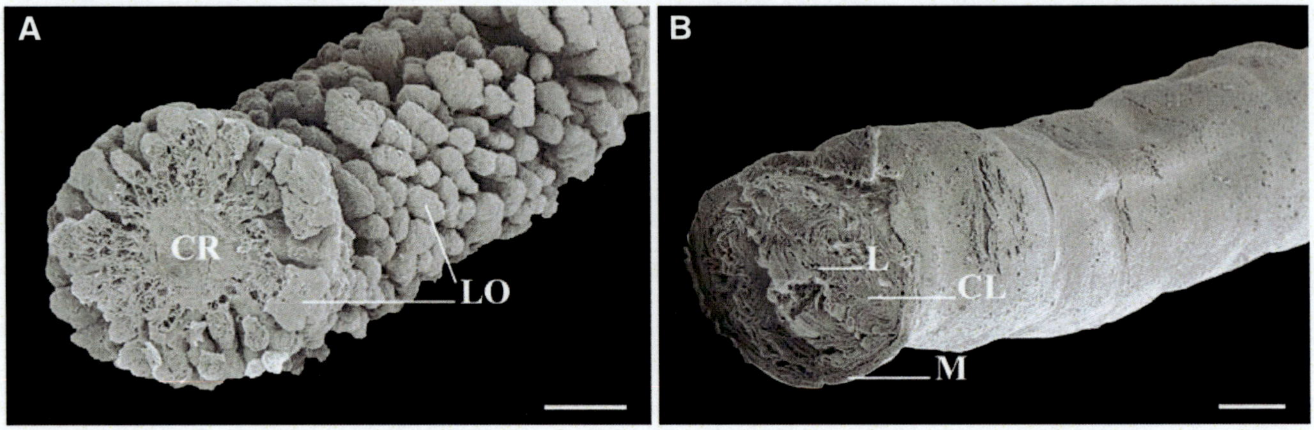

Fig. 6.1 General aspect in SEM of (A) a lobulated tubule from *Actinopyga echinites* and (B) of a smooth tubule of *Holothuria pervicax* (originals). Scale bars = 200 µm. *CR* Central rachis; *CL* connective tissue layer; *L* lumen; *LO* lobules; *M* mesothelium

any surface (e.g., a predator integument), the lengthened tubules become instantly sticky. Finally, the elongated tubules autotomize at the attachment point on the respiratory tree and are left behind as the sea cucumber crawls away (VandenSpiegel and Jangoux, 1987; Flammang, 2006). Lost tubules are then regenerated in a few weeks: 15–18 days in *H. leucospilota*, about 5 weeks in *H. forskali* (VandenSpiegel et al., 2000; Hamel and Mercier, 2000). As only a portion of the tubules are emitted at one time, the total number may suffice for several responses (Hamel and Mercier, 2000).

Cuvierian tubule adhesive strength has been measured in terms of tenacity (i.e., adhesion force per unit surface area). The mean normal tenacity on glass observed in different species of holothuroids varies between 30 and 135 kPa, which is in the range of adhesive strengths described in marine organisms, although these values are among the lowest ones (Flammang et al., 2002). Adhesiveness on glass and other polar surfaces is stronger than on non-polar substrata such as plastics, indicating the importance of polar interactions in the adhesion. Moreover, adhesive forces depend not only on the temperature and the salinity of the sea water but also on the time elapsed after expulsion of the tubules (Flammang et al., 2002).

6.2 Morphology of Cuvierian Tubules

Most of the knowledge on the morphology and fine structure of smooth Cuvierian tubules comes from studies performed on the European species *H. forskali* (Müller et al., 1972; VandenSpiegel and Jangoux, 1987; VandenSpiegel et al., 2000). In terms of morphology, one has to distinguish quiescent tubules, *in situ* before their expulsion, from elongated tubules, in the external medium after their expulsion.

6.2.1 Structure of Quiescent Tubules

Quiescent tubules form a group of cæca (between 200 and 600 in *H. forskali*) attached independently to a basal dilatation of the left respiratory tree called the Cuvierian chamber (VandenSpiegel and Jangoux, 1987). Cuvierian tubules always consist of an outer mesothelium responsible for adhesion, a connective tissue layer enclosing muscle fibers, and an inner epithelium delimiting a narrow central lumen (Fig. 6.1B) (Müller et al., 1972; VandenSpiegel and Jangoux, 1987; VandenSpiegel et al., 2000).

In *H. forskali*, the mesothelium is pseudostratified and is composed of an inner layer of granular cells and an outer layer of peritoneocytes (Fig. 6.3A, B). Granular cells are very flat cells, attached to the basal lamina but not one to another. In longitudinal sections, they present a double-fold arrangement: they first form successive V-shaped structures which are then folded a second time in the same direction (Fig. 6.3A). Granular cells display a Golgi complex and a well-developed RER, both localized close to the nucleus. Their cytoplasm is filled with electron-dense granules (1–2 µm in diameter) limited by a membrane and containing the adhesive material (Fig. 6.3A, B). Peritoneocytes are T-shaped, displaying a flattened apical part lining the cœoelomic cavity and a thin elongated basal part running between the granular cell folds (Fig. 6.3A, B). They bear a single cilium and a few short microvilli. The nucleus and organelles are located exclusively in the apical cytoplasm which also con-

Table 6.1: Summary of the main morphological characteristics of Cuvierian tubules from various species of Holothuriidae (adapted from VandenSpiegel (1993); and Flammang and Plumet, unpubl. obs.)

Genus	Species	Number of tubules	Expulsion elongation adhesion	General morphology	Structure, cell types		
					Mesothelium	Connective tissue layer	Internal epithelium
Holothuria	*H. forskali* *H. impatiens* *H. pervicax* *H. sanctori*	Max. 600	Tubules are expelled, can elongate and are sticky	Smooth, 2–2.5 cm long	T-shaped adluminal cells and granular cells forming V-shaped structures	Collagen fibers organized in helices. Vacuolar cells, neurosecretory cells	Support cells and spherulocytes
	H. hilla *H. leucospilota*	Max. 600	idem	Smooth, 1–2 cm long	idem	idem	Support cells. Absence of spherulocytes
	H. maculosa	Max. 50	idem	Smooth, 3 cm long	T-shaped adluminal cells and large granular cells (no V-shaped structures)	Collagen fibers organized in helices. Absence of vacuolar and neurosecretory cells	Support cells and spherulocytes
	H. mammata	Max. 50	No expulsion, no elongation and not sticky	Smooth, 0.2–0.5 cm long	T-shaped adluminal cells and granular cells forming V-shaped structures	Collagen fibers randomly distributed. Vacuolar cells, neurosecretory cells	Support cells. Absence of spherulocytes
	H. nobilis	Max. 110	idem	Smooth, ramified, 6 cm long	Flattened adluminal cells and large cylindrical granular cells	Collagen fibers randomly distributed. Neurosecretory cells but absence of vacuolar cells	idem
Bohadschia	*B. argus* *B. atra* *B. similis* *B. subrubra* *B. vitiensis*	Max. 400	Tubules are expelled, can elongate and are sticky	Smooth, 7 cm long	T-shaped adluminal cells and granular cells forming V-shaped structures	Collagen fibers organized in helices. Vacuolar cells and neurosecretory cells	idem
	B. marmorata	Max. 400	idem	idem	idem	idem	Support cells and spherulocytes
Pearsonothuria	*P. graeffei*	Max. 400	No expulsion, no elongation but sticky	Smooth, 5 cm long	idem	idem	Support cells. Absence of spherulocytes
Actinopyga	*A. echinites* *A. lecanora* *A. mauritiana* *A. miliaris*	Max. 10	No expulsion, no elongation and not sticky	Lobulated and ramified with a primary trunk and secondary/tertiary branches	No granular cells. Spherules and branches limited by flattened adluminal cells	Collagen fibers randomly distributed. Nodule of vacuolar cells in the center of the lobules. Neurosecretory cells	Cord of epithelial cells. No central lumen

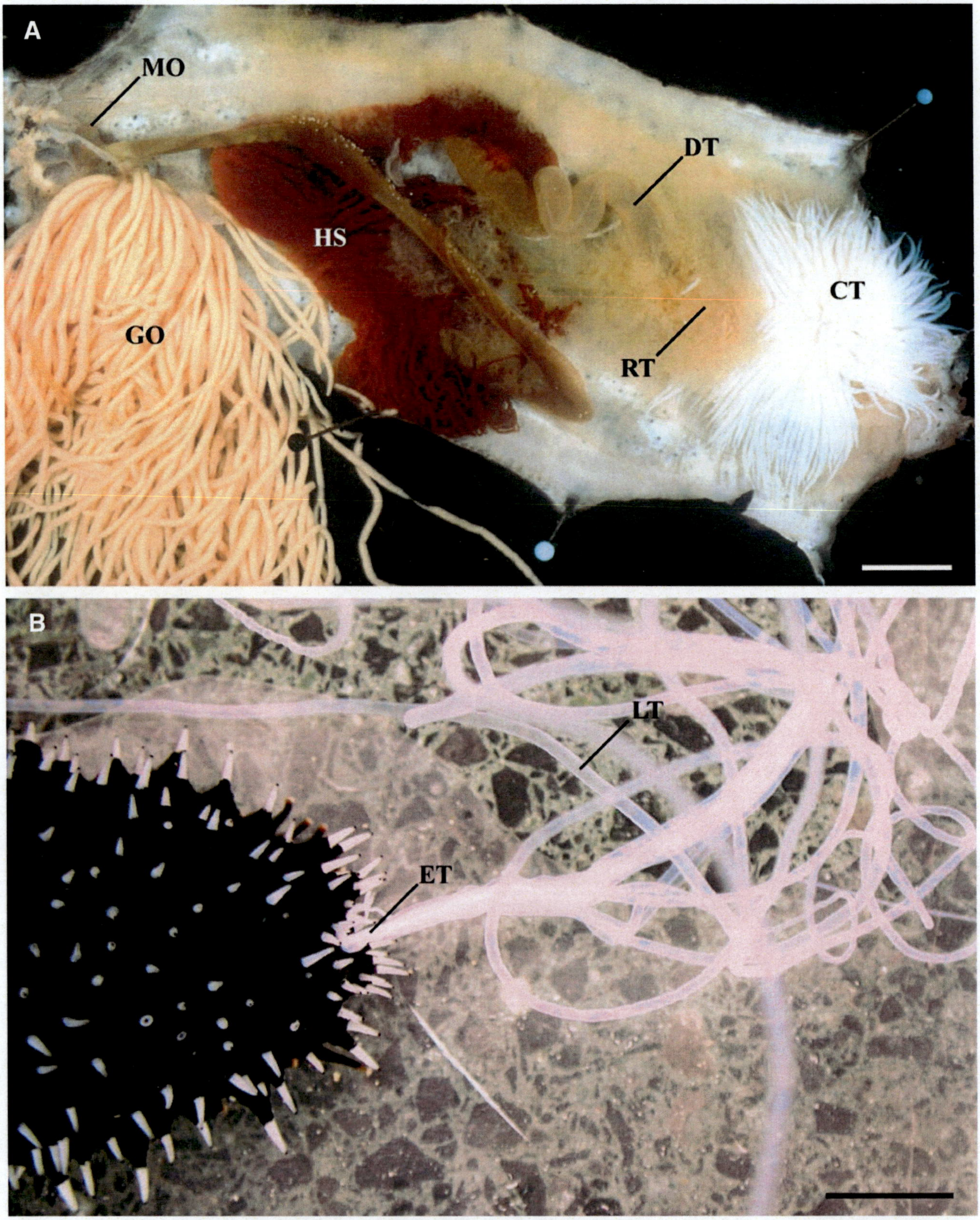

Fig. 6.2 Cuvierian tubules of *H. forskali*. (A) Internal anatomy of a sea cucumber showing the position of the tubules (*CT*), at the basis of the left respiratory tree (*RT*) (from VandenSpiegel, 1993). (B) Ejection of the tubules through the cloacal orifice (original). Scale bars in (A) 2 cm and (B) 1 cm. *DT* Digestive tract; *ET* emitted tubules; *GO* gonad; *HS* hemal system; *LT* lengthened tubules; *MO* mouth

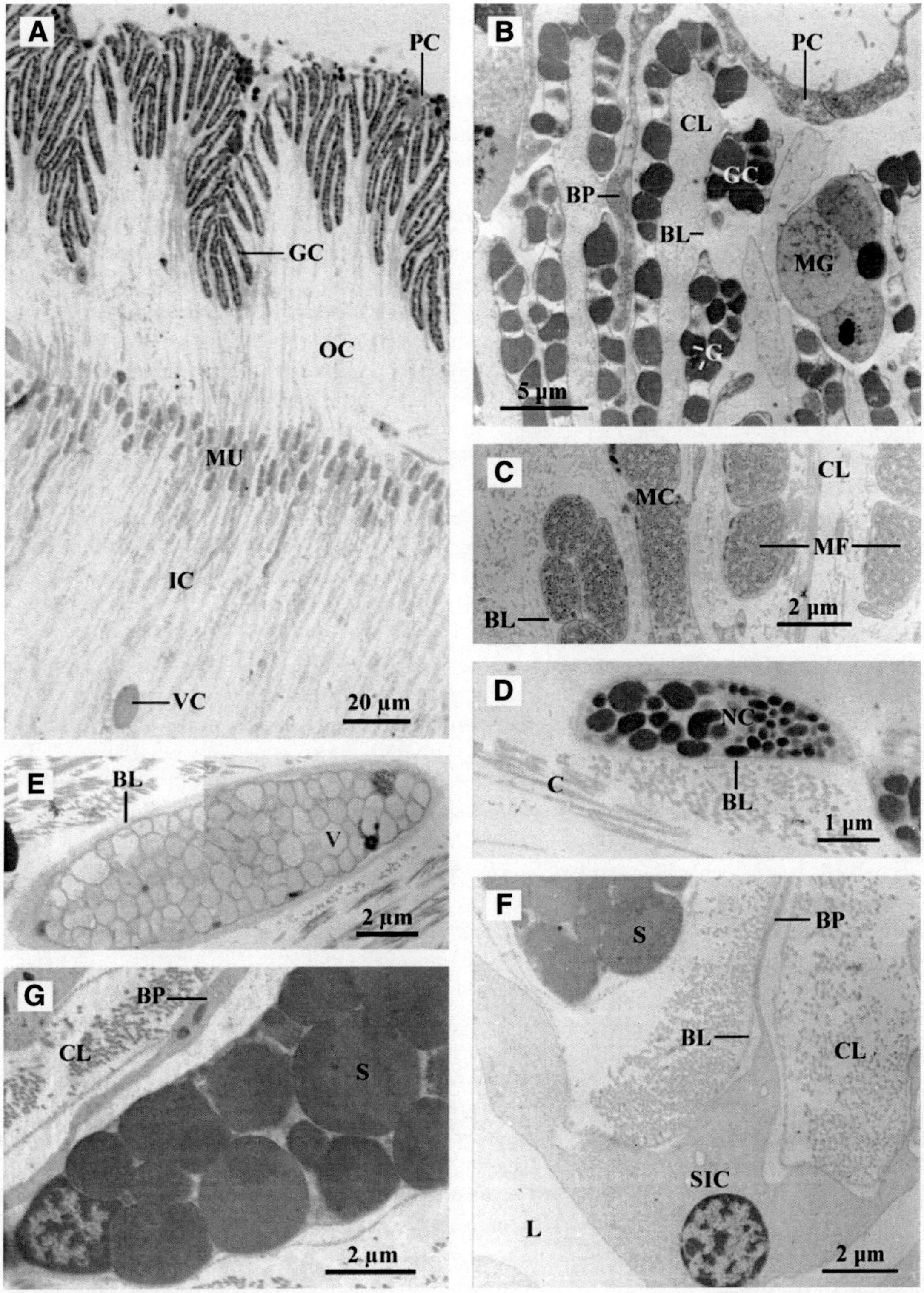

Fig. 6.3 Ultrastructure of the Cuvierian tubules of *Holothuria forskali* (A, E, G: originals; B–D, F: from DeMoor et al., 2003). (A) Longitudinal semi-thin section through the mesothelium and the connective tissue layer; (B) Mesothelium; (C) Muscle layer; (D) Detail of the inner connective tissue layer showing collagen fibers and neurosecretory cell processes; (E) Vacuolar cell; (F) Inner epithelium; (G) Detail of a spherulocyte from the inner epithelium. *BL* Basal lamina; *BP* basal process; *C* collagen; *CL* connective tissue layer; *GC* granular cell; *IC* inner connective tissue layer; *G* granule; *L* lumen; *MC* myocyte; *MF* myofibril; *MG* mucous granule; *MU* muscle; *NC* neurosecretory cell; *OC* outer connective tissue layer; *PC* peritoneocytes; *SIC* support cell of the inner epithelium; *S* spherule; *V* vacuole; *VC* vacuolar cells

tains enlarged mucous vesicles (Fig. 6.3B). The presence of these cells prevents quiescent tubules from sticking one to another and to the other organs (VandenSpiegel and Jangoux, 1987).

The connective tissue is well developed and forms about 90% of the thickness of the tubules. It contains both circular and longitudinal muscle fiber layers that separate the connective tissue layer into a thin outer area, just below the mesothelium, and a thick inner area (Fig. 6.3A, C). In the latter, collagen fibers are organized in up to six imbricated helices parallel to the long axis of the tubules (VandenSpiegel and Jangoux, 1987). The inner area also contains elongated processes from neurosecretory cells that form a network distributed between the collagen fibers (Fig. 6.3D). These cell processes are limited by a basal lamina and filled with numerous electron-dense granules (VandenSpiegel et al., 2000). Two populations of processes can be distinguished based on the size of their granules: cells with small granules (about 0.2 µm in diameter) and cells with larger granules (about 0.5 µm in diameter) (Flammang and Van Dyck, unpubl. obs.). A few vacuolar cells are also scattered in the connective tissue layer, always adjacent to the muscle cells (Fig. 6.3A, E). These cells are either isolated or gathered in small groups and their function is unknown. They are of mesothelial origin and are always limited by a basal lamina (VandenSpiegel et al., 2000).

The inner epithelium is convoluted and lines in the irregular narrow central lumen of the tubules (Fig. 6.3F). It is composed of epithelial cells bearing a few elongated and spaced microvilli, a basal nucleus, a Golgi complex and some mitochondria. These cells are imbricated and attach to each other by means of well-developed septate junctions (VandenSpiegel and Jangoux, 1987). They also present one or two thin basal processes that extend deeply in the underlying connective tissue layer. These basal processes are closely associated to a second cell type, the spherulocytes, that are characterized by large heterogeneous spherules measuring from 1 to 3 µm in diameter (Fig. 6.3G) (VandenSpiegel and Jangoux, 1987; Flammang and Van Dyck, unpubl. obs.).

6.2.2 Structure of Elongated Tubules

Elongated tubules mainly differ from quiescent tubules in that they show a thinner wall surrounding a larger, dilated lumen (VandenSpiegel and Jangoux, 1987). This is due to the forceful injection of water in the tubule lumen. During elongation, peritoneocytes disintegrate and this causes the release of the mucus contained in their vesicles. This mucus covers the tubules and it was proposed that it could prevent them from adhering to each other and to the body wall of the sea cucumber (VandenSpiegel and Jangoux, 1987). After elongation, the mesothelium consists only of unfolded granular cells, which are therefore the outermost on the tubule (VandenSpiegel and Jangoux, 1987). In scanning electron microscopy, granular cells, lying on the basal lamina, are visible at the surface of the tubules (Fig. 6.4A). Numerous filaments (100–200 nm in diameter) forming a dense network are observed close to their granules (Fig. 6.4A). Interestingly, their abundance seems inversely proportional to the number of granules. They are indeed scarce when granules are numerous while they are densely agglomerated when granules are fewer (Flammang and Wattier, unpubl. obs.). Consequently, these filaments probably correspond to the glue. This is corroborated by the observation of glue prints deposited on glass surfaces, showing that the adhesive material forms dense patches composed of intermingled filaments (Fig. 6.4B). The secreted glue is also visible on transmission electron micrographs (Fig. 6.4C, D) but appears as an electron-dense granular rather than a fibrillar material (VandenSpiegel and Jangoux, 1987; Flammang and Wattier, unpubl. obs.). Empty vacuoles are present within the secreted adhesive (Fig. 6.4C, D). Their origin is uncertain but they could be remnants of cellular membranes resulting from the disintegration of peritoneocytes.

The connective tissue layer is thinner and forms no more than 50% of the total wall thickness, due to the considerable elongation and flattening of the collagen helices (VandenSpiegel and Jangoux, 1987). Naturally elongated tubules appear to be lacking in longitudinal muscle fibers whereas circular fibers are still intact and form successive rings along the entire tubule. Intraconnective neurosecretory cell processes (especially those containing small granules) are still present but vacuolar cells, on the other hand are no longer visible (VandenSpiegel and Jangoux, 1987; Van Dyck and Flammang, unpubl. obs.). The inner epithelium is dissociated and discontinuous, the lumen being in places limited only by the basal lamina. Spherulocytes have disintegrated and the contents of their spherules have been released into the tubule lumen during elongation (VandenSpiegel and Jangoux, 1987). There, the spherule contents could act as a cement improving the rigidity of elongated tubules. Indeed, an increase in the tensile strength of the tubules with time after expulsion has been observed (Van Dyck and Flammang, unpubl. obs.).

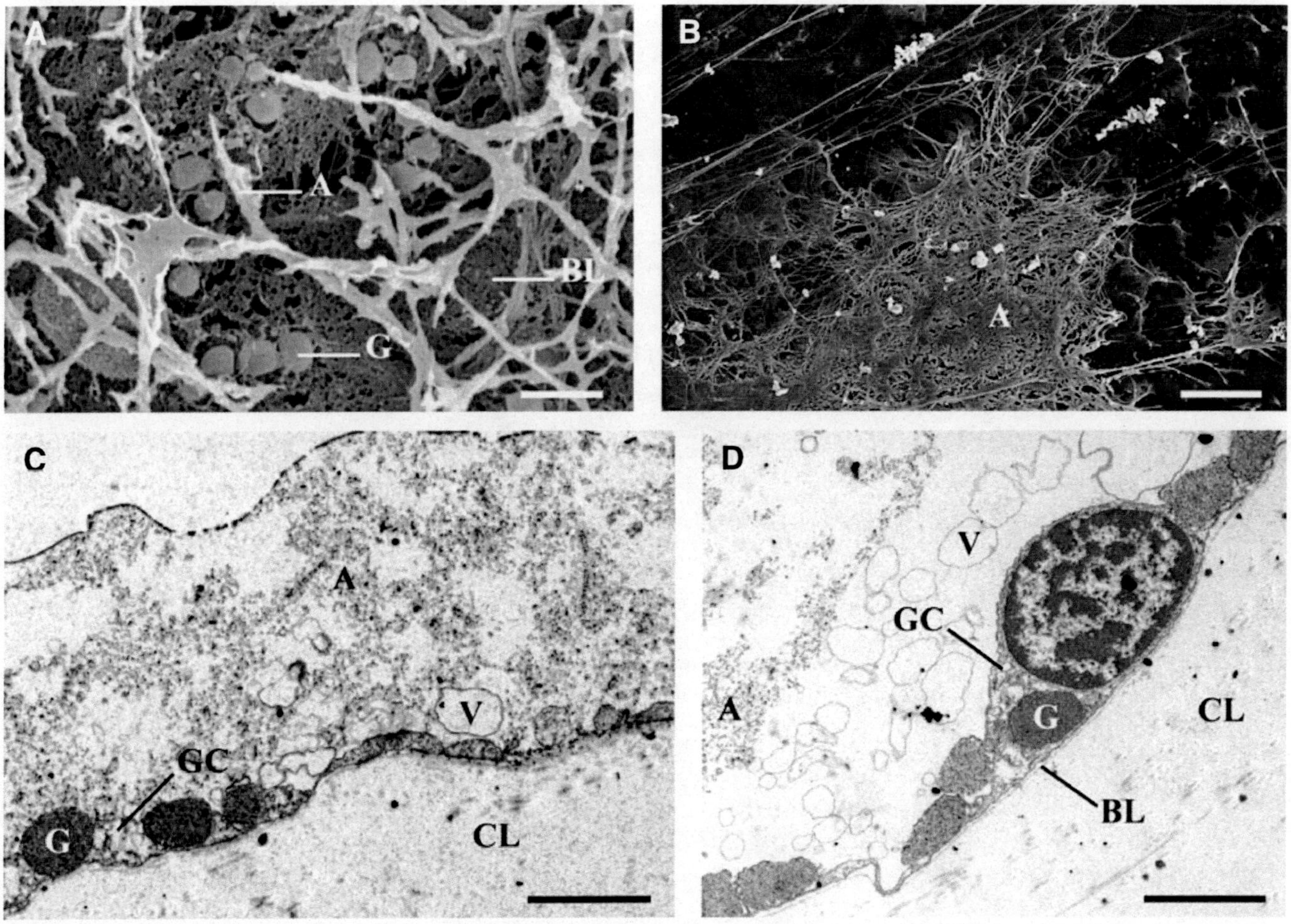

Fig. 6.4 Ultrastructure of the mesothelium in elongated tubules of *Holothuria forskali* (originals). (A) SEM view of the surface of a tubule and the adhesive; (B) SEM view of tubule print material; (C, D) TEM views of the mesothelium and the adhesive. Scale bars for all images 2 µm except for B = 10 µm. *A* Adhesive; *BL* basal lamina; *CL* connective tissue layer; *G* granule; *GC* granular cell; *V* vacuole

6.2.3 Interspecific Diversity in Cuvierian Tubule Morphology

The length of Cuvierian tubules differs among species from a few millimeters in *H. mammata* up to 7 cm in the genus *Bohadschia* while their diameter is usually about 1 mm (VandenSpiegel, 1993; Flammang and Plumet, unpubl. obs.). Table 6.1 summarizes the main characteristics of Cuvierian tubules from various species, evidencing their similarities and differences. As reported in the Introduction, lobulated tubules, characteristic of the genus *Actinopyga*, are radically different from smooth tubules (Fig. 6.1A). They are ramified and covered with small lobules which are composed of a mass of vacuolar cells forming a large central nodule, surrounded by a thin layer of connective tissue and a layer of peritoneocytes. Their mesothelium does not contain granular cells (VandenSpiegel and Jangoux, 1993; Flammang and Plumet, unpubl. obs.).

In smooth tubules, the global structure of the tubules is constant in every species investigated and the organization of the tissue layers is always similar to the one described in *H. forskali*, with a pseudostratified mesothelium, a connective tissue layer enclosing muscle fibers and an inner epithelium (Jourdan, 1883; Endean, 1957; VandenSpiegel and Jangoux, 1988; VandenSpiegel, 1993, 1995; Flammang and Plumet, unpubl. obs.). However, some differences in the fine structure of these layers occur. The genus *Holothuria* presents the highest variability in terms of tubule morphology (Table 6.1). Some species differ only slightly from what was described in *H. forskali* (Jourdan, 1883; Endean, 1957; VandenSpiegel, 1993; Flammang and Plumet, unpubl. obs.). For example, in *H. hilla* and *H. leucospilota*, granular cells make only simple folds (compare Figs. 6.3A and 6.5D) and the inner epithelium lacks spherulocytes (Fig. 6.5C). Differences are more important in *H. maculosa*: the mesothelium is relatively thick and is composed of particularly large granular cells of up to 100 μm long (Fig. 6.5E), apparently not folded along the tubule long axis. These cells show

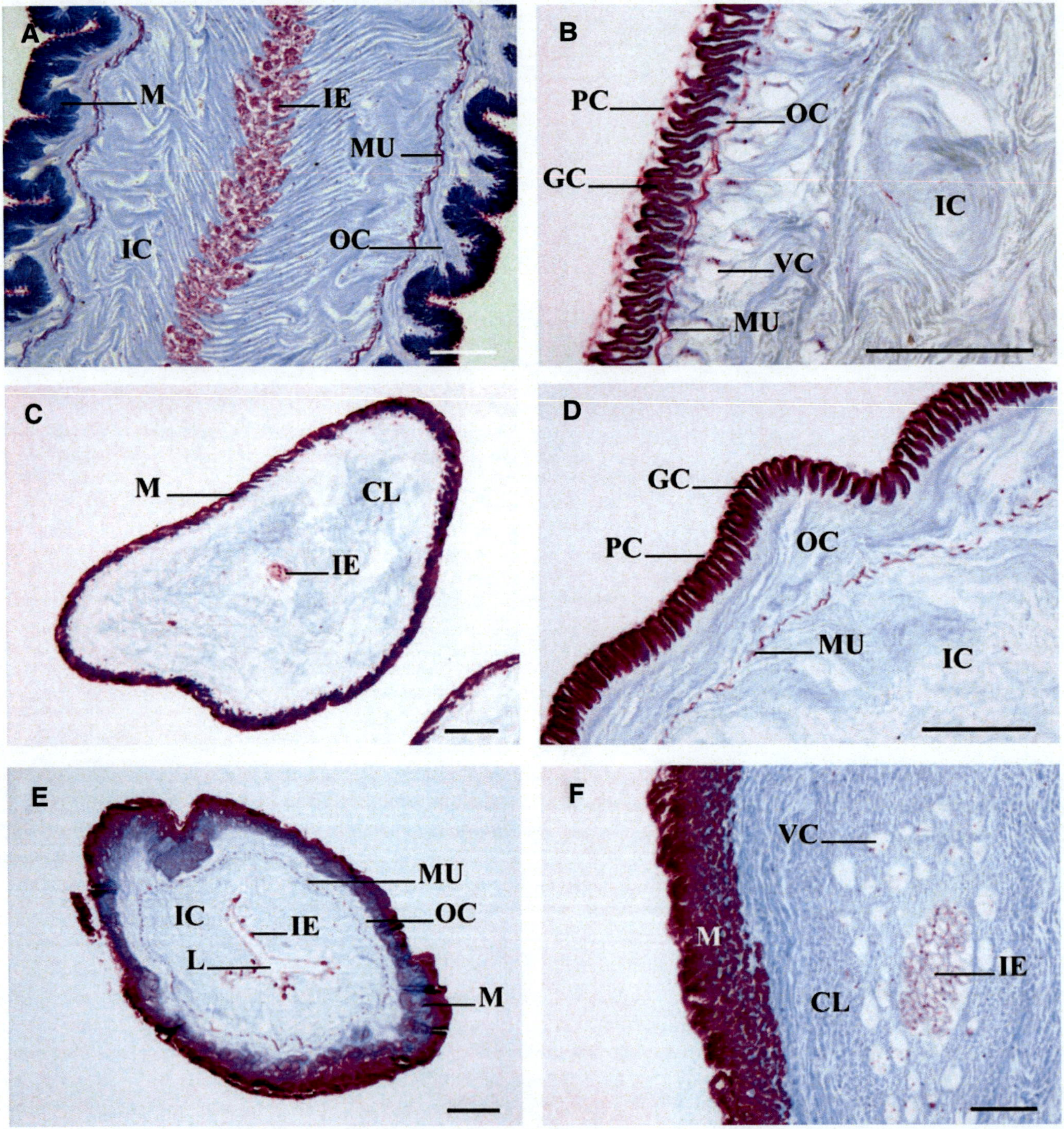

Fig. 6.5 Histology (LM) of the Cuvierian tubules of various sea cucumber species (originals). (A) *Bohadschia marmorata*, (B) *Bohadschia subrubra*, (C) *Holothuria leucospilota*, (D) *Holothuria hilla*, (E) *Holothuria maculosa*, (F) *Pearsonothuria graeffei*. Longitudinal (A, B, D) and transverse (C, E, F) sections stained with Heidenhain's azan trichrome stain. Scale bars = 100 μm. *CL* Connective tissue layer; *GC* granular cells; *IC* internal connective tissue layer; *IE* inner epithelium; *L* lumen; *M* mesothelium; *MU* muscles; *OC* outer connective tissue layer; *PC* peritoneocytes; *VC* vacuolar cells

a very different aspect compared to those of other species with a differentiation of their contents from the cell basal part, where this material appears loosely-packed, to the apical part where it condenses into large polygonal granules (Flammang and Plumet, unpubl. obs.). Cuvierian tubules of *H. mammata* are particularly short, their maximum length reaching only 5 mm, and a random distribution of the collagen fibers is observed in the connective tissue layer (Table 6.1). In this species, tubules are not ejected, not sticky and cannot lengthen (VandenSpiegel and Jangoux, 1988). In *H. nobilis*, adluminal cells are flattened and not T-shaped and lack mucous vesicles, while granular cells are cylindrical and separated by connective tissue processes. Additionally, collagen fibers are not arranged in helices but are rather randomly distributed (VandenSpiegel, 1995; Flammang and Plumet, unpubl. obs.). This species possesses ramified tubules that are not expelled, not sticky and cannot elongate (VandenSpiegel, 1995). In the genus *Bohadschia*, Cuvierian tubules are very similar to those of *H. forskali* (Fig. 6.5A, B). However, in their mesothelium, peritoneocytes are always devoid of mucous vesicles, and the inner epithelium of most species lacks spherulocytes (VandenSpiegel, 1993; Flammang and Plumet, unpubl. obs.). Another morphological characteristic of the tubules in *Bohadschia* spp. is the abundance of vacuolar cells which are located close to the mesothelium (Fig. 6.5B). In *P. graeffei*, the mesothelium is very thick (about 120 μm) but the characteristic folded organization of granular cells is difficult to observe (Flammang and Plumet, unpubl. obs.). Peritoneocytes lack mucous vesicles and spherulocytes are absent. Like in the genus *Bohadschia*, vacuolar cells are abundant, but they are located deeper in the connective tissue layer (Fig. 6.5F). In this species, tubules are never ejected although they can be elongated manually and then become sticky.

6.3 Glue Composition

In *H. forskali*, tubule print material (TPM) – i.e., the secreted adhesive left on the substratum after mechanical detachment of the tubule (Fig. 6.4B) – is composed of 60% protein and 40% neutral carbohydrate (DeMoor et al., 2003). The proteinic nature of the adhesive material is confirmed by the observation that proteolytic enzymes reduce the adhesive strength of Cuvierian tubules in *H. forskali* (Zahn et al., 1973). The amino acid compositions of the protein fraction in *H. forskali*, *H. leucospilota*, *B. subrubra*, and *P. graeffei* indicate that their adhesives are closely related (Flammang, 2006). All are rich in small side-chain amino acids, especially glycine, and in charged and polar amino acids. Only a small fraction of the secreted Cuvierian tubule adhesive (tubule prints) can be extracted using denaturing buffers containing both chaotropic and reducing agents. This soluble fraction contains about ten different proteins with molecular masses ranging from 10 to 220 kDa, but with closely related amino acid compositions, resembling that of the whole adhesive (DeMoor et al., 2003). It was suggested that these different proteins could be polymers of a low molecular weight protein (Flammang, 2006).

Polyclonal antibodies raised against the TPM of *H. forskali* were used to locate the origin of TPM constituents in the tubule. Peritoneocytes and the granules of the granular cells show extensive immunoreactivity, suggesting that their secretions make up the bulk of the adhesive material (DeMoor et al., 2003). These antibodies were tested on 11 other species from the genera *Holothuria*, *Bohadschia*, *Pearsonothuria*, and *Actinopyga* (Flammang and Plumet, unpubl. obs.). In all species, peritoneocytes are labeled (Fig. 6.6). As far as granular cells are concerned, there are more interspecific differences. Granular cells are strongly labeled in all species of the genera *Bohadschia* and *Holothuria* (Fig. 6.6A, C), except in *H. maculosa* and *H. nobilis*. In the former, only the basal part of granular cells is labeled but not the apex (Fig. 6.6B); while in the latter no labeling is observed. In *P. graeffei*, the contents of granular cells are immunoreactive but the labeling is weaker than in *Bohadschia* and *Holothuria* (Fig. 6.6D). Finally, in the lobulated tubules of *Actinopyga echinites* that lack granular cells, only the peritoneocytes are immunoreactive while the vacuolar cells are not labeled.

Histochemical experiments show that granular cells always stain strongly for proteins in general, and for tyrosine residues and disulfide bonds in particular (Endean, 1957; Flammang and Plumet, unpubl. obs.). There are only small variations in staining intensity between species, with the lowest reactivity always detected in *H. nobilis* followed by *P. graeffei* (Flammang and Plumet, unpubl. obs.). On the other hand, in *H. maculosa*, the staining of granular cells follows a base to apex gradient. In *H. forskali*, VandenSpiegel and Jangoux (1987) reported that the secretory granules of granular cells are lipoproteic in nature. However, no lipids were detected in the TPM of this species (DeMoor et al., 2003). In species in which peritoneocytes possess mucous vesicles (see above), these vesicles stain for proteins and for both neutral and acidic mucopolysaccharides (Guislain,

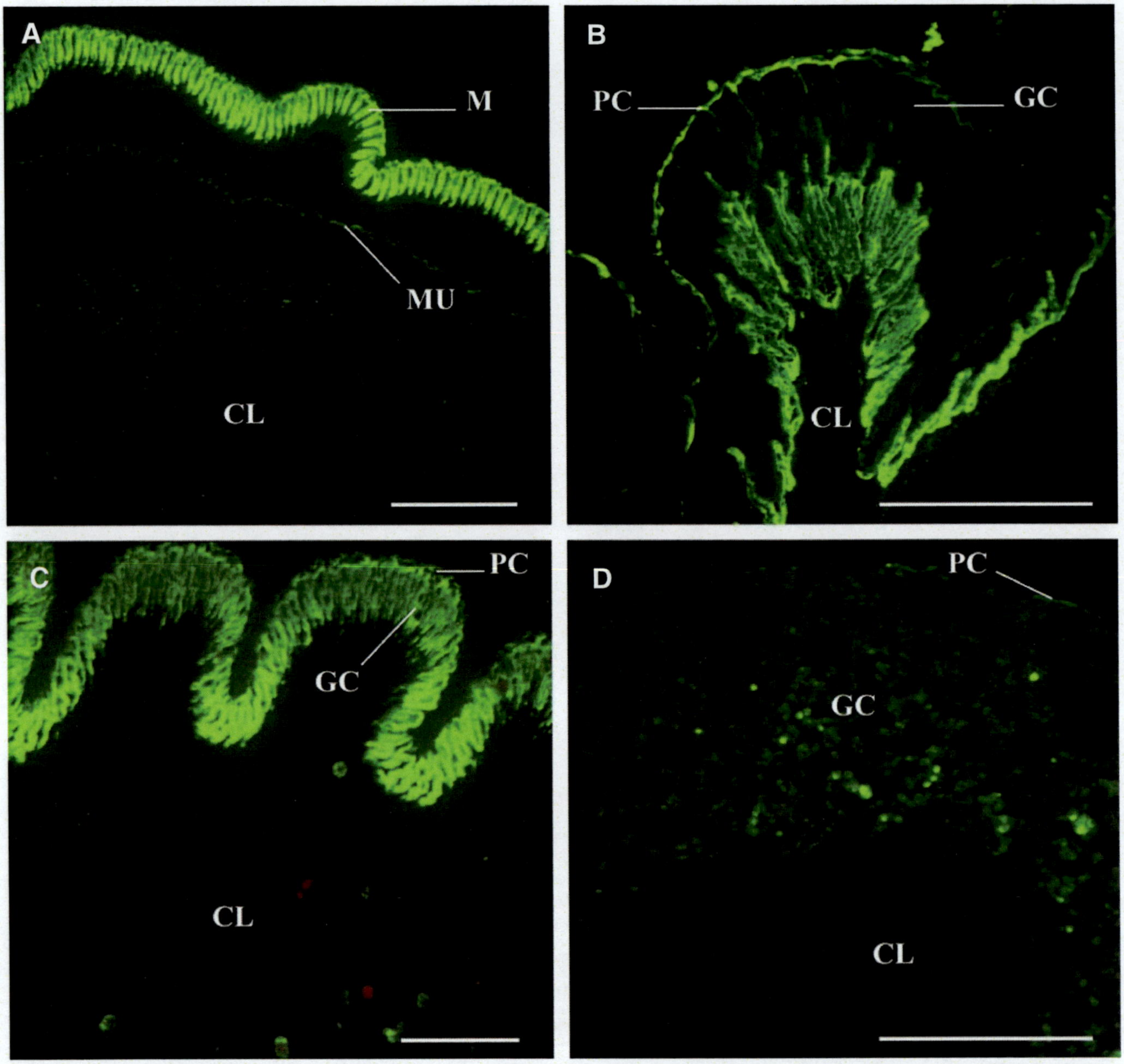

Fig. 6.6 Immunolabeling of Cuvierian tubules using antibodies raised against the tubule print material of *Holothuria forskali* (originals). (A) *Holothuria impatiens*, (B) *Holothuria maculosa*, (C) Bohadschia *marmorata*, (D) *Pearsonothuria graeffei*. Immunoreactive structures are labeled green. Scale bars = 100 μm. *CL* Connective tissue layer; *GC* granular cells; *M* mesothelium; *MU* muscles; *PC* peritoneocytes

1953; VandenSpiegel and Jangoux, 1987; Flammang and Plumet, unpubl. obs.).

In marine adhesives, proteins are usually modified post-translationally and their modifications are of three main types: hydroxylation, phosphorylation and glycosylation (Sagert et al., 2006; Hennebert et al., 2010). Hydroxylation is the most investigated protein modification in adhesive proteins. All the proteins identified in the attachment plaques from the byssus of mussels contain 3,4-dihydroxyphenylalanine (DOPA), a residue formed by the post-translational hydroxylation of tyrosine (Sagert et al., 2006). As granular cells stain strongly for tyrosine residues, a specific DOPA stain (Arnow reaction) was used on tubule sections but no positive reaction was detected, indicating that Cuvierian tubule adhesion is not based on DOPA (Flammang and Wattier,

Table 6.2: Intensity of the lectin labeling in the different tissue layers of the Cuvierian tubules of *Holothuria forskali* (Flammang, Bohem and VanDyck, unpubl. obs.)

Tissue layer and cell type	Lectin reactivity*						
	Con A	LCA	PSA	RCA	GSL I	PNA	UEA I
Mesothelium							
Granular cells	–	–	–	–	–	–	–
Peritoneocytes	+++	+++	+++	+++	–	–	–
Connective tissue layer							
Myocytes	+	+	+	–	–	–	–
Vacuolar cells	++	+	+	–	–	–	–
Neurosecretory cells	++	+	+	++	–	–	–
Inner epithelium							
Spherulocytes	+	+	+	++	++	–	++
Inner epithelial cells	+	+	+	–	–	–	–

*Strong (+++), moderate (++), weak (+), and absent (–).
Con A Concanavaline A; *GSL I* Griffonia (Bandeiraea) simplicifolia lectin I; *LCA* Lens culinaris agglutinin; *PNA* Peanut agglutinin; *PSA* Pisum sativum agglutinin; *RCA* Ricinus communis agglutinin; *UEA I* Ulex europaeus agglutinin I.

unpubl. obs.). Protein phosphorylation occurs in marine adhesives in the form of serine phosphorylation and has been described in two mussel adhesive proteins and in one tube-worm cement protein (Sagert et al., 2006). The use of anti-phosphoserine antibodies on Cuvierian tubules reveals that their adhesive is rich in phosphoserine residues, a strong immunolabeling being detected in the granular cells from the tubules of three different species (*B. subrubra*, *H. forskali*, and *P. graeffei*) (Flammang et al., 2009). Immunoblots and amino acid analyses confirmed the presence of polyphosphoproteins in the adhesive secretion of the Cuvierian tubules from these species (Flammang et al., 2009). Finally, lectins (i.e., proteins that specifically bind to oligosaccharidic structures) were used on histological sections to characterize carbohydrate distribution in the Cuvierian tubules of *H. forskali* (Flammang, Bohem and VanDyck, unpubl. obs.). Seven lectins specific of neutral sugars (Con A, LCA, and PSA specific of mannose containing oligosaccharides; GSL I, PNA, and RCA specific of galactose containing oligosaccharides; and UEA I specific of fucose containing oligosaccharides; Hennebert et al., 2010) were selected in view of the TPM composition (DeMoor et al., 2003). Granular cells of the mesothelium are never labeled with any of the lectins tested (Table 6.2). On the other hand, there is an extensive labeling of the mucous vesicles in peritoneocytes with galactose- and glucose-binding lectins. The other cell types showing some lectin reactivity are vacuolar cells, neurosecretory cells and spherulocytes (Flammang, Bohem and VanDyck, unpubl. obs.).

6.4 Discussion

Smooth Cuvierian tubules occurring in sea cucumbers of the genera *Bohadschia* and *Holothuria* are expelled as sticky white threads that function as a defence mechanism against predators (Hamel and Mercier, 2000; Flammang et al., 2002). In the genus *Pearsonothuria*, the tubules are never ejected although they can be elongated manually and then become sticky. To function properly, sticky Cuvierian tubules must lengthen greatly after expulsion and they must be able to adhere instantaneously upon contact. For this to occur two apparently necessary morphological characteristics are the organization of collagen fibers into helices and the folded arrangement of granular cells in the mesothelium (Table 6.1). Other morphological features such as the presence or absence of mucous vesicles in peritoneocytes, the abundance of vacuolar cells, or the presence or absence of spherulocytes in the inner epithelium vary among species, indicating that they are not prerequisite for elongation and stickiness of the Cuvierian tubules. They may, however, be important for other functions.

In *H. forskali*, the only species in which the composition of the adhesive has been investigated (DeMoor et al., 2003), the tubule print material is composed of both protein and neutral carbohydrate. Histochemistry and lectin labeling suggests that the protein fraction would originate mostly from the granules of granular cells, and the carbohydrate fraction from the mucous vesicles of peritoneocytes. Therefore, in species in which peritoneocytes lack mucous vesicles, like in the genus *Bohadschia*, the carbohydrate fraction could be reduced or even absent. Polyclonal antibodies raised against tubule print material of *H. forskali* are directed mostly toward protein epitopes (DeMoor et al., 2003), and they were used to investigate the variability of Cuvierian tubule adhesive secretions in the different sea cucumber species. Although there was a significant immunolabeling of the peritoneocytes in every species investigated, a strong immunolabeling of the granular cells was only observed in species possessing sticky tubules. This indicates that the adhesives in all the species considered are closely related, probably sharing many identical molecules or, at least many identical epitopes on their constituents. This was confirmed by amino acid analyses of the TPM in *H. forskali*, *H. leucospilota*, *B. subrubra*, and *P. graeffei* (Flammang, 2006). The immunolabeling of the granular cells was however totally absent in *H. nobilis*, a species in which the ramified tubules are never ejected nor become sticky, and break very easily when stretched manually (VandenSpiegel, 1995). Therefore, in this species, the loss of collagen helices and folded granular cells is accompanied by chemical modifications of the granular cell contents.

Acknowledgments

Work supported by the "Service Public de Wallonie – Program Winnomat 2" and by the "Communauté française de Belgique – Actions de Recherche Concertées". Patrick Flammang is a Senior Research Associate of the Fund for Scientific Research of Belgium (F.R.S.–FNRS). This study is a contribution of the "Centre Interuniversitaire de Biologie Marine" (CIBIM; http://www.ulb.ac.be/sciences/biomar/).

References

DeMoor S, Waite JH, Jangoux M, and Flammang P (2003) Characterization of the adhesive from cuvierian tubules of the sea cucumber *Holothuria forskali* (Echinodermata, Holothuroidea). Marine Biotechnology 5: 45–57.

Endean R (1957) The Cuvierian tubules of *Holothuria leucospilota*. Quarterly Journal of Microscopical Science 98: 455–472.

Flammang P (2006) Adhesive secretions in echinoderms: an overview. In: Smith AM and Callow JA (eds) Biological Adhesives. Springer-Verlag, Heidelberg: pp 183–206.

Flammang P, Ribesse J, and Jangoux M (2002) Biomechanics of adhesion in sea cucumber cuvierian tubules (Echinodermata, Holothuroidea). Integrative and Comparative Biology 42: 1107–1115.

Flammang P, Lambert A, Bailly P, and Hennebert E (2009) Polyphosphoprotein-containing marine adhesives. The Journal of Adhesion 85: 447–464.

Guislain R (1953) Recherches histochimiques sur les canaux de Cuvier de *Holothuria impatiens* Forskal. Comptes Rendus de l'Académie des Sciences, Paris 147: 1254–1256.

Hamel JF and Mercier A (2000) Cuvierian tubules in tropical holothurians: usefulness and efficiency as a defence mechanism. Marine and Freshwater Behaviour and Physiology 33: 115–139.

Hennebert E, Wattiez R, and Flammang P (2010) Characterization of the carbohydrate fraction of the temporary adhesive material secreted by the tube feet of the sea star *Asterias rubens*. Marine Biotechnology (accepted for publication).

Jourdan M (1883) Recherches sur l'histologie des Holothuries. Annales du Musée d'Histoire Naturelle de Marseille 1(6): 1–64.

Lawrence JM (2001) Function of eponymous structures in echinoderms: a review. Canadian Journal of Zoology 79: 1251–1264.

Müller WEG, Zahn RK, and Schmid K (1972) The adhesive behaviour in Cuvierian tubules of *Holothuria forskali*. Biochemical and biophysical investigations. Cytobiologie 5: 335–351.

Sagert J, Sun C, and Waite JH (2006) Chemical subtleties of Mussel and Polychaete holdfasts. In: Smith AM and Callow JA (eds) Biological Adhesives. Springer-Verlag, Heidelberg: pp 125–143.

VandenSpiegel D (1993) Morphologie fonctionelle et comparée des organes de défense (tubes de Cuvier) des holothuries (Echinodermata). PhD thesis, Université Libre de Bruxelles: 142 pp.

VandenSpiegel D (1995) Fine structure and behaviour of the Cuvierian organs of the holothuroid *Microthele nobilis* (Echinodermata). In: Emson RH, Smith AB, and Campbell AC (eds) Echinoderm Research. Balkema, Rotterdam: pp 121–127.

VandenSpiegel D and Jangoux M (1987) Cuvierian tubules of the holothuroid *Holothuria forskali* (Echinodermata): a morphofunctional study. Marine Biology 96: 263–275.

VandenSpiegel D and Jangoux M (1988) Les tubes de Cuvier d'Holothuria mammata Grube, 1840 (Holothuroidea, Echinodermata). Annales de la Société Royale Zoologique de Belgique 118: 191–198.

VandenSpiegel D and Jangoux M (1993) Fine structure and behaviour of the so-called Cuvierian organs on the holothuroid genus *Actinopyga* (Echinodermata). Acta Zoologica 74: 43–50.

VandenSpiegel D, Jangoux M, and Flammang P (2000) Maintaining the line of defense: regeneration of Cuvierian tubules in the sea cucumber *Holothuria forskali* (Echinodermata, Holothuroidea). Biological Bulletin 198: 34–49.

Zahn RK, Müller WEG, and Michaelis M (1973) Sticking mechanisms in adhesive organs from a *Holothuria* – Klebemechanismen adhäsiver Organe einer Holothurie. Research in Molecular Biology, 2nd Ed. Akademie der Wissenschaften und der Literatur, Mainz.

7 Adhesion Mechanisms Developed by Sea Stars: A Review of the Ultrastructure and Composition of Tube Feet and Their Secretion

Elise Hennebert

Contents

7.1 Introduction

Like all animals belonging to the phylum Echinodermata, sea stars are characterized by a water-vascular system. This sophisticated hydraulic system consists of a series of interconnected canals: a central ring canal that encircles the gut of the animal and from which arise a single axial canal, the stone canal, communicating with the external seawater through a perforated plate (the madreporite), and five radial canals which extend into each arm of the sea star. The radial canals lead to a multitude of specialized external appendages, the tube feet (Nichols, 1966). According to the sea star species, tube feet may be involved in one or several of the following functions: locomotion, fixation to the substratum, feeding and burrowing. These different functions are allowed by the mobility of the proximal part of the tube foot (the so-called stem) as well as by the attachment of the distal part of the tube foot to the substratum (Flammang, 1996). Tube foot attachment is temporary. Indeed, although tube feet can adhere very strongly to the substratum, they are also able to detach easily and voluntarily before reinitiating another attachment–detachment cycle (Thomas and Hermans, 1985; Flammang, 1996). Suction has long been regarded as a major mean of tube foot attachment in sea stars (Paine, 1926; Smith, 1937; Nichols, 1966). However, a number of more recent observations argue for an adhesive process principally, if not exclusively, mediated by the secretion of an adhesive material (Chaet, 1965; Thomas and Hermans, 1985; Flammang et al., 1994, 1998). For instance, sea star tube feet can adhere very strongly to meshed or perforated substrata, or with only the margin of their distal part, two situations which prevent the use of suction by tube feet (Thomas and Hermans, 1985; Flammang et al., unpubl. obs.).

The comprehension of the complex adhesion mechanisms developed by sea stars requires the complete description of their secretory organs, the tube feet, the elucidation of the mode of formation of the adhesive material, and the characterization of the different molecules constituting it. Tube foot morphology and ultrastructure have been investigated in a large diversity of species while the secreted material has been described and characterized mainly in the species *Asterias rubens* (Flammang, 1996; Flammang et al., 1994, 1998; Santos et al., 2005a).

7.2 Comparative Morphology of Sea Star Tube Feet

Sea star tube feet were originally subdivided into two main categories: those in which the distal part is pointed, the knob-ending tube feet, and those in which the tip is flattened, the disk-ending tube feet (Hyman, 1955; Lawrence, 1987; Clark and Downey, 1992; Flammang, 1996). This classification was later reviewed by Vickery and McClintock (2000) who described three tube foot morphotypes in sea stars: pointed non-sucker, flat-tipped non-suckered and flat-tipped suckered tube feet. These morphotypes were confirmed by Santos et al. (2005a) but were renamed knob-ending, simple disk-ending and reinforced disk-ending tube feet, respectively. Indeed, the previous terminology implied a suction functioning of tube feet for which there is no consensus (see above). Tube feet from all three morphotypes possess a same histological organization. They consist of four tissue layers that are, from the inside to the outside: a myomesothelium surrounding the water-vascular lumen, a connective tissue layer, a nerve plexus, and an epidermis covered by a thin cuticle (Fig. 7.1D–F, Flammang, 1996; Santos et al., 2005a).

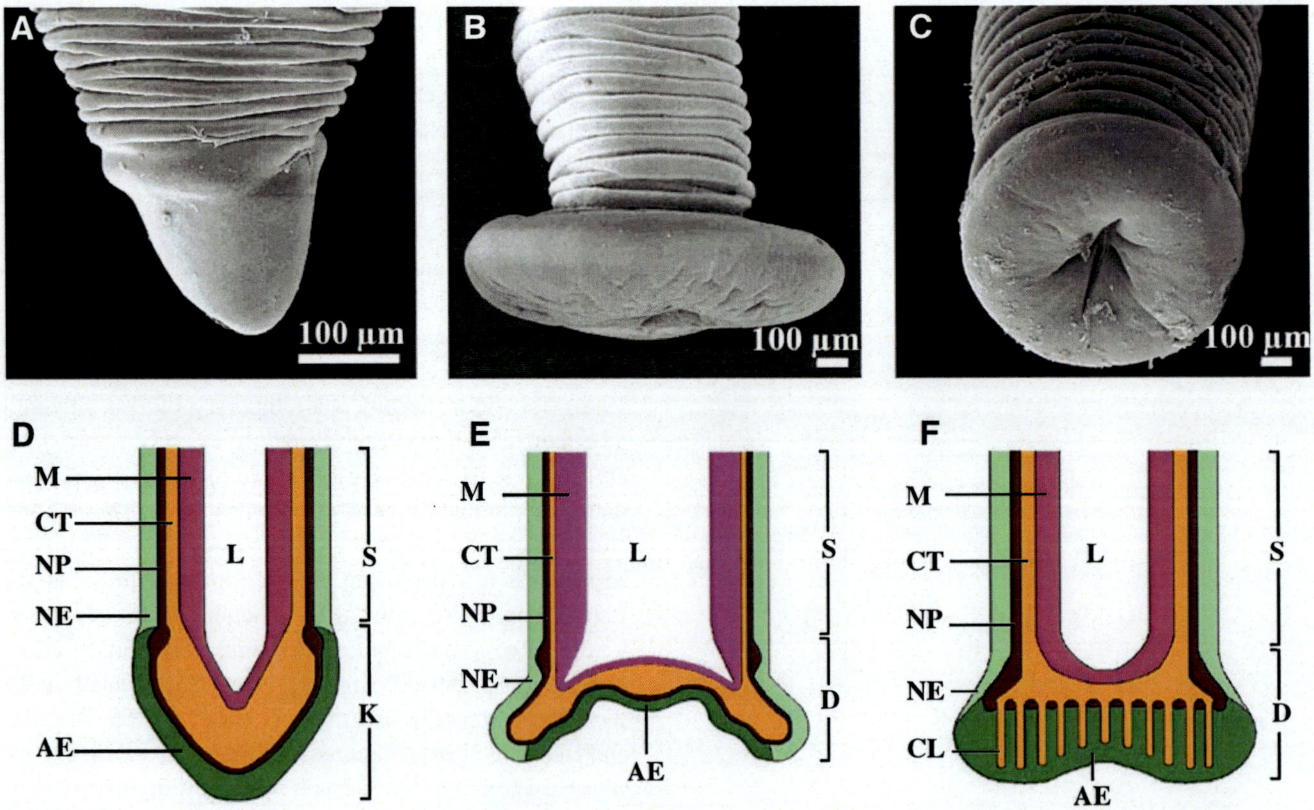

Fig. 7.1 External morphology and histological organization of the three tube foot morphotypes described in sea stars. (A–C) SEM views (originals) of a knob-ending tube foot (*Astropecten irregularis*, A), of a simple disk-ending tube foot (*Mithrodia clavigera*, B) and of a reinforced disk-ending tube foot (*Asterias rubens*, C). (D, E) Schematic drawings of a knob-ending tube foot (D), of a simple disk-ending tube foot (E) and of a reinforced disk-ending tube foot (F). Image from Santos et al. (2005a) and reproduced with permission

7.2.1 Knob-ending Tube Feet

Knob-ending tube feet occur in the order Paxillosida. They are made up of a basal cylindrical stem ending distally with a pointed knob (Fig. 7.1A, Vickery and McClintock, 2000; Santos et al., 2005a; Haesaerts et al., unpubl. obs.). The water-vascular lumen extends into the knob where it tapers off to a point. Except for the myomesothelium, all tissue layers are thicker in the knob than in the stem. The knob epidermis is a tall columnar epithelium consisting mostly of granule-filled secretory cells. The nervous tissue consists of a nerve ring on the proximal side of the knob and a thick basiepithelial nerve plexus underlying the knob epidermis. Both layers lie on a thick layer of loose connective tissue (Fig. 7.1D). Such tube feet allow paxillosids to bury themselves by digging vertically into the sediment (Santos et al., 2005a).

7.2.2 Simple Disk-ending Feet

Simple disk-ending tube feet occur in most species of the order Valvatida (Vickery and McClintock, 2000; Santos et al., 2005a; Haesaerts et al., unpubl. obs.). They consist of a basal cylindrical stem with an apical extremity that is enlarged and flattened to form the so-called disk (Fig. 7.1B). The disk has the same basic structure as the knob of knob-ending tube feet: it encloses the distal extremity of the water-vascular lumen and its epidermis, nerve plexus, and connective tissue layer are thickened compared to their equivalents in the stem (Fig. 7.1E). In most of the species, the disk is much larger than the stem giving the tube foot a flared shape. The water-vascular lumen extends radially into the disk margin (Fig. 7.1E) where it usually presents a scalloped circumference. The disk lumen may even be entirely compartmentalized by radial connective tissue septa, as it is the case in *Archaster typicus*. In the oreasterid species *Culcita schmideliana*, *Pentaceraster mammillatus*, and *Protoreaster lincki*, the large disk is supported by a skeleton made up of numerous calcareous ossicles and spicules located within the connective tissue layer, on the proximal side of the disk lumen. The epidermis of the proximal side of the disk (i.e., the epidermis facing the skeleton) encloses large spumous glandular cells that were not observed in the other valvatid species (Santos et al., 2005a). Simple disk-ending tube feet are used by most of valvatid species mainly for locomotion. The tissular organization of the disk provides the tube foot with a large flat distal surface that can be used in locomotion on soft substrata (e.g., in *P. mammillatus* and *P. lincki*) as well as on hard substrata (e.g., in *Acanthaster planci* or in *C. schmideliana*). Simple disk-ending tube feet are also used for attachment to the substratum in a few valvatid species (e.g., in *A. planci* or in *Linckia laevigata*), and for burrowing in the sediment in *A. typicus*. In this species, the fact that the disk lumen is divided radially by collagenous septa would enable modifications of the disk shape and orientation to allow digging (Santos et al., 2005a).

7.2.3 Reinforced Disk-ending Tube Feet

Reinforced disk-ending tube feet occur in all the investigated species of the orders Velatida, Spinulosa and Forcipulatida, and in a few valvatids (Vickery and McClintock, 2000; Santos et al., 2005a). They have more or less the same external morphology as simple disk-ending tube feet: they are made up of a basal cylindrical stem topped by a flattened disk (Fig. 7.1C). However, the diameter of their disk never greatly exceeds that of the stem. The histological organization of reinforced disk-ending tube feet also differs from that of the two other morphotypes. The lumen does not extend into the disk margin and the disk connective tissue layer sends distally numerous bundles of collagen fibers that insinuate themselves within the adhesive epidermis, reaching up to the cuticle (Fig. 7.1F). This tissular organization varies according to the species considered: from the tube feet of the velatid *Crossaster papposus* which have few, uniformly distributed bundles of collagen fibers, to the tube feet of the forcipulatid *A. rubens* which possess a complex array of radial collagenous laminae (Santos et al., 2005a). Reinforced disk tube feet are used in locomotion, attachment to the substratum, and, in some species, in bivalve opening for feeding (Hennebert et al., 2010a). They appear to be better designed for strong adhesion than simple disk-ending tube feet. Indeed, the collagen fibers present in the disk epidermis may function as tension-bearing structures, transferring the stress due to adhesion from the distal cuticle to the connective tissue of the stem (Santos et al., 2005a, b).

The morphological diversity of tube feet in sea stars appears to be a result of adaptation of species to various habitats, but within limits imposed by the evolutionary lineage (Santos et al., 2005a). For instance, most species inhabiting turbulent environments, where strong attachment to the substratum is required, possess reinforced disk-ending tube feet regardless of their taxonomic position. In this case, adaptation to habitat overrides belonging to a particular order. On the other hand, *A. typicus*, which has the capacity to bury itself completely, is homeomorph with paxillosid species of the genus *Astropecten*, which also bury themselves. This homeomorphy, however, does not extend to the tube feet, which are simple disk-ending in *A. typicus*. In this case, therefore, the taxonomic position of the species seems to supplant habitat constraints (Santos et al., 2005a).

7.3 Ultrastructure of Tube Foot Adhesive Areas

Tube foot adhesive areas are located at the level of the epidermis of functionally important parts of the tube feet, i.e., the whole knob surface in knob-ending tube feet and the distal disk surface in simple disk-ending tube feet, and reinforced disk-ending tube feet (Smith, 1937; Engster and Brown, 1972; Perpeet and Jangoux, 1973; Flammang, 1996; Santos et al., 2005a). These epidermal adhesive areas always consist of four cell categories: adhesive cells, de-adhesive cells, sensory cells, and support cells (Fig. 7.2, see Flammang, 1996 for review). Externally, the epidermis is covered by a multi-layered glycocalyx, the cuticle (Fig. 7.2A, B, D). Its most external layer is the "fuzzy coat" and consists of numerous fine fibrils (McKenzie, 1988; Flammang, 1996; Flammang et al., 1998; Ameye et al., 2000).

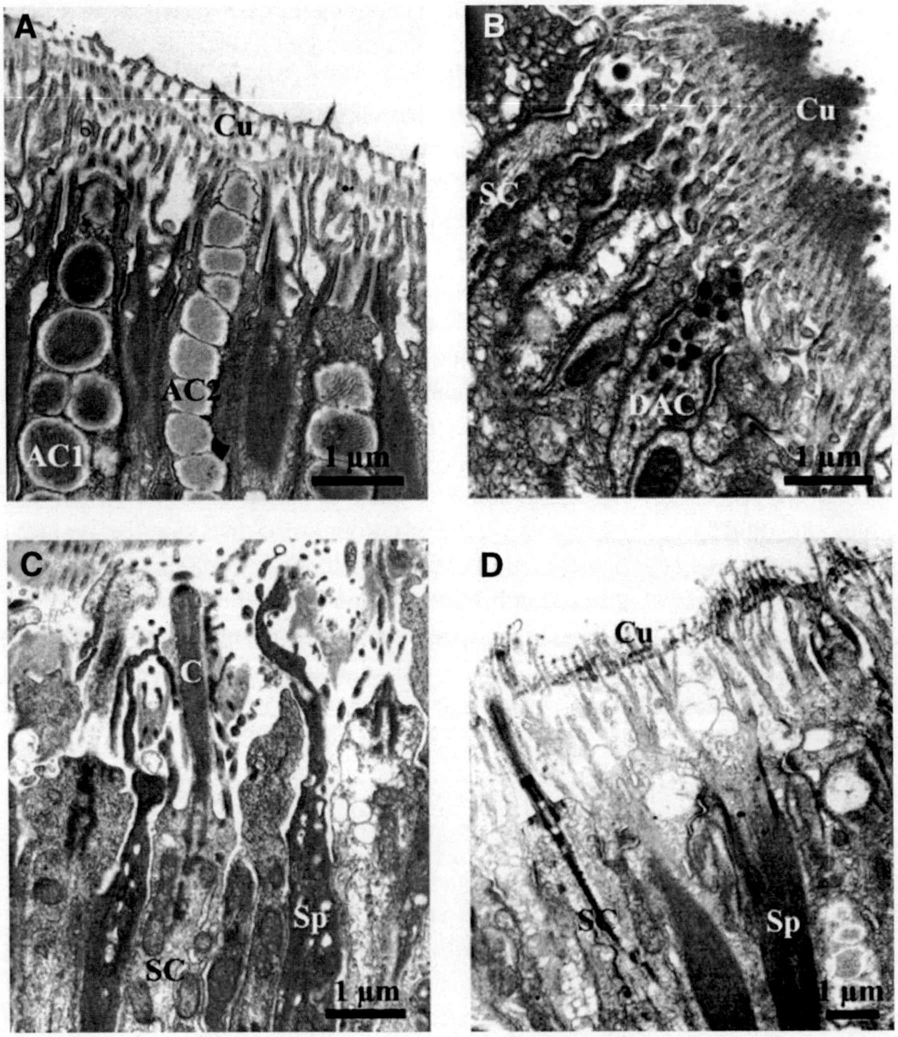

Fig. 7.2 TEM photographs of the four cell categories constituting the tube foot epidermal adhesive areas. (A) Two types of adhesive cells (*Asterias rubens*). Image from Ameye et al. (2000) and reproduced with permission. (B) De-adhesive cell (*Henricia oculata*, original). (C) Sensory cell (*Luidia penangensis*). Image from Flammang (1995) and reproduced with permission. (D) Support cell (*Marthasterias glacialis*, original)

7.3.1 Adhesive Cells

Adhesive cells are flask-shaped. Their enlarged cell bodies are located basally and send out a long apical process that reaches the surface of the tube foot. According to the species, adhesive cells may be of one or two types, which may be distinguished on the basis of the ultrastructure of the secretory granules enclosed in the cytoplasm of both their cell body and apical process (Fig. 7.2A). According to Flammang (1996), two broad groups of granules may be recognized in sea stars. Type 1 granules are ellipsoid and consist of a large electron-dense core made up of a parallel arrangement of fibrils or rods, surrounded by a thin layer of less electron-dense material (Fig. 7.3A–C). Type 2 granules resemble the previous group but with a smaller, more homogeneous core. They are spherical to ellipsoid and are always smaller than type 1 granules (Fig. 7.3D–G). Subcategories exist in these groups, reflecting a high variability in granule ultrastructure between different species (Flammang, unpubl. obs.; Fig. 7.3). In the cell body of adhesive cells, developing granules are closely associated with Golgi membranes and rough endoplasmic reticulum (RER) cisternae, suggesting that these organelles are involved in the synthesis of the granule contents. Granules are then conveyed to the apex of the cell probably with the help of the longi-

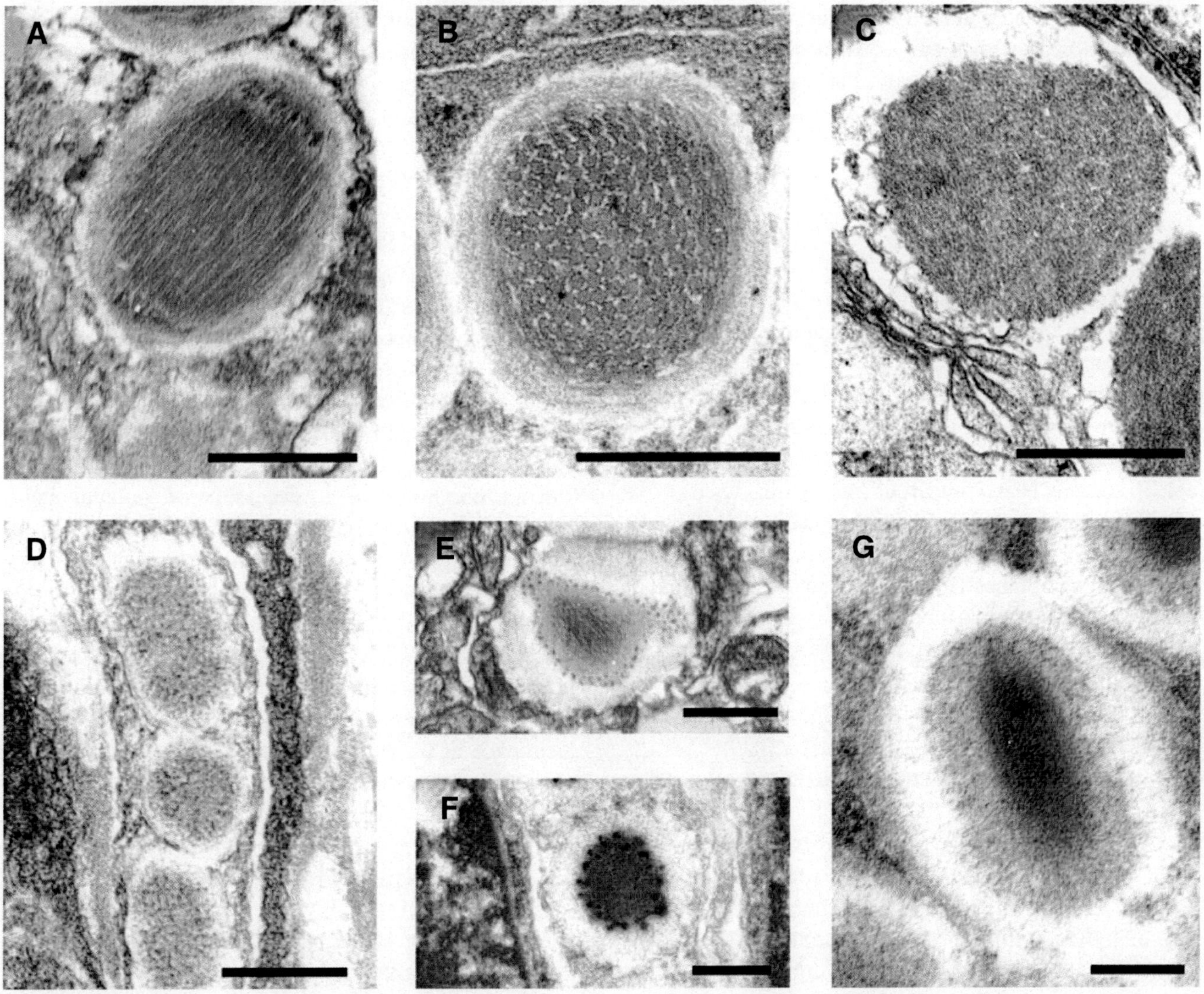

Fig. 7.3 Ultrastructure of the secretory granules of adhesive cells from sea star tube feet. (A, B) Longitudinal and transverse sections through type 1 granules from *Asterias rubens*. Image from Flammang (2006) and reproduced with permission. (C) Type 1 granule from *Luidia ciliaris* (original). (D–F) Type 2 granules from *Asterias rubens*, *Ludia ciliaris* and *Luidia maculata*, respectively (originals). (G) Longitudinal section through a type 2 granule from *Marthasterias glacialis* (original). Scale bar in (A–F) = 0.5 μm, (G) = 0.1 μm

tudinal microtubules occurring in the adhesive cell distal process. They are then extruded through a duct delimited by a ring of microvilli and opening onto the knob or disk surface as a pore (Flammang, 1996).

7.3.2 De-adhesive Cells

De-adhesive cells are narrow and have a centrally-located nucleus. They are filled with small homogeneous electron-dense secretory granules whose ultrastructure is remarkably constant in all the species (Fig. 7.2B). The cytoplasm of these cells also contains numerous RER cisternae, a small Golgi apparatus and longitudinally arranged microtubules. The basal end of these cells is tapered and contacts the nerve plexus, while their apical process ends with a bulge just beneath the cuticle. This bulge is devoid of microvilli and generally bears a short subcuticular cilium, except in paxillosid tube feet (Flammang, 1995, 1996).

7.3.3 Other Cells

Sensory cells are narrow and have a centrally located nucleus (Fig. 7.2C). They are characterized by a single short cilium whose apex protrudes into the cuticle fuzzy coat (Ameye et al., 2000). Sensory cells terminate basally within the nerve plexus (Flammang, 1996).

Support cells have a centrally located nucleus. Their apical surface bears numerous microvilli. Their cytoplasm encloses one longitudinal bundle of intermediate filaments that traverses the cell and joins its apical and basal membranes (Fig. 7.2D, Flammang, 1996). In reinforced disk-ending tube feet, it is these bundles of filaments that connect the connective tissue distal protrusions to the cuticle.

7.4 Structure of the Adhesive Material

In all sea star species investigated to date, after detachment of the tube foot, the adhesive secretion usually remains firmly bound to the substratum as a footprint. The material constituting these footprints can be stained, allowing the observation of their morphology under the light microscope (Chaet, 1965; Thomas and Hermans, 1985; Flammang, 1996; Hennebert et al., 2008). The footprints have the same shape and the same diameter as the distal surface of the tube foot disks (Fig. 7.4A). Various techniques have been used to study the fine structure of the material constituting the footprints and, whatever the method used, the adhesive material always appears as a sponge-like meshwork deposited on a thin homogeneous film (Fig. 7.4; Flammang, 1996; Flammang et al., 1994, 1998; Hennebert et al., 2008). This aspect has been observed in LM, conventional SEM, cryo-SEM, TEM, and AFM; and it does not differ according to whether the footprint has been fixed or not and whether it is observed partially hydrated or dried (Flammang, 1996; Flammang et al., 1998; Hennebert et al., 2008). On the basis of TEM observations, the thickness of the adhesive layer ranges from 1.4 to 7.3 μm, while that of the homogeneous film varies between 0.3 and 3 μm (Fig. 7.6F). The thickness of the adhesive layer may also vary between different areas in a same footprint, giving different aspects to the adhesive material (Flammang et al., 1994; Hennebert et al., 2008). In thin areas, meshes ranging from 1 to 5 μm in diameter are clearly distinguishable; whereas in thick areas, these meshes are obscured because of the accumulation of material (Fig. 7.4B, C; Flammang et al., 1994; Hennebert et al., 2008). In TEM, the meshes do not appear empty as in SEM but filled with a loose electron-lucent material (Fig. 7.6F). Comparison of the images obtained with the different techniques therefore suggests that, during drying (as for instance for sample preparation for SEM), the loose material would collapse, leaving empty mesh components on the thin homogeneous film (Hennebert et al., 2008). At the nanometre scale, both the material forming the meshwork and the homogeneous film are composed of a succession of globular nanostructures (Fig. 7.4B). The diameter of these structures is almost the same for dry footprints observed either with SEM or AFM (50–100 and 50–90 nm, respectively). On the other hand, AFM investigations on partially hydrated footprints reveal a globule diameter ranging between 80 and 200 nm. This slight difference in the size of the globular structures between dry and hydrated footprints suggests once again that the adhesive material may shrink upon drying (Hennebert et al., 2008).

7.5 Composition of Footprint Material

The composition of the material constituting footprints was first investigated by histochemical tests performed on longitudinal sections through tube feet and on footprints. In all the species investigated, these tests revealed the presence of acid mucopolysaccharides in the secretory granules of adhesive cells (Defretin, 1952; Chaet

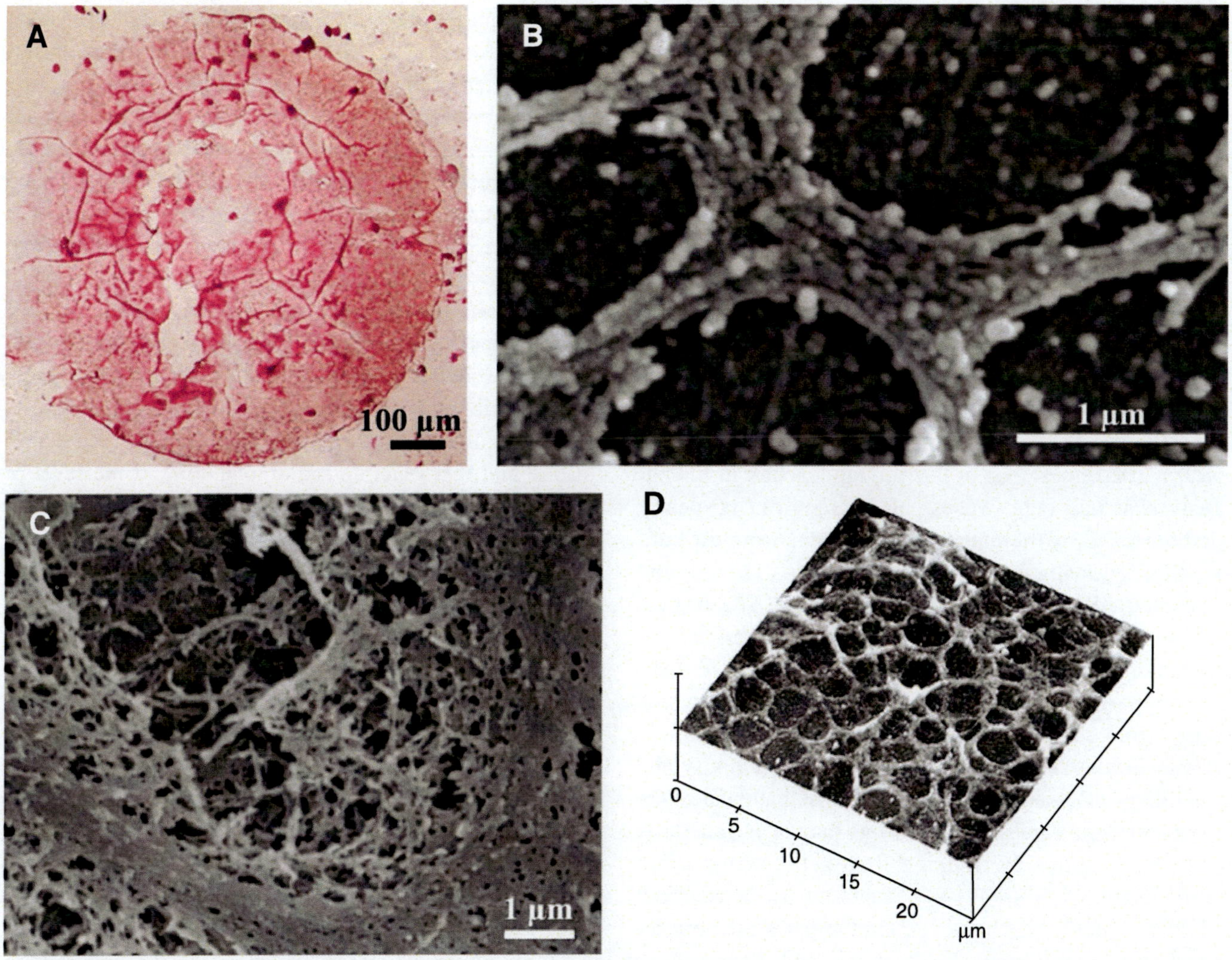

Fig. 7.4 Morphology of sea star footprint material. (A) LM, (B, C) SEM and (D) AFM. (A) Low magnification view of a complete fresh footprint of *Asterias rubens*, stained with a 0.05% aqueous solution of Crystal violet (original). (B) Detailed view of a mesh component in a thin area from a footprint of *A. rubens*. Image from Hennebert et al. (2008) and reproduced with permission. (C) Footprint area presenting different thicknesses of the adhesive material in *Marthasterias glacialis* (original). (D) 3-dimensional view of wet adhesive material of *A. rubens* observed in ambient air. Image from Hennebert et al. (2008) and reproduced with permission

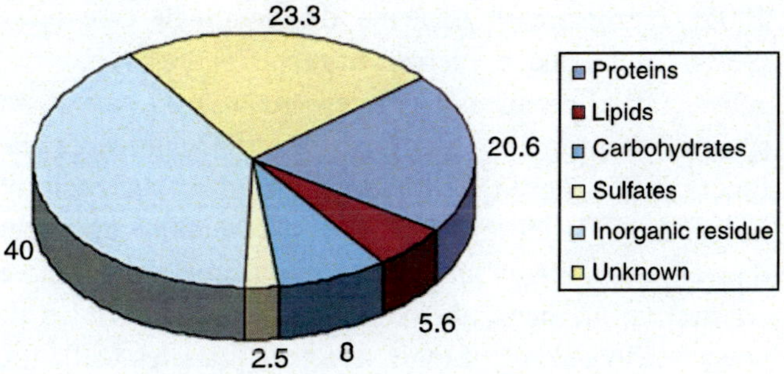

Fig. 7.5 Pie diagram illustrating the biochemical composition of the footprint material of *Asterias rubens*. Data from Flammang et al. (1998) and reproduced with permission of The Journal of Experimental Biology. Values are expressed as percentages of dry mass. The unknown fraction reflects the incomplete recovery when all fractions are added and can be explained by losses or incomplete hydrolysis in the chemical analyses

and Philpott, 1964; Chaet, 1965; Souza Santos and Silva Sasso, 1968; Engster and Brown, 1972). Later, Perpeet and Jangoux (1973) also demonstrated the presence of proteins associated with these mucopolysaccharides in *A. rubens*. This was in accordance with the positive staining obtained for the footprints of *Asterias forbesi* with ninhydrin (Chaet, 1965).

At present, data on the biochemical composition of sea star footprints are available only for *A. rubens*. The water content of the adhesive material has never been measured but, in terms of dry weight, the footprints are made up mainly of proteins (20.6%), carbohydrates (8%), and a large inorganic fraction (about 40%) (Fig. 7.5). This composition is in accordance with the results obtained from histochemical tests (Flammang et al., 1998). Lipids were also detected in footprints, although they have not been detected in the secretory granules of adhesive cells in histochemistry (Perpeet and Jangoux, 1973; Flammang et al., 1998). This lipid fraction might come from the membranes of the adhesive granules or could be a contaminant in the footprint material (Flammang et al., 1998).

The protein moiety contains slightly more polar (55%) than non-polar (45%) amino acids, and among the former, more charged (34%, of which 22% are acidic) than uncharged residues (21%). Moreover, it contains higher levels of glycine, proline, isoleucine, and cysteine than the average eukaryotic proteins (Flammang et al., 1998). The importance of this protein fraction in the adhesiveness and cohesiveness of the footprint is clearly demonstrated by the removal of footprints from the substratum following a treatment with trypsin (Thomas and Hermans, 1985; Flammang, unpubl. obs.). The protein fraction of the footprints was successfully solubilized using a non-hydrolytic method and, after SDS-PAGE analysis, appears to consist of about 25 protein bands which range from 25 to 450 kDa in apparent molecular weight (Hennebert et al., 2010b). Using mass spectrometry and homology-database search, it was shown that most of the minor protein bands correspond to known intracellular proteins, presumably resulting from cell contamination of the samples, while the major protein bands were not identified and might be novel. On the basis of these results, these major bands likely corresponded to the tube foot adhesive proteins and were named the sea star footprint proteins (Sfps) (Hennebert et al., 2010b).

On the basis of colorimetric assays, Flammang et al. (1998) showed that the footprint carbohydrate fraction is made up of neutral sugars (3% of the footprint dry weight), amino sugars (1.5%) and uronic acids (3.5%). The composition of this moiety was further investigated using lectins, molecules which specifically recognize carbohydrate residues. These lectins were used on tube foot histological sections, on footprints, and on extracted and separated adhesive proteins. The results indicate that a fraction of the carbohydrate content of the footprints is linked to two of the Sfps, their outer chains enclosing mannose, galactose, N-acetylgalactosamine, fucose and sialic acid residues. Another fraction or the carbohydrate moiety may be in the form of larger molecules, such as sialylated proteoglycans (Hennebert et al., 2010b).

In addition to glycosylation, sulfation appears to be a modification of the adhesive material. Indeed, sulfate groups have been detected in the footprints of *A. rubens* and account for about 2.5% of their dry weight (Fig. 7.5; Flammang et al., 1998). Currently, however, it is not known whether these groups are attached directly to the proteins or to the carbohydrate residues. Among the other modified amino acids common in marine adhesive proteins (Sagert et al., 2006), DOPA has not been detected in the footprint material while phosphoserine appears to be present in one of the glycosylated Sfps (Hennebert, unpubl. obs.).

The variability of the adhesive secretions from 14 sea star species representing 5 orders and families has been investigated by immunohistochemistry, using polyclonal antibodies raised against the footprint material of *A. rubens*. In every species investigated, there was a very strong and reproducible immunolabeling at the level of the disk adhesive area owing to the labeling of numerous granule-containing adhesive cells, the cuticle being the only other immunoreactive structure. This immunoreactivity seems to be independent of the taxon considered, of the tube foot morphotype or function, of the species habitat, and of the adhesive cell secretory granule ultrastructure. It indicates that sea star adhesives are closely related, probably sharing many identical molecules or, at least, many identical epitopes on their constituents (Santos et al., 2005a; Flammang, 2006). However, differences in the composition of adhesive secretion may exist that are not detected by the antibodies used, and that could account for the differences observed in the structure and function of asteroid tube feet (Santos et al., 2005a).

7.6 A Model for Temporary Tube Foot Adhesion

Based on TEM observations, Flammang et al. (1994, 2005) demonstrated that the different cell categories present in the disk epidermis of tube feet are involved in adhesion and de-adhesion, and function as a duo-gland system, as proposed by Hermans (1983). Indeed, the function of adhesive cells was deduced on the basis of observations made on reinforced disk-ending tube feet that were amputated and fixed while firmly attached to a substratum. In these tube feet, the adhesive cells have always released some of their secretory granules while the de-adhesive cells have not. On the other hand, in tube feet that were cut off and fixed just after they voluntarily became detached from a substratum, de-adhesive cells appear to have released their most apical secretory granules (Flammang, 1996; Flammang et al., 2005). Moreover, antibodies raised against the footprint material of

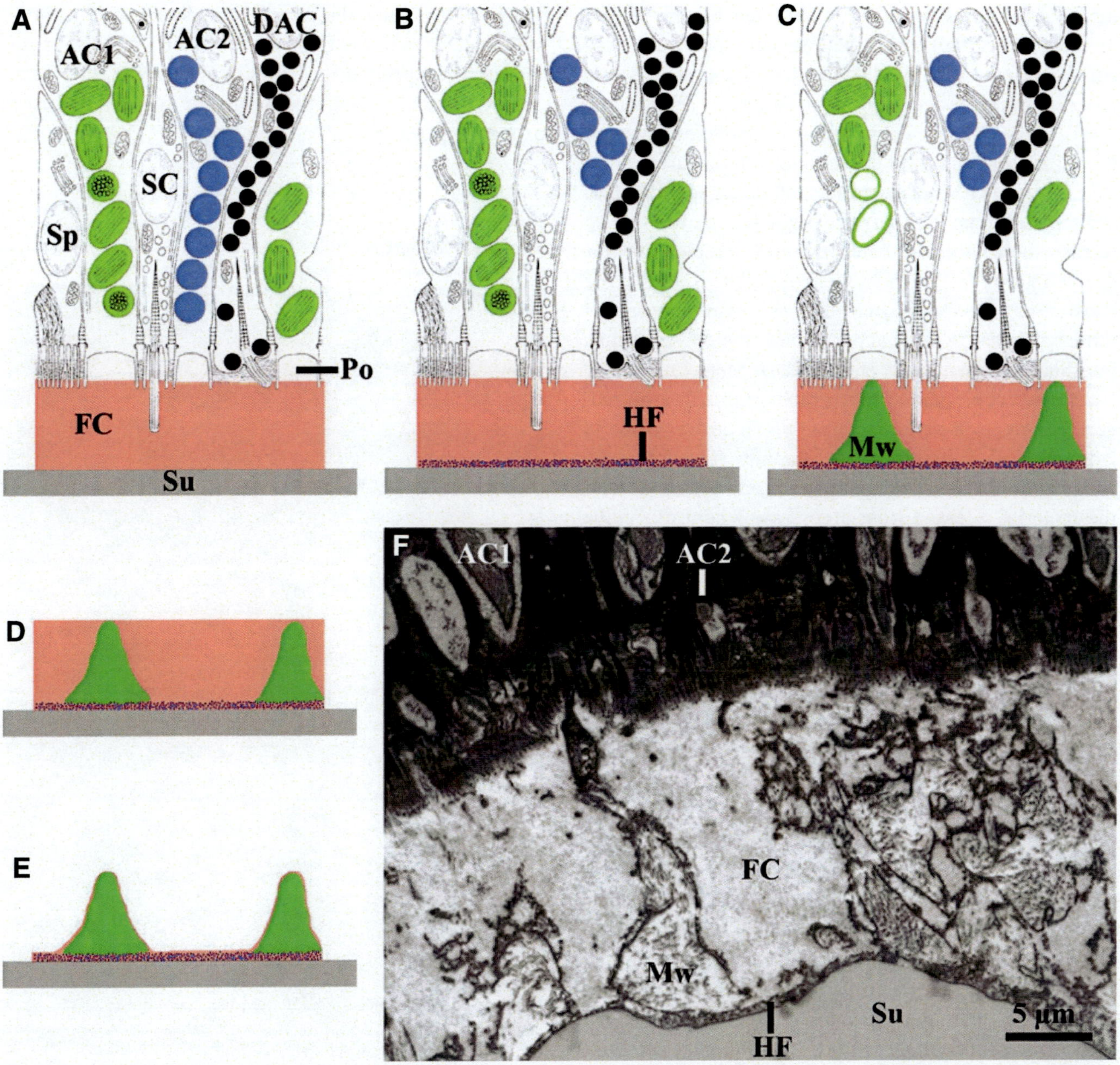

Fig. 7.6 Model for the formation of the adhesive material microstructure in the footprints of *Asterias rubens*. Diagrammatic representations (not to scale) of longitudinal sections through the disk epidermis of a tube foot illustrating the proposed successive steps in the secretion (A–C), deposition (D) and drying (E) of a footprint (see text for details). Image from Hennebert et al. (2008) and reproduced with permission. (F) General view in TEM of a longitudinal section through the disk epidermis of a tube foot attached to a substratum by adhesive material (original), corresponding to step C

A. rubens strongly labelled the adhesive cells and the fuzzy coat, but did not label the de-adhesive cells. These results demonstrate that the footprint material is mainly made up of the secretions of the adhesive cells and also contains elements from the cuticle, while de-adhesive secretions are not incorporated into footprints (Flammang et al., 1998).

A model for the formation of the adhesive material in *A. rubens* may be proposed from all the morphological observations published (Fig. 7.6). When a tube foot attaches to a substratum, it first contacts the surface with its thick fuzzy coat and the compliant disk adhesive pad adapts its distal surface to the substratum profile (Fig. 7.6A, Santos et al., 2005b). Contact with the substratum is presumably detected by the cilium of the sensory cells, and their stimulation could trigger the release of the adhesive material from the two types of adhesive cells *via* the nerve plexus (Flammang et al., 1994). These secretions are delivered through the fuzzy coat and bind the tube foot to the substratum (Fig. 7.6F). TEM observations of attached tube feet suggest that type 2 adhesive cells would be the first to release their contents, being therefore responsible for the formation of the homogeneous film covering the substratum (Fig. 7.6B). Indeed, the material released by these cells has the same appearance as the one constituting this film (Hennebert et al., 2008). Meanwhile, type 1 adhesive cells start to release their content, a heterogeneous electron-dense material which derives directly from the rods constituting the granules (Fig. 7.6C, F). This material seems to expand gradually and fuses with the material released from other type 1 adhesive cells, initiating the formation of a meshwork (Fig. 7.6F). Type 1 cells would be therefore at the origin of the meshwork pattern and the arrangement of their secretory pores on the disk surface could act as a template for the formation of this pattern (Hennebert et al., 2008).

At present, it is not known which sea star footprint proteins constitute the homogeneous layer and which make up the meshwork. However, Sfps may allow adhesion through polar and electrostatic interactions between their amino acid side chains and the substratum surface, while the cohesiveness of the adhesive could be reinforced through intermolecular disulfide bonds (Flammang et al., 1998; Flammang, 2006). In addition, the high proportion of small side-chain amino acids, characteristic of elastomeric proteins which can withstand significant deformation without rupture, may provide compliance to the adhesive (Flammang, 2006; Santos et al., 2009). The carbohydrate fraction of footprints presumably functions in the same way as the protein fraction in adhesion, the functional groups of glycan chains, such as hydroxyls, carboxylates and amines, forming polar and electrostatic interactions with the substratum surface.

The detachment of the tube foot is initiated by the release of the secretory granules by the de-adhesive cells, possibly through a stimulation of their subcuticular cilium (Flammang, 1996). The nature of this de-adhesive material is not known yet but ultrastructural and immunocytochemical data suggest that it might function enzymatically to jettison the fuzzy coat, thereby allowing the tube foot to detach (Flammang et al., 1994, 1998). Once the tube foot is detached, the adhesive material and the fuzzy coat are left on the substratum as a footprint (Fig. 7.6D). Upon drying, the fuzzy coat collapses giving the footprint its characteristic aspect (Fig. 7.6E).

Abbreviations

AC1	type 1 adhesive cell
AC2	type 2 adhesive cell
AE	adhesive epidermis
C	cilium
CL	connective tissue radial lamellae
CT	connective tissue layer
Cu	cuticle
D	disk
DAC	de-adhesive cell
FC	fuzzy coat
HF	homogeneous film
K	knob
L	water-vascular lumen
M	myomesothelium
Mw	meshwork
NE	non-adhesive epidermis
NP	nerve plexus
Po	pore
S	stem
SC	sensory cell
Sp	support cell
Su	substratum

Acknowledgments

The author would like to thank Dr. J. von Byern for providing the opportunity to write this review and Dr. P. Flammang for reviewing the manuscript. This work was supported by a FRIA doctoral grant to E.H., by a FRFC Grant No. 2.4532.07, and by the "Communauté française de Belgique – Actions de Recherche Concertées". This study is a contribution of the "Centre Interuniversitaire de Biologie Marine" (CIBIM, Belgium).

References

Ameye L, Hermann R, Dubois P, and Flammang P (2000) Ultrastructure of the echinoderm cuticle after fast-freezing/freeze substitution and conventional chemical fixations. Microscopy Research and Technique 48(6): 385–393.

Chaet AB (1965) Invertebrate adhering surfaces: secretions of the starfish, *Asterias forbesi*, and the coelenterate, *Hydra pirardi*. Annals of the New York Academy of Sciences 118(24): 921–929.

Chaet AB and Philpott DE (1964) A new subcellular particle secreted by the starfish. Journal of Ultrastructure Research 11: 354–362.

Clark AM and Downey ME (1992) Starfishes of the Atlantic, 1st edn. Natural History Museum Publications: Chapman and Hall, London.

Defretin R (1952) Etude histochimique des mucocytes des pieds ambulacraires de quelques échinodermes. Recueil des travaux de la Station Marine d'Endoume 6: 31–33.

Engster MS and Brown SC (1972) Histology and ultrastructure of the tube foot epithelium in the phanerozonian starfish, *Astropecten*. Tissue and Cell 4(3): 503–518.

Flammang P (1995) Fine structure of the podia in three species of paxillosid asteroids of the genus *Luidia* (Echinodermata). Belgian Journal of Zoology 125(1): 125–134.

Flammang P (1996) Adhesion in echinoderms. In: Jangoux M and Lawrence JM (eds) Echinoderm Studies. A.A. Balkema, Rotterdam: pp 1–60.

Flammang P (2006) Adhesive secretions in echinoderms: An overview. In: Smith AM and Callow JA (eds) Biological Adhesives. Springer-Verlag, Heidelberg: pp 183–206.

Flammang P, Demeulenaere S, and Jangoux M (1994) The role of podial secretions in adhesion in two species of sea stars (Echinodermata). Biological Bulletin 187: 35–47.

Flammang P, Michel A, Cauwenberge AV, Alexandre H, and Jangoux M (1998) A study of the temporary adhesion of the podia in the sea star *Asterias rubens* (Echinodermata, asteroidea) through their footprints. Journal of Experimental Biology 201(Pt 16): 2383–2395.

Flammang P, Santos R, and Haesaerts D (2005) Echinoderm adhesive secretions: From experimental characterization to biotechnological applications. In: Matranga V (ed) Progress in Molecular and Subcellular Biology Subseries Marine Molecular Biology. Springer-Verlag, Berlin.

Hennebert E, Viville P, Lazzaroni R, and Flammang P (2008) Micro- and nanostructure of the adhesive material secreted by the tube feet of the sea star *Asterias rubens*. Journal of Structural Biology 164(1): 108–118.

Hennebert E, Haesaerts D, Duois X, and Flammang P (2010a) Evaluation of the different forces brought into play during tube foot activities in sea stars. Journal of Experimental Biology 213: 1162–1174.

Hennebert E, Wattiez R, and Flammang P (2010b) Characterization of the carbohydrate fraction of the temporary adhesive material secreted by the tube feet of the sea star *Asterias rubens*. Marine Biotechnology, accepted for publication.

Hermans CO (1983) The duo-gland adhesive system. Oceanography and Marine Biology: An Annual Review 21: 283–339.

Hyman LH (1955) The Invertebrates: Echinodermata – The Coelomate Bilateria, vol. IV. McGraw-Hill Book Company, New York.

Lawrence J (1987) A Functional Biology of Echinoderms. Croom Helm, London.

McKenzie JD (1988) Echinoderm surface coats: Their ultrastructure, function and significance. In: Burke RD, Mladenov PV, Lambert P, and Parsley RL (eds) Echinoderm Biology. Balkema, Rotterdam: pp 697–706.

Nichols D (1966) Functional morphology of the water-vascular system. In: Boolootian RA (ed) Physiology of Echinodermata. Interscience Publishers, New York: pp 219–244.

Paine VL (1926) Adhesion of the tube feet in starfishes. Journal of Experimental Zoology 45: 361–366.

Perpeet C and Jangoux M (1973) Contribution á l'etudedes pieds et des ampoules ambulacraires d'*Asterias rubens* (Echinodermata, Asteroides). Forma et functio 6: 191–209.

Sagert J, Sun C, and Waite JH (2006) Chemical subtleties of Mussel and Polychaete holdfasts. In: Smith AM and Callow JA (eds) Biological Adhesives. Springer-Verlag, Heidelberg: pp 125–143.

Santos R, Haesaerts D, Jangoux M, and Flammang P (2005a) Comparative histological and immunohistochemical study of sea star tube feet (Echinodermata, Asteroidea). Journal of Morphology 263(3): 259–269.

Santos R, Gorb SN, Jamar V, and Flammang P (2005b) Adhesion of echinoderm tube feet to rough surfaces. Journal of Experimental Biology 208: 2555–2567.

Santos R, Hennebert E, Coelho AV, and Flammang P (2009) The echinoderm tube foot and its role in temporary underwater adhesion. In: Gorb SN (ed) Functional surfaces in Biology. Adhesion related Phenomena, Vol. 2. Springer-Verlag, Heidelberg: pp 9–41.

Smith JE (1937) The structure and function of the tube feet in certain echinoderms. Journal of the Marine Biological Association of the United Kingdom 22: 345–357.

Souza Santos H and Silva Sasso W (1968) Morphological and histochemical studies on the secretory glands of starfish tube feet. Acta Anatomica 69: 41–51.

Thomas LA and Hermans CO (1985) Adhesive interactions between the tube feet of a starfish, *Leptasterias hexactis*, and substrata. Biological Bulletin 169: 675–688.

Vickery MS and McClintock JB (2000) Comparative morphology of tube feet among the Asteroidea: phylogenetic implications. American Zoologist 40: 355–364.

8 Adhesive Exocrine Glands in Insects: Morphology, Ultrastructure, and Adhesive Secretion

Oliver Betz

Contents

Abstract

A literature survey is provided summarizing the available information on exocrine epidermal glands that produce adhesive secretions in insects. The focus is on both the ultrastructure of the gland cells and the identity and function of the chemical secretion produced by them. Insects employ adhesives for various functions such as tarsal attachment during locomotion, resisting external detachment forces, mating, phoresy and parasitism, egg anchorage, retreat building, self-grooming, prey capture, and active and passive defence. The available studies on the ultrastructure and the secretion of adhesive insect glands cover a broad spectrum of developmental stages and higher taxa, i.e., the Elliplura, the Ephemeroptera, the Polyneoptera, the Acercaria, the Coleoptera, the Amphiesmenoptera, the Hymenoptera, and the Diptera (Table 8.1). Based on this diversity of biological contexts and systematic groups, adhesive structures are found at various tagmata of the body, mainly at the head, the abdomen, and the legs, but also within the thorax in the form of the metapleural glands of ants. Class 1 epidermal cells are the predominant glandular cell type among the adhesive gland systems in insects. With respect to their ultrastructure, the adhesive class 1 cells show features (in terms of their provision with endoplasmic reticulum, Golgi system, free ribosomes, and secretion vesicles and granules) that are either indicative of predominant non-proteinaceous (lipid) or protein secretion. In class 1 cells that are employed in locomotion (i.e., reversible tarsal adhesion to natural substrates such as plant surfaces), lipoidal secretion seems to prevail (although these secretions often appear to be complex mixtures of lipids with proteins and carbohydrates), whereas in the contexts of more permanent body or egg anchorage and of retreat building, protein-based secretion dominates. Oenocyte-

Table 8.1: Survey of studies that deal with the histology / ultrastructure of insect epidermal adhesive glands and / or the chemical nature of the adhesive secretion produced by them (*locomotion* tarsal adhesion to various surfaces, *dashes* mark no data available). Nomenclature of epidermal gland cells is according to Noirot and Quennedey (1974, 1991) and Quennedey (1998). The extensive literature on special wax glands of insects (e.g., scale insects) is not considered, although the extensive waxy secretions of certain insects might entangle the sensilla and mouthparts of potential predators and parasitoids. In the column next to the last ("Details on secretion") larger substance classes (as defined in Sect 8.4) are separated by semicolons

Systematic unit	Observed taxon	Developmental stage	Biological context	Involved body structure	Gland type/exit path	Details of secretion (possible adhesive components only)	References
Protura	Not specified	Juvenile (?), adult	Defence	Abdomen	–/–	Acid; proteinaceous; soluble in alcali	Denis, 1949; Dettner, 2010
Collembola							
Onychiuridae	*Onychiurus armatus, O. latus*	Adult	Defence	Secretory glands (pseudocella) on body surface	Several class 1 cells/canalicules in epicuticular lid of pseudocellus	–	Usher and Balogun, 1966; Rušek and Weyda, 1981
Ephemeroptera							
Leptophlebiidae, Siphlonuridae	*Habrophlebia eldae, Siphlonurus lacustris*	Adult	Egg anchorage in water	Follicle and oviduct cells of female ovaries	Class 1-like follicle and oviduct cells tightly interlocked with egg/*via* apical microvillus border onto chorion	Mucous and fibrous adhesive egg coating	Gaino and Mazzini, 1989, 1990; Gaino and Rebora, 2001
Polymitarcidae	*Asthenopus* sp.	Larva	Retreats in submersed wood	Malpighian tubules (secretion coats and glues together chips of wood within feeding tunnels)	–/*via* Malpighian tubules that open into hind-gut	Protein	Sattler, 1967
Plecoptera							
Perlidae	*Dinocras cephalotes, Perla marginata*	Adult	Egg anchorage in water	Follicle cells within ovaries	Class 1-like follicle cells tightly interlocked with egg/ apical microvillus border onto surface of chorion	Polysaccharides; proteins	Rościszewska, 1991, 1995
Blattodea	Genera *Blatta, Blatella, Diploptera, Ectobius, Eurycotis, Leucophaea, Loboptera, Megaloblatta, Nauphoeta, Nyctibora, Periplaneta, Pseudoderopeltis, Supella*	Larva, adult	Defence	Hypodermal cells of caudal abdominal tergites and cerci	Glandular epithelium made of class 3 units/ cuticular ductule opening toward exterior *via* orifice	Carbohydrate; 90% (glyco-) protein (free amino acids, oligo- and polypeptides of 200–12 kDa); water	Roth and Stahl, 1956; Nayler, 1964; Plattner, 1971; Eisner, 1972; Plattner et al., 1972; Quennedey, 1975b; Brossut and Roth, 1977; Brossut and Sreng, 1980; Ichinosé and Zennyoji, 1980; Blum, 1981; Ichinosé and Kishimoto, 1983; Abed et al., 1995

Isoptera							
Mastotermitidae	*Mastotermes darwiniensis*	Adult soldier	Defence	Paired labial acinous salivary glands made of several lobes	Class 1 (with reservoir)/ extracellular ducts that lead into esophagus	Protein-quinone (benzo- and toluquinone) mixture	Moore, 1968; Eisner, 1970; Blum, 1981; Baker and Wasmsley, 1982; Quennedey, 1984 (and citations therein); Czolij and Slaytor, 1988
Rhinotermitidae	Several subfamilies, e.g., Coptotermitinae	Adult soldier; less developed in other casts	Defence	Frontal head gland	Single-layered glandular epithelium of class 1 cells; in *Coptotermes* mixed with bicellular class 3 cells/common reservoir reaching deeply into abdomen, where secretion of both cell types is mixed, opens via frontal head pore; glandular reservoir enveloped by musculature that helps to discharge secretion	Mixture of lipids (alkanes (C_{22}–C_{27}), (nitro-) alkenes, alcohols, aliphatic and vinyl alkyl ketones, carboxylic acids, fatty acids (e.g., hexacosanoic and lignoceric acid), esters, sphingolipids (ceramides), (keto-) aldehyde compounds; aqueous mucopolysaccharides (based on glucosamines and glucose units); (hydroxy-) benzenes; terpenoids such as mono-, sesqui- (e.g., cadinene, cadinene aldehyde), and diterpenes; (glyco-) proteins	Moore, 1968, 1969; Noirot, 1969; Sannasi, 1969; Eisner, 1970; Quennedey et al., 1973; Vrkoč and Ubik, 1974; Howse, 1975; Prestwich, 1975, 1984 (and citations therein); Prestwich et al., 1975; Quennedey, 1975a, 1984 (and citations therein); Blum, 1981; Deligne et al., 1981; Zalkow et al., 1981; Baker and Wasmsley, 1982; Blum et al., 1982; Grassé, 1982; Prestwich and Collins, 1982; Pasteels et al., 1983; Bagneres et al., 1990; Chuah et al., 1990; Chen et al., 1999; Nelson et al., 2001, 2008; Šobotnik and Hubert, 2003; Šobotnik et al., 2004, 2010a, b; Zhang et al., 2006; Ohta et al., 2007; Piskorski et al., 2007;
Serritermitidae	*Glossotermes oculatus, Serritermes serrifer*	Adult soldier	Defence	Frontal head gland lacking a frontal pore and penetrating deeply into the abdomen	Single-layered glandular epithelium of class 1 cells/self-rupture (autothysis) of frontal gland	–	Costa-Leonardo, 1998b
Termitidae	Macrotermitinae (e.g., *Hypotermes, Macrotermes, Microtermes, Odontotermes*) and Termitinae: *Globitermes*	Adult soldier	Defence	Paired labial acinous salivary glands made of several lobes composed of glandular vacuole, foam, and parietal cells; *Globitermes*: self sacrifice (autothysis)	Class 1/collector canal (supplied by salivary canals of lobes) and reservoir into preoral cavity	(Muco-) polysaccharides (based on β-glucopyranose and (N-acetyl-) β-glucosamine); (benzo- and tolu-) quinones, in some groups in aqueous mixture with proteins (rich in Cys); sesquiterpenes; glycoproteins	Noirot, 1969; Eisner, 1970; Maschwitz et al., 1972; Quennedey, 1973, 1975a, 1984 (and citations therein); Maschwitz and Tho, 1974; Howse, 1975; Wood et al., 1975; Blum, 1981; Baker and Wasmsley, 1982; Pasteels et al., 1983; Prestwich, 1983, 1984; Billen et al., 1989; Plasman et al., 1999; Šobotnik et al., 2010b

(Continued)

Table 8.1: (Continued)

Systematic unit	Observed taxon	Developmental stage	Biological context	Involved body structure	Gland type/exit path	Details of secretion (possible adhesive components only)	References
	Several genera (e.g., *Acanthotermes*, *Macrotermes*) of Macrotermitinae and Termitinae	Adult soldier; less developed in other casts	Defence	Frontal head gland	Single-layered glandular epithelium of class 1 cells; in some Termitinae mixed with bicellular class 3 cells/common reservoir opens *via* frontal head pore	Saturated (straight-chain and methyl-branched) and unsaturated (straight-chain) hydrocarbons (e.g., C_{22}–C_{34} normal and iso-alkanes) mixed with fatty acids, phospholipids and sterols (Macrotermitinae), methylketones, alcohols; terpenoids: mono-, sesqui- and diterpenes (Termitinae)	Cmelik, 1971; Quennedey, 1973, 1975a, 1984 (and citations therein); Howse, 1975; Prestwich, 1975, 1979a, 1984; Evans et al., 1977; Prestwich et al., 1977; Meinwald et al., 1978; Baker and Wasmsley, 1982; Grassé, 1982; Šobotnik et al., 2010a, b
	Nasutitermitinae	Adult soldier; less developed in other casts	Defence	Frontal head gland	Single-layered glandular epithelium of class 1 cells; in many Nasutitermitinae mixed with bicellular class 3 cells/common reservoir, where secretion of both cell types is mixed; opens *via* frontal head pore; in several taxa glandular reservoir enveloped by intrinsic or extrinsic musculature that helps to discharge secretion	Ketones, esters, alcohols, epoxides; hydrogen-bonded bicyclic terpenes and (tricyclic) diterpenes soluted in monoterpene solvent, monoterpenes (dissolved in higher isoprenoids), monoterpene acetates, sesquiterpenes, steroids; monoterpenes having fortifying knockdown effect	Moore, 1964, 1968, 1969; Nutting et al., 1974; Howse, 1975; Prestwich 1975, 1977, 1979a, b, 1983, 1984 (and citations therein); Quennedey, 1973, 1975a, 1984; Meinwald et al., 1978; Prestwich et al., 1979; Braekman et al., 1980, 1984 (and citations therein); Blum, 1981; Deligne et al., 1981; Baker and Wasmsley, 1982; Blum et al., 1982; Grassé, 1982; Pasteels et al., 1983; Costa-Leonardo and De Salvo, 1987; Goh et al., 1990; Costa-Leonardo, 1992, 1998a, 2001; Dettner, 2010; Šobotnik et al., 2010b
Mantophasmatodea							
Austrophasmatidae	*Karoophasma biedouwense*	Adult	Locomotion	Tarsal arolium	Glandular epithelium formed by class 1 cells/glandular reservoir and possibly pore canals in soft cuticle of arolium	–	Eberhard et al., 2009

Ensifera							
Tettigoniidae	*Tettigonia viridissima*	Adult	Locomotion, predation	Tarsal euplantulae	Glandular epithelium probably formed by class 1 cells/possible subcuticular reservoir and pore canals in soft cuticle of euplantula	–	Dewitz, 1884; Henning, 1974
Caelifera							
Acrididae	*Locusta migratoria, Schistocerca gregaria*	Adult	Locomotion	Tarsal euplantulae and arolium	*Schistocerca* arolium: glandular epithelium formed by class 1 cells/ possibly pore canals in soft cuticle of arolium	*Locusta* euplantulae: aliphatic hydrocarbons, alcohols, free C_{16}–C_{20} fatty acids (saturated and unsaturated), glycerides, glycerol; carbohydrates (e.g., glucose); cholesterol; amino acids (peptides?/ proteins?)	Kendall, 1970; Vötsch et al., 2002
Auchenorrhyncha							
Cercopidae	Not identified	Larva	Defence	Anus fluid mixed with secretion from abdominal glands	–/–	Mucopolysaccharides; proteins containing amino acids Gly, Pro, and Ser	Jacobs and Renner, 1988; Li et al., 2008
Fulgoromorpha	2 unidentified spp.	Adult	Fibrous binder in egg mass	–	–/–	Proteinaceous (300-280, 95-15 kDa), containing amino acids (Gly, Thr, Ala, Arg, Val, Pro), glycoproteins	Li et al., 2008
Sternorrhyncha							
Aphidoidea	Most representatives of Aphidoidea, e.g., *Acyrthosiphon pisum, Myzus persicae, Rhopalosiphum padi*	Larva, adult	Defence	Abdominal cornicles	Ensemble of secretory cells similar to class 1 and class 2 within sac-like basal membrane/ holocrine secretion via cuticular valves at apex of cornicle	Triglycerides (with high content of myristic acid and acids of short chain length (C6, C8) emulsified in aqueous medium; sugars (trehalose?); aliphatic sesquiterpenes; amino acids	Edwards, 1966; Strong, 1967; Brown, 1975; Blum, 1981; Foldi-Hope, 1990; Strümpel, 2003
	Megoura viciae	Adult	Locomotion	Tarsal pulvilli	Glandular epithelium of class 1 cells/ reservoir-like porous procuticle and fine filaments or pores	Lipoprotein (?)	Lees and Hardie, 1988

(*Continued*)

Table 8.1: (Continued)

Systematic unit	Observed taxon	Developmental stage	Biological context	Involved body structure	Gland type/exit path	Details of secretion (possible adhesive components only)	References
Heteroptera							
Reduviidae	*Brontostoma colossus, Peirates* sp., *Rhodnius prolixus, Triatoma dimidiata, Zelus leucogrammus*	Adult	Locomotion, mating, predation	Hairy attachment structure (fossula spongiosa) at apex of tibia	Glandular epithelium formed by class 1 or class 3 (Barth, 1953) cells that directly connect to tenent hairs/pore canals (?) in hollow hair shaft	–	Gillett, 1932; Miller, 1942; Barth, 1953; Eisner, 1988; Weirauch, 2007
Pentatomidae	*Graphosoma italicum*	Adult	Locomotion	Tarsal pulvilli	Glandular epithelium of class 1 cells/-	–	Ghazi-Bayat and Hasenfuss, 1980
	Poecilometis sp.	Adult	Egg anchorage	–	–/–	Carbohydrates; high protein content, containing Gly, Ser, and 4-hydroxyproline	Li et al., 2008
Tingidae	Several genera, e.g., *Catoplatus, Copium, Corythucha, Dictyla, Elasmotropis, Galeatus, Lasiacantha, Stephanitis, Tingis*	Larva	Camouflage, defence	Outgrowths of body surface (tubercles) bearing hollow tubular processes	Class 2 (oenocytes) and class 3/subcuticular space connected to exterior via tubular body outgrowths	Unbranched aldehydes, ketones, diketones; chromones, chromanones	Livingstone, 1978; Oliver et al., 1985, 1987, 1990; Lusby et al., 1987; Scholze, 1992; Dettner, 2010
Coleoptera							
Carabidae	*Loricera pilicornis*	Larva	Predation	Galea (bulb-shaped)	Glandular epithelium of class 1 (?) cells lining two central reservoirs/spongy cuticle onto surface of galea (*via* fine pore canals?)	–	Bauer and Kredler, 1988; Betz and Kölsch, 2004
Dytiscidae	*Dytiscus marginalis*	Adult ♂	Mating	Associated with suckers of fore and middle tarsi	Class 1 (with reservoir?)/ outside the actual sucker surface *via* pore canals at base of sucker stalk	–	Blunck, 1912
Staphylinidae – Pselaphinae	*Batrisodes* spp., *Pselaphus* spp.	Larva	Predation, cocoon building	Predation: head gland connected to eversible head organ close to antennal insertion; predation & cocoon building: dorsal body surface of thorax and abdomen	Class 1 (with reservoir)/ cuticular ductule opening at exterior (in cases of head gland, ductule opening at base of adhesive exertile head organ)	–	De Marzo, 1985, 1988
	Bryaxis spp.	Adult	Predation	Maxillary palps	Class 3 (?)/–	–	Schomann et al., 2008

Staphylinidae – Oxytelinae	*Deleaster dichrous*	Adult	Defence	Abdominal glands	Many class 3 units forming an ovoid mass/a large reservoir that opens at base of 10th tergite	Monomeric adhesive containing the monoterpene iridodial	Dettner et al., 1985; Dettner, 1993
Staphylinidae – Steninae	*Stenus* spp.	Adult	Predation	Glands located in head; openings at lateral wall of paraglossae of elongated labium	Many class 3 units arranged in two groups within head/long chitinous ductules within labial tube that open out at pore canals at paraglossae	Lipids; carbohydrates; proteins	Schmitz, 1943; Weinreich, 1968; Kölsch and Betz, 1998; Kölsch, 2000; Betz et al., 2009
	Stenus spp.	Adult	Locomotion	Tarsal tenent hairs	Class 1 (with reservoir)/fine pore canals in hollow tenent hair shaft (?); class 3/canal orifices at ventral and dorsal tarsal cuticular surface	Long hydrocarbons and/or triglycerides, free fatty acids, glycerides (?); proteins	Betz, 2003
Staphylinidae – Staphylininae	*Philonthus marginatus*	Adult	Locomotion, predation	Tarsal tenent hairs	Class 1 (with reservoir)/large pore canals in hollow tenent hair shaft	–	Betz and Mumm, 2001
Coccinellidae	*Coccinella septempunctata*, *Epilachna* spp.	Adult	Locomotion	Tarsal tenent hairs	–/–	Hydrocarbons, triglycerides, fatty acids, alcohols, waxes	Ishii, 1987; Kosaki and Yamaoka, 1996
	Harmonia conformis, *Hippodamia variegata*, *Propylea quatuordecimpunctata*	Adult	Egg anchorage	–	–/–	High protein content, containing Gly and Ser	Li et al., 2008
Chrysomelidae	*Hemisphaerota cyanea*	Adult	Defence, locomotion	Tarsal tenent hairs	–/Hollow tenent hair shaft	C_{22}–C_{29} n-alkanes and n-alkenes (e.g., (Z)-9-pentacosene)	Eisner, 1972; Attygalle et al., 2000; Eisner and Aneshansley, 2000
	Leptinotarsa decemlineata	Adult	Locomotion	Tarsal tenent hairs	Class 1 (with reservoir)/fine pore canals in hollow tenent hair shaft (?); class 3/canal orifices at tarsal cuticular surface	C_{26}–C_{37} mono-, di-, and trimethyl-branched alkanes	Geiselhardt et al., 2010
	Gastrophysa viridula	Adult	Locomotion	Tarsal tenent hairs	–/–	n-alkanes, methyl-branched alkanes (e.g., i-methylheptacosane), 9-alkenes, acetates of long-chain alcohols	Geiselhardt et al., 2009
	Acalymma sp.	Adult	Egg anchorage	–	–/–	High protein content	Li et al., 2008
Curculionidae	*Entimus* sp., *Eupholus* sp.	Adult	Locomotion	Tarsal tenent hairs	Class 1/hollow tenent hair shaft	–	Dewitz, 1884

(Continued)

Table 8.1: (Continued)

Systematic unit	Observed taxon	Developmental stage	Biological context	Involved body structure	Gland type/exit path	Details of secretion (possible adhesive components only)	References
Hymenoptera							
Formicidae	*Iridomyrmex* spp., *Tapinoma* spp., and other Dolichoderinae	Adult	Defence	Pygidial (="anal") glands	–/–	Ketones; quinones; monoterpene iridodial	Pavan and Trave, 1958; Cavill and Ford, 1960; Roth and Eisner, 1962; Blum, 1981; Dettner, 2010
	Crematogaster inflata and other Myrmicinae	Adult	Defence	Metathoracic gland	Class 1 pouring into common reservoir/reservoir opening *via* pore or slit	Non-sacchariferous (?); mixture of phenols	Maschwitz, 1974; Blum, 1981; Buschinger and Maschwitz, 1984; Attygalle et al., 1989
	Formicinae: *Camponotus saundersi* and related spp.	Adult	Defence	Mandibular glands/ self sacrifice (= autothysis)	–/–	Non-proteinaceous	Maschwitz and Maschwitz, 1974; Blum, 1981; Hermann and Blum, 1981
Vespidae	Stenogastrinae (several genera)	Adult	Defence	Dufour's gland of abdomen/building of "ant guards"	Glandular epithelium of class 1 cells lining a central reservoir/common reservoir opening into vagina	Long-chain (un)saturated hydrocarbons (mainly C_{21}–C_{33}), alcohols, fatty acid salts, alkoxyethanol emulsifiers; nectar	Hermann and Blum, 1981; Keegans et al., 1993; Sledge et al., 2000
Sphecidae	*Liris niger*	Adult	Egg anchorage to prey	Dufour's gland of abdomen	Glandular epithelium of class 1 cells lining a central reservoir/common reservoir opening inside sting	Straight-chain unbranched hydrocarbons of medium chain length (mainly pentadecane, (Z)-8-heptadecene), esters; polypeptides (200-14 kDa)	Gnatzy et al., 2004
Apidae	*Bombus terrestris*	Adult ♂	Mating plug	Male's accessory gland	–/–	Fatty acids (palmitic, linoleic, oleic, stearic acid); cyclic peptide (cycloprolylproline)	Baer et al., 2000
Lepidoptera							
Hesperiidae	*Calpodes ethlius*	Larva	Retreat (pupal cocoon)	Several parts (e.g., middle region) of paired labial silk gland	Glandular epithelium of class 1 cells lining a central reservoir/silk glands of both sides unite into unpaired chitinous duct that opens at the labial spinneret	Mucosubstances; proteins (sericin?)	Wiley and Lai-Fook, 1974

Pieridae	*Pieris brassicae*	Adult ♀	Egg anchorage	Abdominal cement gland	–/–	Unsaturated lipoids; phenolic material (?); proteins	Beament and Lal, 1957
Bombycoidae (similar situation probably occurring in many other Ditrysia (e.g., Pyraloidea, Yponomeutoidea)	*Bombyx mori*	Larva	Retreat (pupal cocoon)	Several parts (mainly middle region) of paired labial silk gland	Glandular epithelium of class 1 cells lining a central reservoir/silk glands of both sides unite into unpaired chitinous duct that opens at the labial spinneret	Lipids; polysaccharides; up to 15 adhesive sericin-type (glyco-) proteins (containing at least two types of carbohydrate units) of 400-65 kDa (rich in Ser, Gly, Asp) that wrap up the fibroin fiber core	Yamanouchi, 1922; Voigt, 1965a, b; Sprague, 1975; Gamo et al., 1977; Sinohara, 1979; Akai, 1983, 1998; Abbot, 1990; Chapman, 1998; Fedič et al., 2002; Takasu et al., 2002; Akai et al., 2003; Padamwar and Pawar, 2004; Teramoto and Miyazawa, 2005; Sehnal and Sutherland, 2008; Sehnal, 2008 (and citations therein)
		Adult ♀	Egg anchorage	Abdominal accessory gland of reproductive system (colleterial gland)	–/–	Carbohydrates; proteins of 240, 190, and 30-70 kDa, small amounts of free amino acids; water	Amornsak et al., 1992; Jin et al., 2006
Crambidae	*Diatraea saccharalis*	Larva	Retreat (pupal cocoon)	Middle region of paired labial silk gland	–/–	Lipid, sericin protein	Victoriano et al., 2007
Saturniidae	*Opodiphthera* spp.	Adult ♀	Egg anchorage	Abdominal accessory gland of reproductive system (colleterial gland)	–/opening into colleterial gland reservoir	Treacle-like liquid containing: carbohydrates; high protein content (mainly >500, 140, and 85–90 kDa with sequenced peptides) containing Gly, Ser, Pro, and 4-hydroxyproline	Li et al., 2008
Trichoptera							
Brachycentridae	*Brachycentrus echo*	Larva	Underwater retreat (pupal cocoon)	Labial silk gland	–/–	Polyphosphoproteins (rich in phosphorylated Ser)	Stewart and Wang, 2010

(*Continued*)

Table 8.1: (Continued)

Systematic unit	Observed taxon	Developmental stage	Biological context	Involved body structure	Gland type/exit path	Details of secretion (possible adhesive components only)	References
Limnephilidae	*Neophylax concinnus, Pycnopsyche guttifer*	Larva	Underwater retreat (pupal cocoon)	Several parts of paired labial silk gland	Glandular epithelium of class 1 cells lining a central reservoir/silk glands of both sides unite into unpaired chitinous duct that opens at the labial spinneret	Adhesive sericin-like proteins and other glycoproteins	Engster, 1976a, b
Stenopsychidae	*Stenopsyche marmorata*	Larva	Underwater retreat (pupal cocoon)	Middle and posterior region of paired silk gland	–/–	Adhesive 70 kDa protein Smsp-72k being abundant in Cys and charged residues; also containing amino acids with hydroxyl side-chains (Ser, Thr)	Wang et al., 2010
Diptera							
Chironomidae	*Chironomus tentans, C. thummi*	Larva, (pre-) pupa	Retreat (underwater)	Salivary glands	Single-layered glandular epithelium of class 1 cells lining a central reservoir/common reservoir opening via secretory duct toward mouth	Family of (glycosylated) secretory proteins (Sps); SpI family proteins (1000–750 kDa) are especially Cys rich	Kloetzel and Laufer, 1969, 1970; Dignam and Case, 1990; Case and Wieslander, 1992; Case et al., 1994; Wieslander, 1994; Hoffman et al., 1996; Case and Thornton, 1999; Sehnal and Sutherland, 2008
Simuliidae	*Simulium erythrocephalum, S. niditifrons, S. nölleri, S ornatum., S. rostratum*	Larva, pupa	Body anchorage, locomotion, retreat (underwater)	Salivary gland	Single-layered glandular epithelium of class 1 cells lining a central reservoir/2 common reservoirs uniting into common outlet duct that opens toward mouth	Mucopolysaccharides (?); proteins (mostly 70, 40, and 20 kDa), muco- and glycoproteins (?)	Debot, 1932; MacGregor and Mackie, 1967; Prügel and Rühm, 1994; Kiel and Röder, 2002
Drosophilidae	*Drosophila melanogaster*	Adult ♀	Egg anchorage to fruit	Abdominal accessory (colleterial) glands of reproductive system	Single-layered glandular epithelium of class 1 cells lining a central reservoir/duct connecting to the uterus	Carbohydrates and proteins that are linked to form mucoproteins	Wilson, 1960a, b; Riley and Forgash, 1967
	Drosophila spp.	Larva (3rd instar)	Pupal anchorage to substrate	Salivary glands	Single-layered glandular epithelium of class 1 cells lining a central reservoir/2 common reservoirs uniting into	(Poly-) saccharides (containing glucose, galactose, mannose, and aminosugar glucosamine); Thr-rich (muco-) proteins	Fraenkel and Brookes, 1953; Perkowska, 1963; Bodenstein, 1965; Rizki, 1967; Lane and Carter, 1972; von Gaudecker,

					common outlet duct that opens into floor of pharynx	(15 partly glycosylated fractions ranging from 360 to 8 kDa), encoded by sgs and ng gene families, glycoproteins (e.g., sialomucines?)	1972; von Gaudecker and Schmale, 1974; Zhimulev and Kolesnikov, 1975; Beckendorf and Kafatos, 1976; Korge, 1977; Kress, 1982; Ramesh and Kalisch, 1988; Farkaš and Šutakova, 1998; Graham, 2008
	Drosophila melanogaster	Adult ♂	Mating plug	Seminal fluid (coagulation)	–/–	PEB-me protein (38 kDa) rich in Gly and Pro (central region containing tandem Pro-Ser-Pro-Gly-(Gly/Glu) repeats)	Baer et al., 2000; Lung and Wolfner, 2001; Graham, 2008
Syrphidae	Not specified	Larva	Defence	Salivary gland	–/–	Proteins	Blum, 1981
Calliphoridae	*Calliphora erythrocephala, C. vomitoria*	Adult	Locomotion	Tarsal pulvilli	Glandular epithelium of class 1 cells/reservoir-like spongy endocuticle and pore canal system	Non-polar lipids: hydrocarbons, wax esters, free fatty acids, triglycerides; sterol esters	Bauchhenß and Renner, 1977; Bauchhenß, 1979; Walker et al., 1985
	Lucilia cuprina	Adult	Egg anchorage	–	–/–	Mainly proteinaceous, containing Gly, Leu, and Ala	Li et al., 2008

like class 2 adhesive gland cells have hitherto only been found in the defence systems of Aphidoidea and Tingidae (both Hemiptera). Adhesive class 3 glands are almost always bicellular, consisting of a terminal secretorily active cell and an adjacent canal cell that surrounds the cuticular conducting duct.

The constituents found in insect adhesives belong to aliphatic compounds, to carbohydrates, to phenols, to isoprenoids, to heterocyclic compounds, and to amino acids, peptides, and proteins. Insect adhesives do not consist of one compound only but are highly complex (often emulsion-like) structural and chemical mixtures whose chemical and micromechanical functions are often poorly understood. The possible functional aspects of such mixtures include *(1) the polar and nonpolar interactions of aliphatic compounds* with the substratum, *(2) the in situ differentiation of alkanes and alkenes* at ambient temperatures forming colloid suspensions of solid wax crystals within a liquid matrix, *(3) the non-Newtonian rheological behavior* of colloid- and emulsion-like adhesive fluids, *(4) the lipoid shields* that prevent the aqueous fraction of an adhesive from desiccation and sticking to the walls of the outlet ductule, *(5) those hydrocarbon properties* (e.g., chain length and degree of unsaturation) that are decisive for the adhesive performance, *(6) the rapid hardening of triglyceride-based adhesives* caused by processes other than polymerization, *(7) the water attractance* of the large carbohydrate component of glycoproteins, *(8) the quinone tanning* induced by protein-polymerizing quinones, *(9) the increased wetting properties* toward lipophilic surfaces caused by monoterpenes combined with dissolved diterpenes that retard the rapid vaporization of the monoterpenes, *(10) the production of aqueous lipid-glycoprotein-mucopolysaccharide mixtures*, *(11) those glues based on hydrophilic proteins* (e.g., sericin with high Ser levels) coupling adhesion with high levels of extension and showing extensive hydrogen bonding, ester linkage, and/or ionic linkage, and *(12) the proteinaceous underwater glues* with their high levels of Cys (forming disulfide bonds) and charged amino acids.

Those adhesives that work mechanically might comprise high-molecular compounds containing proteins, terpenes (resins), mixtures of long-chain hydrocarbons and mucopolysaccharides, or waxes. However, defensive adhesive secretions in particular not only function mechanically, but also concomitantly develop a chemical irritant function caused by reactive substances of a low-molecular weight that are mixed within the sticky secretion to produce "toxic glue".

8.1 Introduction

As in other groups of animals, adhesion is a phenomenon that is widely used in the daily life of an insect (Fig. 8.1) being employed not only for (1) reversibly attaching to surfaces during locomotion (e.g., Stork, 1980; Lees and Hardie, 1988; Creton and Gorb, 2007), but also for (2) resisting external detachment forces caused by wind gusts (e.g., Stork, 1980) or attacking predators (e.g., Eisner and Aneshansley, 2000). Moreover, insects need adhesive structures (3) to restrain mating partners during copulation (e.g., Rothschild and Hinton, 1968; Stork, 1981; Voigt et al., 2008), (4) to hold onto the substratum in contexts such as phoresy or parasitism (Gorb, 2001), (5) to anchor their eggs to each other and to specific oviposition sites (e.g., Weber, 1930; Gaino and Rebora, 2001; Gillot, 2002; Li et al., 2008; Voigt and Gorb, 2010), (6) to build retreats (mostly from silk; e.g., Engster, 1976a, b; Akai et al., 2003), and (7) to self-groom (Schwalb, 1961). An additional largely neglected functional aspect of adhesive mechanisms involves (8) prey capture, i.e., several insects hunt with a sticky secretion that is applied *via* special prey capture devices formed by certain body structures (e.g., mouthparts) and that is used to glue a prey animal at the very moment of contact (cf. Betz and Kölsch, 2004). Conversely, a sticky slime can be used by an insect (9) actively to defend itself or its eggs (cf. the alder leaf beetle *Agelastica alni*) against a small attacker (usually an arthropod predator or parasitoid) by mechanically entangling and immobilizing its mouthparts, sensilla, or spiracles (e.g., Ernst, 1959; Eisner, 1972; Pasteels et al., 1983; Scholze, 1992; Blum and Hilker, 2002; Betz and Kölsch, 2004). Moreover, (10) passive defence is possible *via* adhesive secretion, since dust particles adhering to the sticky covering can act as camouflage (e.g., Crowson, 1981). Finally, (11) adhesive secretion is used by several insects to form mating plugs that facilitate sperm movement or to prevent subsequent matings or sperm loss (e.g., Lung and Wolfner, 2001; Dunham and Rudolf, 2009).

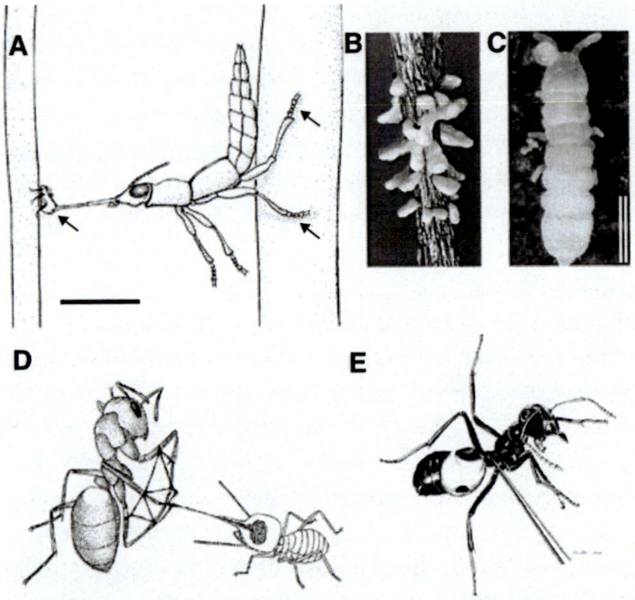

Fig. 8.1 Examples of adhesive prey capture (A) and defence mechanisms (B–E) that involve the exudation of a sticky secretion. (A) *Stenus cicindeloides* staphylinid beetle hunting an aphid in the vegetation with its elongated protrusible labium. The arrows indicate the adhesive interfaces during this action, i.e., the adhesive pads at the tip of the labium (left arrow) and the tarsi (right arrows) that help to keep the beetle attached to the plant surface during prey capture. Bar = 2 mm. Adapted from Betz (1999) and reproduced with permission. (B) Sticky "ant guards" produced by the hover wasp *Parischnogaster jacobsoni*. Adapted from Turillazzi (1991) and reproduced with permission. (C) Springtail of the genus *Onychiurus* releasing a secretion droplet from a pseudocellus. Adapted from Dettner (2010) and reproduced with permission. Bar = 0.5 mm. (D) *Trinervitermes* termite soldier entangling an ant with its sticky defence secretion produced by the frontal head gland and discharged *via* the elongated rostrum. Adapted from Quennedey (1975a) and reproduced with permission. (E) After a worker of *Camponotus saundersi* is seized with a pair of forceps, it contracts its abdominal wall violently, finally bursting open to release a secretion from its hypertrophied mandibular glands. Adapted from Buschinger and Maschwitz (1984) and reproduced with permission

In their review on the role of adhesion in prey capture and predator defence in arthropods, Betz and Kölsch (2004) argue that adhesive prey capture devices, in evolutionary terms, have the advantage that they may not require particularly advanced sensory and neuromuscular mechanisms that assure the exact control of closing movements, as are necessary in clamp-like raptorial legs or mandibles (e.g., Just and Gronenberg, 1999). This might be of special relevance for inert life forms that are physiologically limited with respect to their sensory or motile performance. Using prey capture devices covered with glue, such predators can catch their prey merely by contacting it and sticking to it. Conversely, in the context of predator avoidance, sticky defence agents immobilize the mouthparts or sensilla of an attacker. As in predation, this is advantageous, especially for slow-moving insects (Pasteels et al., 1983). Solely chemically acting defence secretions are well known to select specialists that easily evolve mechanisms of countering the efficiencies of these substances (Dettner, 2007). Since the employment of defensive glue is a physical defence strategy, i.e., adhesion works mostly mechanically and not physiologically as in purely chemically acting defence secretions, counterstrategies might have been much more difficult for predators to evolve against it. Thus, adhesive secretions should be equally effective toward both specialist and generalist predators.

Although many adhesive devices are so-called "dry adhesives" that are chiefly based on van der Waals interaction (e.g., Gorb, 2008), the phenomena reviewed here employ supplementary adhesive fluids produced by glandular systems underlying external adhesive cuticular structures. These structures can be assigned to "hairy" systems differentiated into an array of tenent hairs or, alternatively, "smooth" systems that form pliable pad-like structures with flat surfaces (Beutel and Gorb, 2001; Gorb and Beutel, 2001). Whereas the external morphology and the resulting adhesive performance of both hairy and smooth adhesive devices in insects have been investigated in numerous empirical and theoretical studies (e.g., Gorb, 2001; Scherge and Gorb, 2001; Persson, 2007), relatively little information is available on the ultrastructure of the glue-producing cells and the chemical nature of the adhesive secretion produced by them.

The goal of the present contribution is to provide a literature survey of the structure and function of glandular systems in insects involved in the production of adhesive fluids. This includes both the production of the adhesive fluid and its transport and final discharge to the exterior. Functionally connected to this subject is the chemical nature and action of the adhesive secretion in insects. Although we are still largely ignorant of both these aspects, a summary of the relevant findings over the last few decades should stimulate future research into this central functional aspect of insect adhesion. This survey only considers studies in which dependable data on the histology or ultrastructure of the glands and/or the chemical nature of their secretion have been acquired. The numerous cases in which adhesive body structures have been observed without exploring further structural or chemical details have been omitted.

A special category of insect glands is formed by single- or multicellular wax glands. These produce true waxes, i.e., esters of long-chain alcohols and fatty acids, although other lipoid and non-lipoid components such as resins, proteins, and amino acids might also be included (Waku and Foldi, 1984). Both the wax glands and their secretion show tremendous diversity among both the Auchenorrhyncha (e.g., Weber, 1930; Liang and Wilson, 2002; Liang and O'Brien, 2002; Liang and Jiang, 2003; Strümpel, 2010) and the Sternorrhyncha (especially the Coccina; e.g., Brown, 1975; Miller and Kosztarab, 1979; Foldi, 1981, 1991; Byrne and Bellows, 1991; Gill, 1992; Foldi and Lambdin, 1995; Bährmann, 2002). Wax glands are also present in several larval Coleoptera, Hymenoptera (mainly "Symphyta"), Lepidoptera, and adult Aculeata (Hymenoptera) (Peters, 2010). Although wax coverings might not only protect from desiccation, but also help in the avoidance of predators and parasitoids by coating their body structures (Waku and Foldi, 1984), direct observations or experiments concerning this function remain scarce. Hence, by building their own category, wax glands are not considered in this review and are generally reviewed elsewhere (Meinwald et al., 1975; Jackson and Blomquist, 1976; Strümpel, 1983; Pope, 1984; Waku and Foldi, 1984; Peters, 2010).

Several herbivorous insects, e.g., certain Reduviidae (Heteroptera), Diprionidae, and Pergidae (both "Symphyta"), do not produce adhesives on their own but obtain substances such as resin terpenes and ethereal oils from their plant diet. These insects can cover themselves or their eggs with these substances, or, on assault, regurgitate the sequestered glue, which sometimes contains repellent mono- and sesquiterpenes, onto the attacker (Eisner, 1972, 1988; Morrow et al., 1976; Dettner, 2010).

8.2 Function and Distribution of Adhesive Glands in Insects

The available studies on the ultrastructure and the secretion of adhesive insect glands cover a broad spectrum of higher taxa, i.e., the Elliplura, the Ephemeroptera, the Polyneoptera, the Acercaria, the Coleoptera, the Hymenoptera, the Amphiesmenoptera, and the Diptera (Table 8.1). Structures for use in repelling attackers or temporarily or permanently adhering to a substratum or a mating partner have been found in four developmental stages of an insect, i.e., the egg, the larvae, the pupae, and the adult. For instance, the males of dytiscid beetles have specialized fore (and middle) tarsi that employ an adhesive secretion in combination with prominent bowl-shaped suckers of various sizes (cf. fig. 4.19 in Gorb, 2001). The tarsal glands provide an additional adhesive secretion that flows onto the inner sucker surface and adds to the attachment force caused by the sucker (Blunck, 1912). Moreover, some insects have developed sticky prey capture devices or use adhesive glue for cocoon building.

Based on this diversity of biological contexts, adhesive structures are found at various tagmata of the body, mainly at the head, the abdomen, and the legs, but also within the thorax in the form of the metapleural glands of ants (Table 8.1). Adhesive glands of the head can involve the mouthparts and the antennae, the labial salivary glands, or specific head glands as seen in many termites. In some groups of termites, both the labial and the head glands form extremely large reservoirs that extend far into the abdomen (cf. Chapman, 1998).

With regard to the legs, adhesive glands are chiefly associated with the tarsi onto which they discharge an adhesive that is used for reversibly attaching to the substratum during locomotion (e.g., Gorb, 2001). In insects with raptorial legs, adhesive tarsi might also be used for prey capture (Betz and Kölsch, 2004). In some Reduviidae, adhesive glands are not restricted to the tarsi, since they sometimes extend toward the tibia (e.g., Barth, 1953).

Abdominal adhesive glands can be recruited from highly different glandular systems of the abdomen (Table 8.1): (1) the epithelium underlying the tergites and the cerci; (2) specific abdominal gland systems associated with the female reproductive system (Dufour's gland in Aculeata, colleterial, and cement glands in Lepidoptera (functionally equivalent glands have been described in the Hemiptera; Weber, 1930), accessory reproductive glands in Diptera); (3) other abdominal glands (pygidial glands in Formicidae and Staphylinidae); (4) specific tubes associated with the 5th and 6th abdominal segments (the cornicles of aphids that do not only release an alarm pheromone but also a viscous secretion).

Finally, an extreme case of defence has been independently evolved in some Termitinae (Termitidae) and Formicinae (Formicidae), in which individuals sacrifice themselves upon attack by bursting their body wall; this leads to the release of large quantities of sticky slime from cephalic glands that are enlarged into the abdomen (autothysis; e.g., Maschwitz and Maschwitz, 1974; Hermann and Blum, 1981; Buschinger and Maschwitz, 1984; Prestwich, 1984; Costa-Leonardo, 2004).

In several cases, muscles are associated with the gland systems used in defence and either control their opening (Aphididae) or help to spray the secretion forcefully toward an assailant (e.g., Costa-Leonardo and De Salvo, 1987).

8.3 Histological and Ultrastructural Characteristics of Adhesive Glands in Insects

The morphology and ultrastructure of insect exocrine adhesive glands have to be viewed in the broader context of insect epidermal glands that are found in many body regions of an insect, e.g., the tegument of both the tergum and sternum of the body segments, the pygidium, the legs, the mouthparts, and the genital apparatus (Peters, 2010). In general, epidermal glands and their secretions can take a tremendous diversity of functions in the contexts of (1) protection from adverse environmental conditions (e.g., toxic chemicals) and microbial contamination (e.g., Baker et al., 1979; Leger, 1991), (2) regulation of water balance (e.g., Beament, 1962; de Renobales et al., 1991), (3) intra- and interspecific communication *via* pheromones and alelochemicals (e.g., Percy and Weatherston, 1974; Percy-Cunningham and MacDonald, 1987; de Renobales et al., 1991; Blomquist et al., 1993; Howard, 1993), (4) repulse of predators and parasitoids (e.g., Dettner, 1987, 1989; Eisner, 1970, 1972), (5) construction of dwellings and other mechanical supporting structures (e.g., threads of spermatophores; e.g., Alberti and Storch, 1976; Bitsch, 1990; Akai, 1998; Eberhard and Krenn, 2003; Sutherland et al., 2010), and (6) making food accessible (e.g., Ribeiro, 1987).

According to Noirot and Quennedey (1974, 1991) and Quennedey (1975b, 1998), epidermal gland cells can be classified into three categories in accordance with their cell structure and the route followed by the secretion on crossing the cuticular barrier toward the exterior (Fig. 8.2): (1) Class 1 cells, the simplest type, are single epidermal cells that are regularly integrated into the epidermis that directly adjoins the outer cuticle. Their secretion has to cross this cuticle in order to be released; the cuticle can be perforated by pores that are the openings of pore canals crossing the various cuticular layers. As a specialization, type 1 cells might apically possess a deeply invaginated reservoir-like extracellular space bounded by a microvillus border (e.g., Skilbeck and Anderson, 1994; Betz and Mumm, 2001; Betz, 2003) (Fig. 8.3A). (2) Class 2 cells are also single cells, but lie deeper within the epidermis, so that their apical side does not reach the outer cuticle. Consequently, its secretion has to be transferred to the surrounding epidermis cells before it can be conveyed to the external layer. A cell type belonging to this category is the oenocyte (e.g., Gnatzy, 1970; Romer 1975, 1980) that, among others, is involved in the secretion of the epicuticular wax layer. (3) Class 3 units are compound glandular units consisting of one to several secretory cells plus one or two canal cells that surround a conducting duct formed by internalized cuticle (e.g., Kölsch, 2000; Betz, 2003). This duct has two functions, i.e., it takes up the secretion from the inner secretory cells (the receiving canal) and then conducts it toward the outside (the conducting canal). The transfer of the secretion from the terminal cell into the receiving canal takes place in the so-called end-apparatus, which is formed by the internalized extracellular space of the terminal cell delimited by a continuous microvillus border. Class 3 units show a certain degree of interspecific structural variability (cf. fig. 22 in Quennedey, 1998). For instance, an additional so-called intercalary cell might be inserted between the terminal cell and the canal cell (cf. ic in Fig. 8.3C). If this cell is secretorily active, it might form an extracellular space and microvillus border similar to those of the terminal cell. Another modification of class 3 glands occurs if the canal cell is entirely enclosed by the gland cell (Quennedey, 1998).

In general, hundreds of gland cells and glandular units belonging to categories 1 or 3 might be aggregated to form whole gland organs that can discharge much larger amounts of a complex secretion *via* a common reservoir into which the single cells release their secretory products. In this case, the glandular epithelium might be deeply invaginated forming a sac-like structure beneath the cuticle (Quennedey, 1998).

Being derivatives of the epidermis, adhesive exocrine glands can be assigned to one of the three gland types displayed in Fig. 8.2. Table 8.1 provides an overview of case studies performed on insect adhesive glands that, in

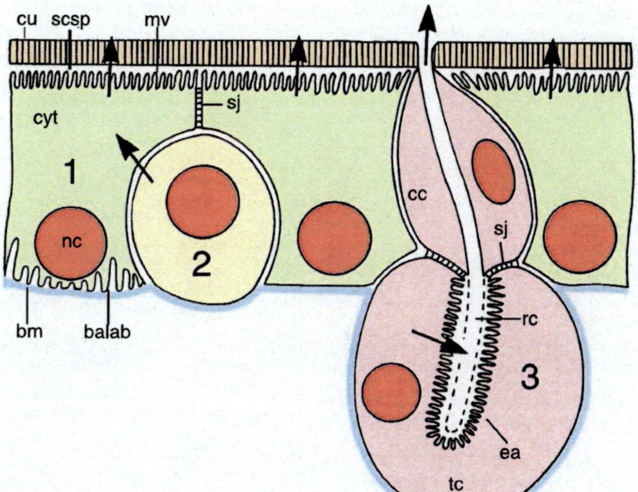

Fig. 8.2 The three types of epidermal gland cells found in insects. Whereas class 1 cells (*1*) directly adjoin the cuticle of the body surface, class 2 cells (*2*) (e.g., the epidermal oenocytes) do not. In class 3 cells (*3*) the contact toward the exterior is formed by a cuticular duct enveloped by a canal cell. Additional secretorily active intercalary cells are sometimes inserted between the terminal cell and the canal cell. The arrows are indicative of the pathway along which the secretion is discharged. This figure was redrawn and modified after Quennedey (1975b)

many cases, provide sufficient structural data to classsify the investigated glands with respect to the above-mentioned system of Noirot and Quennedey (1974, 1991) and Quennedey (1975b, 1998). In several cases, a neuronal innervation of these cells has been observed, which helps, especially in defence glands, to discharge the secretion upon an exterior stimulus (e.g., Quennedey, 1984).

In the following, the histological and ultrastructural characteristics of the adhesive gland cells of the investigated systems of Table 8.1 are surveyed. In order to discover features they may have in common, this discussion is subdivided according to the biological contexts in which these gland systems are employed. It also includes the involved outlet structures and paths to the exterior.

8.3.1 Glands Employed in Locomotion

Adhesive structures used in locomotion are mostly restricted to the tarsus and the tibia, where they occur in the form of smooth (arolium, pulvillus, and euplantula) or hairy (pulvillus, fossula spongiosa, and tenent hairs) systems (cf. fig. 9.3 in Gorb, 2001). Interestingly, the larvae of the Simuliidae employ a different structure for their under water locomotion: they produce "adhesive secretion pads" on the substratum *via* their labial glands (e.g., Reidelbach and Kiel, 1990). Simuliid larvae use their labial gland secretion not only for locomotion, but also for body anchorage and for building an underwater retreat. Hence, their ultrastructure will be treated below in the context of the labial glands of other dipterans.

In all investigated cases, the adhesive secretion is produced by clusters of class 1 cells that are integrated into a glandular epithelium made of uniform cylindrical class 1 epidermis cells. In tarsal hairy systems, single tube-shaped class 1 cells are directly connected to the base of a tenent hair shaft (e.g., Betz and Mumm, 2001; Betz, 2003; Geiselhardt et al., 2010) (Fig. 8.3A). Additional class 3 glands are found in the tarsi of some coleopterans (Betz, 2003; Geiselhardt et al., 2010). Since they are not directly associated to single tenent hairs, it remains unclear as to whether they fulfill a special function in the context of tarsal adhesion. Adhesive is poured directly into the hollow shafts of the tenent hairs in hairy tarsal systems (e.g., Betz and Mumm, 2001; Betz, 2003; Geiselhardt et al., 2010) but is discharged into subcuticular or spongy cuticular reservoirs in smooth tarsal systems (e.g., Bauchhenß, 1979; Lees and Hardie, 1988). The final passage through the outer cuticular wall is assumed to proceed *via* a system of extremely fine pore canals and

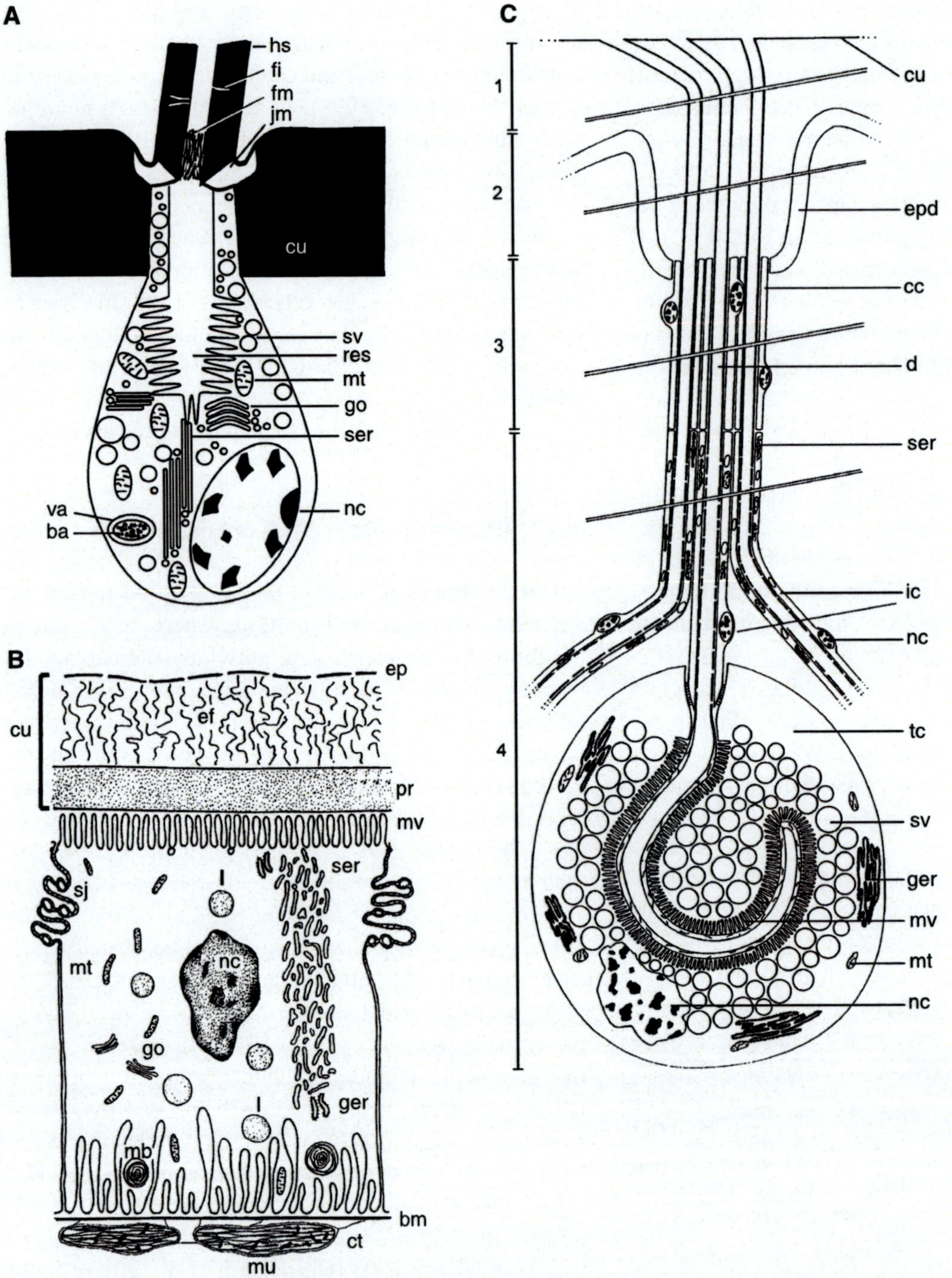

Fig. 8.3 Diagrams of the ultrastructure of selected examples typical of adhesive epidermal glandular cell units occurring in insects. (A) Class 1 cell underlying a tenent seta at the bottom of the tarsus of the staphylinid beetles *Stenus* spp. This cell possesses a deeply invaginated reservoir-like extracellular space bounded by a microvillus border. Adapted from Betz (2003) and reproduced with permission. (B) Class 1 cell found in the glandular epithelium of the frontal weapon of the termite *Armitermes euamignathus*. Adapted from Costa-Leonardo (2001) and reproduced with permission. Although it has not as yet been confirmed that the secretion produced by this species is adhesive, its overall ultrastructure resembles very closely the adhesive cells that form the frontal gland of other Nasutitermitinae (cf., images in Grassé, 1982). (C) Class 3 gland unit supplying the prey capture apparatus of the staphylinid beetles *Stenus* spp. (cf. Figs. 8.1A, 8.4). Adapted from Kölsch (2000) and reproduced with permission. Because of their considerable length, the ductules and cells are drawn with interruptions, indicated by the angled parallel lines. The ductules (only three are shown) coming from the respective terminal cells are accompanied by individual intercalary cells and combine into a bundle within the head. In the proximal half of the labium, the ductules are surrounded by canal cells (Sect. 3 as drawn left to the Fig. 8.3C). Once the ductules have crossed the epidermis, they lie in the subcuticular space in a deep indentation of the epidermis (Sect. 2 as drawn left to the Fig. 8.3C)

epicuticular filaments of a few nanometers in diameter (Lees and Hardie, 1988), a system that is presumed to be a general pathway of lipids in insect cuticles (e.g., Locke, 1961; Wigglesworth, 1985). Only in a few cases have larger draining pores been established by transmission electron microscopy (TEM), viz., in the adhesive setae of the staphylinid *Philonthus marginatus* (Betz and Mumm, 2001) or the syrphid *Episyrphus balteatus* (Gorb, 1998).

With regard to the ultrastructure of the class 1 gland cells, their nuclei are often large (several micrometers in diameter) and lie in the basal half of the cell (Kendall, 1970; Betz and Mumm, 2001; Betz, 2003; Eberhard et al., 2009), so that the cells attain a polar structure. Secretorily active cells are usually characterized by their large number of large mitochondria with an ovoid or spherical shape, an extensive endoplasmic reticulum (ER), Golgi complexes, and secretion vesicles of various shapes and contrast (electron-lucent *versus* electron-dense; Bauchhenß and Renner, 1977; Bauchhenß, 1979; Lees and Hardie, 1988; Betz and Mumm, 2001; Betz, 2003; Eberhard et al., 2009). Bauchhenß (1979) also describes coated vesicles (associated with the microvillus border), lipid droplets, and dense bodies, multivesicular bodies, and lamellar bodies associated with the Golgi complex.

Depending on the system, the endoplasmic reticulum (ER) can be of the granular (GER) or the smooth (SER) type. Usually, only one of these types prevails within a cell; SER is prevalent in the gland cells supporting the tenent hairs of the staphylinid beetles described by Betz and Mumm (2001) and Betz (2003), whereas GER prevails in the tarsal pulvilli and arolia described by Lees and Hardie (1988), Bauchhenß and Renner (1977), Bauchhenß (1979), and Eberhard et al. (2009). Abundant GER (often in association with electron-dense secretion vesicles) suggests that proteins are secreted by the gland cells. Moreover, free ribosomes might be abundant in the cytosol (Lees and Hardie, 1988). Contrary to this, abundant SER suggests non-proteinaceous secretion such as lipoids (e.g., Quennedey, 1998; Eberhard et al., 2009).

Junctions between the class 1 cells within a glandular epithelium comprise, on their apical side, belt desmosomes and septate junctions (Eberhard et al., 2009). On their basal side, the cells might be connected to the extracellular basal membrane *via* hemidesmosomes (Eberhard et al., 2009). The basal cell membrane can be deeply infolded, developing a prominent basal labyrinth that may facilitate the transport of precursors of the secretion from the hemolymph toward the cytoplasm (Bauchhenß, 1979). The apical side of the cell is usually differentiated into a microvillus border (Bauchhenß and Renner, 1977; Bauchhenß, 1979; Lees and Hardie, 1988; Eberhard et al., 2009), which often deeply invaginates into the cell to form an extensive reservoir for the secretion that directly passes into the hollow shaft of the adhesive setae (Betz and Mumm, 2001; Betz, 2003; Geiselhardt et al., 2010) (Fig. 8.3A). Evidence for the exocytosis of vesicles into the reservoir has been detected at these microvilli (Bauchhenß and Renner, 1977; Bauchhenß, 1979; Eberhard et al., 2009). Within the reservoir, the secretion is sometimes visible forming a matrix containing fibrils or flakes (Betz, 2003). Betz (2003) has detected bacteria that are scattered throughout the cytoplasm of these gland cells (Fig. 8.3A). The enclosure of the bacteria in cell vacuoles, i.e., in an controlled environment, suggests that they might have some endosymbiontic function for the cell (cf. Locke, 1984).

8.3.2 Glands Employed in Prey Capture

Although adhesive prey capture devices are widespread among insects (Betz and Kölsch, 2004), only few studies deal in greater detail with the chemical nature of the adhesive and/or the ultrastructure of the adhesive-producing glands. In several cases, insects merely use their fore tarsi to clasp their prey. In these cases, the involved tarsal gland structures are identical to those described above in the context of locomotion. In addition, several cases are known in carabid and staphylinid beetles in which other body structures have been modified into highly advanced prey capture organs. A well-investigated case is the staphylinid beetle genus *Stenus*, the members of which have a unique protrusible labium, which is one of the most specialized prey capture structures among insects (Figs. 8.1A and 8.4). Its general structure and function have been elucidated in several studies (Schmitz, 1943; Weinreich, 1968; Bauer and Pfeiffer, 1991; Betz, 1996, 1998a, b; Kölsch and Betz, 1998; Kölsch, 2000; reviewed in Betz and Kölsch, 2004). When not in use, the labium is withdrawn back into the head, where it is wrapped by a connecting membranous tube. In order to capture prey, the beetles rapidly protrude their prementum from this resting position within just 3–5 ms. The prey adheres to the sticky cushions and is seized by the mandibles after immediate retraction of the prementum.

The elongated prementum carries, on its apex, two paraglossae, which are modified into membranous sticky cushions (Fig. 8.4B). Within the prementum, bundles of ductules, which transport an adhesive secretion produced by special secretory glands in the forehead, lead to the

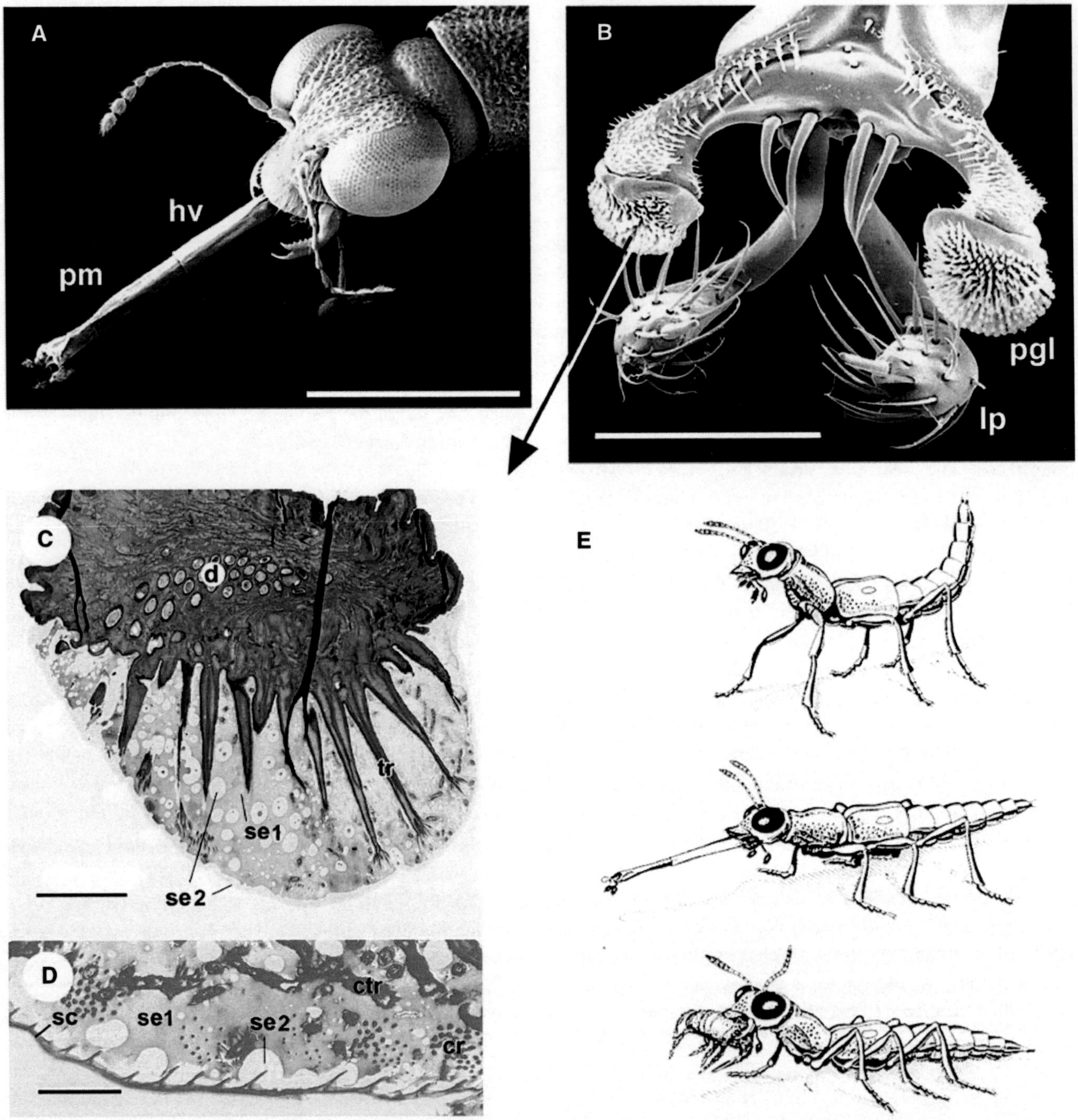

Fig. 8.4 The adhesion-capture apparatus of *Stenus* spp. (A, B) Scanning electron microscopic (SEM) images; (C, D). Transmission electron microscopic (TEM) images. A, B *S. comma*; adapted from Bauer and Pfeiffer (1991) and reproduced with permission; C, D *Stenus* spp.; adapted from Kölsch (2000) and reproduced with permission; E *S. comma*; adapted from Weinreich (1968) and reproduced with permission. (A) Head with protruded sticky rod (labium); (B) Dorso-frontal view of the apex of the labium with the paraglossae modified into sticky pads; (C) Ultrathin section through a sticky pad (paraglossa) of *S. juno* showing tenent trichomes deeply immersed in a multiphasic adhesive secretion. Presumed lipid phase (colored pink in the image) is emulsified into a presumably watery (protein and/or carbohydrate containing) phase (gray in the image). The pink-colored lipid droplets within the emulsion appear to flow together by coalescence toward the periphery of the secretion, which might serve as a protective surface film preventing the desiccation of the underlying aqueous phase; (D) Close-up of the periphery of the sticky pad showing a scale of a collembole (*sc*) wetted by the presumed lipoid phase (colored pink) of the secretion; (E) Prey-capture sequence in *S. comma*. After approaching the prey to a critical distance (top), the prey-capture apparatus is rapidly protruded (middle) and the glued prey is withdrawn to the mandibles (bottom). Scale bar in (A) = 1 mm, (B) = 100 μm, (C) = 10 μm, (D) = 2 μm

sticky cushions. These class 3 secretory glands consist of three cells (Fig. 8.3C). Two cells (the terminal and the intercalary cell; tc and ic in Fig. 8.3C) are secretorily active; the third cell is the canal cell (cc in Fig. 8.3C) that secretes the glandular ductule (Kölsch, 2000). The terminal cell shows a zonation into an ecto- and an endoplasm. Typical organelles of the ectoplasm are the ovoid nucleus, the extensive GER, and the interposed mitochondria, whereas the endoplasm is filled with secretory vesicles that are assumed to have a proteinaceous nature (Weinreich, 1968; Kölsch, 2000). These vesicles are internally secreted *via* an end-apparatus that consists of an extracellular secretion reservoir (formed by microvilli) and a receiving canal. According to its size, the terminal cell probably contributes the major part of the adhesive secretion. The terminal cell connects to the smaller intercalary cell, which possesses only a thin cytoplasmic layer around the ductule. Although intercalary cells do not necessarily have to be secretorily active (Sreng and Quennedey, 1976), they clearly contribute to the formation of the adhesive in *Stenus* beetles, as characterized by SER, mitochondria, and light-gray staining droplets that presumably have a lipoidal nature (Kölsch, 2000). These cells add their secretion directly into the receiving canal without the mediation of an internal microvillus border. In this way, the intercalary cells modify the proteinaceous (and presumably carbohydraceous according to Betz et al., 2009) secretion of the terminal cell, by probably adding a lipoid component to it. The intercalary cell lies adjacent to a canal cell that lines the canal during its passage through the outer cuticle. The interplay between both the terminal and the intercalary cells in delivering various adhesive components finally leads to a secretion consisting of at least two immiscible phases found in the ductules and on the paraglossae (Kölsch, 2000; Betz et al., 2009) (Fig. 8.4C, D).

Adhesive glands are also employed for prey capture in the megadiverse staphylinid subfamily Pselaphinae. In the adults of two *Bryaxis* species, Schomann et al. (2008) have found that these beetles employ their apparently sticky maxillary palps to fix prey at the moment of contact. Corresponding glandular structures (probably class 3) have been detected in the interior of the palps. Characteristic cell structures are the large nucleus, the secretion vesicles filled with electron-dense material (probably protein), and extensive GER. Golgi complexes and mitochondria are present only in low numbers.

Pselaphine larvae, in many taxa, have a pair of specific organs (eversible glands?) that can be rapidly protruded out of the head by hemolymph pressure, like the finger of a glove (De Marzo 1984, 1985, 1988). The structure and function of these sac-like membranous organs have been more intensely studied in the genera *Pselaphus* and *Batrisodes* by De Marzo (1985, 1988): they appear to arise from the articular membrane connecting the antenna with the head capsule and function in capturing prey such as springtails. The prey sticks to their terminal part, which is differentiated into a number of hair-like trichomes. Specific head glands produce a sticky secretion that is led, *via* cuticular ductules, toward the exterior at the base of the protrusible organs. From here, it probably spreads across the surface of the protrusible organ providing it with a sticky surface. The cephalic glands are characterized by an extensive ER together with darkly staining secretion vesicles that are secreted into a large intracellular reservoir joined by microvilli (cf. fig. 19 in De Marzo, 1985). The same ultrastructural features are found in the much larger dermal thoracic and abdominal glands of *Batrisodes oculatus*, which produce a sticky secretion used for both prey capture and cocoon building.

8.3.3 Glands Employed in Defence

Other than prey capture, adhesiveness is used in a converse way, i.e., for predator avoidance. Two basic mechanisms are employed. First, predation can be avoided by firm temporary or permanent adhesion to the substratum, so that the prey is not detachable by the predator. This mechanism has been demonstrated by Attygalle et al. (2000) in the foot adhesion system of the cassidine chrysomelid beetle *Hemisphaerota cyanea*. When assaulted by a predator (e.g., an ant), the beetles press their tarsi down flatly, which dramatically increases their adhesion with the surface. In this way, they can withstand lateral and vertical pulling forces (as exerted by the attacking predator) of many times their body mass.

Second, predators are avoided by the exudation of a sticky secretion, which immobilizes the appendages or the sensilla of the opponent. Studies dealing with the ultrastructure of the involved gland cells exclusively consider this second context. In this case, repellents or toxins are sometimes added to the secretions, although they mainly work mechanically rather than chemically. Depending on the taxon and the system, all three gland types (class 1, 2, or 3) can be involved, either singly or as a mixture of two types (Table 8.1). The most common are class 1 cells, which have been described in the defence glands of certain taxa of springtails, termites (often mixed with class

3 units), aphids, ants, and wasps. Class 3 units (made of a secretorily active terminal cell and a canal cell) form the exclusive gland types found in the abdominal defence system of the terga of the last abdominal segments of roaches in which the viscous secretion is transported to the exterior *via* cuticular ductules formed by canal cells (Table 8.1). Class 3 units have also been described forming the defensive glands in staphylinid beetles of the genus *Deleaster* (Dettner et al., 1985; Dettner, 1993).

In the cockroach *Blatta orientalis*, the ultrastructure of the secretorily active cell of the bicellular tergal abdominal class 3 gland units comprises a large basally lying nucleus (with a prominent nucleolus), extensive GER, free ribosomes, mitochondria, and an extensive Golgi apparatus with associated electron-dense secretion granules of probably proteinaceous content (Plattner, 1971; Plattner et al., 1972). These granules are internally secreted *via* apocrine secretion at the microvillus border of a prominent end-apparatus (Plattner, 1971). At their base, these cells have a weakly developed basal labyrinth that is underlaid by a basal lamina and a collagen layer. Tracheoles and nerves are associated with the glandular epithelium (Plattner et al., 1972). In addition to the described "normal" class 3 gland cells, other class 3 gland cells sporadically occur that lack GER and an electron-dense secretion. These cells contain numerous mitochondria, Golgi apparatus, and electron transparent vesicles. They are assumed to produce a non-proteinaceous secretion that helps to adjust the viscosity of the adhesive produced by the other gland cells in order to optimize its performance (Plattner et al., 1972). Conditions that deviate with regard to the ultrastructural details of exocrine abdominal glands in other blattodean taxa are documented in Brossut and Sreng (1980).

The largest amount of data on adhesive glandular defence systems in insects is available for the Isoptera (Table 8.1). Termites employ various cephalic gland systems in defence, i.e., the labial salivary gland, the mandibular gland, and the frontal head gland (Quennedey, 1975b). Depending on the taxon, both the labial salivary gland and the frontal head gland of termite soldiers produce a viscous secretion to entangle the appendages of an attacker. The amount of defensive glue can be considerable, since many individual soldiers often work together to drive away an intruder (e.g., Roth and Eisner, 1962). The dominant cells are of the class 1 type. They form the only cell type in the labial salivary glands, whereas in the frontal head glands of some Rhinotermitidae and Termitidae (Table 8.1), bicellular class 3 units are scattered between the class 1 cells. In some cases (e.g., in *Prorhinotermes*, cf. Šobotnik et al., 2004), these class 3 units are separated from the glandular epithelium of class 1 cells, emitting their secretion not directly into the reservoir but further distally into the efferent duct of the gland.

Labial salivary gland: The defensive secretion of these glands has been established as being viscous (apart from its toxic and/or irritant effect) in the investigated soldiers of the Mastotermitidae and Termitidae (Macrotermitinae and Termitinae; Table 8.1). According to Quennedey (1984), the secretory labial gland cells in the monospecific Mastotermitidae are class 1 cells that are characterized by a large nucleus, free ribosomes, extensive GER, well-developed mitochondria, and a Golgi apparatus that secretes large electron-lucent vesicles. The apical plasma membrane is differentiated into a microvillus border; the basal cell membrane forms a basal labyrinth that increases the surface contact with the hemocoel. The basal cell membrane is faced by a basal membrane that is penetrated by tracheoles and neurosecretory axons (Quennedey, 1984), which suggest that the discharge of the salivary secretion is under nervous control. Lateral cell–cell connections are septate junctions. The descriptions of the labial gland cells of the Mastotermitidae given by Czolij and Slaytor (1988) are more detailed than those provided by Quennedey (1984). These authors describe three morphologically distinct cell types (probably class 1 cells), two of them ("central cells" and "peripheral cells") showing a deep intracellular reservoir (called "intracellular ductile" by the authors) bordered by microvilli. The third cell type ("storage cell") contains a large storage vacuole filled with osmophilic homogeneous granular material (Czolij and Slaytor, 1988). The "central cells" and the "peripheral cells" contain numerous vacuoles with electron-dense or electron-lucent material. "Central cells" and "storage cells" are both characterized by extensive GER indicative of the extensive protein production of these cells. Further ultrastructural details including the close association with axons resemble the descriptions of Quennedey (1984).

The fine structure of the large labial gland cells of *Macrotermes bellicosus* soldiers (Termitidae) is characterized by abundant clear vesicles, few mitochondria, free ribosomes, and a moderately developed GER (Billen et al., 1989). These features indicate the production of proteinaceous components in addition to the quinones previously described for these glands in other termitid species.

Frontal head gland: In these gland systems, a clear distinction can be made between class 1 and class 3 units in terms of their ultrastructure. Whereas in the class 1 cells, SER prevails, the ER of the class 3 units is of the granular type (GER; Quennedey, 1984).

With regard to the class 1 secretory cells of the frontal gland (Fig. 8.3B), a characteristic cytological feature in many taxa is the differentiation of their apical plasma membrane into microvilli (often including microfilaments), which increase the cell surface covered by the secretion (Quennedey, 1984; Costa-Leonardo, 1992). The cells are often extensively interdigitated laterally and connect to each other in their apical part by desmosomes followed by septate junctions. Internally, the desmosomes serve as attachment sites for microtubules that might also attach to hemidesmosomes adhering to the basement membrane (Quennedey, 1984; Šobotnik et al., 2004). A further specialization of these cells is their outer cuticle that shows epicuticular pores or even develops into large extracellular spaces forming subcuticular reservoirs that store the secretory products continuously produced by the underlying glands (cf. fig. 5.16 in Quennedey, 1984; Costa-Leonardo, 1992, 2001). The basal cell membrane facing the hemocoel in many taxa forms a well-developed basal labyrinth and is lined by a sometimes thick basal lamina that in some cases contains collagen fibers. The cytological features of the class 1 cells are partly taxon- and even caste-specific (e.g. Quennedey et al., 1973; Quennedey, 1975b, 1984; Costa-Leonardo, 1998a; Šobotnik et al., 2004). Depending on the investigated taxon, they comprise a large basal irregular or round nucleus, (moderately) abundant mitochondria and microtubules, SER (not found in the Prorhinotermitinae species investigated by Šobotnik et al. (2004)), Golgi bodies (not found in the Prorhinotermitinae species investigated by Šobotnik et al. (2004)), free ribosomes, glycogen granules, lysosomes, and myeloid bodies, all of which are considered to constitute the glandular secretion (Quennedey, 1984). The rich development of SER is indicative of the production of non-proteinaceous molecules (e.g., terpenes). This secretion has the appearance of large electron-dense vesicles in the Coptotermitinae, Prorhinotermitinae (both Rhinotermitidae), and Nasutitermitinae (Termitidae). Costa-Leonardo (2001) assumes a lipid nature of the large and spherical droplets observed in *Armitermes* soldiers (Nasutitermitinae) (Fig. 8.3B). In Psammotermitinae, secretory precursors have been observed that are collected from the hemocoel and finally channeled *via* the intercellular spaces (Quennedey, 1984). Šobotnik et al. (2004) have observed that the secretory vesicles in various *Prorhinotermes* castes change over time, beginning as small electron-dense granules, taking up lipid components in their intermediate stages, and finally developing into large electron-lucent vacuoles after the lipids have dissolved. In *Glossotermes* and *Serritermes* soldiers (Serritermitidae), heterogeneous granules and vesicles occur with various electron densities (Costa-Leonardo, 1998b; Šobotnik et al., 2010a).

With regard to the class 3 glands, the large Golgi bodies in the terminal cells of Coptotermitinae produce clear vesicles with a flocculent content (probably saccharides such as glycogen) that is also visible in both the extracellular space of the end-apparatus and the subcuticular space, where it mixes with the electron-dense material of the class 1 cells (Quennedey, 1984). In Prorhinotermitinae, the secretory cell contains GER, numerous Golgi apparatus, and abundant mitochondria. The GER probably produces the electron-dense vesicles, whereas the electron-lucent vesicles are released from the Golgi apparatus (Šobotnik et al., 2004). In Termitinae, the oval shaped secretory cell has a large basal nucleus and well-developed Golgi bodies that secrete numerous large clear vesicles with a granular content. This secretion is also found in the extracellular spaces of the end-apparatus and the glandular reservoir, where it is mixed with the small and clear vesicles of the class 1 cells (Quennedey, 1984). In Nasutitermitinae, the secretorily active terminal class 3 units are ultrastructurally characterized by numerous glycogen granules, free ribosomes, and large vesicles with a granular content that originates from the highly developed Golgi bodies (Quennedey, 1984; Šobotnik et al., 2004).

The secretion of the various class 1 cells and class 3 units are mixed in the sub- or intracuticular spaces and/or the larger glandular reservoirs prior to discharge, allowing the formation of complex structural mixtures and/or chemical reactions between their components.

Aphids discharge, from their cornicles, a secretion that contains not only alarm pheromones, but also waxes that block the mouthparts of assailants (e.g., Hesse and Doflein, 1943; Blum, 1981; Strümpel, 2003). The cornicle wax arises from class 1 and class 2 gland cells, the latter resembling oenocytes (Edwards, 1966; cf. fig. 1 in Foldi-Hope, 1990). Mature secretory cells are characterized by abundant SER and mitochondria, rare GER, numerous electron-lucent vacuoles with a volatile (probably pheromone-like) content, and a large osmophilic inclusion of lipid. In response to an external stimulus, the secretory cells burst within the cornicles, so that the droplike secretion is discharged in a holocrine manner (Foldi-Hope, 1990).

8.3.4 Glands Employed in Body Anchorage

Although widespread among insects, gland cells used by female insects to glue their eggs to the substratum during oviposition have rarely been studied at the ultrastructural level (Table 8.1). These glands are associated with the reproductive system, where they are formed by the follicle cells within the ovaries or by accessory glands of the reproductive system (e.g., Dufour's gland in apocrite Hymenoptera; cement and colleterial glands in Lepidoptera and Diptera). In all investigated cases, the involved gland cells are of the class 1 type.

The follicle cells that produce the underwater egg glue in certain mayflies and stoneflies (Table 8.1) closely interlock with the plasma membrane of the oocyte *via* the microvilli of their apical cell pole (Gaino and Rebora, 2001). They are laterally attached to each other *via* septate junctions and desmosomes (described for perlid stoneflies by Rościszewska, 1995) and are generally characterized by their rich GER, Golgi elements, secretory vesicles, mitochondria, and granular cytoplasm (Gaino and Mazzini 1989, 1990; Rościszewska, 1995). Having deposited the egg chorion layers, the follicle cells discharge an adhesive coat (containing mucous materials that are later enriched by fibrous material from the oviduct cells) enveloping each egg. This material stems from electron-dense precursor granules occuring in the cytoplasm of the follicle cells (Gaino and Rebora, 2001). Paracrystalline bodies are visible within the mucous material (Gaino and Mazzini, 1989).

The class 1 cells that form the glandular epithelium of the Dufour's gland in sphecid digger wasps have been described by Gnatzy et al. (2004); these class 1 cells show an apical microvillus border and basal outwardly facing protrusions. They are laterally connected *via* septate junctions and desmosomes. The underlying basal membrane is accompanied by tracheoles and axons. The cytoplasm contains a basal nucleus, dense SER, scattered GER, abundant free ribosomes, electron-lucent vesicles, and mitochondria of sometimes considerable size. A distinct Golgi apparatus has not been observed in these cells. The inner side of the gland lumen is lined with a cuticle, so that the secretory material has to pass this cuticular layer before entering the lumen. Since no pores or ductules are present, the secretion transits *via* the fibrillar network of the procuticle and the epicuticle (cf. fig. 7 of Gnatzy et al., 2004).

Simuliid and drosophilid dipterans use the secretion of their labial glands to anchor themselves to the substrate. Whereas simuliid larvae employ this secretion for underwater locomotion and to avoid their being swept away by the current (e.g., Prügel and Rühm, 1994), drosophilid third instar larvae need the glue to anchor their puparium to an underlying dry surface. The class 1 cells that form the glandular epithelium of the salivary glands of simuliid larvae have been described ultrastructurally by MacGregor and Mackie (1967): they show an apical microvillus border and a moderately developed basal labyrinth underlaid by a thick basal membrane. The lateral contact surfaces of adjoining cells are interdigitated and attach to each other *via* septate junctions. The cytoplasm contains a spherical nucleus, extensive GER, free ribosomes, Golgi complexes, secretion granules (consisting of a dark and a light component), and microtubules. These ultrastructural features indicate that the major components of the secretion are proteinaceous. The secretion is probably concentrated in the Golgi regions, formed into secretion granules, and passed out of the cell into the lumen of the gland by exocytosis (MacGregor and Mackie, 1967).

The ultrastructural appearance of the class 1 cells that form the salivary glands in *Drosophila* larvae (Rizki, 1967; Lane and Carter, 1972; von Gaudecker, 1972; Farkaš and Šutakova, 1998) is highly similar to that of the described simuliid larvae. In addition, fields of glycogen occur in the cytoplasm of the *Drosophila* gland cells. The adhesive secretion granula mature in specialized "corpus cells" that arise at the middle of the third instar shortly before the puparium is developed (Lane and Carter, 1972; von Gaudecker, 1972; von Gaudecker and Schmale, 1974). The secretion granula are composed of (1) electron-dense peripheral material that disappears upon discharge of the secretion into the gland lumen, (2) a fibrous component that shows a paracrystalline structure (probably the mucoprotein), and (3) a foamy substance. They are enclosed by a smooth enveloping membrane that apically merges into the microvillus border on their discharge into the gland lumen *via* exocytosis (von Gaudecker, 1972; Farkaš and Šutakova, 1998). According to Kress (1982), the synthesis of larval glue protein 1 occurs in three successive steps: a precursor protein is formed that is modified by two subsequent steps of glycosylation, one involving the attachment of hexosamine in the GER and the other occurring in the Golgi apparatus, where the terminal hexoses are added.

8.3.5 Glands Employed in Retreat Building

Adhesion is also involved in the construction of retreats made of proteinaceous silk. Insects produce silks in highly different contexts such as supporting sperm, covering eggs, or building retreats and cocoons (e.g., Sehnal and Akai, 1990; Akai, 1998; Sehnal and Sutherland, 2008). The involved silks are produced by a variety of dermal glands, Malpighian tubuli, colleterial reproductive glands, the gut, and the labial glands (Akai, 1998; Sehnal and Sutherland, 2008). The mechanism of silk secretion, silk composition, and silk structure has been examined most systematically in Lepidoptera (especially domesticated *Bombyx mori*), being stimulated by the interests of commercial sericulture (Padamwar and Pawar, 2004). The silk filament used by these groups for cocoon building is largely built of proteins called fibroins and coated by glue-type proteins known as sericins (Akai, 1998; Sehnal and Sutherland, 2008). Studies that focus on the chemical identification of the adhesive sericins and/or investigate the ultrastructure of the involved glands have mainly focused on a few groups of Lepidoptera and Trichoptera (Table 8.1). In all these cases, certain regions of the paired labial silk glands produce the glue-like silk components. Yamanouchi (1922), Voigt (1965a, b), Akai (1983, 1998), and Akai et al. (2003) describe the histology and ultrastructure of the glandular epithelium (consisting of class 1 cells) that forms the middle part of the labial silk gland of the silkworm *Bombyx mori*, which is responsible for the production of the sericin-containing silk glue: similar conditions have been found in the hesperiid lepidopteran *Calpodes ethlius* (Wiley and Lai-Fook, 1974). In *Bombyx mori*, the gland cells show a basal labyrinth underlaid by a thick basal membrane that contains fibers (probably collagen). The apical cell membrane is differentiated into microvilli. The cytoplasm contains a ramified nucleus with numerous nucleoli, a few free ribosomes, Golgi complexes (with increased activity during the spinning process), many mitochondria, and an extensive cistern-like GER. Across the entire cytoplasm, extensive fields of vacuoles occur with an electron-dense fine particulate granular content, i.e., the precursor sericin secretion. These vacuoles (sericin globules) stem from the Golgi apparatus and increase and merge toward the apical cell pole, where the secretion matures into the final adhesive secretion that is loose and filiform. The content of these vacuoles is finally exocytosed *via* the apical microvillus border of the cell into a silk layer zone. From here, the sericin passes a perforated cuticular membrane and is finally transferred into the central lumen of the gland, where it is added to the fibroin mass being transferred from the posterior silk gland (cf. fig. 7 in Voigt, 1965a, b; fig. 19 in Akai, 1998). The various divisions of the silk glands produce somewhat different types of sericin, so that the appearance of the content of the sericin vacuoles within the cytoplasm varies (Akai, 1983, 1998).

In two limnephilid trichopteran larvae, i.e., members of the sister group of the Lepidoptera, the ultrastructural appearance of the silk glands, which also secrete a sericin-like protein, is similar to that described above for lepidopterans (Engster, 1976a, b).

In chironomid and simuliid dipterans, the salivary glands produce proteinaceous substances that have glue-like properties and that are used to build viscous underwater housing tubes. The ultrastructure of the salivary glands of Simuliidae larvae has been considered above (in the context of body anchorage). Similar to these simuliid larvae, the main salivary gland cells ("large cells") of *Chironomus* larvae (Chironomidae) show structural details characteristic of the secretion of exportable proteins, i.e., (1) an abundant compact GER (in lamellar or random order), (2) well-developed Golgi complexes (comprising numerous peripheral Golgi vesicles that might function as shuttle carriers, a large number of moderately electron-dense vesicles containing flocculent, fibrous material, and membrane-bound secretory granules), and (3) an accumulation of dense secretory granules in the apical part of the cell (Kloetzel and Laufer, 1969, 1970). Other cytological features comprise a large nucleus, a basement membrane, a basal labyrinth, multivesicular bodies, and a well-developed apical microvillus border. The dense secretory granules associated with the Golgi system concentrate at the microvillus region together with Golgi complexes and coated vesicles of unknown function. The dense secretory granules are probably proteinaceous and appear to be synthesized by the ribosomes of the GER, followed by accumulation and condensation in the Golgi regions and finally expulsion into the saliva in the gland lumen. The association of a large population of mitochondria within the basal labyrinth suggests an uptake of proteins from the hemolymph in addition to *de novo* synthesis within the gland cell. This double origin of the dense secretory granules has been verified by radioautographic studies (Kloetzel and Laufer, 1970). Additional gland cell types described by the authors appear to contribute a distinctive type of secretory material to the saliva produced by the main gland cells.

8.3.6 Conclusions on the Ultrastructural Characteristics of Adhesive Glands in Insects

In conclusion, the foregoing literature survey on the ultrastructural characteristics of adhesive glands in insects suggests that class 1 cells are the predominant glandular cell type among the adhesive gland systems in insects. This predominance is understandable, since class 1 cells are simply derived from regular epidermis cells that form the subcuticular epithelium all over the insect body. The modifications that might make the gland cells different concern (1) their diverse dimension and shape, (2) the organization of the overlying cuticle, which has to develop perforations for the passage of the secretion (although lipids do not necessarily require open channels to permeate the cuticle (e.g., Seifert and Heinzeller, 1989), which might explain the lack of pores in several cases), and (3) the development of extracellular spaces (including subcuticular reservoirs) allowing the storage of the secretion (Quennedey, 1998). An example for a class 1 gland cell are the elongated single gland cells that support the tarsal tenent setae of beetles. They have an apical cell membrane, being differentiated into a border of long slender microvilli, that is deeply invaginated to form a voluminous reservoir for the secretion (Fig. 8.3A). Other features in common within class 1 adhesive cells are their well-developed basal labyrinth (being indicative of the uptake of substances from the adjoining hemocoel), their underlying basal membrane (sometimes provided with (collagen) fibers and axons), and their apical microvillus border that greatly increases the surface available for the exocytosis of the adhesive material toward the exterior, the subcuticular reservoirs, or a central glandular lumen. The basal labyrinth might be accompanied by an especially dense accumulation of mitochondria as established in the larval salivary glands of chironomid dipterans (Kloetzel and Laufer, 1969); such an organization resembling "mitochondrial pumps" is observed in some absorbing epithelia of insects and is indicative of an intense uptake of secretion precursors and/or components from the hemocoel.

If organized in a glandular epithelium, the adjoining class 1 cells adhere to each other by belt desmosomes and/or septate junctions. Such glandular epithelia may be enveloped by intrinsic or extrinsic muscle cells that help to squirt the secretion (e.g., in several Nasutitermitinae: Costa-Leonardo and De Salvo, 1987).

According to their ultrastructure, the adhesive class 1 cells show features (in terms of their provision with ER, Golgi system, free ribosomes, and secretion vesicles and granules) that are either indicative of predominant non-proteinaceous (lipid) or protein secretion. Indeed, a comparison of the literature on the ultrastructure of the cells and the chemical identity of the secretion has revealed close correspondence in this respect. Interestingly, in class 1 cells that are employed in locomotion (i.e., reversible tarsal adhesion to natural substrates such as plant surfaces), lipoidal secretion seems to prevail (although these secretions often appear to be complex mixtures of lipids with proteins and carbohydrates), whereas in the contexts of more permanent body or egg anchorage and retreat building, protein-based secretion dominates (Table 8.1), as is reflected by the specific ultrastructure of these cells. Only in the digger wasp *Liris niger* Gnatzy et al. (2004) established that the secretion used for egg anchorage is not only proteinaceous, but also contains a prominent lipid component. Class 1 cells employed in active defence (comprising the active discharge of a viscous secretion toward the assailant) have predominantly been investigated in termites. In as much as such cell types are part of the labial salivary glands (in Mastotermitidae and Termitidae), they are characterized by GER and many free ribosomes, being indicative of proteinaceous secretion, whereas in the frontal head glands of termites, SER predominates, being indicative of non-proteinaceous secretion. Together with the special appearance of the secretory vesicles, these differences are indicative of proteinaceous adhesive on the one hand, and a non-proteinaceous lipid-based secretion on the other. This is in good accordance with the available chemical data on this taxon (Table 8.1).

Oenocyte-like class 2 adhesive gland cells have hitherto only been found in the defence systems of Aphidoidea and Tingidae (both Hemiptera).

Adhesive class 3 glands are almost always bicellular, consisting of a terminal secretorily active cell and an adjacent canal cell that surrounds the cuticular conducting duct. Only in the stick-capture apparatus of staphylinid *Stenus* beetles have tricellular glands been established, consisting of two secretory cells (providing a proteinaceous and a lipoidal component, respectively) and one canal cell (Fig. 8.3C). Whereas class 3 units may be involved in the modification of the tarsal adhesive secretion in beetles (Betz, 2003; Geiselhardt et al., 2010), they have mainly been found in defence systems in which they might accompany class 1 glands (as found in the frontal glands of several termites) and might add a proteinaceous component to their secretion.

8.4 Chemical Identity and Functional Aspects of Insect Adhesive Secretion

Insect glues are still a largely unexplored source of adhesives and might have a large biomimetic potential. For example, a significant clinical need exists for strong, elastic, and biocompatible adhesives that work in moist environments, and potential medical applications lie in the areas of wound healing, surgical repair, or dentistry (Graham, 2008). Only a few insect adhesive secretions have been analyzed in molecular detail (Li et al., 2008), so that the modes of action of their compounds (including the relevance of their concentrations in the adhesive liquid) in the adhesive process are poorly understood. Notwithstanding, Graham (2008) has provided an overview on general aspects, relating specific chemical parameters of selected bioadhesives to their biomechanic properties and function. He depicts most bioadhesives as relying on polymers, i.e., carbohydrates and proteins, to achieve the necessary adhesive and cohesive strength. Strong cohesion can be achieved either by providing preformed long-chain molecules of high-molecular mass or by linking smaller units *in situ*. The latter glue types have the advantage of low initial viscosity, which facilitates their engagement with the surface (Graham, 2008). Whereas permanent attachment can often be ascribed to covalent bonds, reversible adhesives (as necessary for prey capture or locomotion) are expected to rely on noncovalent interactions. For instance, intermolecular cross-linking *via* covalent bonds or high-affinity metal coordination might be expected to contribute to the high cohesive strength of many irreversible biological adhesives such as those found in the mantid ootheca and in the adhesives used to anchor marine algae or mussels (Graham, 2008).

Natural adhesives used by plants and animals are composed of only a few basic components, i.e., proteins, polysaccharides, polyphenols, and lipids (including terpenes and terpene resins) and their various combinations (e.g., glycoproteins, proteoglycanes, phenolic proteins, phenolic polysaccharides; Ambsdorf and Peter, 1992; Onusseit, 2004). Focusing on adhesive defence secretions in insects, Dettner (2010) stresses that they have diverse chemical identities. Added to this, natural adhesives do not consist of one compound only but are highly complex (emulsion-like) structural and chemical mixtures (Table 8.1) whose chemical and micromechanical functions are often poorly understood. Those adhesives that work mechanically might comprise high-molecular compounds containing proteins, terpenes (resins), mixtures of long-chain hydrocarbons and mucopolysaccharides, or waxes (Pasteels et al., 1983; Dettner, 2010). However, defensive adhesive secretions in particular not only function mechanically, but also concomitantly develop a chemical irritant function caused by reactive substances of a low-molecular weight that are mixed within the sticky secretion to produce a "toxic glue" (e.g., Moore, 1969; Dettner, 2010). For instance, highly odorous benzoquinone and other quinones emitted by the primitive termite *Mastotermes darwiniensis* show tanning reactions with salivary proteins but show, at the same time, a repellent action that contributes to the effectiveness of the secretion (Eisner, 1970). Moreover, in social insects (e.g., termites) and aphids, alarm pheromones might be mixed within the sticky secretion.

In the following, the adhesive compounds found in insects will be categorized according to the organization of their carbon skeleton (cf., Walter and Francke, 2004). In several cases listed in Table 8.1, the chemical components have been identified in greater detail (e.g., the hydrocarbons in the tarsal secretion of beetles or the terpenoid classes found in termites) and can be elicited from the cited original literature. The constituents found in insect adhesives belong to aliphatic compounds, to carbohydrates, to phenols, to isoprenoids, to heterocyclic compounds, and to amino acids, peptides, and proteins.

8.4.1 Aliphatic Compounds

Aliphates occur in great diversity within insect adhesive secretion (Table 8.1). According to their chemical constitution, they are classified into alkanes, alkenes, and alkines. The following compounds have been established (Table 8.1): straight-chain and mono-, di-, and trimethyl-branched alkanes (C_{21}–C_{37}) and (nitro-)alkenes (C_{22}–C_{29}), alcohols, alkoxyethanol, aliphatic aldehydes, aliphatic and vinyl alkyl ketones, carboxylic acids, saturated and unsaturated fatty acids (e.g., hexacosanoic and lignoceric acid), fatty acid salts, esters (e.g., acetates of long-chain alcohols, (tri-)glycerides (e.g., with high content of myristic acid and C_6–C_8 acids), waxes, phospholipids), glycerol, epoxides, sphingolipids (ceramides), diketones, and (keto-)aldehyde compounds.

Aliphatic compounds form a major constituent of the glandular secretion of the tarsal locomotion organs in insects but are also significantly involved in the formation of defence secretions such as those occurring in termites, aphids, and hymenopterans (Table 8.1). Despite the nonpolarity of the hydrocarbons that make up a major part of the tarsal liquid, this secretion is well suited to wet substrates of high surface polarity, as long as these surfaces show high free surface energies (cf. McFarlane and Tabor, 1950). Even waxy plant surfaces with extremely low surface polarities might still be wettable by nonpolar aliphatic compounds on condition that the surface polarities of both the adhesive and the substratum match closely (e.g., Wu, 1973; Lüken et al., 2009), i.e., in order to wet substrates with low surface polarities, the adhesive should contain only a low (if any) content of polar components such as fatty acids, esters, alcohols, and aldehydes. Added to this, apolar aliphates keep the ventral surface of the tarsi water-repellent and thus prevent the insect from being immersed in water under wet conditions.

The frequently established general chemical congruence between tarsal and epicuticular aliphates (e.g., Kosaki and Yamaoka, 1996; Betz, 2003; Geiselhardt et al., 2009) makes it comprehensible to consider the tarsal secretion as a derivative of the outer semi-liquid free lipid layer of the general body cuticle (Hasenfuss, 1977, 1999; Attygalle et al., 2000) that is discharged at the tarsal surface in a concentrated manner. Hence, the vast literature on insect cuticular lipids (e.g., Gilbert, 1967; Lockey, 1980, 1985, 1988, 1991; Blomquist and Dillwith, 1985; de Renobales et al., 1991; Gibbs and Crowe, 1991; Noble-Nesbitt, 1991; Buckner, 1993; Howard, 1993; Nelson, 1993; Gibbs, 1995, 2002; Gibbs and Pomonis, 1995; Gibbs et al., 1995, 1998) should be a usable source for deriving properties of the tarsal adhesive secretion. According to this literature, the primary function of the apolar material on the outer surface of insect cuticle is to prevent desiccation by restricting water loss. Further functions comprise pheromonal communication, the obstruction of microorganism invasion, self-cleaning, and mechanical protection. If the tarsal liquid is viewed as a part of the overall lipid film, then adhesion has to be considered an additional possible function of this film. Indeed, the (lipoidal) body coating of certain insects (e.g., the larvae of the staphylinid *Stenus* beetles) is known to possess sticky properties and hence can be used for prey capture (Betz and Kölsch, 2004).

Depending on the melting temperature of the involved aliphatic compounds, the epicuticular lipid layer of insects may be liquid even at ambient temperatures (Hasenfuss, 1977, 1999; Geiselhardt et al., 2010). The same holds for the tarsal secretion that is deposited as liquid droplets on solid surfaces (e.g., Gorb, 2001). According to the presumed steady flow of these liquids across the cuticle, the liquid layers of both these compartments might actually intermingle, which would further contribute to their overall chemical resemblance. The melting temperatures (T_m) of hydrocarbons increase with chain length and the ratio of n-alkanes to methyl-branched alkanes (Gibbs, 2002). In 21–40 carbon n-alkanes, the addition of one C unit increases T_m by 1–3°C, whereas, depending on its exact position within the molecule, the insertion of a double bond, an ester bond, or a methyl branch might decrease T_m by 20–50°C (Gibbs and Pomonis, 1995; Gibbs, 2002). The increased fluidity over a range of temperatures is important not only for spreading the lipid over the cuticular surface, but also for its transport through the cuticular canal pore system toward the exterior (Buckner, 1993).

Mixtures between different n-alkanes or between n-alkanes and branched alkanes form mixed crystals that melt at the weighted average of the single components. In contrast, mixtures of waxes with alkanes and alkenes do not form mixed crystals, since alkanes and alkenes crystallize separately, i.e., alkenes melt earlier ($T_m < 25$°C in alkenes with <30 carbon atoms), resulting in solid alkane crystals suspended in a liquid alkene matrix (cf., Geiselhardt et al., 2010). Considering these physicochemical aspects, the (semi-)liquid nature of the aliphatic tarsal secretion in insects not only might function in maximizing the contact area toward the substratum by filling out its surface irregularities (e.g., Drechsler and Federle, 2006; Persson, 2007), but might also entail further interesting properties. The *in situ* phase differentiation of alkanes and alkenes at ambient temperatures forming a colloid suspension of solid wax crystals within a liquid matrix (Geiselhardt et al., 2010) might induce rate-dependent viscosity changes caused by non-Newtonian shear strains. Shear thickening (dilatant) behavior might help in instantaneous resistance to sudden pull-off forces caused by blasts of wind or tremors, whereas shear thinning might facilitate the release of the tarsi from the substratum during rapid locomotion. Non-Newtonian viscosity shifts are also possible properties of emulsions (Barnes, 1994; Bibette et al., 1999), which consist of a watery phase being dispersed in an oily continuous phase or *vice versa*. Several case studies have shown that arthropod adhesives used in locomotion or prey capture are complex mixtures of aliphatic lipids, carbohydrates, and proteins (e.g., Ishii, 1987; Kosaki and Yamaoka, 1996; Benkendorff et al., 1999; Attygalle et al., 2000; Vötsch et al., 2002; Betz, 2003). Functional analyses of this aspect are currently in progress and have established first indications for shear thickening in smooth systems (Drechsler and Federle, 2006; Drechsler, 2008; Dirks et al., 2009), whereas no deviation from Newtonian rheology has been established in hairy systems (Abou et al., 2010).

An additional advantage of metastable emulsions consisting of both a hydrophobic and a hydrophilic phase is their increased flexibility toward substrates of different surface energy and polarity. Moreover, by fine-tuning the relative proportion of each phase within an adhesive emulsion, the animals might even be able to adapt the viscosity to the current demand. Furthermore, lipid droplets flowing together by coalescence toward the outer margin of the secretion film (Fig. 8.4C, D) might serve as a protective film against the desiccation of the underlying aqueous proteinaceous and/or sugar-containing phase. In the sticky pads of the staphylinid *Stenus* beetles, Kölsch (2000) has observed that, although the lipid droplets are small around the opening of the chitinous ductules that transport the sticky secretion onto the surface of the pads, they merge into larger drops toward the exterior (Fig. 8.4C, D). This mechanism probably helps to keep the sticky pads ready for prey capture during the long periods in which the beetles do not make use of them and might function in analogy to the lachrymal fluid in the mammalian eye in which a lipid secretion is released from the Meibomian glands onto the eye surface, where it protects the subjacent aqueous layer from evaporation (Jester et al., 1981; Yokoi et al., 1999). A similar function has been observed in fruit flies (Tephritidae), where each droplet of an aqueous secretion of the tergal glands is covered by a thin film of a waxy material that retards its evaporation (Evans, 1967). In the process of cleaning itself, the fly appears to wipe these secretions over the surface of its body. In members of the phylum Onychophora, which also use a sticky secretion for prey capture and defence, the lipids (plus the surfactant nonylphenol) may act as an external lining to prevent the secretion sticking to the walls of the slime glands and their reservoir (Benkendorff et al., 1999). Furthermore, Benkendorff et al. (1999) hypothesize that these compounds lower the surface tension of the secretion and therefore facilitate the ejection of the slime.

A protective effect from desiccation has also been presumed in the persistently sticky "ant guards" produced by certain stenogastrine wasps (Keegans et al., 1993). The ball of emulsion placed on eggs of these wasps consists of the contents of Dufour's glands (alkenes and alkoxyethanols in *Liostenogaster flavolineata* and alkenes and a fatty acid salt for *Parischnogaster jacobsoni*, respectively), plus fructose and water (both stemming from the nectar collected by the female). Alkoxyethanols have been particularly noted as being useful in inhibiting the evaporation of water by acting as a monomolecular surface film (Deo et al., 1965; Trapeznikov and Zaozerskaya, 1972). They may have a similar function in *L. flavolineata* by reducing water loss from the ball of emulsified nectar and helping to keep the ball in a semi-liquid state for the growing larva. Mixtures of hydrocarbons (chiefly 9-tricosene), honey, water, and either octadecyloxyethanol or sodium palmitate give highly stable emulsions (Keegans et al., 1993).

Whether the aliphatic compounds found in the defence secretion of termites (mainly Coptotermitinae and Macrotermitinae; cf. Table 8.1) have similar functions in helping to prevent the aqueous phase from sticking to the walls of the outlet structures and keeping the final outlet pores clean remains speculative. The frontal glands of *Macrotermes goliath* soldiers produce large amounts of normal and isoalkanes (C_{22}–C_{34}) in addition to free fatty acids, phospholipids, and sterols. Saturated straight chain and methyl branched and unsaturated alkanes have also been found in the frontal gland secretion of *M. subhyalinus* (Cmelik, 1971; Baker and Wasmsley, 1982). Blum et al. (1982), Howard and Blomquist (1982), and Prestwich (1984) suggest that the use of fatty acids as an ant repellent is a widespread phenomenon among insects. However, in *Coptotermes formosanus*, the identified free fatty acids lignoceric acid and hexacosanoic acid presumably do not contribute to soldier unpalatability. Their functional role remains to be clarified.

The possible combined function of the water-repellent lipid fraction as a protective shield that prevents the aqueous glue fraction from drying out and sticking to the walls of the outlet ductules rewards further investigation.

Notwithstanding the physicochemical properties of suspension- and emulsion-like adhesives, the properties of the pure hydrocarbon molecules might also largely influence their adhesive performance. This might be attributable to viscosity and surface tension effects and to molecular re-orientations and intermolecular attraction of the hydrocarbon chains in thin liquid films (Geiselhardt, 2009). For instance, in hydrocarbon molecules, viscosity increases with chain length and decreases with degree of unsaturation and number of double bonds and methyl branches, the position of double bonds and methyl branches having a modifying effect on this behavior (Geiselhardt, 2009). Indeed, Geiselhardt et al. (2009) have experimentally shown that supplementation of unsaturated components (e.g., (Z)-alkenes) to the adhesive tarsal secretions of the potato beetle *Leptinotarsa decemlineata* brings about a significant reduction of friction forces.

A special mechanism of the hardening of a triglyceride-based adhesive secretion that is not based on a polymerization process has been established in aphids, whereby the hardening of the wax-like cornicle secretion droplets occurs within seconds after they are discharged to the exterior (Edwards, 1966; Strong, 1967; Eisner, 1970). In this system, globules within the body constitute the lipid material that is emulsified in an aqueous medium. The globules coalesce and harden as soon as sufficient water has evaporated to supersaturate the solutes contained therein. This evaporation has been hypothesized to cause some supercooling of the droplet, a process that aids in the rapid crystallization of the liquid material. Alternatively, "the cornicle wax" has been suggested to be in a stable liquid-crystalline state within the aphid and to change to the solid crystal phase upon contact with a seeding nucleus. Since the melting point of the wax (37.5–48°C depending on species) is higher than ambient temperatures, crystallization by seeding can occur under natural conditions. Because of this delayed crystallization, the waxes can literally form a cast around the mouthparts of a predator and impede any further movement.

8.4.2 Carbohydrates

Compared with other substance classes within the adhesive secretions of insects, little analytical effort has been concentrated into the characterization of carbohydrates, so that in many cases, only general indications regarding their occurrence have been presented. Often, only simple histochemical detections of "carbohydrates", "(poly-)saccharides", or "mucosubstances" have been performed. In general, carbohydrates are subdivided into three major categories, i.e., monosaccharides (e.g., pentoses and hexoses), oligosaccharides (consisting of 2–6 monosaccharides), and polysaccharides (i.e., higher molecular saccharides). The few studies that make more detailed statements (Table 8.1) name glucose, trehalose (presumed), and (muco-)polysaccharides (containing glucose, galactose, mannose, β-glucopyranose, and/or (N-acetyl-β-) glucosamine) as constituents of insect adhesives.

In Coptotermitinae (Rhinotermitidae), carbohydrates form an important component of the adhesive defence secretion. Alarmed *Coptotermes* spp. emit a milk-like suspension of lipids (largely n-alkanes C_{22}–C_{27}) in aqueous mucopolysaccharides based on glucosamine and glucose units from a large abdominal reservoir of their frontal gland; this material subsequently hardens to a colorless-resilient film on the attacker (Moore, 1968; Eisner, 1970; Baker and Wasmsley, 1982; Chuah et al., 1990). No chemical reaction seems to be involved, since the latex can be reconstituted with the addition of water. Mucopolysaccharides are also part of the defence glue discharged by the labial glands of certain Termitidae (Table 8.1). For instance, in *Pseudacanthotermes spinger* (Macrotermitinae), the soldiers release a defensive salivary gland secretion that operates in two different ways: first, by immobilizing the foe with an aqueous solution of a polysaccharide that, on evaporation of the solvent, forms an enrobing solid film, and second, by simultaneously applying toxic macrolactones (Plasman et al., 1999).

Amino sugars such as glucosamine and galactosamine are repetitive components of the polysaccharides of proteoglycans (i.e., heavily glycosylated glycoproteins) that are constituents of the extracellular matrix and that impart highly viscoelastic properties to the connective tissue of animals. Similarly, if present in the adhesive secretions of insects, carbohydrates in the form of amino sugars or mucopolysaccharides (= glycosaminoglycans) often appear to be covalently bonded to polypeptides, forming glycoproteins. Whereas in other organisms, evidence exists that sugars (e.g., the sugar N-acetyl-D-glucosamine of the holdfast system of the bacterium *Caulobacter crescentus*; Tsang et al., 2006) found in glycoproteins are responsible for the stickiness of the secreted glues (Lee, 2010), only a few studies have expanded ideas on the possible functions of sugar components in insect adhesives. Vötsch et al. (2002) suggest that the water-soluble glucose-containing carbohydrate polymers present in the tarsal liquid of the locust might be involved in physicochemical processes that increase the viscosity of the continuous phase of the established emulsion.

According to Graham (2008), the water-binding capacity of the large carbohydrate component in the glycoproteins of the *Drosophila* larval glue (used for the anchorage of the pupa) accounts for their stickiness. Water attractance *via* the hydration of charged molecule groups are a general feature of many glycoproteins and this contributes to the stickiness of these components.

8.4.3 Aromatic Compounds

This substance class comprises benzol (= benzene) and its derivatives and substances with several condensed benzene rings (Walter and Francke, 2004). Phenolic substances and (benzo- and tolu-) quinones have mainly been detected in the adhesive defence secretion of termites and ants, although phenols are also thought to be present in the secretion used in egg anchorage in pierid butterflies (Table 8.1).

Being widespread among the arthropods, quinones are nonspecific toxicants and repellents that easily penetrate the cuticle and denature proteins (e.g., Eisner, 1970; Prestwich, 1979a; Blum, 1981). At the same time, quinones induce quinone-tanning, i.e., an oxidative process that leads to the production of quinone-tanned glues by polymerizing proteins (e.g., Waite, 1990; Graham, 2008). One widely investigated example of quinone-tanning involves dihydroxyphenylalanine (Dopa)- and hydroxyproline-based reactions in which the hydroxyl groups of these posttranslationally modified amino acids enhance the potential for hydrogen bonding and other interactions.

Similar tanning actions appear to be widespread among termites in which such compounds are used to produce toxic glues that are squirted toward an arthropod assailant (Table 8.1). In *Mastotermes darwiniensis* (Mastotermitidae), the soldiers combine p-benzoquinone and traces of toluquinone with salivary proteins, resulting in a mixture that is initially mobile but sets to a dark rubber-like material that immobilizes the enemy. At the same time, the glue-like material possibly retains a toxic or irritant substance (e.g., aggressive quinones and terpenes) for a longer period of time on the surface of an arthropod adversary (Moore, 1968; Eisner, 1970; Baker and Wasmsley, 1982). Similarly to *Mastotermes*, the soldiers of *Macrotermes carbonarius* (Termitidae: Macrotermitinae) produce an aqueous solution of toluquinone with a small amount of benzoquinone. The secretion is pumped from massive labial glands into wounds produced by the mandibles (Baker and Wasmsley, 1982). The same is true for some *Odontotermes* species that emit, from the labial gland, an aqueous brown secretion that becomes sticky on exposure to the air. This secretion has been found to be a mixture of benzoquinone and protein (Wood et al., 1975; Baker and Wasmsley, 1982).

8.4.4 Isoprenoids (Terpenes and Steroids)

Isoprenoids are composed of isopren (C_5H_8) units. Insect adhesives contain a broad spectrum of such substances, including monoterpenes (e.g., iridodial), sesquiterpenes (e.g., cadinene and cadinene aldehyde), diterpenes, triterpenes, and steroids (e.g., cholesterol; Table 8.1). Terpenoids and steroids have predominantly been established in termites as a compound of their defensive glue secretion (Table 8.1). In general, according to their toxicity, they play an important role in insect defence systems (Dettner, 2010). In termites, terpenes, and steroids have been established in Rhinotermitidae and Termitidae in which they possibly occur in complex mixtures together with aliphatic compounds, carbohydrates, aromatic compounds, and (glyco-) proteins (Table 8.1). According to their diversity, substances secreted from the defensive glands can be used for various purposes, i.e., (1) as glues that become sticky after air exposure, (2) as contact poisons, (3) as alarm pheromones attracting soldiers and repelling workers, (4) as irritants causing repellence, deterrence, or disorientation, (5) as greases that prevent wounds caused by mandibular bites from healing, and (6) as antibacterial and antifungal agents (e.g., Prestwich, 1979a; Deligne et al., 1981; Quennedey, 1984; Roisin, 2000; Reinhard et al., 2003; Šobotnik et al., 2010b).

The most highly evolved form of defence in termites is used by the nasute soldiers (Termitidae: Nasutitermitinae) that totally avoid any contact with the enemy. Their head capsule has developed into an elongated rostrum called the nasus through which a gluey viscous secretion is ejected (Fig. 8.1D); the secretion is produced by the frontal head gland. In *Nasutitermes* termites, this spray is a viscous entangling agent, an irritant, and a topical poison (Eisner et al., 1976). Since the secretion is forcefully sprayed, viscous resistance to rapid flow must be critical in this system. Whereas in this group, the monoterpenes are mainly considered toxicants, irritants, and alarm pheromones (e.g., Nutting et al., 1974; Howse, 1975; Eisner et al., 1976; Baker and Wasmsley, 1982), they also seem to play an important role in the formation of the glue (e.g., Vrkoč et al., 1973; Prestwich, 1979a; Deligne et al., 1981; Šobotnik et al., 2010b). Prestwich (1979b, 1983) suggests a model according to which the hydrophobic monoterpenes (e.g., α-pinene, β-pinene, limonene, and terpineole according to Moore, 1968; Eisner, 1970; Vrkoč et al., 1973) are good wetting agents for the lipophilic cuticular surface of an attacking insect whose cuticular waxes might even be dissolved by the monoterpenes. During this wetting process, the monoterpenes effectively display their irritant and toxic effect on the assailant. The evaporation of the monoterpenes is probably reduced by the presence of moderate amounts of diterpenes (e.g., structurally unusual bi- and tricyclic diterpenes in *Trinervitermes* according to Meinwald et al. (1978)), thus enhancing the stickiness of the secretion. Prestwich (1979b) suggests that the role of the diterpenes is to provide the high viscosity required of a good glue. Meinwald et al. (1978) stress that such glues remain sticky for several days. This model is analogous to pine resin, in which diterpenes with one or two polar groups are dissolved in monoterpene hydrocarbons (Prestwich, 1979b).

Notably, the glue produced by the nasute soldiers is advanced in that it is a tractable glue that can be reconstituted when dried out merely by adding solvent (Prestwich, 1984). This is in contrast to the intractable glue produced, by *Coptotermes* termites, from an aqueous lipid-glycoprotein-mucopolysaccharide mixture. This material rapidly becomes sticky on exposure to air (possibly by polymerization) and cannot be reconstituted once dried out (Prestwich, 1984).

Other examples of monoterpene-based glues involve the monoterpene iridodial that exclusively functions as an entangling agent, since it exhibits no insecticidal properties (Blum and Hermann, 1978; Dettner et al., 1985; Dettner, 2010). In the oxyteline staphylinid beetle *Deleaster dichrous* and the ants *Tapinoma* spp., low-molecular compounds such as ketones and quinones have been found in addition to the monoterpene iridodial, which polymerizes when exposed to air to form a sticky butter- or glasslike substance. At the same time, the viscosity of the secretion prevents the vaporization of reactive (carbonylic) substances (Trave and Pavan, 1956; Dettner et al., 1985; Dettner, 1993, 2010).

8.4.5 Heterocyclic Compounds

Heterocyclic compounds contain at least one element other than carbon, such as sulfur, oxygen, or nitrogen within a ring structure. Such compounds have only been established in the adhesive secretion of the heteropteran Tingidae, in the form of chromones and chromanones (Table 8.1).

8.4.6 Amino Acids, Peptides, and Proteins

Amino acids, peptides, and proteins are almost always present in the adhesive secretion of insects (Table 8.1). Aliphatic amino acids have been frequently determined; they probably not only form parts of peptides and proteins, but may also occur in the form of free amino acids in the secretion. Amino acids that have been found in especially high levels in adhesive protein-based secretions are glycine, alanine, valine, leucine, serine, threonine, cysteine, arginine, aspartic acid, proline, and posttranslationally modified 4-hydroxyproline or phosphorylated serine giving rise to polyphosphoproteins. In addition to their amino acid content, oligo- and polypeptides within the adhesive secretion have been characterized with regard to their molecular sizes that span a range from >500 to 8 kDa, although most peptides are in the range below 300 kDa (Table 8.1). The glue-like sericin-type proteins in the silk of ditrysian Lepidoptera (mainly analyzed in *Bombyx mori*) have been especially well-investigated. To date, 15 different sericin types have been identified that are partly glycosylated with diverse carbohydrate units (Table 8.1). Other than pure proteins, conjugated proteins occur in adhesive insect secretion and comprise glycoproteins and mucopolysaccharides, lipoproteins (presumed), and (poly-)phosphoproteins (e.g., those containing high levels of phosphorylated serine).

Protein glues are of special interest mainly because of the possibility of cloning the encoding genes for their key components and expressing the bioadhesive proteins by recombinant means (Li et al., 2008). Notwithstanding, despite of their widespread occurrence in insect glues, only a few studies are available in which peptides and proteins have been characterized in greater molecular detail. These studies comprise adhesive secretions used in more permanent junctions such as egg and pupal anchorage to the substratum and cocoon-like retreat building (cf., Table 8.1).

8.4.6.1 Proteins Employed in Egg Anchorage

Li et al. (2008) have investigated the saturniid gum moths *Opodiphthera* spp. in closer detail and established that a treacle-like protein-based liquid egg attachment glue is stored in the reproductive gland (colleterial gland) reservoirs of gravid females. The viscous fluid sets quickly to form a highly elastic hydrogel that binds newly laid eggs to the substratum and to each other. In general, egg anchorage proteins in insects contain high levels of Gly, Ser, and/or Pro (Li et al., 2008), which is consistent with our knowledge that structural, adhesive, and elastic proteins are typically rich in Gly, Pro, Ser, and/or Gln (Graham, 2008). According to Graham (2008), tandem repeats of Pro- and Gly-rich motifs can form a range of specialized poly-Pro/β-turn conformations called Pro-β helices. The dipeptide Pro-Gly is particularly likely to constitute the center of a type II β-turn, and tandems of this type constitute a β-turn spiral (nanospring) that is considered highly elastic.

The proteins in bioadhesive secretions are often extensively glycosylated (Graham, 2008), which often enhances their solubility. Moreover, highly glycosylated proteins such as mucins are known to have adhesive properties (Beeley, 1985). Other subunits of an adhesive protein improve adhesion by their elastin-like elastic properties (e.g., Choresh et al., 2009). They form long dissipative bonds that elongate over a long distance during pull-off, so that the elastic energy is dissipated within the glue rather than being used to break the interfacial adhesion bonds toward the substratum (cf., Persson, 2007). Such coupling of adhesion with extension is a common design principle of natural adhesives (Lee, 2010). In Araneae, Sahni et al. (2010) have established that the micron-sized glue droplets are composed of an aqueous coating of salts surrounding nodules made of physically and chemically cross-linked glycoproteins. The pull-off forces of these glycoproteins are highly rate-dependent, suggesting that the spider glue droplets behave as a viscoelastic solid, and that the elasticity of the glue is critical in enhancing adhesion caused by specific adhesive ligands. At low extension rates, similar to the movements of trapped insects, glycoproteins deform like an ideal elastic rubber, which is essential in retaining the insects trapped in the web. At high extension rates, the adhesive forces are dramatically enhanced, because of high viscous effects, making it easier for the threads of capture silk to hold on to fast-flying insects when they initially impact webs (Sahni et al., 2010).

In the fibrous binder found in the egg masses of planthoppers, Li et al. (2008) have found unusually high proportions of both Val and Pro. Several insect hydrogels contain 4-hydroxyproline, which in some cases is actually the most abundant amino acid. This modified amino acid is also found in many other protein-based bioadhesives (e.g., Benkendorff et al., 1999; Sagert et al., 2006).

8.4.6.2 Proteins Employed in Terrestrial Cocoon Building

Much work has been performed on the sericin-like (glyco-) proteins that function in cocoon building in Lepidoptera. The silk produced by the labial glands of ditrysian lepidopteran larvae has two main components, i.e., the highly elastic fibroin and the gelatinous sericin (Akai, 1984). Whereas the term fibroin is used to describe the protein of the solid fibers of lepidopteran cocoons, the cement that coats the fibers and that frequently sticks them together is another protein called sericin. Sericins are a family of at least six glycoproteins functioning to glue the fibroin threads (Fedič et al., 2002). The sericins of *Bombyx mori* range in size from 400 to 65 kDa. Sericin is characterized by its unusually high serine content (up to 30%), which gives it a high hydrophilicity and a sensitivity to chemical modification (Teramoto and Miyazawa, 2005). The sericin domains that are responsible for the special properties of these proteins, such as their stickiness, substrate adherence, and mechanical properties (Teramoto and Miyazawa, 2005), remain to be identified (Sehnal, 2008). However, according to Graham (2008), a propensity of β-style hydrogen bonding is considered the basis of the silk-gluing properties of sericins. The sericin content in silkworm (*Bombyx mori*) cocoons varies from 19 to 28% and gradually declines during cocoon spinning. It is usually maximal in the outer layer, becomes progressively lower in the central layers, and is minimal in the most-inner cocoon layer that is spun last. According to Sehnal (2008), the rheology of silk material inside the gland and during spinning leads to a layered distribution of different sericins in the spun-out silk fiber, and these layers solidify successively. The outer sericin layer remains sticky for some time and serves as a glue in cocoon construction. Sericin proteins stick to surfaces, and since they contain positively charged residues such as arginine or lysine, they provide ideal coating, in biomedical applications, for the attachment of cells with a predominantly negatively charged glycocalyx (Gotoh et al., 1998).

8.4.6.3 Proteins Employed in Underwater Retreat Building

Although uncertainty remains as to whether the peripheral layer of limnephilid trichopteran silks is sericin, it may serve an equivalent adhesive function (Engster, 1976b). The reducing ends of the sugars in the gum might bind to the amino groups of fibroin asparagine, or the amino groups of any galactosamine in the gum might hydrogen-bond to the carboxyl groups of fibroin glutamic or aspartic acid residues. The difficulty in separating the two components of *Pycnopsyche* silk might be attributable to a combination of extensive hydrogen-bonding and ionic linkages between the fibroin and the silk gum molecules. According to Chen and Cyr (1970), hydrogen-bonding is especially important in wet adhesives. If serine is the major amino acid in sericin-like silk gum, and if the arginine is mainly from fibroin, the hydroxyl groups of serine might hydrogen-bond with the guanidium groups of arginine residues. The serine hydroxyl groups might also form ester links with the aspartic or glutamic acids of fibroin (Hunt, 1970). This last type of linkage has also been suggested for several lepidopteran silks (Seifter and Gallop, 1966). The carbohydrate residues of either component, but more likely those of the silk gum, possibly also help bind the two silk components together.

Stewart and Wang (2010) have found that the underwater silk of the caddisfly *Brachycentrus echo* (Brachycentridae) possesses a high content of phosphorylated serines (pS) that have also been identified in marine underwater bioadhesives. The mussel foot protein mfp-5 at the interface of the adhesive plaque and the substrate contains several pS residues that might promote interfacial adhesion to calcareous materials (Waite and Qin, 2001). The same is true for the adhesive secreted in Echinodermata (Flammang et al., 2009) and the marine annelid *Phragmatopoma californica* (Stewart et al., 2004) (Chapter 10, p. 169). In this annelid, polyphosphoproteins may promote interfacial adhesion to mineral substrates but probably also play an important role in cohesion and the triggered setting of the glue. Polymeric phosphates are widely known as adhesion promoters in the coatings industry and in dentistry. Hence, the caddisfly silk phosphates might also function to promote adhesion to wet substrates in an aquatic environment. The phosphates attached to the serines are negatively charged, whereas other amino acids in the protein are positively charged. Such chains of proteins with alternating regions of positive and negative charges line up in parallel with positive and negative charges attracting each other (Stewart and Wang, 2010).

In stenopsychid caddisflies, Wang et al. (2010) have established a novel protein (Smsp-72k) that exists on the peripheral layer of a silk filament and has an adhesive function, on the basis that the high content of cysteine residues and charged amino acids in this protein is shared by many potential and known adhesive proteins that are secreted underwater. This is in contrast to silks of terrestrial animals such as the silkworm and spider (Wang et al., 2010).

The underwater silks of the chironomid midge *Chironomus tentans* (Chironomidae) also contain large amounts of cysteine (Cys) caused by multiple tandem copies of Cys-containing motifs (Dignam and Case, 1990; Case and Wieslander, 1992; Case et al., 1994; Wieslander, 1994; Case and Thornton, 1999). According to Case and Wieslander (1992), the secretory proteins form a random 3D network primarily attributable to the large number of contacts made between numerous spI molecules. The Cys residues form intramolecular or intermolecular disulfide bonds with a Cys in another domain, resulting in a network of parallel bundles. Cysteine-rich proteins are also typical for other underwater adhesive proteins such as the *Mytilus galloprovincialis* foot protein 2 (Mgfp2), a component of the adhesive plaque of the mussel. Similarly, the barnacle *Megabalanus rose* cement protein (Mrcp-20k) also has a high content of cysteines and charged amino acids; it serves to attach to rocks and other surfaces underwater. Further evidence comes from many cell adhesion proteins that can promote cell adhesion through their cysteine-rich domains (e.g., Zigrino et al., 2007). Moreover, in these systems, cysteine is assumed to play a role in maintaining the topology of charged amino acids on the molecular surface by the formation of intra- and intermolecular disulfide bonds (Smith et al., 1995). Generally, the solubility of proteins in water increases with the content of hydrophilic amino acids (Teramoto and Miyazawa, 2005). However, such disulfide cross-links (e.g., toward the L-fibroins of the silk) might render the above-mentioned Smsp-72k of stenopsychid caddisflies (Wang et al., 2010) insoluble, despite its generally hydrophilic nature.

The hydroxyl groups of amino acids with hydroxyl side-chains (serine, threonine) play a role in removing the weak boundary-water layer in the *Mytilus* byssus protein (Waite, 1987). They are also a major component of the larval salivary glues used to anchor the puparia of *Drosophila* spp. to the substrate (Ramesh and Kalisch, 1988; Swida et al., 1990) and are of general importance in wet adhesives (Chen and Cyr, 1970). In the above-mentioned caddisfly silk, the hydroxyl groups of Ser and Thr might be able to hydrogen-bond with certain amino acids (Arg, Asp, Glu) of the fibroin component of the silk (Engster, 1976b) and thus link the peripheral Smsp-72k to the filament region protein H-fibroin.

8.4.7 Other Systems

Female sphecid wasps such as representatives of the genus *Liris* use glue to attach their eggs firmly to the outside of their paralyzed prey. Gnatzy et al. (2004) suggest that the proteinaceous material of the content of Dufour's gland is used in this context. The established additional lipophilic material might be involved in the gluing process, thereby softening or diluting the secretion so that it is easier to apply.

The identity and function of proteins in the tarsal adhesive secretion of insects has not as yet been investigated, and studies are confined to the general detection of peptides and single amino acids in the secretion (cf., Table 8.1). Vötsch et al. (2002) ascribe the increase of the viscosity of the fluid and their possible function as emulsifiers and/or surfactants to these proteins.

Abbreviations

ba	bacterium
balab	basal labyrinth
bm	basal membrane
cc	canal cell

cr	cross section of ramifications of adhesive trichome
ct	connective tissue
ctr	cross section of adhesive trichome
cu	cuticle
cyt	cytoplasm
d	chitinous ductules that transport the secretion toward the exterior
ea	end-apparatus
ef	epicuticular filaments
ep	epicuticle
epd	epidermis
fi	filament-like pore canals
fm	fibrillar matrix
ger	granular endoplasmic reticulum
go	Golgi apparatus
hs	hair shaft
hv	connecting membrane
ic	intercalary cell
jm	joint membrane
l	lipid secretion
lp	labial palps
mb	multilamellar body
mt	mitochondrion
mu	musculature
mv	microvilli
nc	nucleus
pgl	paraglossae (= sticky pads)
pm	prementum
pr	procuticle
rc	receiving canal
res	secretion reservoir
sc	scale of a collembole
scsp	subcuticular space
se1	presumed watery protein/carbohydrate containing phase of secretion
se2	presumed lipoid phase of secretion (colored pink)
ser	smooth-surfaced endoplasmic reticulum
sj	septate junctions
sv	secretory vesicle
tc	terminal cell
tr	trichome
va	vacuole

Acknowledgments

I wish to thank my wife Dr. Heike Betz and both the editors Dr. Janek von Byern and Dr. Ingo Grunwald for their editorial help with the manuscript. Gabriela Schmid-Kloss re-drew Figs. 8.1A and 8.2. I thank Dr. Theresa Jones who corrected the English of the manuscript.

References

Abbot A (1990) Bioadhesives: potential for exploitation. Science Progress 74: 131–146.

Abed D, Farine JP, Francois J, and Brossut R (1995) Protein and amino acid composition of the tergal gland secretions of *Blatta orientalis* and *Eurycotis floridana* (Dictyoptera: Blattidae). Chemoecology 5/6(1): 19–25.

Abou B, Gay C, Laurent B, Cardoso O, Voigt D, Peisker H, and Gorb SN (2010) Extensive collection of femtolitre pad secretion droplets in the beetle *Leptinotarsa decemlineata* allows nanolitre microrheology. Journal of the Royal Society of Interface, eprint arXiv:1002.2235.

Akai H (1983) II: The structure and ultrastructure of the silk gland. Experientia 39: 443–449.

Akai H (1984) The ultrastructure and functions of the silk gland cells of *Bombyx mori*. In: King RC and Akai H (eds) Insect Ultrastructure, Vol. 2. Plenum Press, New York: pp 323–359.

Akai H (1998) Silk glands. In: Harrison FW and Locke M (eds) Microscopic Anatomy of Invertebrates, Vol. 11A: Insecta. Wiley, New York: pp 219–253.

Akai H, Hakim RS, and Kristensen NP (2003) Labial glands, silk and saliva. In: Kristensen NP (ed) Handbook of Zoology 4: Arthropoda, Insecta. Part 36: Lepidoptera: Moths and Butterflies, Vol. 2: Morphology, Physiology and Development. Walter de Gruyter, Berlin: pp 377–388.

Alberti G and Storch V (1976) Transmissions- und rasterelektronenmikroskopische Untersuchung der Spinndrüsen von Embien (Embioptera, Insecta). Zoologischer Anzeiger 197: 179–186.

Ambsdorf J and Peter MG (1992) Strategien und Techniken des Klebstoffeinsatzes in der Natur – Anregungen für die Technik. Universität Kiel: Studie im Auftrag des Ministeriums für Natur, Umwelt und Landesentwicklung des Landes Schleswig-Holstein: 82 pp.

Amornsak W, Noda T, and Yamashita O (1992) Accumulation of glue proteins in the developing colleterial glands of the silkworm, *Bombyx mori*. Journal of Sericultural Science of Japan 61(2): 123–130.

Attygalle AB, Aneshansley DJ, Meinwald J, and Eisner T (2000) Defense by foot adhesion in a chrysomelid beetle (*Hemisphaerota cyanea*): characterization of the adhesive oil. Zoology 103: 1–6.

Attygalle AB, Siegel B, Vostrowsky O, Bestmann HJ, and Maschwitz U (1989) Chemical composition and function of metapleural gland secretion of the ant, *Crematogaster deformis Smith* (Hymenoptera: Myrmicinae). Journal of Chemical Ecology 15(1): 317–328.

Baer B, Maile R, Schmid-Hempel P, Morgan ED, and Jones GR (2000) Chemistry of a mating plug in bumblebees. Journal of Chemical Ecology 26(8): 1869–1875.

Bagneres AG, Clément JL, Blum MS, Severson RF, Joulie C, and Lange C (1990) Cuticular hydrocarbons and defensive compounds of *Reticulitermes flavipes* (Kollar) and *R. santonensis* (Feytaud): polymorphism and chemotaxonomy. Journal of Chemical Ecology 16(12): 3213–3244.

Bährmann R (2002) Die Mottenschildläuse. Aleyrodina. Die Neue Brehm-Bücherei Bd. 644. Pflanzensaugende Insekten Band 2. Westarp Wissenschaften-Verlagsgesellschaft mbH, Hohenwarsleben.

Baker R and Wasmsley S (1982) Soldier defense secretions of the South American termites *Cortaritermes silvestri*, *Nasutitermes* sp. n.d. and *Nasutitermes kemneri*. Tetrahedron 38(13): 1899–1910.

Baker JE, Sukkestad DR, Nelson DR, and Fatland CL (1979) Cuticular lipids of larvae and adults of the cigarette bettle, *Lasioderma serricorne*. Insect Biochemistry 9: 603–611.

Barnes HA (1994) Rheology of emulsions – a review. Physicochemical and Engineering Aspects 91: 89–95.

Barth R (1953) Die Fangdrüsen an den Beinen von *Zelus* (*Diplocodus*) *leucogrammus* (Perty 1834) (Heteroptera, Reduviidae). Zoologische Jahrbücher: Abteilung für Anatomie und Ontogenie der Tiere 73: 325–336.

Bauchhenß E (1979) Die Pulvillen von *Calliphora erythrocephala* (Diptera, Brachycera) als Adhäsionsorgane. Zoomorphologie 93: 99–123.

Bauchhenß E and Renner M (1977) Pulvillus of *Calliphora erythrocephala* Meig (Diptera; Calliphoridae). International Journal of Insect Morphology and Embryology 6(3/4): 225–227.

Bauer T and Kredler M (1988) Adhesive mouthparts in a ground beetle larva (Coleoptera, Carabidae, *Loricera pilicornis* F.) and their function during predation. Zoologischer Anzeiger 221(3/4): 145–156.

Bauer T and Pfeiffer M (1991) Shooting springtails with a sticky rod: the flexible hunting behaviour of *Stenus comma* (Coleoptera; Staphylinidae) and the counter-strategies of its prey. Animal Behaviour 41: 819–828.

Beament JWL (1962) The surface properties of insects – some evolutionary and ecological implications. Proceedings of the Linnean Society of London 173: 115–119.

Beament JWL and Lal R (1957) Pennetration through the egg-shell of *Pieris brassicae* (L.). Bulletin of Entomological Research 48: 109–125.

Beckendorf SK and Kafatos FC (1976) Differentiation in the salivary glands of *Drosophila melanogaster*: characterization of the glue proteins and their developmental appearance. Cell 9(3): 365–373.

Beeley JG (1985) Glycoprotein and proteoglycan techniques. In: Burdon RH and van Knipperberg PH (eds) Laboratory Techniques in Biochemistry and Molecular Biology. Elsevier, Amsterdam: pp 105–108.

Benkendorff K, Beardmore K, Gooley AA, Packer NH, and Tait NN (1999) Characterisation of the slime gland secretion from the peripatus *Euperipatoides kanangrensis* (Onychophora: Peripatopsidae). Comparative Biochemistry and Physiology, Part B 124: 457–465.

Betz O (1996) Function and evolution of the adhesion-capture apparatus of *Stenus* species (Coleoptera, Staphylinidae). Zoomorphology 116: 15–34.

Betz O (1998a) Comparative studies on the predatory behaviour of *Stenus* spp. (Coleoptera: Staphylinidae): the significance of its specialized labial apparatus. Journal of Zoology 244: 527–544.

Betz O (1998b) Life forms and hunting behaviour of some central European *Stenus* species (Coleoptera, Staphylinidae). Applied Soil Ecology 9: 69–74.

Betz O (1999) A behavioural inventory of adult *Stenus* species (Coleoptera: Staphhylinidae). Journal of Natural History 33: 1691–1712.

Betz O (2003) Structure of the tarsi in some *Stenus* species (Coleoptera, Staphylinidae): external morphology, ultrastructure, and tarsal secretion. Journal of Morphology 255: 24–43.

Betz O and Kölsch G (2004) The role of adhesion in prey capture and predator defence in arthropods. Arthropod Structure & Development 33: 3–30.

Betz O and Mumm R (2001) The predatory legs of *Philonthus marginatus* (Coleoptera, Staphylinidae): functional morphology and tarsal ultrastructure. Arthropod Structure & Development 30: 77–97.

Betz O, Koerner L, and Gorb SN (2009) An insect's tongue as the model for two-phase viscous adhesives? Adhesion 3: 32–35.

Beutel RG and Gorb SN (2001) Ultrastructure of attachment specializations of hexapods (Arthropoda): evolutionary patterns inferred from a revised ordinal phylogeny. Journal of Zoological Systematics and Evolutionary Research 39: 177–207.

Bibette J, Leal-Calderon F, and Poulin P (1999) Emulsion: basic principles. Reports on Progress in Physics 62: 969–1033.

Billen J, Joye L, and Leuthold RH (1989) Fine structure of the labial gland in *Macrotermes bellicosus* (Isoptera, Termitidae). Acta Zoologica 70(1): 37–45.

Bitsch J (1990) Ultrastructure of the phallic glands of the firebrat, *Thermobia domestica* (Packard) (Thysanura: Lepismatidae). International Journal of Insect Morphology and Embryology 19(2): 65–78.

Blomquist GJ and Dillwith JW (1985) Cuticular lipids. In: Kerkut GE and Gilbert LI (eds) Insect Physiology, Biochemistry and Pharmacology. Pergamon Press, Oxford: pp 117–154.

Blomquist GJ, Tillman-Wall JA, Guo L, Quilici DR, Gu P, and Schal C (1993) Hydrocarbon and hydrocarbon-derived sex pheromones in insects: biochemistry and endocrine regulation. In: Stanley-Samuelson DW and Nelson DR (eds) Insect Lipids: Chemistry, Biochemistry and Biology. University of Nebraska Press, Lincoln: pp 317–351.

Blum MS (1981) Chemical Defenses of Arthropods. Academic Press, New York.

Blum MS and Hermann HR (1978) Venoms and venom apparatuses of the Formicidae: Dolichoderinae and Aneuretinae. In: Bettini S (ed) Arthropod Venoms. Handbook of Experimental Pharmacology. Springer-Verlag, Berlin: pp 871–894.

Blum MS and Hilker M (2002) Chemical protection of insect eggs. In: Hilker M and Meiners T (eds) Chemoecology of Insect Eggs and Egg Deposition. Wiley-Blackwell, Berlin: pp 61–90.

Blum MS, Jones TH, Howard DF, and Overal WL (1982) Biochemistry of termite defenses: *Coptotermes*, *Rhinotermes* and *Cornitermes* species. Comparative Biochemistry and Physiology 71B(4): 731–733.

Blunck H (1912) Beitrag zur Kenntnis der Morphologie und Physiologie der Haftscheiben von *Dytiscus marginalis* L. Zeitschrift für wissenschaftliche Zoologie 100: 459–492.

Bodenstein D (1965) The postembryonic developmental of *Drosophila*. In: Demerec M (ed) Biology of *Drosophila*. Hafner Publishing Company, New York: pp 275–367.

Braekman JC, Daloze D, Dupont A, Pasteels JM, and Josens G (1984) Diterpene composition of defense secretion of four West African *Trinervitermes* soldiers. Journal of Chemical Ecology 10(9): 1363–1370.

Braekman JC, Daloze D, Dupont A, Pasteels JM, Tursch B, Declerco JP, Germain G, and van Meerssche M (1980) Secotrinervitane, a novel bicyclic diterpene skeleton from a termite soldier. Tetrahedron Letters 21: 2761–2762.

Brossut R and Roth LM (1977) Tergal modifications associated with abdominal glandular cells in the Blattaria. Journal of Morphology 151: 259–298.

Brossut R and Sreng L (1980) Ultrastructure comparee des glandes exocrines abdominales des Blattaria (Dictyoptera). International Journal of Insect Morphology and Embryology 9: 199–213.

Brown KS (1975) The chemistry of aphids and scale insects. Chemical Society Reviews 4: 263–288.

Buckner JS (1993) Cuticular polar lipids of insects. In: Stanley-Samuelson DW and Nelson DR (eds) Insect Lipids: Chemistry, Biochemistry and Biology. University of Nebraska Press, London: pp 227–270.

Buschinger A and Maschwitz U (1984) Defensive behavior and defensive mechanisms in ants. In: Hermann HR (ed) Defensive Mechanisms in Social Insects. Praeger Publishers, New York: pp 95–150.

Byrne DN and Bellows TS (1991) Whitefly biology. Annual Review of Entomology 36: 431–457.

Case ST, Powers J, Hamilton R, and Burton MJ (1994) Silk and silk proteins from two aquatic insects. In: Kaplan D, Adams WW, Farmer B, and Viney C (eds) Silk Polymers: Materials Science and Biotechnology, Vol. 544. American Chemical Society, Washington D.C.: pp 80–90.

Case ST and Thornton JR (1999) High molecular mass complexes of aquatic silk proteins. International Journal of Biological Macromolecules 24: 89–101.

Case ST and Wieslander L (1992) Secretory proteins of *Chironomus* salivary glands: structural motifs and assembly characteristics of a novel biopolymer. In: Case ST (ed) Structure, Cellular Synthesis and Assembly of Biopolymers. Springer-Verlag, Berlin: pp 187–226.

Cavill GWK and Ford DL (1960) The chemistry of ants. III. Structure and reactions of iridodial. Australian Journal of Chemistry 13: 296–310.

Chapman RF (1998) The Insects: Structure and Function, 4th Ed. Cambridge University Press, Cambridge.

Chen JL and Cyr CN (1970) Composition producing adhesion through hydration. In: Manley RS (ed) Adhesion in Biological Systems. Academic Press, New York: pp 163–181.

Chen J, Henderson G, and Laine RA (1999) Lignoceric acid and hexacosanoic acid: major components of soldier frontal gland secretions of the formosan subterranean termite (*Coptotermes formosanus*). Journal of Chemical Ecology 25(4): 817–824.

Choresh O, Bayarmagnai B, and Lewis RV (2009) Spider web glue: two proteins expressed from opposite strands of the same DNA sequence. Biomacromolecules 10: 2852–2856.

Chuah CH, Goh SH, and Tho YP (1990) Chemical defense secretions of some species of Malaysian Rhinotermitidae (Isoptera, Rhinotermitidae). Journal of Chemical Ecology 16(3): 685–692.

Cmelik SHW (1971) Composition of the lipids from the frontal gland of the major soldiers of *Macrotermes goliath*. Insect Biochemistry 1: 439–445.

Costa-Leonardo AM (1992) The secretory epithelium of the frontal gland in *Velocitermes* sp. soldiers (Isoptera, Termitidae). Naturalia 17: 99–109.

Costa-Leonardo AM (1998a) Morphology of the frontal weapon in the soldier caste of *Cornitermes cumulans* (Isoptera, Termitidae). Revista Brasileira de Entomologia 41(2–4): 195–197.

Costa-Leonardo AM (1998b) The frontal weapon of the termite soldiers *Serritermes serrifer* (Isoptera, Serritermitidae). Ciencia e Cultura 50: 65–67.

Costa-Leonardo AM (2001) The frontal weapon of the termite *Armitermes euamignathus* Silvestri (Isoptera, Termitidae, Nasutitermitinae). Revista Brasileira de Zoologia 18(2): 411–419.

Costa-Leonardo AM (2004) A new interpretation of the defense glands of neotropical *Ruptitermes* (Isoptera, Termitidae, Apicotermitinae). Sociobiology 44(2): 1–12.

Costa-Leonardo AM and De Salvo CR (1987) A comparative study of the frontal glands in three species of Brazilian termite soldiers (Isoptera, Termitidae). Revista Brasileira de Entomologia 31(3): 465–471.

Creton C and Gorb SN (2007) Sticky feet: from animals to materials. MRS Bulletin 32: 466–472.

Crowson RA (1981) The Biology of the Coleoptera. Academic Press, London.

Czolij RT and Slaytor M (1988) Morphology of the salivary glands of *Mastotermes darwiniensis* Froggatt (Isoptera: Mastotermitidae). International Journal of Insect Morphology and Embryology 17(3): 207–220.

De Marzo L (1984) Osservazioni etologiche sulle larve di *Batrisodes oculatus* Aubé (Coleoptera, Pselaphidae). Frustula Entomologica 7–8: 501–506.

De Marzo L (1985) Organi erettili e ghiandole tegumentali specializzate nelle larve di *Batrisodes oculatus* Aubé: studio morfoistologico (Coleoptera, Pselaphidae). Entomologica (Bari) 20: 125–145.

De Marzo L (1988) Comportamento predatorio, nelle larve di *Pselaphus heisei* Herbst (Coleoptera, Pselaphidae). Attidel XV. Congresso Nazionale Italiano di Entomologia, L'Aquila 1988: 817–824.

de Renobales M, Nelson DR, and Blomquist GJ (1991) Cuticular lipids. In: Binnington K and Retnakaran A (eds) Physiology of the Insect Epidermis. CSIRO Publishing, East Melbourne Victoria: pp 240–251.

Debot L (1932) L'appareil séricigene et les glandes salivaires de la larve de *Simulium*. La Cellule 41: 205–216.

Deligne J, Quennedey A, and Blum MS (1981) The enemies and defense mechanisms of termites. In: Hermann HR (ed) Social Insects II. Academic Press, New York: pp 1–76.

Denis R (1949) Sous-classe des aptérygotes. In: Grassé PP (ed) Traité de Zoologie IX. Masson, Paris: pp 111–275.

Deo AV, Kulkarni SB, Gharpurey MK, and Biswas AB (1965) Specific resistance of the monolayers of *n*-long chain alcohols and *n*-long chain alkoxy ethanols to the evaporation of substrate water molecules. Indian Journal of Chemistry 3: 200–204.

Dettner K (1987) Chemosystematics and evolution of beetle chemical defenses. Annual Review of Entomology 32: 17–48.

Dettner K (1989) Chemische Ökologie – Ein interdisziplinäres Forschungsgebiet zwischen Biologie und Chemie. Zeitschrift für Umweltchemie und Ökotoxikologie 4: 29–36.

Dettner K (1993) Defensive secretions and exocrine glands in free-living staphylinid beetles – their bearing on phylogeny (Coleoptera: Staphylinidae). Biochemical Systematics and Ecology 21(1): 143–162.

Dettner K (2007) Gifte und Pharmaka aus Insekten – ihre Herkunft, Wirkung und ökologische Bedeutung. Entomologie heute 19: 3–28.

Dettner K (2010) Insekten als Nahrungsquelle, Abwehrmechanismen. In: Dettner K and Peters W (eds) Lehrbuch der Entomologie, Teil 2, 2nd Ed. Spektrum Akademischer Verlag, Heidelberg: pp 555–559.

Dettner K, Schwinger G, and Wunderle P (1985) Sticky secretion from two pairs of defensive glands of the rove beetle *Deleaster dichrous* (Grav.) (Coleoptera: Staphylinidae). Gland morphology, chemical constituents, defensive functions, and chemotaxonomy. Journal of Chemical Ecology 11(7): 859–883.

Dewitz H (1884) Ueber die Fortbewegung der Thiere an senkrechten glatten Flächen vermittelst eines Secretes. Pflügers Archiv für die Gesamte Physiologie des Menschen und der Tiere 33: 440–481.

Dignam SS and Case ST (1990) Balbiani ring 3 in *Chironomus tentans* encodes a 185-kDa secretory protein which is synthesized throughout the fourth larval instar. Gene 88: 133–140.

Dirks JH, Clemente CJ, and Federle W (2009) Insect tricks: two-phasic foot pad secretion prevents slipping. Journal of the Royal Society Interface 7: 587–593.

Drechsler P (2008) Mechanics of adhesion and friction in stick insects and tree frogs. PhD thesis, Bayerische Julius-Maximilians-Universität, Würzburg: 149 pp.

Drechsler P and Federle W (2006) Biomechanics of smooth adhesive pads in insects: influence of tarsal secretion on attachment performance. Journal of Comparative Physiology A 192(11): 1213–1222.

Dunham AE and Rudolf VHW (2009) Evolution of sexual size monomorphism: the influence of passive mate guarding. Journal of Evolutionary Biology 22(7): 1376–1386.

Eberhard SH and Krenn HW (2003) Salivary glands and salivary pumps in adult Nymphalidae (Lepidoptera). Zoomorphology 122: 161–167.

Eberhard MJB, Pass G, Picker MD, Beutel R, Predel R, and Gorb SN (2009) Structure and function of the arolium of Mantophasmatodea (Insecta). Journal of Morphology 270: 1247–1261.

Edwards JS (1966) Defence by smear: supercooling in the cornicle wax of aphids. Nature 211: 73–74.

Eisner T (1970) Chemical defense against predation in arthropods. In: Sondheimer E and Simeone JB (eds) Chemical Ecology. Academic Press, New York: pp 157–217.

Eisner T (1972) Chemical ecology: on arthropods and how they live as chemists. Verhandlungen der Deutschen Zoologischen Gesellschaft 65: 123–137.

Eisner T (1988) Insekten als fürsorgliche Eltern. Verhandlungen der Deutschen Zoologischen Gesellschaft 81: 9–17.

Eisner T and Aneshansley DJ (2000) Defense by foot adhesion in a beetle (*Hemisphaerota cyanea*). Proceedings of the National Academy of Sciences 97(12): 6568–6573.

Eisner T, Kriston I, and Aneshansley DJ (1976) Defensive behavior of a termite (*Nasutitermes exitiosus*). Behavioral Ecology and Sociobiology 1: 83–125.

Engster MS (1976a) Studies on silk secretion in the Trichoptera (F. Limnephilidae) I. Histology, Histochemistry, and ultrastructure of the silk glands. Journal of Morphology 150: 183–212.

Engster MS (1976b) Studies on silk secretion in the Trichoptera (F. Limnephilidae) II. Structure and amino acid composition of the silk. Cell and Tissue Research 169: 77–92.

Ernst E (1959) Beobachtungen beim Spritzakt der *Nasutitermes*-Soldaten. Revue Suisse de Zoologie 66: 289–295.

Evans DA, Baker R, Briner PH, and McDowell PG (1977) Defensive secretions of some African termites. Proceedings of the 8th International Congress of the International Union for the Study of Social Insects IUSSI, Wageningen: pp 46–47.

Evans JJT (1967) The integument of the Queensland fruit fly *Dacus tryoni* (Frogg.) I. The tergal glands. Zeitschrift für Zellforschung 81: 18–33.

Farkaš R and Šutakova G (1998) Ultrastructural changes of *Drosophila* larval and prepupal salivary glands cultured in vitro with ecdysone. In Vitro Cellular & Developmental Biology – Animal 34(10): 813–823.

Fedič R, Zurovec M, and Sehnal F (2002) The silk of Lepidoptera. Journal of Insect Biotechnology and Sericology 71: 1–15.

Flammang P, Lambert A, Bailly P, and Hennebert E (2009) Polyphosphoprotein-containing marine adhesives. Journal of Adhesion 85(8): 447–464.

Foldi I (1981) Ultrastructure of the wax-gland system in subterranean scale insects (Homoptera, Coccoidea, Margarodidae). Journal of Morphology 168: 159–170.

Foldi I (1991) The wax glands in scale insects: comparative ultrastructure, secretion, function and evolution (Homoptera: Coccoidea). Annales de la Société Entomologique de France 27: 163–188.

Foldi I and Lambdin P (1995) Ultrastructural and phylogenetic assessment of glands in pit scales (Homoptera: Coccoidea). International Journal of Insect Morphology and Embryology 24(1): 35–49.

Foldi-Hope I (1990) Ultrastructure des cornicules de *Rhopalosiphum padi* L. (Homoptera: Aphidae). International Journal of Insect Morphology and Embryology 19: 25–33.

Fraenkel G and Brookes VJ (1953) The process by which the puparia of many species of flies become fixed to a substrate. Biological Bulletin 105: 442–449.

Gaino E and Mazzini M (1989) Chorionic adhesive material of the egg of the mayfly *Habrophlebia eldae* (Ephemeroptera, Leptophlebiidae): morphology and synthesis. Italian Journal of Zoology 56(4): 291–298.

Gaino E and Mazzini M (1990) Follicle cell activity in the ovarioles of *Habrophlebia eldae* (Ephemeroptera: Leptophlebiidae). Transactions of the American Microscopical Society 109(3): 300–310.

Gaino E and Rebora M (2001) Synthesis and function of the fibrous layers covering the eggs of *Siphlonurus lacustris* (Ephemeroptera, Siphlonuridae). Acta Zoologica 82: 41–48.

Gamo T, Inokuchi T, and Laufer H (1977) Polypeptides of fibroin and sericin secreted from the different sections of the silk gland in *Bombyx mori*. Insect Biochemistry 7: 285–295.

Geiselhardt SF (2009) Chemical, morphological and biomechanical aspects of tarsal attachment in beetles. PhD Thesis, Albert-Ludwigs-Universität, Freiburg, Fakultät für Biologie: 135 pp.

Geiselhardt SF, Geiselhardt S, and Peschke K (2009) Comparison of tarsal and cuticular chemistry in the leaf beetle *Gastrophysa viridula* (Coleoptera: Chrysomelidae) and an evaluation of solid-phase microextraction and solvent extraction techniques. Chemoecology 19: 185–193.

Geiselhardt SF, Lamm S, Gack C, and Peschke K (2010) Interaction of liquid epicuticular hydrocarbons and tarsal adhesive secretion in *Leptinotarsa decemlineata* Say (Coleoptera: Chrysomelidae). Journal of Comparative Physiology A 196: 369–378.

Ghazi-Bayat A and Hasenfuss I (1980) Zur Herkunft der Adhäsionsflüssigkeit der tarsalen Haftlappen bei den Pentatomidae (Heteroptera). Zoologischer Anzeiger 204(1/2): 13–18.

Gibbs A (1995) Physical properties of insect cuticular hydrocarbons: model mixtures and lipid interactions. Comparative Biochemistry and Physiology 112B(4): 667–672.

Gibbs A (2002) Lipid melting and cuticular permeability: new insights into an old problem. Journal of Insect Physiology 48: 391–400.

Gibbs A and Crowe JH (1991) Intra-individual variation in cuticular lipids studied using fourier transform infrared spectroscopy. Journal of Insect Physiology 37(10): 743–748.

Gibbs A and Pomonis JG (1995) Physical properties of insect cuticular hydrocarbons: the effects of chain length, methyl-branching and unsaturation. Comparative Biochemistry and Physiology 112B(2): 243–249.

Gibbs A, Kuenzli M, and Blomquist GJ (1995) Sex- and age-related changes in the biophysical properties of cuticular lipids of the housefly, *Musca domestica*. Archives of Insect Biochemistry and Physiology 29: 87–97.

Gibbs A, Louie AK, and Ayala JA (1998) Effects of temperature on cuticular lipids and water balance in a desert *Drosophila*: is thermal acclimation beneficial? Journal of Experimental Biology 201: 71–80.

Gilbert LI (1967) Lipid metabolism and function in insects. Advances in Insect Physiology 4: 69–211.

Gill RJ (1992) The morphology of whiteflies. In: Gerling D (ed) Whiteflies: Their Bionomics, Pest Status, and Management. Intercept Ltd., Andover: pp 13–46.

Gillett JD and Wigglesworth VB (1932) The climbing organ of an insect, *Rhodnius prolixus* (Hemiptera; Reduviidae). Proceedings of the Royal Society B: Biological Sciences 111: 364–375.

Gillot C (2002) Insect accessory reproductive glands: key players in production and protection of eggs. In: Hilker M and Meiners T (eds) Chemoecology of Insect Eggs and Egg Deposition. Blackwell Publishing, Berlin: pp 37–59.

Gnatzy W (1970) Struktur und Entwicklung des Integuments und der Oenocyten von *Culex pipiens* L. (Dipt.). Zeitschrift für Zellforschung 110: 401–443.

Gnatzy W, Voknandt W, and Schulz S (2004) Dufour gland of the digger wasp *Liris niger*: structure and developmental and biochemical aspects. Cell and Tissue Research 315: 125–138.

Goh SH, Chuah CH, Vadiveloo J, and Tho YP (1990) Soldier defense secretions of Malaysian free-ranging termite of the genus *Lacessititermes*. Journal of Chemical Ecology 16(2): 619–630.

Gorb SN (1998) The design of the fly adhesive pad: distal tenent setae are adapted to the delivery of an adhesive secretion. Proceedings of the Royal Society B: Biological Sciences 265: 747–752.

Gorb SN (2001) Attachment Devices of Insect Cuticle. Kluwer Academic Publishers, Dordrecht.

Gorb SN (2008) Biological attachment devices: exploring nature's diversity for biomimetics. Philosophical Transactions of the Royal Society A: Mathematical, Physical and Engineering Sciences 366: 1557–1574.

Gorb SN and Beutel R (2001) Evolution of locomotory attachment pads of hexapods. Naturwissenschaften 88: 530–534.

Gotoh Y, Tsukada M, and Minoura N (1998) Effect of the chemical modification of the arginyl residue in *Bombyx mori* silk fibroin on the attachment and growth of fibroblast cells. Journal of Biomedical Materials Research 39(3): 351–357.

Graham LD (2008) Biological adhesives from nature. In: Wnek GE and Bowlin GL (eds) Encyclopedia of Biomaterials and Biomedical Engineering, 2nd Ed. Informa Healthcare USA Inc., New York: pp 236–253.

Grassé PP (1982) Termitologia. Anatomie, physiologie, biologie, systematique des termites. Tome I. Masson, Paris, New York.

Hasenfuss I (1977) Die Herkunft der Adhäsionsflüssigkeit bei Insekten. Zoomorphologie 87: 51–64.

Hasenfuss I (1999) The adhesive devices in larvae of Lepidoptera (Insecta, Pterygota). Zoomorphology 119: 143–162.

Henning B (1974) Morphologie und Histologie der Tarsen von *Tettigonia viridissima* L. (Orthoptera, Ensifera). Zeitschrift für Morphologie der Tiere 79: 323–342.

Hermann HR and Blum MS (1981) Defense mechanisms in the social Hymenoptera. In: Hermann HR (ed) Social Insects, Vol. II. Academic Press, New York: pp 77–197.

Hesse R and Doflein F (1943) Tierbau und Tierleben in ihrem Zusammenhang betrachtet. Gustav Fischer, Jena.

Hoffman RT, Schmid ER, and Case ST (1996) A cell-specific glycosylated silk protein from *Chironomus thummi* salivary glands. Journal of Biological Chemistry 271(16): 9809–9815.

Howard RW (1993) Cuticular hydrocarbons and chemical communication. In: Stanley-Samuelson DW and Nelson DR (eds) Insect Lipids: Chemistry, Biochemistry and Biology. University of Nebraska Press, Lincoln, London: pp 179–226.

Howard RW and Blomquist GJ (1982) Chemical ecology and biochemistry of insect hydrocarbons. Annual Review of Entomology 27: 149–172.

Howse PE (1975) Chemical defences of ants, termites and other insects: some outstanding questions. Proceedings of the IUSSI Symposium (Dijon): pp 23–40.

Hunt S (1970) Polysaccharide-Protein Complexes in Invertebrates. Academic Press, London.

Ichinosé T and Kishimoto M (1983) The comparative morphology of the caudal secretory hypodermis of three species of cockroaches, *Periplaneta fuliginosa* Serv., *P. japonica* Karny, and *P. americana* L. (Orthoptera: Blattidae), with regard to their development and rearing densities. Applied Entomology and Zoology 18(2): 191–199.

Ichinosé T and Zennyoji K (1980) Defensive behavior of the cockroaches, *Periplaneta fuliginosa* Serville and *P. japonica* Karny (Orthoptera: Blattidae) in relation to their viscous secretion. Applied Entomology and Zoology 15(4): 400–408.

Ishii S (1987) Adhesion of a leaf feeding ladybird *Epilachna vigintioctomaculata* (Coleoptera: Coccinellidae) on a vertically smooth surface. Applied Entomology and Zoology 22(2): 222–228.

Jackson LL and Blomquist GJ (1976) Insect waxes. In: Kolattukudy PE (ed) Chemistry and Biochemistry of Natural Waxes. Elsevier, Amsterdam: pp 201–233.

Jacobs W and Renner M (1988) Biologie und Ökologie der Insekten. 2nd Ed. Gustav Fischer Verlag, Stuttgart.

Jester JV, Nicolaides N, and Smith RE (1981) Meibomian gland studies: histologic and ultrastructural investigations. Investigative Ophthalmology & Visual Science 20(4): 537–547.

Jin YX, Jiang YH, and Xu MK (2006) Proteome analysis of the silkworm (*Bombyx mori*. L.) colleterial gland during different development stages. Archives of Insect Biochemistry and Physiology 61: 42–50.

Just S and Gronenberg W (1999) The control of mandible movements in the ant *Odontomachus*. Journal of Insect Physiology 45: 231–240.

Keegans SJ, Morgan ED, Turillazzi S, Jackson BD, and Billen J (1993) The Dufour gland and the secretion placed on eggs of two species of social wasps, *Liostenogaster flavolineata* and *Parischnogaster jacobsoni* (Vespidae: Stenogastrinae). Journal of Chemical Ecology 19: 279–290.

Kendall MD (1970) The anatomy of the tarsi of *Schistocerca gregaria* Forskal. Zeitschrift für Zellforschung 109: 112–137.

Kiel E and Röder T (2002) Gelelectrophoretic studies on labial gland secretions of immature blackflies (Simuliidae, Diptera). Archives of Insect Biochemistry and Physiology 61: 42–50.

Kloetzel JA and Laufer H (1969) A fine-structural analysis of larval salivary gland function in *Chironomus thummi* (Diptera). Journal of Ultrastructure Research 29: 15–36.

Kloetzel JA and Laufer H (1970) Developmental changes in fine structure associated with secretion in larval salivary glands of *Chironomus*. Experimental Cell Research 60: 327–337.

Kölsch G (2000) The ultrastructure of glands and the production and function of the secretion in the adhesive capture apparatus of *Stenus* species (Coleoptera: Staphylinidae). Canadian Journal of Zoology 78: 465–475.

Kölsch G and Betz O (1998) Ultrastructure and function of the adhesion-capture apparatus of *Stenus* species (Coleoptera, Staphylinidae). Zoomorphology 118: 263–272.

Korge G (1977) Larval saliva in *Drosophila melanogaster*: production, composition, and relationship to chromosome puffs. Developmental Biology 58: 339–355.

Kosaki A and Yamaoka R (1996) Chemical composition of footprints and cuticula lipids of three species of lady beetles. Japanese Society of Applied Entomology and Zoology 40: 47–53.

Kress H (1982) Biochemical and ontogenetic aspects of glycoprotein synthesis in *Drosophila virilis* salivary glands. Developmental Biology 93: 231–239.

Lane NJ and Carter YR (1972) Puffs and salivary gland function: the fine structure of the larval and prepupal salivary glands of *Drosophila melanogaster*. Wilhelm Roux' Archiv 169: 216–238.

Lee H (2010) Biomaterials: intelligent glue. Nature 465: 298–299.

Lees AD and Hardie J (1988) The organs of adhesion in the aphid *Megoura viciae*. Journal of Experimental Biology 136: 209–228.

Leger RJS (1991) Integument as a barrier to microbial infections. In: Binnington K and Retnakaran A (eds) Physiology of the Insect Epidermis. CSIRO Publishing, East Melbourne Victoria: pp 284–306.

Li D, Huson MG, and Graham LD (2008) Proteinaceous adhesive secretions from insects, and in particular the egg attachment glue of *Opodiphthera* sp. moths. Archives of Insect Biochemistry and Physiology 69: 85–105.

Liang AP and Jiang GM (2003) Wax-exuding cuticular pores in *Oeclidius nanus* Van Duzee (Hemiptera: Fulgoromorpha: Kinnaridae). Journal of Entomological Science 38(2): 247–253.

Liang AP and O'Brien LB (2002) External morphology of the wax glands of *Epiptera woodworthi* (Hemiptera: Fulgoromorpha: Achilidae). Southwestern Entomologist 27: 209–215.

Liang AP and Wilson MR (2002) Wax-secreting cuticular structures in nymphs of *Scolops abnormis* Ball (Hemiptera: Fulgoromorpha: Dictyopharidae). Journal of the Kansas Entomological Society 75(2): 132–137.

Livingstone D (1978) On the body outgrowths and the phenomenon of "sweating" in the nymphal instars of Tingidae (Hemiptera: Heteroptera). Journal of Natural History 12: 377–394.

Locke M (1961) Pore canals and related structures in insect cuticle. Journal of Biophysical and Biochemical Cytology 10: 589–616.

Locke M (1984) The structure and development of the vacuolar system in the fat body of insects. In: King RC and Akai H (eds) Insect Ultrastructure, Vol. 2. Plenum Press, New York: pp 151–197.

Lockey KH (1980) Insect cuticular hydrocarbons. Comparative Biochemistry and Physiology 65B: 457–462.

Lockey KH (1985) Insect cuticular lipids. Comparative Biochemistry and Physiology 81B(2): 263–273.

Lockey KH (1988) Lipids of the insect cuticle: origin, composition and function. Comparative Biochemistry and Physiology, Part B 89(4): 595–645.

Lockey KH (1991) Insect hydrocarbon classes: implications for chemotaxonomy. Insect Biochemistry 21(1): 91–97.

Lüken D, Voigt D, Gorb SN, and Zebitz CP (2009) Die Tarsenmorphologie und die Haftfähigkeit des Schwarzen Batatenkäfers *Cylas puncticollis* (Boheman). Mitteilungen der Deutschen Gesellschaft für allgemeine und angewandte Entomologie 17: 109–113.

Lung O and Wolfner MF (2001) Identification and characterisation of the major *Drosophila melanogaster* mating plug protein. Insect Biochemistry and Molecular Biology 31: 543–551.

Lusby WR, Oliver JE, Neal JW, and Heath RR (1987) Isolation and identification of the major component of setal exudates from *Corythucha ciliata*. Journal of Natural Products 50(6): 1126–1130.

MacGregor HC and Mackie JB (1967) Fine structure of the cytoplasm in salivary glands of *Simulium*. Journal of Cell Science 2: 137–144.

Maschwitz U (1974) Vergleichende Untersuchungen zur Funktion der Ameisenmetathorakaldrüse. Oecologia 16: 303–310.

Maschwitz U and Maschwitz E (1974) Platzende Arbeiterinnen: Eine neue Art der Feindabwehr bei sozialen Hautflüglern. Oecologia 14: 289–294.

Maschwitz U and Tho YP (1974) Chinone als Wehrsubstanzen bei einigen orientalischen Macrotermitinen. Insectes Sociaux 21(3): 231–234.

Maschwitz U, Jander R, and Burkhardt D (1972) Wehrsubstanzen und Wehrverhalten der Termite *Macrotermes carbonarius*. Journal of Insect Physiology 18: 1715–1720.

McFarlane JS and Tabor D (1950) Adhesion of solids and the effect of surface films. Proceedings of the Royal Society of London Series A, Mathematical, Physical and Engineering Sciences 202: 224–243.

Meinwald J, Prestwich GD, Nakanishi K, and Kubo I (1978) Chemical ecology: studies from East Africa. Science 199: 1167–1173.

Meinwald J, Smolanoff J, Chibnall AC, and Eisner T (1975) Characterization and synthesis of waxes from homopterous insects. Journal of Chemical Ecology 1: 269–274.

Miller CE (1942) On the structure of the legs in Reduviidae (Rhynchota). Proceedings of the Royal Entomological Society of London Part A 17: 49–58.

Miller DR and Kosztarab M (1979) Recent advances in the study of scale insects. Annual Review of Entomology 24: 1–27.

Moore BP (1968) Studies on the chemical composition and function of the cephalic gland secretion in Australian termites. Journal of Insect Physiology 14: 33–39.

Moore BP (1969) Biochemical studies in termites. In: Krishna K and Weesner FM (eds) Biology of Termites, Vol. 1. Academic Press, New York: pp 407–432.

Moore I (1964) A new key to the subfamilies of the nearctic Staphylinidae and notes on their classification. The Coleopterists Bulletin 18(3): 83–91.

Morrow PA, Bellas TE, and Eisner T (1976) *Eucalyptus* oils in the defensive oral discharge of Australian sawfly larvae (Hymenoptera: Pergidae). Oecologia 24: 193–206.

Nayler LS (1964) The structure and function of the posterior abdominal glands of the cockroach *Pseudoderopeltis bicolor* (Thunb.). Journal of the Entomological Society of South Africa 27(1): 62–66.

Nelson DR (1993) Methyl-branched lipids in insects. In: Stanley-Samuelson DW and Nelson DR (eds) Insect Lipids: Chemistry, Biochemistry and Biology. University of Nebraska Press, Lincoln: pp 271–315.

Nelson LJ, Cool LG, Forschler BT, and Haverty MI (2001) Correspondence of soldier defense secretion mixtures with cuticular hydrocarbon phenotypes for chemotaxonomy of the termite genus *Reticulitermes* in North America. Journal of Chemical Ecology 27(7): 1449–1479.

Nelson LJ, Cool LG, Forschler BT, and Haverty MI (2008) Cuticular hydrocarbons and soldier defense secretions of *Reticulitermes* in Southern California: a critical analysis of the taxonomy of the genus in North America. Journal of Chemical Ecology 34: 1452–1475.

Noble-Nesbitt J (1991) Cuticular permeability and its control. In: Binnington K and Retnakaran A (eds) Physiology of the Insect Epidermis. CSIRO Publishing, East Melbourne Victoria: pp 252–283.

Noirot CH (1969) Glands and secretions. In: Krishna K and Weesner FM (eds) Biology of Termites. Academic Press, New York: pp 89–123.

Noirot CH and Quennedey A (1974) Fine structure of insect epidermal glands. Annual Review of Entomology 19: 61–80.

Noirot CH and Quennedey A (1991) Glands, gland cells, glandular units: some comments on terminology and classification. Annales de la Société Entomologique de France 27(2): 123–128.

Nutting WL, Blum MS, and Fales HM (1974) Behavior of the North American termite *Tenuirostritermes tenuirostris* with special reference to the soldier frontal gland secretion, its chemical composition, and use in defense. Psyche 81: 167–177.

Ohta M, Matsuura F, Henderson G, and Laine RA (2007) Novel free ceramides as components of the soldier defense gland of the Formosan subterrannean termite (*Coptotermes formosanus*). Journal of Lipid Research 48: 656–664.

Oliver JE, Lusby WR, and Neal JW (1990) Exocrine secretions of the andromeda lace bug *Stephanitis takeyai* (Hemiptera: Tingidae). Journal of Chemical Ecology 16(7): 2243–2252.

Oliver JE, Neal JW, and Lusby WR (1987) Phenolic acetogenins secreted by rhododendron lace bug, *Stephanitis rhododendri* Horvath (Hemiptera: Tingidae). Journal of Chemical Ecology 13(4): 763–769.

Oliver JE, Neal JW, Lusby WR, Aldrich JR, and Kochansky JP (1985) Novel components from secretory hairs of azalea lace bug *Stephanitis pyrioides* (Hemiptera: Tingidae). Journal of Chemical Ecology 11(9): 1223–1228.

Onusseit H (2004) Klebstoffe der Natur. Biologie in unserer Zeit 34(5): 307–314.

Padamwar MN and Pawar AP (2004) Silk sericin and its applications: a review. Journal of Scientific & Industrial Research 63: 323–329.

Pasteels JM, Gregoire JC, and Rowell-Rahier M (1983) The chemical ecology of defense in arthropods. Annual Review of Entomology 28: 263–289.

Pavan PM and Trave R (1958) Etudes sur les Formicidae IV Sur le venin du dolichoderide *Tapinoma nigerrimum* Nyl. Insectes Sociaux V(3): 299–308.

Percy JE and Weatherston J (1974) Gland structure and pheromone production in insects. In: Birch M (ed) Frontiers of Biology: Pheromones, Vol. 32. North-Holland Publishing Co., Amsterdam: pp 11–34.

Percy-Cunningham JE and MacDonald JA (1987) Biology and ultrastructure of sex pheromone-producing glands. In: Prestwich GD and Blomquist GJ (eds) Pheromone Biochemistry. Academic Press, Orlando: pp 27–75.

Perkowska E (1963) Some characteristics of the salivary gland secretion of *Drosophila virilis*. Experimental Cell Research 32: 259–271.

Persson BNJ (2007) Biological adhesion for locomotion: basic principles. Journal of Adhesion Science and Technology 21(12–13): 1145–1173.

Peters W (2010) Integument. In: Dettner K and Peters W (eds) Lehrbuch der Entomologie, Teil 1, 2nd Ed. Spektrum Akademischer Verlag, Heidelberg: pp 1–46.

Piskorski R, Hanus R, Vasickova S, Cvacka J, Šobotnik J, Svatos A, and Valterova I (2007) Nitroalkenes and sesquiterpene hydrocarbons from the frontal gland of three *Prorhinotermes* termite species. Journal of Chemical Ecology 33: 1787–1794.

Plasman V, Daloze D, Braekman JC, Connetable S, Robert A, and Bordereau C (1999) New macrolactones form the defensive salivary secretion of soldiers of the African termite *Pseudacanthotermes spiniger*. Tetrahedron Letters 40: 9229–9232.

Plattner H (1971) Funktionelle Organisation eines Protein-sezernierenden Insektenepithels. Naturwissenschaften 58(6): 325–326.

Plattner H, Salpeter M, Carrel JE, and Eisner T (1972) Struktur und Funktion des Drüsenepithels der postabdominalen Tergite von *Blatta orientalis*. Zeitschrift für Zellforschung 125: 45–87.

Pope RD (1984) Visible insect waxes: form, function and classification. Antenna 8: 4–8.

Prestwich GD (1975) Chemical analysis of soldier defensive secretions of several species of East African termites. Proceedings of the IUSSI Symposium (Dijon) 1975: 149–152.

Prestwich GD (1977) Isotrinervi-2β-ol. Structural isomers in the defense secretions of allopatric populations of the termite *Trinervitermes gratiosus*. Experientia 34(6): 682–684.

Prestwich GD (1979a) Chemical defense by termite soldiers. Journal of Chemical Ecology 5(3): 459–480.

Prestwich GD (1979b) Interspecific variation in the defence secretions of *Nasutitermes* soldiers. Biochemical Systematics and Ecology 7: 211–221.

Prestwich GD (1983) Chemical systematics of termite exocrine secretions. Annual Review of Ecology, Evolution, and Systematics 14: 287–311.

Prestwich GD (1984) Defense mechanisms of termites. Annual Review of Entomology 29: 201–232.

Prestwich GD and Collins MS (1982) Chemical defense secretions of the termite soliders of *Acorhinotermes* and *Rhinotermes* (Isoptera, Rhinotermitinae): Ketones, vinyl ketones and β-ketoaldehydes derived from fatty acids. Journal of Chemical Ecology 8(1): 147–161.

Prestwich GD, Bierl BA, Devilbiss ED, and Chaudhury MFB (1977) Soldier frontal glands of the termite *Macrotermes subhyalinus*: Morphology, chemical composition, and use in defense. Journal of Chemical Ecology 3(5): 579–590.

Prestwich GD, Kaib M, Wood WF, and Meinwald J (1975) 1,13-tetradecadien-3-one and homologs: new natural products isolated from *Schedorhinotermes* soldiers. Tetrahedron Letters 52: 4701–4704.

Prestwich GD, Lauher JW, and Collins MS (1979) Two new tetracyclic diterpenes from the defense secretion of the Neotropical termite *Nasutitermes octopilis*. Tetrahedron Letters: 3827–3830.

Prügel MR and Rühm W (1994) Die Labialdrüsen der Kriebelmückenlarven (Simuliidae, Diptera). Mitteilungen Hamburgisches Zoologisches Museum und Institut 91: 127–146.

Quennedey A (1973) Observations cytologiques et chimiques sur la glande frontale des termites. Proceedings of the VIII International Congress of IUSSI London: pp 324–326.

Quennedey A (1975a) La guerre chemique chez les termites. La Recherche 54(6): 274–276.

Quennedey A (1975b) Morphology of exocrine glands producing pheromones and defensive substances in subsocial and social insects. Proceedings of the IUSSI Symposium (Dijon): pp 1–21.

Quennedey A (1984) Morphology and ultrastructure of termite defense glands. In: Hermann H (ed) Defensive Mechanism in Social Insects. Praeger Publishers, New York: pp 151–200.

Quennedey A (1998) Insect epidermal gland cells: ultrastructure and morphogenesis. In: Harrison FW and Locke M (eds) Microscopic Anatomy of Invertebrates. Vol. 11A: Insecta. Wiley, New York: pp 177–207.

Quennedey A, Brule G, Rigaud J, Dubois P, and Broussut R (1973) La glande frontale des soldats de *Schedorhinotermes putorius* (Isoptera): analyse chimique et fonctionnement. Insect Biochemistry 3: 67–74.

Ramesh SR and Kalisch WE (1988) Glue proteins in *Drosophila nasuta*. Biochemical Genetics 26(7/8): 527–541.

Reidelbach J and Kiel E (1990) Observations on the behavioural sequences of looping and drifting by blackfly larvae (Diptera: Simuliidae). Aquatic Insects 12(1): 49–60.

Reinhard J, Quintana A, Sreng L, and Clement JL (2003) Chemical signals inducing attraction and alarm in European *Reticulitermes termites* (Isoptera, Rhinotermitidae). Sociobiology 42(3): 675–691.

Ribeiro JMC (1987) Role of saliva in blood-feeding by arthropods. Annual Review of Entomology 32: 463–478.

Riley RC and Forgash AJ (1967) *Drosophila melanogaster* eggshell adhesive. Journal of Insect Physiology 13: 509–517.

Rizki TM (1967) Ultrastructure of the secretory inclusions of the salivary gland cell in *Drosophila*. Journal of Cell Biology 32: 531–534.

Roisin Y (2000) Diversity and evolution of caste patterns. In: Abe T, Bignell DE, and Higashi M (eds) Termites: Evolution, Sociality, Symbioses, Ecology. Kluwer Academic Publishers, Dordrecht: pp 95–119.

Romer F (1975) Morphologie und Ultrastruktur der larvalen Oenocyten von *Tenebrio molitor* L. (Insecta, Coleoptera) bei Larve, Puppe und Imago. Zeitschrift für Morphologie der Tiere 80: 1–40.

Romer F (1980) Histochemical and biochemical investigations concerning the function of larval oenocytes of *Tenebrio molitor* L. (Coleoptera, Insecta). Histochemistry 69: 69–84.

Rościszewska E (1991) Ultrastructural and histochemical studies of the egg capsules of *Perla marginata* (Panzer, 1799) and *Dinocras cephalotes* (Curtis, 1827) (Plecoptera: Perlidae). International Journal of Insect Morphology and Embryology 20(4/5): 189–203.

Rościszewska E (1995) Oogenesis of stone flies. Development of the follicular epithelium and formation of the eggshell in ovaries of *Perla marginata* (Panzer) and *Perla pallida* Guérin (Plecoptera: Perlidae). International Journal of Insect Morphology and Embryology 24(3): 253–271.

Roth LM and Eisner T (1962) Chemical defenses of arthropods. Annual Review of Entomology 7: 107–136.

Roth LM and Stahl WH (1956) Tergal and cercal secretion of *Blatta orientalis* L. Science 123: 798–799.

Rothschild M and Hinton HE (1968) Holding organs on the antennae of male fleas. Proceedings of the Royal Entomological Society of London Part A 43(7–9): 105–107.

Rušek J and Weyda F (1981) Morphology, ultrastructure and function of pseudocelli in *Onychiurus armatus* (Collembola: Onychiuridae). Revue d'Écologie et de Biologie du Sol 18(1): 127–133.

Sagert J, Sun C, and Waite JH (2006) Chemical subtleties of mussel and polychaete holdfasts. In: Smith AM and Callow JA (eds) Biological Adhesives. Springer-Verlag, Heidelberg: pp 125–143.

Sahni V, Blackledge TA, and Dhinojwala A (2010) Viscoelastic solids explain spider web stickiness. Nature Communications 1: 19. doi: 10.1038/ncomms1019:1-$.

Sannasi A (1969) Morphology, histology and histochemistry of the frontal gland of soldier termite *Rhinotermes magnificus silvestri* (Isoptera, Rhinotermitidae). Cellule 67: 367–375.

Sattler W (1967) Über die Lebensweise, insbesondere das Bauverhalten neotropischer Eintagsfliegen-Larven (Ephemeroptera, Polymitarcidae). Studies on Neotropical Fauna and Environment 5: 89–110.

Scherge M and Gorb SN (2001) Biological Micro- and Nanotribology. Nature's Solutions. Springer-Verlag, Berlin.

Schmitz G (1943) Le labium et les structures bucco-pharyngiennes du genre *Stenus* Latreille. Cellule 49: 291–334.

Scholze W (1992) Sekretschwitzen bei Netzwanzenlarven: Zur Morphologie, Chemie und biologischen Bedeutung von Integumentbildungen mit sekretorischer Funktion bei Netzwanzenlarven (Heteroptera, Tingidae). PhD Thesis, Universität Bayreuth, Fakultät für Biologie, Chemie und Geowissenschaften.

Schomann A, Afflerbach K, and Betz O (2008) Predatory behaviour of some Central European pselaphine beetles (Coleoptera: Staphylinidae: Pselaphinae) with description of relevant morphological features of their heads. European Journal of Entomology 105: 889–907.

Schwalb H (1961) Beiträge zur Biologie der einheimischen Lampyriden *Lampyris noctiluca* Geoffr. und *Phausis splendidula* Lec. und experimentelle Analyse ihres Beutefang- und Sexualverhaltens. Zoologische Jahrbücher. Abteilung für Systematik, Ökologie und Geographie der Tiere 88: 399–550.

Sehnal F (2008) Prospects of the pratical use of silk sericins. Entomological Research Bulletin 38: 1–8.

Sehnal F and Akai H (1990) Insect silk glands: their types, development and function, and effects of environmental factors and morphogenetic hormones on them. International Journal of Insect Morphology and Embryology 19(2): 79–132.

Sehnal F and Sutherland TD (2008) Silks produced by insect labial glands. Prion 2(4): 145–153.

Seifert P and Heinzeller T (1989) Mechanical, sensory and glandular structures in the tarsal unguitractor apparatus of *Chironomus riparius* (Diptera, Chironomidae). Zoomorphology 109: 71–78.

Seifter S and Gallop PM (1966) The structure proteins. In: Neurath H (ed) The Proteins, Vol. 4. Academic Press, New York: pp 155–458.

Sinohara H (1979) Glycopeptides isolated from sericin of the silkworm, *Bombyx mori*. Comparative Biochemistry and Physiology, Part B 63: 87–91.

Skilbeck CA and Anderson M (1994) The fine structure of glandular units on the antennae of two species of the parasitoid *Aleochara* (Coleoptera: Staphylinidae). International Journal of Insect Morphology and Embryology 23(4): 319–328.

Sledge MF, Fortunato A, Turillazzi S, Francescato E, Hashim R, Moneti G, and Jones GR (2000) Use of Dufour's gland secretion in nest defence and brood nutrition by hover wasps (Hymenoptera, Stenogastrinae). Journal of Insect Physiology 46: 753–761.

Smith SV, Correia JJ, and Case ST (1995) Disulfide bonds in a recombinant protein modeled after a core repeat in an aquatic insect's silk protein. Protein Science 4: 945–954.

Šobotnik J and Hubert J (2003) The morphology and ontogeny of the exocrine glands of *Prorhinotermes simplex* (Isoptera: Rhinotermitidae). Acta Societatis Zoologicae Bohemicae 67: 83–98.

Šobotnik J, Weyda F, Hanus R, Kyjakova P, and Doubsky J (2004) Ultrastructure of the frontal gland in *Prorhinotermes simplex* (Isoptera: Rhinotermitidae) and quantity of the defensive substance. European Journal of Entomology 101: 153–163.

Šobotnik J, Bourguignon T, Hanus R, Weyda F, and Roisin Y (2010a) Structure and function of defensive glands in soldiers of *Glossotermes oculatus* (Isoptera: Serritermitidae). Biological Journal of the Linnean Society 99: 839–848.

Šobotnik J, Jirošová A, and Hanus R (2010b) Chemical warfare in termites. Journal of Insect Physiology 56(9): 1012–1021.

Sprague KU (1975) The *Bombyx mori* silk proteins: characterization of large polypeptides. Biochemistry 14(5): 925–931.

Sreng L and Quennedey A (1976) Role of a temporary ciliary structure in the morphogenesis of insect glands: an electron microscope study of the tergal glands of male *Blattella germanica* L. (Dictyoptera, Blattellidae). Journal of Ultrastructure Research 56: 78–95.

Stewart RJ and Wang CS (2010) Adaptation of caddisfly larval silks to aquatic habitats by phosphorylation of H-fibroin serines. Biomacromolecules 11: 969–974.

Stewart RJ, Weaver JC, Morse DE, and Waite JH (2004) The tube cement of *Phragmatopoma californica*: a solid foam. Journal of Experimental Biology 207: 4727–4734.

Stork NE (1980) Experimental analysis of *Chrysolina polita* (Chrysomelidae: Coleoptera) on a variety of surfaces. Journal of Experimental Biology 88: 91–107.

Stork NE (1981) The structure and function of the adhesive organs on the antennae of male *Harpocera thoracica* (Fallen) (Miridae; Hemiptera). Journal of Natural History 15: 639–644.

Strong FE (1967) Observations on aphid secretions. Annual Report of the Entomological Society of America 60(3): 668–673.

Strümpel H (1983) Handbuch der Zoologie: Homoptera (Pflanzensauger), 4 (28). Walter de Gruyter, Berlin.

Strümpel H (2003) Ordnung Sternorrhyncha, Pflanzenläuse. In: Gruner HE and Dathe HH (eds) Lehrbuch der Speziellen Zoologie. Band I: Wirbellose Tiere. 5. Teil: Insecta. Gustav Fischer Verlag, Stuttgart: pp 366–401.

Strümpel H (2010) Die Zikaden – Auchenorrhyncha. Die Neue Brehm-Bücherei Bd. 668. Pflanzensaugende Insekten Band 6. Westarp Wissenschaften-Verlagsgesellschaft mbH, Hohenwarsleben.

Sutherland TD, Young JH, Weisman S, Hayashi CY, and Merrit DJ (2010) Insect silk: one name, many materials. Annual Review of Entomology 55: 171–188.

Swida U, Lucka L, and Kress H (1990) Glue protein genes in *Drosophila virilis*: their organization, developmental control of transcription and specific mRNA degradation. Development 108: 269–280.

Takasu Y, Yamada H, and Tsubouchi K (2002) Isolation of three main sericin components from the cocoon of the silkworm, *Bombyx mori*. Bioscience, Biotechnology, and Biochemistry 66(12): 2715–2718.

Teramoto H and Miyazawa M (2005) Molecular orientation behavior of silk sericin film as revealed by ATR infrared spectroscopy. Biomacromolecules 6: 2049–2057.

Trapeznikov AA and Zaozerskaya LA (1972) Viscosity of hexadecanol and oxyethylated octadecanol monolayers and the effect of a wave on it. Kolloid Zeitschrift 34: 579–584.

Trave R and Pavan M (1956) Veleni degli insetti: principi estratti dalla formica *Tapinoma nigerrimum* Nyl. La chimica e l'industria 38: 1015–1019.

Tsang P, Li G, Brun YV, Freund LB, and Tang JX (2006) Adhesion of single bacterial cells in the micronewton range. Proceedings of the National Academy of Sciences 103: 5764–5768.

Turillazzi S (1991) The Stenogastrinae. In: Ross KG and Matthews RW (eds) The Social Biology of Wasps. Comstock Publishing Associates, Ithaca: pp 74–98.

Usher MB and Balogun RA (1966) A defence mechanism in *Onychiurus* (Collembola, Onychiuridae). Entomologist 102: 237–238.

Victoriano E, Pinheiro O, and Gregorio EA (2007) Histochemical and ultrastructural evidence of lipid secretion by the silk gland of the sugarcane borer *Diatraea saccharalis* (Fabricius) (Lepidoptera: Crambidae). Neotropical Entomology 36(5): 707–711.

Voigt WH (1965a) Zur funktionellen Morphologie der Fibroin- und Sericinsekretion der Seidendrüse von *Bombyx mori* L. I. Mitteilung: Der proximale Abschnitt der Seidendrüse. Zeitschrift für Zellforschung 66: 548–570.

Voigt WH (1965b) Zur funktionellen Morphologie der Fibroin- und Sericin-Sekretion der Seidendrüse von *Bombyx mori* L. II. Mitteilung: Der mediale Abschnitt der Seidendrüse. Zeitschrift für Zellforschung 66: 571–582.

Voigt D and Gorb SN (2010) Egg attachment of the asparagus beetle *Crioceris asparagi* L. (Coleoptera, Chrysomelidae) to the crystalline waxy surface of the plant *Asparagus officinalis* L. (Asparagaceae). Proceedings of the Royal Society B: Biological Sciences 277: 895–903.

Voigt D, Schuppert JM, Dattinger S, and Gorb SN (2008) Sexualdimorphismus der Haftfähigkeit an rauen Oberflächen bei *Leptinotarsa decemlineata* Say (Coleoptera, Chrysomelidae). Mitteilungen der Deutschen Gesellschaft für allgemeine und angewandte Entomologie 16: 431–434.

von Gaudecker B (1972) Der Strukturwandel der larvalen Speicheldrüse von *Drosophila melanogaster*. Zeitschrift für Zellforschung 127: 50–86.

von Gaudecker B and Schmale EM (1974) Substrate-histochemical investigations and ultrahistochemical demonstrations of acid phosphatase in larval and prepupal salivary glands of *Drosophila melanogaster*. Cell and Tissue Research 155: 75–89.

Vötsch W, Nicholsen G, Müller R, Stierhof YD, Gorb SN, and Schwarz U (2002) Chemical composition of the attachment pad secretion of the locust *Locusta migratoria*. Insect Biochemistry and Molecular Biology 32: 1605–1613.

Vrkoč J and Ubik K (1974) 1-nitro-trans-1-pentadecene as the defensive compound of termites. Tetrahedron Letters 15: 1463–1464.

Vrkoč J, Ubik K, Dolejs L, and Hrdy I (1973) On the chemical composition of frontal gland secretion in termites of the genus *Nasutitermes* (*N. costalis* and *N. ripertii*; Isoptera). Acta Entomologica Bohemoslovaca 70: 74–80.

Waite JH (1987) Nature's underwater adhesive specialist. International Journal of Adhesion and Adhesives 7: 9–14.

Waite JH (1990) The phylogeny and chemical diversity of quinone-tanned glues and varnishes. Comparative Biochemistry and Physiology, Part B 97: 19–29.

Waite JH and Qin X (2001) Polyphosphoprotein from the adhesive pads of *Mytilus edulis*. Biochemistry 40(9): 2887–2893.

Waku Y and Foldi I (1984) The fine structure of insect glands secreting waxy substances. In: King RC and Akai H (eds) Insect Ultrastructure, Vol. 2. Plenum Press, New York: pp 303–322.

Walker G, Yule AB, and Ratcliffe J (1985) The adhesive organ of the blowfly, *Calliphora vomitoria*: a functional approach (Diptera: Calliphoridae). Journal of Zoology 205: 297–307.

Walter W and Francke W (2004) Beyer Walter Lehrbuch der Organischen Chemie, 24th Ed. S. Hirzel Verlag, Stuttgart.

Wang Y, Wang H, Zhao T, and Nakagaki M (2010) Characterization of a cysteine-rich protein specifically expressed in the silk gland of caddisfly *Stenopsyche marmorata* (Trichoptera: Stenopsychidae). Bioscience, Biotechnology, and Biochemistry 74(1): 108–112.

Weber H (1930) Biologie der Hemipteren. Eine Naturgeschichte der Schnabelkerfe, Bd. 91, Heft 4. Julius Springer-Verlag, Berlin.

Weinreich E (1968) Über den Klebfangapparat der Imagines von *Stenus* Latr. (Coleopt., Staphylinidae) mit einem Beitrag zur Kenntnis der Jugendstadien dieser Gattung. Zeitschrift für Morphologie und Ökologie der Tiere 62: 162–210.

Weirauch C (2007) Hairy attachment structures in Reduviidae (Cimicomorpha, Heteroptera), with observations on the fossula spongiosa in some other Cimicomorpha. Zoologischer Anzeiger 246: 155–175.

Wieslander L (1994) The balbiani ring multigene family: coding repetitive sequences and evolution of a tissue-specific cell function. Progress in Nucleic Acid Research and Molecular Biology 48: 275–313.

Wigglesworth VB (1985) The transfer of lipid in insects from the epidermal cells to the cuticle. Tissue and Cell 17(2): 249–265.

Wiley MJ and Lai-Fook J (1974) Studies on the silk glands of *Calpodes ethlius* Stoll (Lepidoptera, Hesperiidae). Journal of Morphology 144: 297–322.

Wilson BR (1960a) Some chemical components of *Drosophila melanogaster* egg shell. I. Amino acids. Annals of the Entomological Society of America 53: 170–173.

Wilson BR (1960b) Some chemical components of *Drosophila melanogaster* egg shell. II. Amino sugars and elements. Annals of the Entomological Society of America 53(6): 732–735.

Wood WF, Truckenbrodt W, and Meinwald J (1975) Chemistry of the defensive secretion from the African termite *Odontotermes badius*. Annals of the Entomological Society of America 68(2): 359–360.

Wu S (1973) Polar and nonpolar interactions in adhesion. Journal of Adhesion 5: 39–55.

Yamanouchi M (1922) Morphologische Beobachtung über die Seidensekretion bei der Seidenraupe. Journal of Agricultural College Sapporo 10: 1–49.

Yokoi N, Mossa F, Tiffany JM, and Bron AJ (1999) Assessment of meibomian gland function in dry eye using meibometry. Archives of Ophthalmology 117: 723–729.

Zalkow LH, Howard RW, Gelbaum LLT, Gordon MM, Deutsch WM, and Blum MS (1981) Chemical ecology of *Reticulitermes flavipes* (Kollar) and *R. virginicus* (Banks) (Rhinotermitidae). Chemistry of the soldier cephalic secretions. Journal of Chemical Ecology 7(4): 717–731.

Zhang S, Mo J, Teng L, Cheng M, and Cheng J (2006) Inter-colonial variation in the compositions of the frontal gland secretion of *Coptotermes formosanus* (Isoptera: Rhinotermitidae). Sociobiology 47(2): 553–561.

Zhimulev IF and Kolesnikov NN (1975) Synthesis and secretion of mucoprotein glue in the salivary gland of *Drosophila melanogaster*. Wilhelm Roux' Archiv 178: 15–28.

Zigrino P, Steiger J, Fox JW, Löffek S, Schild A, Nischt R, and Mauch C (2007) Role of ADAM-9 disintegrin-cysteine-rich domains in human keratinocyte migration. Journal of Biological Chemistry 282(42): 30785–30793.

9 Mechanisms of Adhesion in Adult Barnacles

Anne Marie Power, Waltraud Klepal, Vanessa Zheden, Jaimie Jonker, Paul McEvilly and Janek von Byern

Contents

9.1 General Introduction

Barnacles belong to the phylum Crustacea (following the taxonomy of Newman, 1987), which makes them segmented animals with jointed limbs, an exoskeleton that periodically moults, and a complex lifecycle involving metamorphosis between larval and adult forms. The group of crustaceans to which barnacles belong, the Cirripedia, has a unique larval form – the cyprid. This life history stage is adapted to locate a spot on which to permanently settle, develop, grow, and survive for the rest of its life. Barnacles have a worldwide distribution and various lifestyles, from parasitic species on the gills of decapod crustaceans to free-living groups. The free-living groups are adapted to permanently attach *via* cement onto other living organisms, rocks or man-made materials, and barnacle "fouling" on marine installations and cargo ships is increasingly of economic concern (Adamson and Brown, 2002). Within the free-living barnacles, a further division is recognized between acorn (Order Sessilia) and stalked (Order Pedunculata) forms. Certain stalked species are termed "pleustonic" due to a lifestyle at the air/water interface (see Bainbridge and Roskell, 1966) and these are the species which will be emphasized in this chapter (Fig. 9.1A–C).

Barnacles use glue to perform underwater adhesion in marine habitats. Temporary adhesion is used for exploration by the cyprid prior to committing itself permanently to one spot (for more information see Nott, 1969; Walker and Yule, 1984; Matsumura et al., 1998b; Dreanno et al., 2006a). Once a spot has been chosen by the cyprid, permanent glue is produced. This is a low viscosity fluid initially (Dougherty, 1990) that hardens or "cures" over the course of 1–3 hours (Yule and Walker, 1987) to form "cement". Tensile strength measurements of cement average $9.3 \times 10^5\,Nm^{-2}$ in adult barnacles and these measurements have demonstrated barnacle cement to be stronger

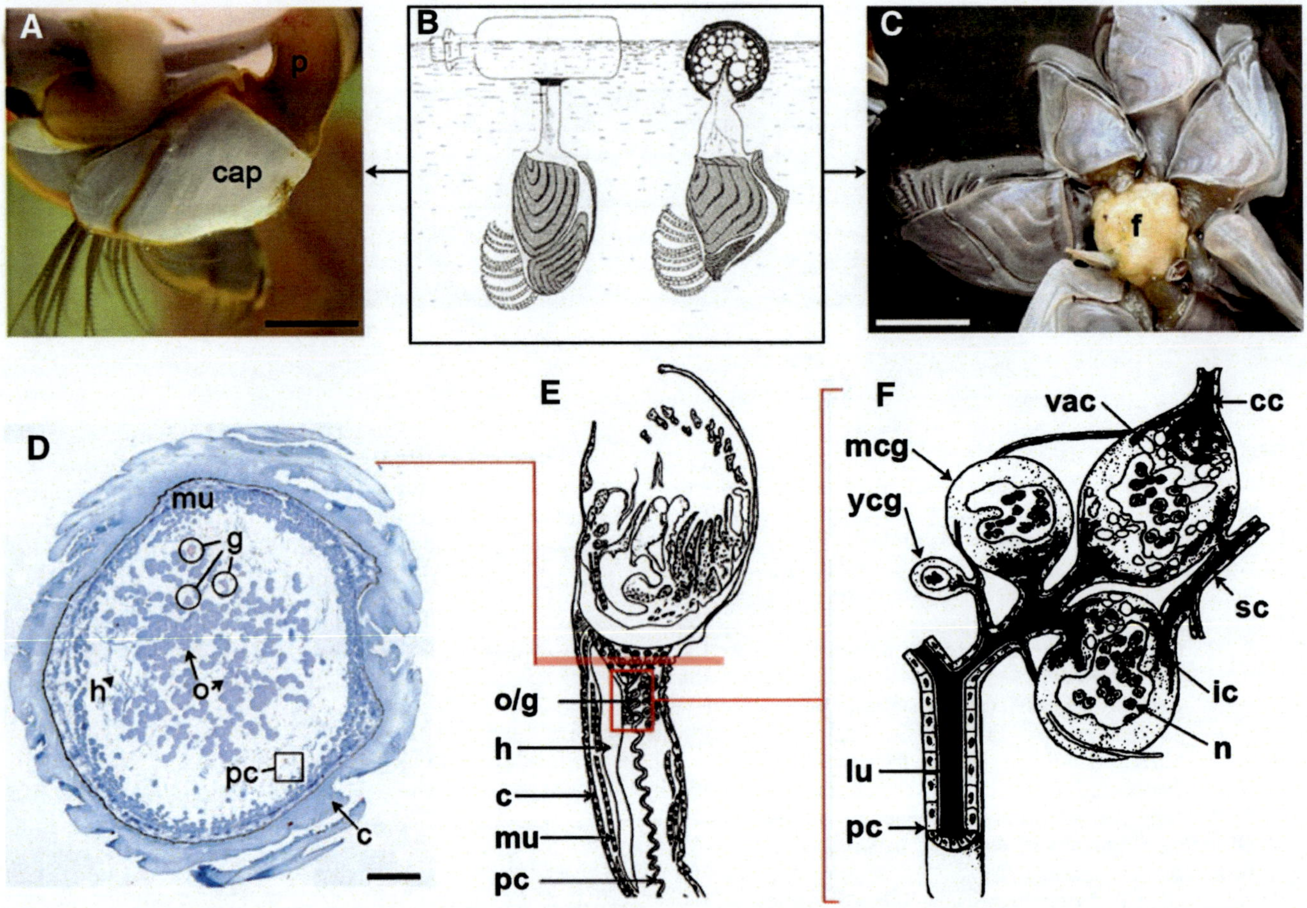

Fig. 9.1 (A) *Lepas anatifera* showing capitulum (*cap*) and peduncle (*p*), scale bar 1 cm; (B) pleustonic species *L. anatifera* attached to glass and *Dosima fascicularis* with glue float; (C) *D. fascicularis* with float (*f*), scale bar 1 cm; (D) transverse section of peduncle in *L. anatifera* stained using AZAN (Kiernan, 1999) showing position of the cuticle lining of the peduncle (*c*), circular and longitudinal muscle layers (*mu*), ovarioles (*o*), hemocoelic space (*h*) and glue gland cells (*g*), scale bar 500 µm; (E) schematic of glue apparatus in *L. anatifera* including the position of the ovarioles/glue glands (*o/g*) in the peduncle and principal canal (*pc*); (F) schematic of detailed glue glands in *L. anatifera* including mature cement gland (*mcg*), young cement gland (*ycg*), lumen (*lu*) of the principal canal, vacuole (*vac*), collector canal (*cc*), secondary canal (*sc*), intracellular canal (*ic*), large nucleus with numerous nucleoli (*n*). Schematic in B is reproduced with permission from Ankel (1962) and drawings in E and F are reprinted with permission of Lacombe and Liguori (1969)

than mussel ($5–9 \times 10^5$ Nm^{-2}) or limpet ($1–3 \times 10^5$ Nm^{-2}) adhesive, although less so than commercial dental adhesives (Yule and Walker, 1984; Nakajima et al., 1995). As with other glues, the adhesive strength in barnacles relates to the strength and number of chemical bonds which must be severed in order to detach the animal (Crisp, 1972). The main innovation in natural underwater adhesives, compared with synthetic alternatives, is the ability of natural adhesives to displace the bound water layer on the substratum and to maintain a stable bond under various levels of humidity. Other important processes for adhesion underwater are spreading, coupling with various materials, curing and resisting biodegradation (e.g., Yule and Walker, 1987; Waite et al., 2005; Kamino, 2008). Understanding the mechanism(s) involved is important since this could potentially lead to interesting surgical applications, or alternatively, could be used to prevent fouling.

The ingenuity of barnacle glue is, above all, demonstrated by the huge range of materials with different surface properties to which they attach. This includes materials with high (e.g. glass) and low (Teflon) surface free energies (Cheung et al., 1977). A drifting lifestyle in pleustonic species means that they are adapted to attach to a variety of floating objects made of different materials; *Lepas anatifera* (Fig. 9.1A, B) has been documented to attach to driftwood, glass bottles, plastic boxes, and seaweed (Boëtius, 1952) as well as to fur and feathers, e.g., various seal species, macaroni penguins, and wandering albatross (Baldridge, 1977; Arnbom and Lundberg, 1995; Barnes et al., 2004; Setsaas and Bester, 2006). Minchin (1996) found that the most common substratum to which stalked barnacles were attached, was tar, followed by plastic and finally natural debris. Another pleustonic species *Dosima* (previously known as "*Lepas*") *fascicularis* displays variation on this theme by either attaching to floating debris (e.g., fucoid algae, feathers, cork, and coke pieces – Boëtius, 1952) or adopting a free-floating lifestyle *via* secretion of a natural buoyancy aid made from cement (Fig. 9.1B, C). Such a flotation device is necessary whether or not the animal is attached to debris like wood or seaweed because these can decay and sink eventually. The size of the float has been correlated to the size of the barnacle (i.e., capitulum length) (Boëtius, 1952).

Some progress has been made in detailing the structure of glue-producing cells and the molecular components of cement in acorn barnacles (e.g., Cheung et al., 1977; Dougherty, 1996; Kamino et al., 1996; Phang et al., 2006; Mori et al., 2007; Urushida et al., 2007; Kamino, 2008; Dickinson et al., 2009; Sullan et al., 2009). However this information is still almost completely lacking in stalked barnacles which may differ somewhat from acorn barnacles due to their different lifestyles. This chapter will review past and recent developments in detailing the morphology of glue gland structures and the hypothesized adhesive mechanisms in barnacles. Where possible, we will emphasize stalked barnacles *L. anatifera* and *D. fascicularis* and present some preliminary results on the glandular structure in these species.

9.2 Peduncular Structure and the Adult Glue Apparatus

9.2.1 Structural Differences Between Acorn and Stalked Barnacles

The adult glue "apparatus" includes the protein producing cells, the canals that drain them and any supporting tissues. Almost no work has been carried out on the glue apparatus of stalked barnacle species, with the notable exception of a single study on *L. anatifera* (Lacombe and Liguori, 1969). In contrast, the glue apparatus in acorn barnacles has been described by a succession of studies (Nott, 1969; Nott and Foster, 1969; Lacombe, 1970; Walker 1970, 1971; Cheung and Nigrelli, 1972; Walker and Yule, 1984). The structural differences between acorn and the typical stalked barnacle groups are fairly minor because the stalk is equivalent to the mantle region at the base of an acorn barnacle. In stalked species this has been greatly stretched to produce the long stalk or "peduncle" (Anderson, 1994). Glue gland cells in acorn species occur basally in the mantle tissue, near the substratum, whereas in stalked barnacles they are located toward the capitular end of the stalk (Fig. 9.1D, E). Instead of flowing down a short distance toward the substratum, as is the case in acorn barnacles, the glue must travel *up* the long stalk (i.e., against gravity) in several stalked species, including *L. anatifera* and *D. fascicularis*. This is due to the inverted position of these stalked species in the water column (Fig. 9.1B). Generally there is a close proximity between glue gland cells and the ovarioles in both stalked and acorn barnacles (Lacombe and Liguori, 1969; Lacombe, 1970; Fyhn and Costlow 1976, 1977; Adamson and Brown, 2002; Fig. 9.1D, E) – but see Lacombe (1970) for a slight difference in *Semibalanus balanoides*.

9.2.2 Gland Cells (Acorn and Stalked Barnacles)

In acorn barnacles glue production is *via* unicellular gland cells. These cells were round or oval and up to 200 μm diameter (Fyhn and Costlow, 1976), although sizes were variable depending on maturity of gland cells and species (Walker, 1970). Glue gland nuclei were approximately half the diameter of small (50 μm) gland cells but became larger and more irregular in larger cells (Walker, 1970).

Glue production in stalked barnacle *L. anatifera* is also *via* unicellular glands (Lacombe and Liguori, 1969). A schematic of the glandular structure reproduced from Lacombe and Liguori (1969) is given in Fig. 9.1F. These authors showed that gland cells in *L. anatifera* appeared singly, or in small groups and a "fine connective tissue membrane was visible around them". Gland cells were about 30–40 μm in diameter in adult *Lepas* individuals (8–10 cm total length) (Lacombe and Liguori, 1969). One of the most noteworthy aspects of these cells was the very large nucleus which contained numerous nucleoli, surrounded by "fine chromatin granules" (Lacombe and Liguori, 1969) (Fig. 9.1F).

We have recently collected data on the glue apparatus in stalked barnacles *L. anatifera* and *D. fascicularis* which agree with Lacombe's and Liguori's (1969) scheme in major respects for *L. anatifera*. We present the first report of the glue apparatus in *D. fascicularis*. Like the previous study in *L. anatifera*, gland cells occurred singly or in small groups scattered among the ovarioles and surrounded by connective tissue (Fig. 9.3A). Fibrocytes were seen surrounding individual gland cells but not surrounding clusters of cells. Large irregular nuclei with many nucleoli were present in mature cells in both *L. anatifera* and *D. fascicularis* (Figs. 9.2A; 9.3B). The mature gland cells in *L. anatifera* were a bit larger in the present study compared with the equivalent descriptions in Lacombe and Liguori (1969). Our preliminary findings suggest that these range from 50 to 150 µm in diameter in *L. anatifera* and 100 to 200 µm in *D. fascicularis*.

The glue gland cells typically reached their greatest density near the capitular margin of the peduncle; they were easily distinguished from ovarioles, as they stained red under AZAN compared to ovarioles (blue). The glands cells also stained positive for acid protein (Alcian Blue pH 1.0 and 2.5) but were negative for neutral to basic proteins (Biebrich scarlet at pH 6.0, 8.5, 9.5, and 10.5) and for neutral sugars (PAS). The latter results indicated a lack of mucopolysaccharides in the glandular cytoplasm. However, the drainage canals were weakly PAS positive, as was the connective tissue surrounding the glandular cell and possibly the cell membrane. Ovarioles were strongly PAS positive (Fig. 9.2B). The positive reaction for acidic proteins may be connected to a regulated protein secretion pathway (Sect. 9.3.4).

9.2.3 Canal System in Stalked Barnacles

Individual gland cells in *L. anatifera* and *D. fascicularis* each possess intracellular canals. Intracellular canals are formed within the cytoplasm and appear to be associated with accumulating "vacuoles". We use "vacuole" in the sense that, these structures clearly correspond to the "vacuoles" of Lacombe and Liguori (1969) (Fig. 9.1F), but the precise structure of the vacuoles as well as the mode of secretion is not yet clear (see Sect. 9.3.3). Dense collections of vacuoles and intracellular canal regions were superimposed in *L. anatifera* (Fig. 9.2C, E). Vacuoles were less apparent in *D. fascicularis* and were smaller than in *L. anatifera*. It remains unclear whether the above-mentioned histochemical tests (Sect. 9.2.2) react positively to the vacuoles' content or other material present within the cytoplasm because the small size of the vacuoles allowed no clear assignment.

Ultrastructural investigations of intracellular canals showed that they are lined by a membrane with microvilli (especially in *L. anatifera*). Intracellular canals lead first into a cellular "collector canal region" (Fig. 9.2A, C) and then into secondary canals (Fig. 9.2E) that finally drain into paired principal canals. Collecting canals therefore have direct continuity with intracellular canals within the cell and secondary canals outside the cell (Fig. 9.2A, E, F). Several canals appear to drain each cell of *L. anatifera*, in accordance with the observations of Lacombe and Liguori (1969); a similar observation was made by Lacombe (1970) in the acorn barnacle *S. balanoides*. It has been proposed that the number of collector canals draining a cell may be a factor of cell size whereby larger gland cells are increasingly encircled by collector canals, which are enfolded in the plasma membrane of the gland cell, and these increasingly branch around the cell as it increases in size (Fyhn and Costlow, 1976).

Glue delivery to the substratum is ultimately achieved by a pair of large cuticularized ducts, called principal canals, running parallel to the peduncle (Fig. 9.1E). A cuticle lining was present on principal canals and this rested on an epithelial cell layer, generally one cell thick, in *L. anatifera* (Fig. 9.2G–I) and *D. fascicularis* (Fig. 9.3G). Intracellular and secondary canals lack a cuticle lining (Figs. 9.2F, G; 9.3D). A similar observation was made by Lacombe (1970) in acorn barnacles. It was possible in some cases to see a reticulated substance that appears to be glue within the lumen of principal canals in both *D. fascicularis* (Fig. 9.3G) and *L. anatifera*. To date, this substance has not appeared in secondary canals.

Kugele and Yule (2000) showed that pores leave the principle ducts at the basis of the peduncle in *L. anatifera* and release glue to the substratum. More new pores appear to develop off the principal canals as the animal grows and the basis widens. The basal perimeter extends in the direction of growth and the largest/newest pores are always at the leading edge of growth (Kugele and Yule, 2000).

9.2.4 Canal System in Acorn Barnacles and "Secondary Glue" Production

The drainage canals of acorn barnacles are similar to that described in stalked species – with a major difference. In acorn barnacles, rather than possessing just two principal canals, many such canals are present. These further communicate with circular canals in the basal plate (which can either be calcareous or membranous) and glue is delivered to the perimeter of the base by radial canals. The radial canals increase in number as the animal grows. Most barnacle species are capable of dispatching new glue after being detached or after suffering an injury,

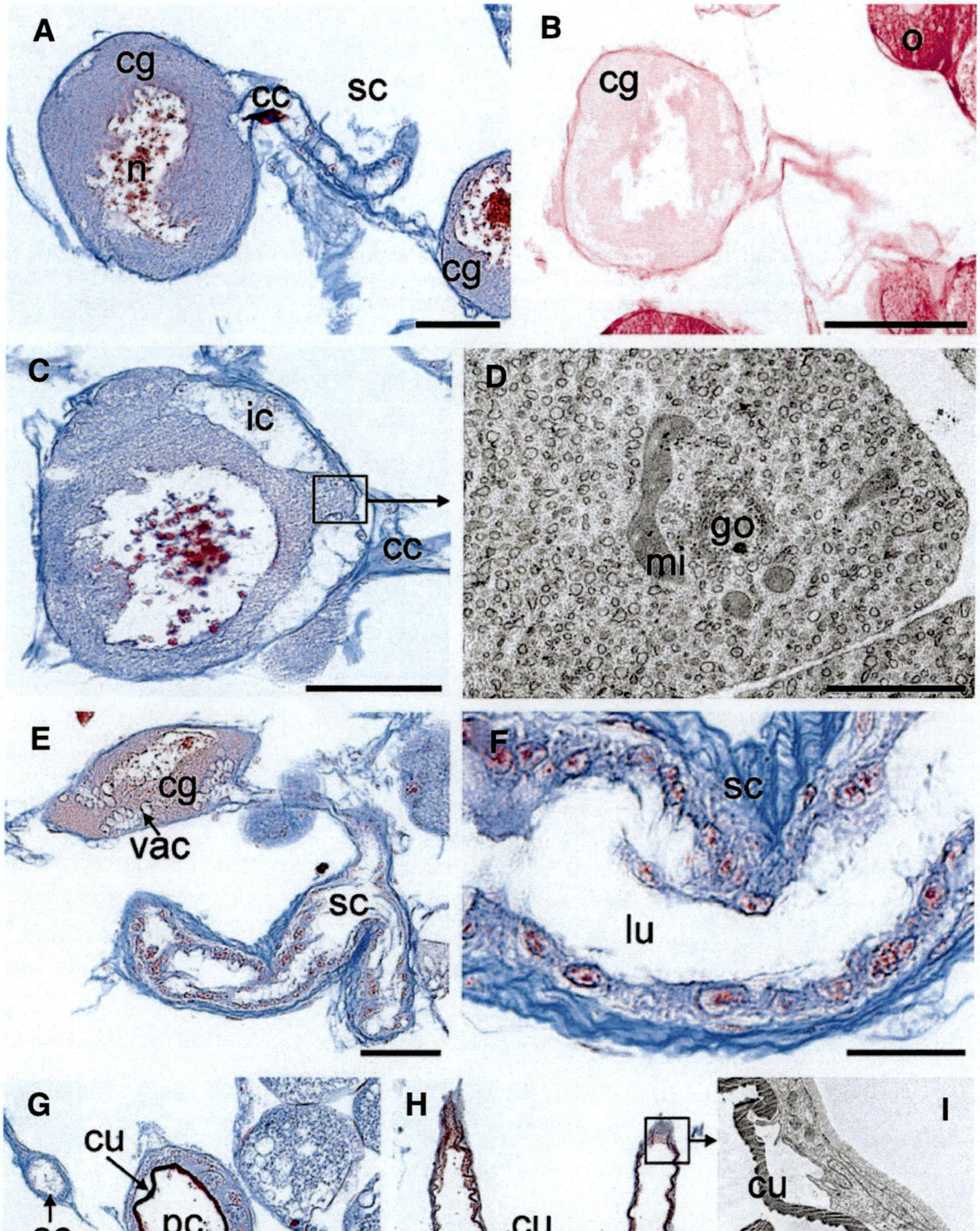

Fig. 9.2 Detail of glue-producing glandular system in *L. anatifera*. (A) Glue gland (*cg*) showing nucleus (*n*), collector canal (*cc*) and secondary canal (*sc*); (B) glue gland and ovariole (*o*) stained negative/positive respectively with PAS (Böck, 1989); (C) glue gland showing intracellular canal (*ic*) and collector canal (*cc*); (D) ultrastructure of region adjacent to intracellular canals (corresponding to marked region in C) showing mitochondria (*mi*) and Golgi (*go*); (E) glue gland with vacuoles (*vac*) adjacent to intracellular canals; (F) lumen (*lu*) of secondary canal; (G) transverse section of secondary canals (*sc*) and principal canal (*pc*), the latter is lined with cuticle (*cu*); (H) longitudinal section of principal canals showing cuticle lining; (I) ultrastructure detail of cuticle lining on principal canal (inset from H). A, C, E–H stained with AZAN (Kiernan, 1999). Scale bar in (A) = 50 μm, (B) = 100 μm, (C) = 50 μm, (D) = 2 μm, (E) = 50 μm, (F) = 20 μm, (G) = 100 μm, (H) = 30 μm, (I) = 2 μm

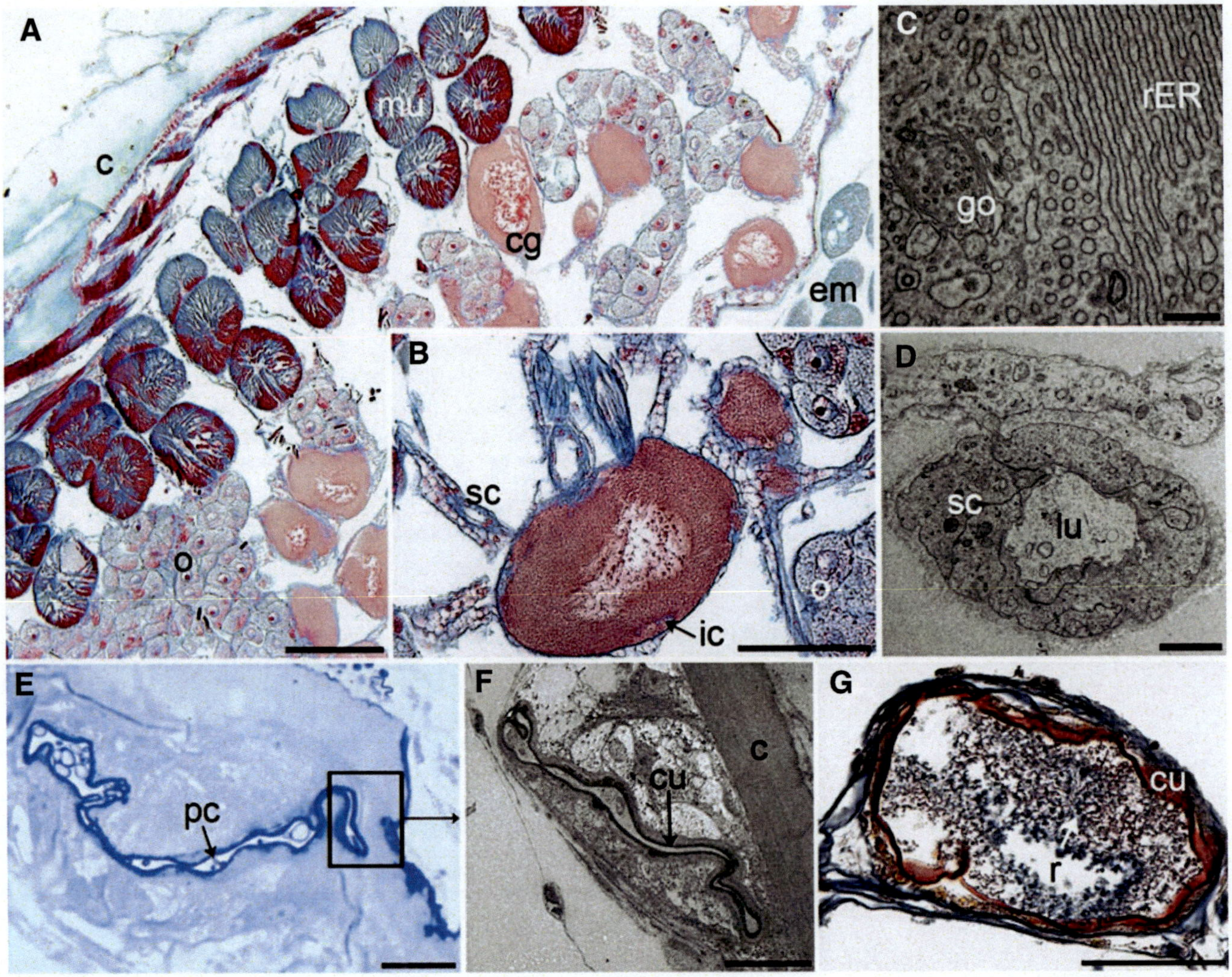

Fig. 9.3 Detail of glue-producing glandular system in *D. fascicularis*. (A) transverse section of peduncle showing the peduncular cuticle (*c*), muscle layer (*mu*), glue gland (*cg*), ovarioles (*o*), embryos (*em*) lying in the capitulum; (B) detail of glue gland cell showing secondary canal (*sc*), intracellular canal (*ic*), ovariole; (C) ultrastructure of gland cell with Golgi (*go*) and rough endoplasmic reticulum (*rER*); (D) ultrastructure of secondary canal (*sc*) showing single-celled lining and lumen (*lu*); (E) principal canal (*pc*) with cuticular lining; (F) ultrastructure of cuticle (*cu*) lining on principal canal that is lying adjacent to the peduncular cuticle (*c*); (G) reticulated substance (*r*) in cuticle-lined principal canal. A, B, G are stained with AZAN (Kiernan, 1999) and E is stained with Toluidine Blue (Trump et al., 1961). Scale bar in (A) = 20 μm, (B) = 100 μm, (C) = 0.5 μm, (D) = 2 μm, (E) = 20 μm, (F) = 20 μm, (G) = 50 μm

such as a crack in the adhesive plaque (this is sometimes called "secondary glue") (Saroyan et al., 1970). The secondary glue does not only appear at the perimeter of the basis but can also apparently be dispatched over central portions of the basis (Saroyan et al., 1970). These authors postulated that a flushing mechanism must be employed to ensure that glue delivery pores are not plugged by previous hardened secretions; however, no such mechanism has been demonstrated to date. The same authors did present some evidence to suggest that "new cement ducts" could be generated, if required. We have also observed secondary glue production in stalked species following injury of the adhesive plaque, but this topic awaits further investigations.

9.2.5 Movement of Liquid Glue in the Canal System

As mentioned above, acorn barnacles may either have a membranous or a calcareous basis by which they are cemented to the substratum. Lateral calcareous plates, which enclose the body of the animal, rest on top of this basis. The lateral plates are held down by muscles and not glue, allowing the barnacle base to grow outward from the center which was the point of original larval attachment. Muscle fibers that are involved in attachment to lateral plates (and opercular valves) stretch across the mantle in acorn species and come into close proximity to gland cells (Saroyan et al., 1970). This has been suggested to create pressure differences inside the mantle tissue and aid movement of liquid glue in some of the larger acorn barnacle species (Saroyan et al., 1970).

Stalked species generally only have membranous bases. Direct muscular support of the glue apparatus has not been observed to date in *L. anatifera* and *D. fascicularis*. However, hydrostatic pressure of up to 33 kNm^{-2} is generated in the peduncle of these species (Crenshaw, 1979). This pressure is generated in the stalk by the action of dense circular and longitudinal muscles surrounding blood sinuses (Burnett, 1972) and the presence of a tough inflexible epidermis on the stalk. The principal glue drainage canals lie close to the muscle layer, adjacent to the tough inflexible epidermis (Fig. 9.1D). It is conceivable that hydrostatic pressure in the peduncle might aid in moving glue, against gravity, up the stalk toward the attachment site. In the present study, nerves were observed near the muscles in the peripheral stalk area and these may permit voluntary control of hydrostatic pressure and hence, controlled movement of glue.

9.2.6 Cuticular Origins of the Glandular Apparatus

It now appears that, with one possible exception (Walker, 1970), most species of barnacle, both stalked and acorn, possess principal canals with a cuticle lining. It had been hypothesized that membranous-based barnacles secrete glue into principal canals with cuticle linings, while in calcareous-based species, the entire canal system is non-cuticularized (Walker, 1970). However, several barnacle species with calcareous bases were subsequently shown to possess a chitinous (cuticle) lining on principal drainage canals (Lacombe, 1970).

The issue of a cuticular lining on canal walls is an interesting one. Adult glue canals are proposed to arise ectodermally; specifically, these may arise during development from invaginations of hypodermal cells in the external mantle wall. These invaginations are suggested to progressively enlarge, ramify and become cuticle-lined as the chitin of the mantle epithelium spreads over the lumen of the entire canal system (Lacombe, 1966). Part of this canal system, the secondary canal walls, have been proposed to give rise to the gland cells *via* a process of enlargement of an epithelial cell on the secondary canal wall lining (Lacombe and Liguori, 1969; Lacombe, 1970). In light of this ectodermal origin of the adult canal system, it is interesting to note that the accumulation of glue secretions within gland cells may be associated with the moulting cycle. Fyhn and Costlow (1976) suggested that cement accumulation reaches its maximum around moulting; expression of glue proteins is also associated with moulting (see Sect. 9.4.8). Several histochemical studies have suggested that the adhesive mechanism in barnacle glue bears the chemical hallmarks of the hardening of proteins in the arthropod exoskeleton (see Sect. 9.4.7), though this is only one of several hypotheses. The characterization of cyprid *temporary* adhesive as one or more cuticular glycoproteins (Matsumura et al., 1998a; Dreanno et al., 2006b) strengthens this contention.

Adult barnacles develop from a metamorphosed cyprid larva and therefore the adult adhesive apparatus also develops from this point. Whether or not the cyprid glue apparatus is homologous to the adult glue system has been debated. Cyprid *permanent* adhesive is produced in a large multicellular gland (Walker, 1971; Cheung and Nigrelli, 1972; Ödling et al., 2008). Because the cyprid gland was suggested to disappear during metamorphosis (Bernard and Lane, 1962), some authors believed that adult and larval glue apparatus are independent of one another (Lacombe, 1966, 1970; Cheung and Nigrelli, 1972). Other studies claim, however, that the gland(s) producing permanent larval glue remain active throughout life because after metamorphosis these structures are regenerated to produce the adult glands (Walley, 1969; Walker, 1973; Yule and Walker, 1987; Anderson, 1994). Specifically, it was suggested that the cyprid glands break down and adult gland cells arise from remnants of the "de-differentiated" cyprid cement gland plus its cuticle-lined drainage canals (Walley, 1969; Walker, 1973). Whether adult glue glands arise from external mantle epithelium (i.e., the hypothesis of Lacombe, 1966) or from the remnants of the cyprid gland and its canals (i.e., Walley, 1969; Walker, 1970), cuticle-lined ducts are involved in either case. Taken together, these lines of evidence suggest that the molecular building blocks involved in adhesion might be linked to exoskeleton formation, albeit in ways that have yet to be unambiguously elucidated.

9.3 Glue Production at Cellular Level in Adult Barnacles

The process of glue protein production within glandular cells in barnacles most likely conforms to a generic cellular pathway dubbed "regulated protein secretion" that was described in relation to mussel foot proteins by Waite et al. (2005). It is worthwhile to summarize Waite et al.'s (2005) scheme: the process starts in the cytoplasm of the secretory cell, where translation of messenger RNA is initiated. Signal peptides cause this process to proceed to the endoplasmic reticulum (ER) and the products of translation are directed into the lumen, where enzymes can covalently modify certain residues in the protein during or post-translation. As protein synthesis is completed, nascent protein becomes membrane bound in specialized areas of the ER and appear as vesicles. These migrate to and merge with the Golgi apparatus, where proteins are sorted according to their fate. Proteins to be secreted by regulated pathways, bud off the trans- portion of the Golgi as vesicles, and fuse with one another to form large vacuoles. Vacuoles serve as reservoirs for accumulating protein and as a place where protein condensation can begin. Protein condensation (i.e., covalent joining of –H and –OH containing groups on amino acids to form dipeptides or polypeptides) occurs at a pH of about 5.5. This entire process shows that the steps in the production of a glue protein for secretion might take place in different parts of the glue-producing cell. As we shall see, some variation exists between barnacle species in this respect.

9.3.1 Glue Secretion Pathways in Acorn Barnacles

Glue synthesis and accumulation zones have been identified within cells in some acorn barnacle species (Lacombe and Liguori, 1969; Walker, 1970) but not in others (Lacombe and Liguori, 1969; Walker, 1970; Cheung and Nigrelli, 1972). Walker (1970) observed that red and blue AZAN staining reactions were patchy inside cells, he then distinguished that the same areas of cells that had stained blue with AZAN also stained stronger for RNA (using methyl green pyronin), i.e., these were "synthesis" zones. Synthesis zones also had dense rough ER (rER), mitochondria and Golgi and only few scattered vacuoles under electron microscopy (EM). "Accumulation" zones were defined as positive for proteins (especially thiol (–SH) containing groups using D.D.D. stain) – and were in close proximity to the cell drainage system. Accumulation zones were also shown to contain aggregated membrane-bound vesicles under EM and only small amounts of rER or Golgi (Walker, 1970).

9.3.2 Glue Secretion Pathways in Stalked Barnacles

In the present study there was no clear cellular distinction into protein synthesis and accumulation zones in *L. anatifera* or *D. fascicularis*, as revealed by light microscopy and EM. No patchiness was seen under AZAN staining (Figs. 9.2A, C, E; 9.3B) and, as yet, no clear substructuring of regions where rER, Golgi and mitochondria were concentrated. Ultrastructure of *D. fascicularis* gland cells clearly showed Golgi with vesicles and rER sometimes arranged in stacks in the cytoplasm (Fig. 9.3C). In *L. anatifera*, large amounts of dense rER, some of which was quite swollen, and high numbers of mitochondria and Golgi (Fig. 9.2D) were seen throughout the cytoplasm of the cement gland cell. However, accumulations of transport vesicles were seen around the intracellular canals in *L. anatifera* and the electron density of the vesicles implied that these contained protein. This indicates that the intracellular canals may be important for export of glue components from the cell. Functional zones were also lacking in previous work on *L. anatifera* (Lacombe and Liguori, 1969), although a region rich in "ergastoplasm" (taken to mean rER) was suggested to be located near the cell membrane.

9.3.3 Basis Type and Mode of Glue Discharge (Acorn and Stalked Barnacles)

For acorn barnacles, Walker (1970) appears to have been correct in suggesting slight differences in the mode of discharge of secretion from the cement cells in membranous-based barnacles (*Elminius modestus* and *Semibalanus balanoides*) compared with calcareous-based species (*Balanus hameri*). He suggested that in membranous-based species, the secretion passes from intracellular canals to the drainage canals but in calcareous-based species, the secretion passes directly into the drainage canals without an intracellular structure being involved (Walker, 1970).

Where intracellular canals are present in acorn barnacles, these were associated with secretion "accumulation zones" and many vacuoles were found abutting onto the intracellular canals in these zones (Walker, 1970). By this arrangement, secretory material presumably passes into the lumen of intracellular canals, having been first "packaged" into a vacuole.

Where intracellular canals are lacking, other means are necessary. Contact between accumulation zones and nuclear membranes, rather than intracellular canals, has been suggested to have a role in glue discharge (Lacombe, 1970). Alternatively, in *B. hameri*, the extracellular canal cells in immediate contact with the gland cell were filled with vacuoles whose size and contents were similar to those inside the gland (Walker, 1970), i.e., the vacuoles passed directly from the gland cytoplasm to the drainage apparatus through the limiting membranes of each. In both of these examples, protein secretions appear to have been packaged inside vacuoles prior to drainage, whether or not intracellular canals were present.

But alternative glue drainage mechanisms have also been described. For example, in the acorn barnacles *Balanus tintinnabulum* and *B. eburneus*, no vacuoles were present, instead, "granules of secretory material" were described (nuclear fast red- and AZAN-stained). At the accumulation zone of the gland, these granules apparently passed through a "fine membrane" into the lumen of a collecting canal (Lacombe, 1970). Similarly, Lindner and Dooley (1972) described movement of melanin particles which are proposed by-products of adhesion (see Sect. 9.4.7) inside collector canals. Walker (1970) believed that vacuole contents were the final secretory product of the cell in acorn barnacles. However a more precise definition of these vacuoles is required, as is the elucidation of other secretory means in situations where vacuoles are not apparent.

By contrast with Walker (1970) observations in acorn barnacles, stalked species such as *L. anatifera* and *D. fascicularis* possessed no distinct functional zones. *L. anatifera* and *D. fascicularis* are membranous-based and appear to utilize the intracellular canals for glue secretion – as indicated by the presence of microvilli and transport vesicles associated with these structures. The precise significance of the many "vacuoles" in the intracellular canal area in stalked barnacle glands (see Sect. 9.2.3) awaits further investigations. Merocrine secretion (*via* exocytosis of secretory vesicles from Golgi – Aumüller et al., 1999) is assumed to be responsible but the precise transcytosis pathways of protein components and the composition and involvement of vacuoles or other vesicles in this process is unknown. The appearance of reticulated substance in the principal canals (e.g., Fig. 9.3G) means that even if vacuoles are the final secretory products of cells (Walker, 1970), some further modification of vacuole contents may occur post-expulsion from the cell.

9.3.4 Regulation of Protein Secretion

Gradients or switches in pH or isoelectric point (pI) within gland cells may be an important control mechanism in the regulation of protein secretion. A change in "polarity" of staining reaction in *B. tintinnabulum* was suggested between synthesis and accumulation zones (nuclear fast red and napthol green) (Lacombe and Liguori, 1969). This polarity change might aid transport of glue proteins across limiting membranes *via* increasing solubility of these proteins. Cytoplasm components of gland cells in acorn barnacles (*B. crenatus*) ranged from pI 3.9 to 9, but these included many components for which a low pI was reported (Lindner and Dooley, 1972). Sometimes pH and pI were associated with elevated levels of a particular amino acid, e.g., Lindner and Dooley (1972) found that high local concentrations of tryptophan in the cytoplasm were associated with pI 6. Similarly for enzyme activity, phenoloxidase activity was demonstrated by several methods to coincide with pI 6 (Lindner and Dooley, 1972). This enzyme was most concentrated around the collector canals and various staining methods (Cu^{++} ions, activators, and enzyme inhibitors) showed melanin particles to apparently enter and move through these canals (Lindner and Dooley, 1972). As seen in Sect. 9.2.2 of the present study, preliminary findings show that proteins in gland cells of both *L. anatifera* and *D. fascicularis* fall in the acidic range around pH 1 to 2.5.

Besides transport, pH and pI also aid in protein condensation and complex coacervation. Complex coacervation can happen when proteins combine in solution to form soluble aggregates. It involves phase separation of liquids to a more dense (coacervate) phase and a less dense (equilibrium) phase. It is pH dependent and typically produces a low viscosity coacervate with low surface tension; so it can be very useful for secretion and coalescence of coacervates in seawater, or spreading and gelation over substrata or around particulates (Stewart et al., 2004; Waite et al., 2005). Crosslinking of the soluble protein aggregates can be triggered by pH or temperature (Waite et al., 2005). It is not known how, if at all, complex coacervate formation would take place in barnacle bulk proteins although it might contrast with the equivalent process in mussel foot proteins (Lim et al., 2010). This is because these components are extremely hydrophobic in barnacles but not in mussels (or tubeworms) (Kamino, 2008).

9.4 Glue Composition and Molecular Adhesion

The gross composition of glue in barnacles is widely accepted to almost exclusively be comprised of protein (Kamino 2006, 2008), with the presence of small amounts of lipid and carbohydrate having been debated in early studies (Lindner and Dooley, 1972). Slightly higher lipid content in some studies was attributed to contamination of the analyzed glue with ovarian material (Crisp, 1972). Adult hardened cement from *L. anatifera* contains ~96% protein; but for *D. fascicularis* the protein fraction in cement may be lower (~75.9%) (Walker and Youngson, 1975). Observations of newly secreted glue show that this starts off as a low viscosity secretion and solidifies later over a period of minutes (Saroyan et al., 1970) to hours (Cheung et al., 1977; Dougherty, 1990). Unlike mussel cement, barnacle adhesive is macroscopically uniform throughout the plaque (Kamino, 2008) although on closer examination it has been shown to be composed of multiple layers that are progressively more elastic from top layers to sub layers (Sun et al., 2004).

9.4.1 Involvement of Physical Adhesive Forces

Crisp (1972) discussed the possibility that the mechanism of adhesion in adult barnacles included physical adhesion by Stèfan forces. The design of the acorn barnacle basis allows for adherence to the substratum, which is so close, that the basis can easily undercut paint or coatings designed to prevent fouling. The cement layer is very thin in some cases, perhaps 5 μm (Saroyan et al., 1970). The thinness of the adhesive layer was suggested to implicate Stèfan-type forces (Crisp, 1972). It was later shown that barnacles are too resistant to forces sliding them across the attachment surface for Stèfan-type adhesion to be a major factor (Dougherty, 1990). Recent work has characterized individual proteins in the cement complex (see Kamino 2006, 2008 for reviews) and has argued that chemical interactions, rather than physical forces, explain adhesion in barnacles.

9.4.2 Cement Solubility

The reagents which can solubilize the hardened cement give an idea what sort of chemical adhesion may be operating. In early studies, Lindner and Dooley (1972) found that hardened barnacle cements were resistant to solubilization with thioglycollate which they believed ruled-out disulfide (–S–S–) bonds as a crosslinking mechanism. The various solvents to which hardened adult glue was resistant also included salts, strong acid/alkali hydrolysis and sodium hypochlorite (although these did solubilize larval glue) (Lindner and Dooley, 1972). But 90% solubility was achieved in other studies with non-proteolytic means, i.e., heat treatment with denaturants (7 M guanidine hydrochloride) and reductants (0.5 M dithiothreitol – DTT) (Kamino et al., 2000). In fact, once certain bulk proteins are solubilized, other protein components can come into solution (Kamino, 2008). Barnes and Blackstock (1976) claimed to completely solubilize hardened *D. fascicularis* cement with a combination of sodium dodecyl sulfate (SDS) and mercaptoethanol (Me). The Me was speculated to break disulfide bonds, with SDS responsible for weaker linkages (Barnes and Blackstock, 1976). However this success has not been found in other studies with the same reagents. In fact, Kamino et al. (2000) showed that SDS and Me only rendered one protein soluble, Mrcp-68k (k = protein size in kilodaltons). Nonetheless, the effectiveness of the reductant DTT showed that disulfide links were an important source of stability within the protein complex (Kamino et al., 2000), although it could not be determined whether these bonds were intermolecular or intramolecular. One anomaly with this theory was the low cysteine (Cys) residues in some of the DTT-solubilized bulk proteins (Kamino et al., 2000; Kamino, 2006) – Cys being a source of disulfide crosslinks. The low Cys content in one such protein, Mrcp-100k, was explained due to the abundant hydrophobic residues in this protein, which perhaps resulted in a hydrophobic barrier that hides disulfide links within the protein (Kamino et al., 2000). But Cys was also quite low (2.7%) in amino acid analyses of total cement (Lindner and Dooley, 1972). In a couple of studies where ½ cystine was reported, values were even lower (0.4–1.10) (Walker and Youngson, 1975; Barnes and Blackstock, 1976). Cys was also low in sequenced proteins, apart from in Mrcp-20k. In Mrcp-20k, Cys probably occurred as cystine, i.e., this protein possessed *intra*molecular rather than *inter*molecular disulfide links (Kamino, 2006).

9.4.3 Cement Proteins in Acorn Barnacles

Following advances in solubilizing barnacle adhesive plaques, it was shown that acorn barnacles possess at least ten cement proteins, six of which have already been characterized (Kamino, 2008). Of the six characterized cement proteins, five are novel, i.e., have no significant homologs on published databases (Kamino, 2008). Three of these proteins Mrcp-100k, Mrcp-68k, and Mrcp-52k were responsible for 90% of the multiprotein complex by weight. Putative functions were suggested for some proteins, e.g., Mrcp-19k may be a "multi-surface coupling", rich in certain amino acids (Serine, Threonine, Alanine, Glycine, Valine, and Lysine) and adsorbing to diverse materials, perhaps *via* hydrogen bonds, electrostatic or hydrophobic interactions. Calcite-specific adsorption may be achieved *via* another cement protein (Mrcp-20k) which is rich in Cys (in intramolecular disulfide form) and charged amino acids like Aspartic Acid, Glutamic Acid, and Histidine (Mori et al., 2007). Some of the most insoluble bulk proteins are strongly hydrophobic and appear to be disulfide linked, but it was suggested that these links were probably intramolecular, i.e., were involved in conferring "shape" to protein molecules; with intermolecular noncovalent bonding being more important for cement curing (Kamino, 2006). Intermolecular noncovalent interaction occurs in hydrophobic interactions but also features strongly in "self-assembly" of beta sheet protein structures such as amyloid-like structures. Barnacle bulk proteins, particularly Mrcp-100k, appears to possess alternating hydrophobic and hydrophilic residues characteristic of beta sheet (amyloid) structures (Kamino et al., 2000). Atomic Force measurements and staining have also suggested cross-beta sheet structures in adhesive plaques of *Balanus amphitrite*, although total content of amyloid(s) were low (<5%) (Sullan et al., 2009).

9.4.4 Cement Proteins in Stalked Barnacles

Very little information is known as yet on this topic. Barnes and Blackstock (1976) suggested that a considerable fraction of hardened *D. fascicularis* cement comprised weakly bonded (e.g., hydrogen-bonded) material. SDS-PAGE analysis gave nine protein bands, 30% of which were small molecular weights <10k. The size range of other protein bands fell into a range 20–90k. However, these results are, at best, partial analyses of the protein components, given the solubility issues outlined above.

9.4.5 Cement *Versus* Uncured Glue

Most studies (acorn barnacles) have examined cured glues (or "cements") but few studies have attempted to harvest uncured glues and analyse their composition (Cheung et al., 1977; Dougherty, 1996; Dickinson et al., 2009). This is advantageous because the insolubility of cured glues presents analytical challenges. Further, collection of uncured or liquid cement presents opportunities to examine activation and inhibition of the curing process. Different authors use different terminologies, e.g., uncured glues are termed "liquid cement" (Cheung et al., 1977) or "cement precursor secretion" (CPS) (Dougherty, 1996) or "secondary glue" (Saroyan et al., 1970). Whatever the terminology, the newly secreted substance exhibited adhesive properties (Dougherty, 1990) and apparently does not differ from primary glue, either histochemically (Saroyan et al., 1970) or in amino acid and peptide composition (Kamino, 2006). *Chthamalus fragilis* liquid cement contained up to 22 proteins ranging in size from 14k to 200k, but apart from examining their size, these proteins were not characterized (Dougherty, 1989). The activation of curing (either by protein crosslinking or self-assembly of cross-beta sheets) in the protein complex is not yet well understood. Curing of the liquid cement took place in air as well as in seawater which means that it was unlikely that this was activated by a pH shift (e.g., on entering seawater) (Cheung et al., 1977; Dougherty, 1990).

Enzyme activation of the polymerization process was not inhibited by serine or cysteine protease inhibitors, indicating that proteolytic cleavage by such enzymes was unimportant in achieving polymerization (Dougherty, 1996). Similarly, trypsin inhibition did not prevent polymerization (Cheung et al., 1977; *but see* Dickinson et al., 2009). Instead Zinc-metalloprotease (an exo-carboxypeptidase) played a role in the transformation from liquid to solid cement, but the target protein was not identified (Dougherty, 1996). It is also interesting, in light of one hypothesis (Sect. 9.4.7), that a range of phenoloxidase inhibitors were tested and did not prevent polymerization of liquid cement (Cheung et al., 1977). An enzyme has been isolated from barnacle cement (though this was not confirmed by northern blot to be a cement protein). This molecule (16k) shared 47% homology with lysozyme and hence had a possible role in anti-biodegradation rather than in proteolytic activation (Kamino, 2006). Regarding the intermolecular noncovalent interactions viewed by some authors as important adhesive mechanisms (Sect. 9.4.3), self-assembly cross-beta sheet formation can easily be stimulated by pH and salt concentration (Kamino, 2008) but, conversely, the mechanisms that prevent these from forming *in vivo* are not clear. The only other enzymatic activity reported was alkaline glycerophosphatase in the collector canals and cement (Arvy et al., 1968).

9.4.6 Post-translation Modifications and Comparison with Other Adhesive Models

Evidence to date suggests that barnacles display a contrasting adhesive system compared to mussel and tubeworm. Little or no post-translational modifications have been found so far in individual barnacle cement proteins (Kamino, 2008) but this is not the case in mussels or tubeworm in which the adhesive mechanism involves several post-translational modifications, particularly Dopa (3,4-dihydroxyphenl-alanine) as important components. Dopa is thought to be involved in at least two major adhesive functions in mussels (i) intermolecular crosslinking of bulk adhesive proteins, and (ii) bonding with surface materials. The first function provides cohesive strength between glue proteins and the second provides adhesive strength between the plaque and the surface materials. The degree of oxidation of the Dopa residues seems to determine which of these functions takes precedence, i.e., Dopa oxidized to Dopa-quinone (see Sect. 9.4.7) may be involved in crosslinking proteins as well as adsorption, but only to certain surface materials, meanwhile unoxidized Dopa bonds to other materials (Waite, 2002; Lee et al., 2006). No Dopa has thus far been isolated from barnacles, which means another mechanism is responsible in these animals. Dopa is an o-dihydroxy derivative of the monophenolic amino acid tyrosine (Tyr), so presence of this amino acid could indicate a mussel-like adhesive mechanism in barnacles; however this does not appear to be the case since Tyr is not present in high quantities in hydrolyzed barnacle cements (2.1–3.7%) (Lindner and Dooley, 1972; Walker and Youngson, 1975; Barnes and Blackstock, 1976). Presence of Tyr by histochemical staining was either found to be absent or only weakly positive from glue gland cytoplasm and nucleoli (Lindner and Dooley, 1972).

9.4.7 Quinone-type Crosslinking

Despite an absence of Dopa in barnacle cements, a quinone-type crosslinking has, for years, been one of the most credited adhesive mechanisms hypothesized for barnacles (Crisp, 1972; Lindner and Dooley, 1972; Barnes and Blackstock, 1976; Phang et al., 2006). The quinone crosslinking mechanism is akin to the process that brings about hardening of proteins in the arthropod cuticle after moulting by the formation of quinone-amine bonds in the cuticle. In this mechanism, bonding of the terminal amino groups in the protein result in lengthening of the protein chains and further bonds involving the amino groups on the protein side chains create crosslinks between individual protein chains (Crisp, 1972). To demonstrate this type of adhesive mechanism in barnacles, several components should be identified: a phenolic precursor molecule, phenoloxidase-catalyzed o-dihydroxyphenols which form semiquinones and free amino groups to bind with the quinones. Evidence to support this mechanism were as follows: banding patterns after Electronparamagnetic Resonance spectroscopy that were similar to free radicals (i.e., semiquinones) and the discovery of high Lysine (Lys) residues (8.6%) as a source of free amino groups in the hardened cement (Lindner and Dooley, 1972). Further, phenoloxidase (PO) was demonstrated by introducing Dopa into tissues, as well as by other methods; the former reaction was strongly positive in cyprid antennules, in the cytoplasm of adult secreting cells (at pI 6), the secretion collector canals and in the hardened cement (Lindner and Dooley, 1972). However it should be noted that the significance of PO is disputed by other evidence (Cheung et al., 1977 – see Sect. 9.4.5). Unlike the findings of Lindner and Dooley (1972) Lys was rather low in some studies (2.8–3.9% weight) (Walker and Youngson, 1975; Barnes and Blackstock, 1976). Difficulties with a quinone crosslinking mechanism also include a failure to identify the phenolic precursor molecule of quinones (Lindner and Dooley, 1972). A lack of pigmentation (melanin) is generally reported in barnacle cements (although a weak melanin positive reaction was reported in primary hardened cement by Lindner and Dooley, 1972). This is relevant because melanin is an o-hydroyphenol derivative and hence a by-product of the quinone-type crosslinking reaction. Acorn barnacle glue has been described as "white and rubbery" by Saroyan et al. (1970) implying little or no melanin content. Neither is melanin apparent in stalked barnacle glue, certainly on initial polymerization, when the clear fluid becomes an opaque white color (pers. obs.).

9.4.8 Possible Implications of Moulting and Hemolymph Systems

Despite some undoubtedly problematic issues, it is tempting to believe that there is a link between the chemistry of adhesion and the post-moult hardening of the barnacle exoskeleton. It appears that glue glands originate ectodermally (see Sect. 9.2.6) so that the molecular components of adhesion and exoskeleton hardening may both arise from very similar tissues. Glue accumulation in cells appeared to be related to the intermoult cycle showing that this was co-incident with the deposition of new exoskeleton (Fyhn and Costlow, 1976). Cement protein gene expression (which occurs in the basis of the animal in the general vicinity of cement glands) also seemed to increase just after moulting (Kamino, 2006), i.e., when the new exoskeleton is being secreted and hardening. In addition, temporary larval glue has been characterized as a cuticular glycoprotein (Matsumura et al., 1998a; Dreanno et al., 2006b). A recent study has suggested what initially seems to be a contrasting mechanism at work, i.e., that barnacle adhesion shares similarities with the vertebrate blood clotting cascade (Dickinson et al., 2009). Barnacle adhesion was suggested to involve a proteolytic activation of enzymes (by a trypsin-like serine protease) and transglutaminase cross-linking of fibrous proteins (Dickinson et al., 2009). While this is intriguing, how the proposed source of transglutaminase (TGase) to catalyze part of this cascade enters inside the glandular apparatus is not immediately clear; also, serine protease activation of glue protein precursors is disputed by other work (Dougherty, 1996). Perhaps hemocytes, a probable source of TGase (Wang et al., 2001), could migrate from the hemolymph across the walls of the secondary canals of the glue apparatus. Saroyan et al. (1979) did describe a histochemical change in glue protein secretion after this had left the gland and was making its way through the glue drainage system. Specifically, a Mallory's trichrome staining reaction switched from negative in the secondary canals to a positive reaction in the principle canals. In this regard, it is perhaps of note that a "hilum" (or notch) appears at the junction between secondary and primary ducts in some species (Lacombe, 1970). Alternatively, the glue harvesting method of Dickinson et al. (2009) could have caused a wound, resulting in contamination of the liquid cement by hemolymph, which is why they recorded the cascade that they described. The hemolymph clotting response in crustaceans utilizes a simple coagulation process involving crosslinking aggregates of soluble clotting proteins by TGase (released from hemocytes) in the presence of Ca^{++} (Wang et al., 2001; Cheng et al., 2008). Under some conditions, crosslinking in crayfish hemolymph can even happen *via* a prophenoloxidase (PPO) activating system (Theopold et al., 2004). It includes a PO proenzyme of 76k and activation of this proenzyme by a serine protease (Aspán et al., 1995). The activation system is produced in hemocytes and held in secretory granules and, interestingly, an enzyme inhibiting PO activity has also been isolated (Söderhäll et al., 2009). Both of these hemolymph clotting systems share aspects of Dickinson's et al. (2009) proposed adhesion mechanism. Hemolymph clotting and adhesion may not necessarily be completely unrelated. Invertebrates are increasingly being viewed as displaying opportunism in their approach to blood clotting mechanisms with high variation in clot components that will be used within and between arthropod classes, for example. Proteins with different functions have been known to be recruited into the clotting system (Theopold et al., 2004). One of these components, the PPO system, is also involved in invertebrate immunity as a recognition system, as well as in wound healing, and in cuticle formation (Söderhäll et al., 2009 and references therein). It is possible that the PPO system is also a component of the barnacle adhesive system and its activation (see Sect. 9.4.7), but this is only one of several hypotheses and requires further work.

9.5 Conclusions

Characterization of glue proteins has begun in acorn barnacles, aided enormously by a breakthrough in solubilizing most of the cement, though a small fraction still remains inert. Despite these promising advances, the adhesive mechanism of the overall multiprotein complex is far from being understood, although several contending possibilities have been proposed. Questions still remain about the similarity (or otherwise) of permanent larval glues and adult glues; on the one hand, evidence suggests similar adhesive mechanisms of quinone crosslinking (Walker, 1971; Lindner and Dooley, 1972; Phang et al., 2006), but on the other hand, adult and permanent larval glues clearly differ somewhat in terms of solubility (Lindner and Dooley, 1972). Whether adult and larval glues are produced from the same apparatus is also unclear, but in any case, the adult glands probably have ectodermal origins involving cuticle-lined epithelia. This resonates with evidence toward a quinone-tanning explanation, although other evidence disputes this. Alternatively, there is strong evidence that noncovalent interaction (cross-beta sheet structures) provides cohesive strength in certain bulk cement proteins, e.g., Mrcp-100k; which works in combination with other proteins such as Mrcp-19k and Mrcp-20k that are involved in surface coupling (Kamino, 2008). Precise characterization and complete functional significance of several acorn barnacle glue proteins is still underway. Even with full characterization of the protein components, advancing our understanding of the functions and assembly of separate elements to provide all the features necessary for underwater adhesion is a long way off. How the glue components are packaged within the gland, the secretion pathways or changes in the glue components from the point of production to the point of expulsion are not understood at present. All protein chemistry to date has also been in acorn species, and the best-studied species (*Megabalanus rosa*; *Balanus amphitrite;* and *Balanus improvisus*) have calcareous bases. In the evolutionary sense, pedunculate barnacles are more ancient compared with acorn species (Moyse, 1987). It is perhaps unlikely that the mechanisms of adhesion have been adapted substantially by acorn barnacles from their pedunculate ancestors, although the hydrodynamic regime acting as a force on a large animal with stalk will be far stronger than that acting on smaller acorn barnacles occupying surface boundary layers. Further, the glue modified to become a "float" in *D. fascicularis* enlarges over time and only sometimes performs an adhesive function, which perhaps warrants individual analysis in this species. Many stalked barnacle species also have a membranous basis and so calcite-adsorbent proteins (e.g., Mrcp-20k – Mori et al., 2007) might not be as important as in calcareous-based species. However other proteins may be universal: Mrcp-100k is suggested to be a protein common to all barnacle species because it is present in both calcareous- and membranous-based barnacle species (e.g., *Semibalanus cariosus*) (Kamino, 2008). Other differences may exist, e.g., in Lys values, which appear to be doubled or even trebled in acorn barnacles compared with *L. anatifera* and *D. fascicularis* (Walker and Youngson, 1975). The glue apparatus in stalked species is similar to that of acorn barnacles in many respects and preliminary results show that glandular cytoplasm contains acidic proteins (pH 1 to 2.5) in both *L. anatifera* and *D. fascicularis*. Functional secretion "synthesis" and "accumulation" zones are lacking in *L. anatifera* and *D. fascicularis* glands. A previous suggestion of a difference in the cuticularization of the glue drainage apparatus according to the type of basis (calcareous or membranous) (Walker, 1970) appears to be unfounded, although there does seem to be a correlation between the presence or absence of intracellular canals and basis type (Lacombe and Liguori, 1969; Lacombe, 1970; Walker, 1970; present study). Where they occur, intracellular canals appear to be an important area for exporting glue components from the cell, but variations on this theme are also necessary in species lacking intracellular canals. The precise composition and functional significance of vacuoles (where present) is unknown in all barnacle species and awaits further investigations.

Acknowledgments

The authors would like to thank Alex Dufort for the use of photographs of live animals. We also thank Dave Duffy, Norbert Cyran, and Dorothea Stuebing for their help in sample collection and fixation. This research was carried out with the support of a Beaufort Marine Research Award under the Irish National Development Plan 2007–2013; a Research Frontiers Award from Science Foundation Ireland (09RFPMTR2311) and Austrian Science Foundation (P21767-B17).

References

Adamson L and Brown N (2002) Anti-fouling systems. Focus on IMO: 1–32.

Anderson DT (1994) Barnacles – Structure, Function, Development and Evolution, 1st Ed. Chapman & Hall, London.

Ankel WE (1962) Das Märchen von den Entenmuscheln. Natur und Museum 92: 207–218.

Arnbom T and Lundberg S (1995) Notes on *Lepas australis* (Cirripedia, Lepadidae) recorded on the skin of southern elephant seal (*Mirounga leonina*). Crustaceana 68: 655–658.

Arvy L, Lacombe D, and Shimony T (1968) Studies on the biology of barnacles: alkaline phosphatase activity histochemically detectable in the cement apparatus of the Balanidae (Crustacea-Cirripedia). American Zoologist 8: 817.

Aspán A, Huang TS, Cerenius L, and Söderhäll K (1995) cDNA cloning of prophenoloxidase from the freshwater crayfish *Pacifastacus leniusculus* and its activation. Proceedings of the National Academy of Sciences of the United States of America 92(4): 939–943.

Aumüller G, Wilhelm B, and Seitz J (1999) Apocrine secretion – fact or artifact? Annals of Anatomy 181(5): 437–446.

Bainbridge V and Roskell J (1966) A re-description of the larvae of *Lepas fascicularis* Ellis and Solander with observations on the distribution of *Lepas* nauplii in the north-eastern Atlantic. In: Barnes H (ed.) Some Contemporary Studies in Marine Science. Allen and Unwin Ltd., London: pp 67–81.

Baldridge A (1977) The barnacle *Lepas pacifica* and the alga *Navicula grevillei* on Northern elephant seals, *Mirounga angustirostris*. Journal of Mammalogy 58(3): 428–429.

Barnes DKA, Warren NL, Webb K, Phalan B, and Reid K (2004) Polar pedunculate barnacles piggy-back on pycnogona, penguins, pinniped seals and plastics. Marine Ecology Progress Series 284: 305–310.

Barnes H and Blackstock J (1976) Further observations on the biochemical composition of the cement of *Lepas fascicularis* Ellis & Solander; Electrophoretic examination of the protein moietis under various conditions. Journal of Experimental Marine Biology and Ecology 25: 263–271.

Bernard FJ and Lane CE (1962) Early settlement and metamorphosis of the barnacle *Balanus amphritrite niveus*. Journal of Morphology 110(1): 19–39.

Böck P (1989) Romeis Mikroskopische Technik, 17th Ed. Urban und Schwarzenberg, München.

Boëtius J (1952) Some notes on the relation to the substratum of *Lepas anatifera* L. and *Lepas fascicularis* E.et S. Oikos 4(II): 112–117.

Burnett BR (1972) Aspects of the circulatory system of *Pollicipes polymerus* J.B. Sowerby (Cirripedia: Thoracica). Journal of Morphology 136: 79–108.

Cheng W, Tsai IH, Huang CJ, Chiang PC, Cheng CH, and Yeh MS (2008) Cloning and characterization of hemolymph clottable proteins of kuruma prawn (*Marsupenaeus japonicus*) and white shrimp (*Litopenaeus vannamei*). Developmental & Comparative Immunology 32(3): 265–274.

Cheung PJ and Nigrelli RF (1972) Histochemical analysis of the fluid and the solid state of the adhesive materials produced by the pre- and postmetamophosed cyprids of *Balanus eburneus* Gould. Zoologica: Scientific Contributions of the New York Zoological Society 57: 79–95.

Cheung PJ, Ruggieri GD, and Nigrelli RF (1977) A new method for obtaining barnacle cement in the liquid state for polymerization studies. Marine Biology 43: 157–163.

Crenshaw DG (1979) Hydrostatic support of the pedunculate barnacle *Pollicipes polymerus*. Comparative Biochemistry and Physiology A 62: 423–425.

Crisp DJ (1972) Mechanisms of adhesion of fouling organisms. In: Acker RF, Brown BF, DePalma DR, and Iverson WP (eds) Proceedings of the 3rd International Congress on Marine Corrosion and Fouling. Northwestern University Press, Gaithersburg, Maryland: pp 691–709

Dickinson GH, Vega IE, Wahl KJ, Orihuela B, Beyley V, Rodriguez EN, Everett RK, Bonaventura J, and Rittschof D (2009) Barnacle cement: a polymerization model based on evolutionary concepts. Journal of Experimental Biology 212(Pt 21): 3499–3510.

Dougherty WJ (1989) Strength of adhesion of adult barnacles, *Chthamalus fragilis*, reattached to polystyrene. American Zoologist 29: A278.

Dougherty WJ (1990) Barnacle adhesion: reattachment of the adult barnacle *Chthamalus fragilis* Darwin to polystyrene surfaces followed by centrifugational shearing. Journal of Crustacean Biology 10(3): 469–478.

Dougherty WJ (1996) Zinc metalloprotease activity in the cement precursor secretion of the barnacle, *Chthamalus fragilis* Darwin. Tissue and Cell 28(4): 439–447.

Dreanno C, Kirby RR, and Clare AS (2006a) Locating the barnacle settlement pheromone: spatial and ontogenetic expression of the settlement-inducing protein complex of *Balanus amphitrite*. Proceedings of the Royal Society of London, Series B: Biological Sciences 273(1602): 2721–2728.

Dreanno C, Kirby RR, and Clare AS (2006b) Smell feet are not always a bad thing: the relationship between cyprid footprint protein and the barnacle settlement pheromone. Biology Letters 2: 423–425.

Fyhn UEH and Costlow JD (1976) A histological study of cement secretion during the intermolt cycle in barnacles. Biological Bulletin 150(1): 47–56.

Fyhn UEH and Costlow JD (1977) Histology and histochemistry of the ovary and oogenesis in *Balanus amphitrite* L. and *B. eburneus* Gould (Cirripedia, Crustacea). Biological Bulletin 152: 351–359.

Kamino K (2006) Barnacles Underwater Attachment. In: Smith AM and Callow JA (eds) Biological Adhesives. Springer-Verlag, Heidelberg: pp 145–166.

Kamino K (2008) Underwater adhesive of marine organisms as the vital link between biological science and material science. Marine Biotechnology 10: 111–121.

Kamino K, Inoue K, Maruyama T, Takamatsu N, Harayama S, and Shizuri Y (2000) Barnacle cement proteins – importance of disulfide bonds in their insolubility. Journal of Biological Chemistry 275(35): 27360–27365.

Kamino K, Odo S, and Maruyama T (1996) Cement proteins of the acorn barnacle, *Megabalanus rosa*. Biological Bulletin 190(3): 403–409.

Kiernan JA (1999) Histological and Histochemical Methods: Theory and Practice, 3rd Ed. Butterworth Heinemann, Oxford.

Kugele M and Yule AB (2000) Active relocation in lepadomorph barnacles. Journal of the Marine Biological Association of the United Kingdom 80: 103–111.

Lacombe D (1966) Glândulas de cimento e seus canais em *Balanus tintinnabulum*. Nota Técnica No. 32, Publição Instituto de Pesquisas da Marinha, Rio de Janeiro.

Lacombe D (1970) A comparative study of the cement glands in some balanid barnacles (Cirripedia, Balanidae). Biological Bulletin 139: 164–179.

Lacombe D and Liguori VR (1969) Comparative histological studies of the cement apparatus of *Lepas anatifera* and *Balanus tinntinnabulum*. Biological Bulletin 137: 170–180.

Lee H, Scherer NF, and Messersmith PB (2006) Single-molecule mechanics of mussel adhesion. Proceedings of the National Academy of Sciences of the United States of America 103(35): 12999–13003.

Lim S, Choi YS, Kang DG, Song YH, and Cha HJ (2010) The adhesive properties of coacervated recombinant hybrid mussel adhesive proteins. Biomaterials 31: 3715–3722.

Lindner E and Dooley CA (1972) Chemical bonding in cirriped adhesive. In: Acker RF, Brown BF, DePalma DR, and Iverson WP (eds) Proceedings of the 3rd International Congress on Marine Corrosion and Fouling. Northwestern University Press, Gaithersburg, Maryland: pp 653–673.

Matsumura K, Nagano M, and Fusetani N (1998a) Purification of a larval settlement-inducing protein complex (SIPC) of the barnacle, *Balanus amphitrite*. Journal of Experimental Zoology 281: 12–20.

Matsumura K, Nagano M, Kato-Yoshinaga Y, Yamazaki M, Clare AS, and Fusetani N (1998b) Immunological studies on the settlement-induced protein complex (SIPC) of the barnacle *Balanus amphitrite* and its possible involvement in larva–larva interactions. Proceedings of the Zoological Society of London, Series B 265: 1825–1830.

Minchin D (1996) Tar pellets and plastics as attachment surfaces for lepadid cirripedes in the North Atlantic Ocean. Marine Pollution Bulletin 32: 855–859.

Mori Y, Urushida Y, Nakano M, Uchiyama S, and Kamino K (2007) Calcite-specific coupling protein in barnacle underwater cement. FEBS Journal 274(24): 6436–6446.

Moyse J (1987) Larvae of lepadomorph barnacles. Crustacean Issues 5. Barnacle Biology: 329–362.

Nakajima M, Sano H, Burrow MF, Tagami J, Yoshiyama M, Ebisu S, Ciucchi B, Russell CM, and Pashley DH (1995) Tensile bond strength and SEM evaluation of caries-affected dentin using dentin adhesives. Journal of Dental Research 74(10): 1679–1688.

Newman WA (1987) Evolution of cirripedes and their major groups. Barnacle Biology Crustacean Issues 5: 3–42.

Nott JA (1969) Settlement of barnacle larvae: surface structure of the antennular attachment disc by scanning electron microscopy. Marine Biology 2: 248–251.

Nott JA and Foster BA (1969) On the structure of the antennular attachment organ of the cypris larva of *Balanus balanoides* (L.). Philosophical Transactions of the Royal Society of London, Series B: Biological Sciences 256: 115–134.

Ödling K, Albertsson C, Russell JT, and Mårtensson LGE (2008) An *in vivo* study of exocytosis of cement proteins from barnacle *Balanus improvisus* (D.) cyprid larva. Journal of Experimental Biology 209: 956–964.

Phang IY, Aldred N, Clare AS, Callow JA, and Vancso GJ (2006) An *in situ* study of the nanomechanical properties of barnacle (*Balanus amphitrite*) cyprid cement using atomic force microscopy (AFM). Biofouling 22(3–4): 245–250.

Saroyan JR, Lindner E, and Dooley CA (1970) Repair and reattachment in the barnacle as related to their cementing mechanism. Biological Bulletin 139: 333–350.

Setsaas TH and Bester MN (2006) Goose barnacle (*Lepas australis*) infestation of the Subantarctic fur seal (*Arctocephalus tropicalis*). African Zoology 41(2): 305–307.

Söderhäll I, Wu C, Novotny M, Lee BK, and Söderhäll K (2009) A novel protein acts as a negative regulator of prophenoloxidase activation and melanization in the freshwater crayfish *Pacifastacus leniusculus*. Journal of Biological Chemistry 284(10): 6301–6310.

Stewart RJ, Weaver JC, Morse DE, and Waite JH (2004) The tube cement of *Phragmatopoma californica*: a solid foam. Journal of Experimental Biology 207(Pt 26): 4727–4734.

Sullan RM, Gunari N, Tanur AE, Chan Y, Dickinson GH, Orihuela B, Rittschof D, and Walker GC (2009) Nanoscale structures and mechanics of barnacle cement. Biofouling 25(3): 263–275.

Sun Y, Guo S, Walker GC, Kavanagh CJ, and Swain GW (2004) Surface elastic modulus of barnacle adhesive and release characteristics from silicone surfaces. Biofouling 20(6): 279–289.

Theopold U, Schmidt O, Söderhäll K, and Dushay MS (2004) Coagulation in arthropods: defence, wound closure and healing. Trends in Immunology 25(6): 289–294.

Trump BF, Smuckler EA, and Benditt EP (1961) A method for staining epoxy sections for light microscopy. Journal of Ultrastructure Research 5: 343–348.

Urushida Y, Nakano M, Matsuda S, Inoue N, Kanai S, Kitamura N, Nishino T, and Kamino K (2007) Identification and functional characterization of a novel barnacle cement protein. FEBS Journal 274(16): 4336–4346.

Waite JH (2002) Adhesion á la Moule. Integrative and Comparative Biology 42: 1172–1180.

Waite JH, Anderson NH, Jewhurst S, and Sun C (2005) Mussel adhesion: finding the tricks worth mimicking. Journal of Adhesion 81: 297–317.

Walker G (1970) The histology, histochemistry and ultrastructure of the cement apparatus of three adult sessile barnacles, *Elminius modestus*, *Balanus balanoides* and *Balanus hameri*. Marine Biology 7: 239–248.

Walker G (1971) A study of the cement apparatus of the cypris larva of the barnacle *Balanus balanoides*. Marine Biology 9: 205–212.

Walker G (1973) The early development of the cement apparatus in the barnacle, *Balanus balanoides* (L.) (Crustacea: Cirripedia). Journal of Experimental Marine Biology and Ecology 12: 305–314.

Walker G and Youngson A (1975) The biochemical composition of *Lepas anatifera* (L.) cement (Crustacea: Cirripedia). Journal of the Marine Biological Association of the United Kingdom 55: 703–707.

Walker G and Yule AB (1984) Temporary adhesion of the barnacle cyprid: the existence of an antennular adhesive secretion. Journal of the Marine Biological Association of the United Kingdom 64: 679–686.

Walley LJ (1969) Studies on the larval structure and metamorphosis of *Balanus balanoides* (L.). Philosophical Transactions of the Royal Society of London, Series B: Biological Sciences 256: 237–280.

Wang R, Liang Z, Hall M, and Söderhäll K (2001) A transglutaminase involved in the coagulation system of the freshwater crayfish, *Pacifastacus leniusculus*. Tissue localisation and cDNA cloning. Fish & Shellfish Immunology 11(7): 623–637.

Yule AB and Walker G (1984) The adhesion of the barnacle, *Balanus balanoides*, to slate surfaces. Journal of the Marine Biological Association of the United Kingdom 64: 147–156.

Yule AB and Walker G (1987) Adhesion in barnacles. Barnacle Biology Crustacean Issues 5: 389–402.

10 Morphology of the Adhesive System in the Sandcastle Worm, *Phragmatopoma californica*

Ching Shuen Wang, Kelli K. Svendsen and Russell J. Stewart

Contents

10.1 Introduction

The marine Sandcastle worm (*P. californica*) and related species live in composite mineralized tubes for shelter. They gather the mineral phase for free from the environment as sand grains and seashell bits with a crown of ciliated tentacles. The captured mineral particles are conveyed for inspection to the building organ – a pincer-shaped pair of dexterous palps in front of the mouth (Fig. 10.1). A dab of proteinaceous adhesive (Jensen and

Fig. 10.1 *Phragmatopoma californica.* At the left side of the photograph the tentacles and operculum of a sandcastle worm protrude from the end of a tube rebuilt with white 0.5 μm zirconium oxide beads. The worm on the right has been removed from its tube. The larger arrow indicates the building organ. The smaller arrow indicates the ventral shield region. Scale bar: 5 mm

Morse, 1988) is secreted from the building organ onto suitable particles as they are pressed onto the end of the tube. The major protein components of the adhesive are a group of heterogeneous proteins, referred to as Pc3x, characterized by serial runs of 10–14 serine residues punctuated with single tyrosine residues (Zhao et al., 2005). Phosphorylation of more than 90% of the serines (Stewart et al., 2004) makes the Pc3 proteins polyacidic ($pI<3$). Other potential protein components identified biochemically (Waite et al., 1992) and by sequencing random cDNAs from an adhesive gland library (Endrizzi and Stewart, 2009) are generally polybasic with predicted pIs greater than 9. Amino acid analysis of secreted glue revealed that, in total, close to 50% of the adhesive protein residues are charged when serine phosphorylation is taken into account. The adhesive also contains Mg^{2+} and Ca^{2+} and a large fraction of the tyrosines are post-translationally hydroxylated to form 3,4-dihydroxyphenylalanine (DOPA), a residue shared with the adhesive plaque proteins of the mussel (Waite and Tanzer, 1981). Phosphates and *o*-dihydroxyphenols are well-known adhesion promoters.

The secreted glue is initially fluid as evident from its crack filling capability and the contact angles of the adhesive on glass substrates (Fig. 10.3D). Within 30 s after secretion the fluid adhesive sets into a force-bearing solid foam (Stevens et al., 2007) (Fig. 10.3) and covalently cures over the next several hours into material with the consistency of shoe leather. The foam is a mix of open and closed cells, covered by a non-porous skin, and often associated with silky fibers (Fig. 10.3E). The skin and perhaps lining of the pores appear to be denser than the matrix of the adhesive (Fig. 10.3F). The set and cure of the adhesive was proposed to be triggered by the pH differential between secretory granules (pH ~5) and seawater (pH 8.2); the set by insolubilization of the divalent cations and polyphosphate, the cure by accelerated oxidation and subsequent crosslinking of some of the DOPA residues (Stewart et al., 2004).

The high charge content and segregation of opposite charges into separate proteins suggested a model in which complex coacervation of polyacidic and polybasic adhesive proteins drove formation of the initial cohesive fluid phase of the secreted adhesive (Stewart et al., 2004). Complex coacervates phase separate from aqueous solutions of oppositely charged polyelectrolytes when the conditions (pH, charge ratios, and ionic strength) are such that the solution is near electrical neutrality (Bungenberg de Jong, 1949). Complex coacervates are dense, cohesive, water-immiscible fluids with low interfacial tension and as such are ideal vehicles for water-borne underwater bioadhesives. Indeed, copying the charge characteristics and ratios of the adhesive proteins with synthetic polyelectrolytes resulted in adhesive complex coacervates that qualitatively replicated several features of the natural underwater adhesive, namely interfacial adhesion to wet substrates, underwater deliverability, and controlled setting reactions (Shao et al., 2009; Shao and Stewart, 2010).

Although the Sandcastle worm has already provided significant insights it has much more to teach us about making underwater adhesives. The sandcastle glue is comprised of separated active components that when mixed just before application react to form bonds – superficially analogous to two-part epoxy adhesives found in hardware stores. Little is known about which components of the multi-part adhesive are segregated into which compartments, when or how the separated components are mixed prior to bonding, the chemical details of the mechanisms that initiate hardening of the adhesive, and when or how water is removed or segregated from the fluid adhesive to produce a solid foam underwater. To address these and other questions morphological analyses of the adhesive gland and secretion process were initiated and preliminary findings are described here. These studies provide a map to investigate the detailed organization of the adhesive system with specific protein and nucleic acid probes. They are also the first step in studying the physiology of glue secretion. The knowledge, of course, will guide development of more sophisticated synthetic analogs of the Sandcastle worm adhesive.

10.2 Sandcastle Worm Morphology

Early morphological studies of the adhesive secretion process were conducted by Jean Vovelle in three related worm species that build composite tubes: *Sabellaria alveolata* (Vovelle, 1965) in the sub-order Sabellida that builds reef-like colonies of conjoined tubes; and *Lagis koreni* (Vovelle and Grasset, 1976) and *Petta pusilla* (Vovelle, 1979) in the sub-order Terebellida, family Pectinariidae, that build solitary conical tubes. They are commonly known as Honeycomb worms and Ice Cream Cone worms, respectively, for their tube-building habits. To summarize Vovelle's observations, all three species had clustered glandular tissue situated around the coelomic cavity of the first three parathoracic metameres. The glands contained two major secretory cell types distinguished by the appearance of their densely packed secretory granules; "homogeneous" granules appeared to

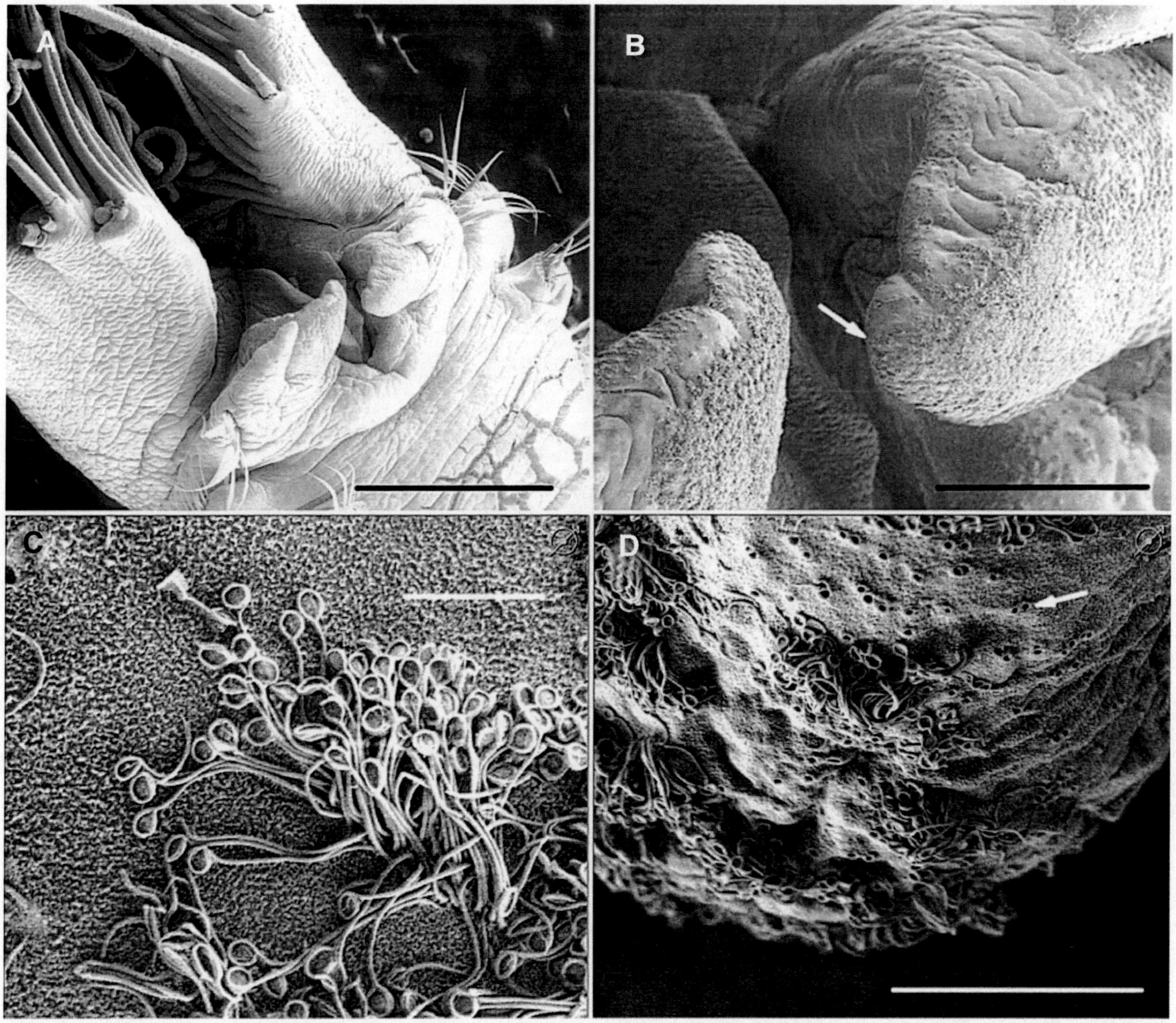

Fig. 10.2 SEM of the Sandcastle building organ. (A) The dexterous and multi-functional building organ; (B) Close up of the heavily ciliated building organ. (C) The area indicated by the arrow in B. Paddle-ended cilia occur in dense clusters. (D) The tip of the building organ palps are dotted with small pores that often occur in pairs. Scale bar in (A) 1 mm; (B) 200 μm; (C) 5 μm; and (D) 20 μm

have a uniform composition, "heterogeneous" granules contained dense inclusions. Both types of granules exited intact from their respective secretory cells through narrow cellular extensions that extended in bundles to the building organ. The granules remained intact and in separate channels until evacuating from the building organ through clustered pores, which in *P. pusilla* was described as a "strainer". Vovelle also reported the presence of Mg, Ca, and P in the heterogeneous granules of *S. alveolata* (Gruet et al., 1987) and *L. koreni* (Truchet and Vovelle, 1977).

10.2.1 The Building Organ

As the name implies, the building organ is the principle appendage used to carry out the detailed work of tube building. The Sandcastle worm, as mason, feels around the end of the tube with the fingers of the building organ evaluating where next to place a particle and the required size and shape. Particles delivered by the tentacles are turned and rotated between the fingers to check the size and it seems surface chemistry or texture. Tubes rebuilt with chicken egg shell fragments, for example, always

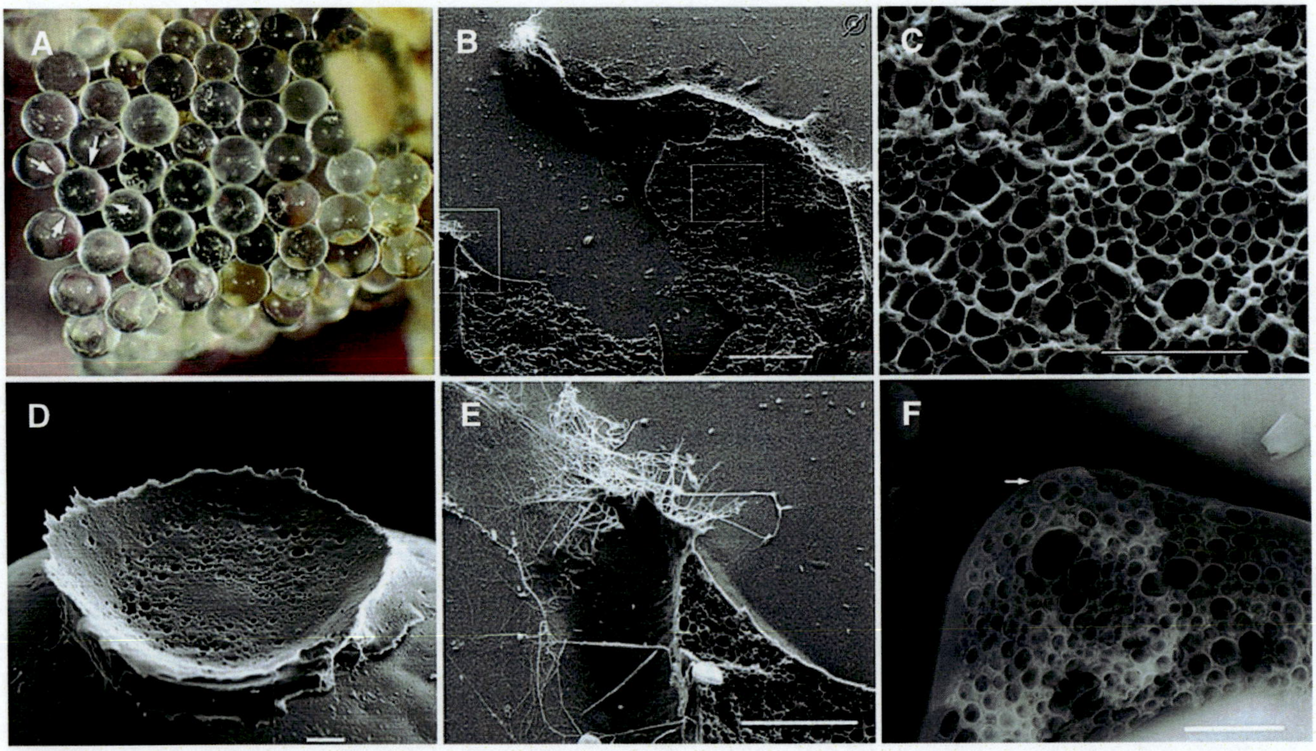

Fig. 10.3 SEM of sandcastle adhesive. (A) A tube partially rebuilt with glass beads. The glue was applied only at the four contact points (arrows); (B) Sandcastle worms placed on coverslips glue glass beads to the surface. The glue fractured when the bead was pried away; (C) The foamy interior in the right box in B; (D) A spot of glue left on a glass bead; (E) Threads and non-porous skin layer on glue; (F) Foamy interior imaged with backscattered electron detector. Distinct layers on the surface (arrow) and lining the pores are visible. Scale bar in (B and D) 50 μm; (C and E) 15 μm; (F) 5 μm

have the outside of the eggshell oriented toward the inside of the tube. When the appropriate particle is in position the worm calculates where to apply the adhesive most effectively. With uniform round particles, like glass beads, the glue is applied precisely at the points where stacked spheres make contact – only at contact points and generally at all contact points (Fig. 10.3A). After observing the placement of numerous particles it appears the worm feels the contact point between particles from the side with the building organ and injects adhesive at that point. In addition to providing the muscle, protractor and compass, the building organ also provides sensory information required for precise and efficient sandcastle construction.

Scanning electron microscopy (SEM) revealed dense clusters of paddle-shaped cilia, which may play a role in sensory functions, sporadically carpeting the surface of the building organ (Fig. 10.2B–D). Clustered pores were observed in the building organ surface that may be the exit points for adhesive components (Fig. 10.2D). At ~0.5 μm in diameter the pores are much smaller than the major 2–4 μm adhesive granules.

10.2.2 The Adhesive Gland

The gross structure of the adhesive gland was revealed with Arnow stain on cryosectioned worms (Fig. 10.4). Arnow's reagent becomes bright red after reacting with DOPA (Arnow, 1937). The main glandular region is apparent in the body cavity, the building organ is uniformly stained, and the so-called ventral shield [the layer of milky white tissue on the ventral surface of the first three segments (Fig. 10.1)] is stained lighter red and in a more punctate pattern. The red striations below the building organ in Fig. 10.4C are channels leading from the adhesive gland to the building organ. A more detailed overview of the adhesive gland is provided by observing the autofluorescence of the adhesive precursors (Fig. 10.5). Secretory cells densely packed with fluorescent secretory granules are distributed around the body cavity of the three parathoracic segments. Individual granules exit from the packed secretory cells and move in single-file clusters through channels toward the building organ (Fig. 10.5C). The granules are still intact when they arrive at the building organ, which appears to be laced around its entire

Fig. 10.4 Distribution of DOPA-containing proteins in the Sandcastle worm. (A) Cryosection through the second parathoracic segment stained with Arnow's reagent. The adhesive gland stains dark red. The ventral shield and building organ stain lighter red; (B) Coronal cryosection unstained; (C) Same section as B stained with Arnow's reagent

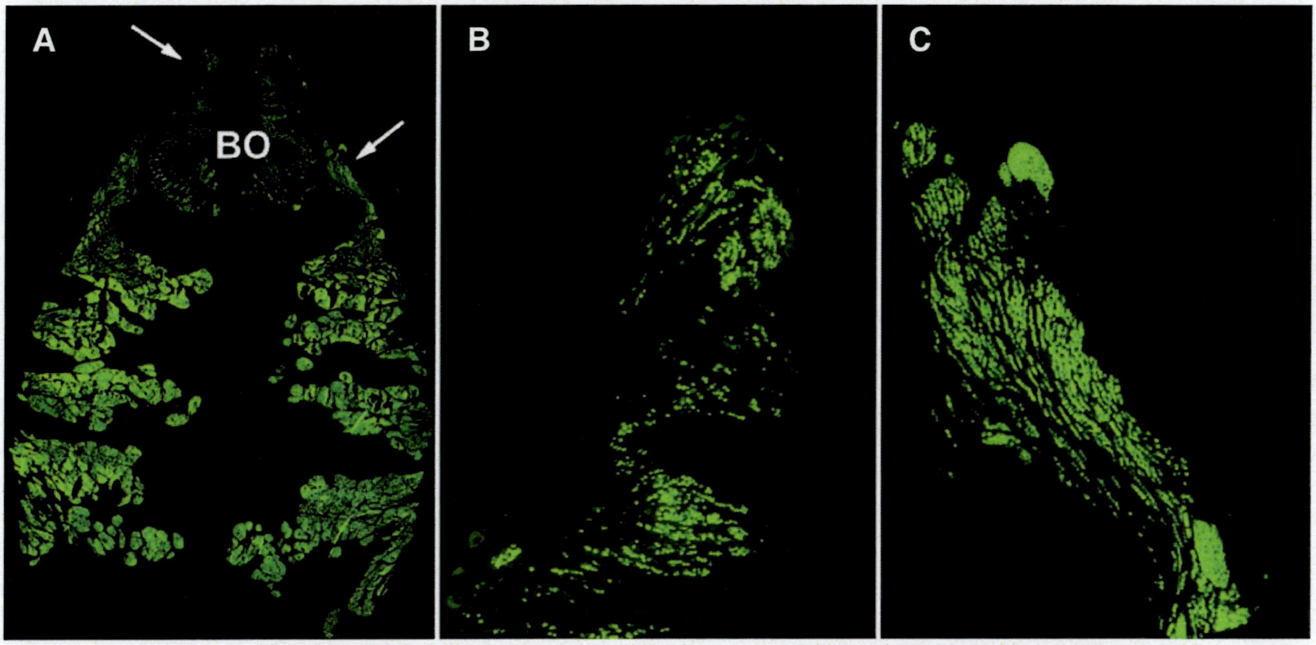

Fig. 10.5 Auto-fluorescence of the sandcastle worm secretory system. (A) Secretory cells in the tri-partite adhesive gland are intensely fluorescent due to the densely packed secretory granules. The channels leading from the secretory cells are apparent as well as radially distributed channels around the building organ (*BO*). (B) The building organ region identified by right arrow in A. Granules are lined up in channels and positioned a few microns from the surface. (C) Area of channel identified by left arrow in A. Granules are in single file lanes in a broad channel leading to the building organ

circumference with radial channels (Fig. 10.5B). The granules move toward the surface between the clefts of the puckered building organ. Most of the intact granules seem to be parked several microns from the surface of the building organ in the unstimulated condition. Although granules are positioned around the entire building organ, the shape of the glue spots on glass beads and slides (Fig. 10.3) suggests the adhesive is released from localized areas of the building organ.

Like the adhesive precursors, the final set and cured sandcastle glue is autofluorescent over the entire visible spectrum (Stevens et al., 2007). The origin of this autofluorescence has not been studied in detail but a reasonable assumption has been that extensive conjugation resulting from oxidative crosslinking between DOPA residues would create fluorescent structures. If this were the mechanism it suggests the adhesive is at least partially DOPA-crosslinked before secre-

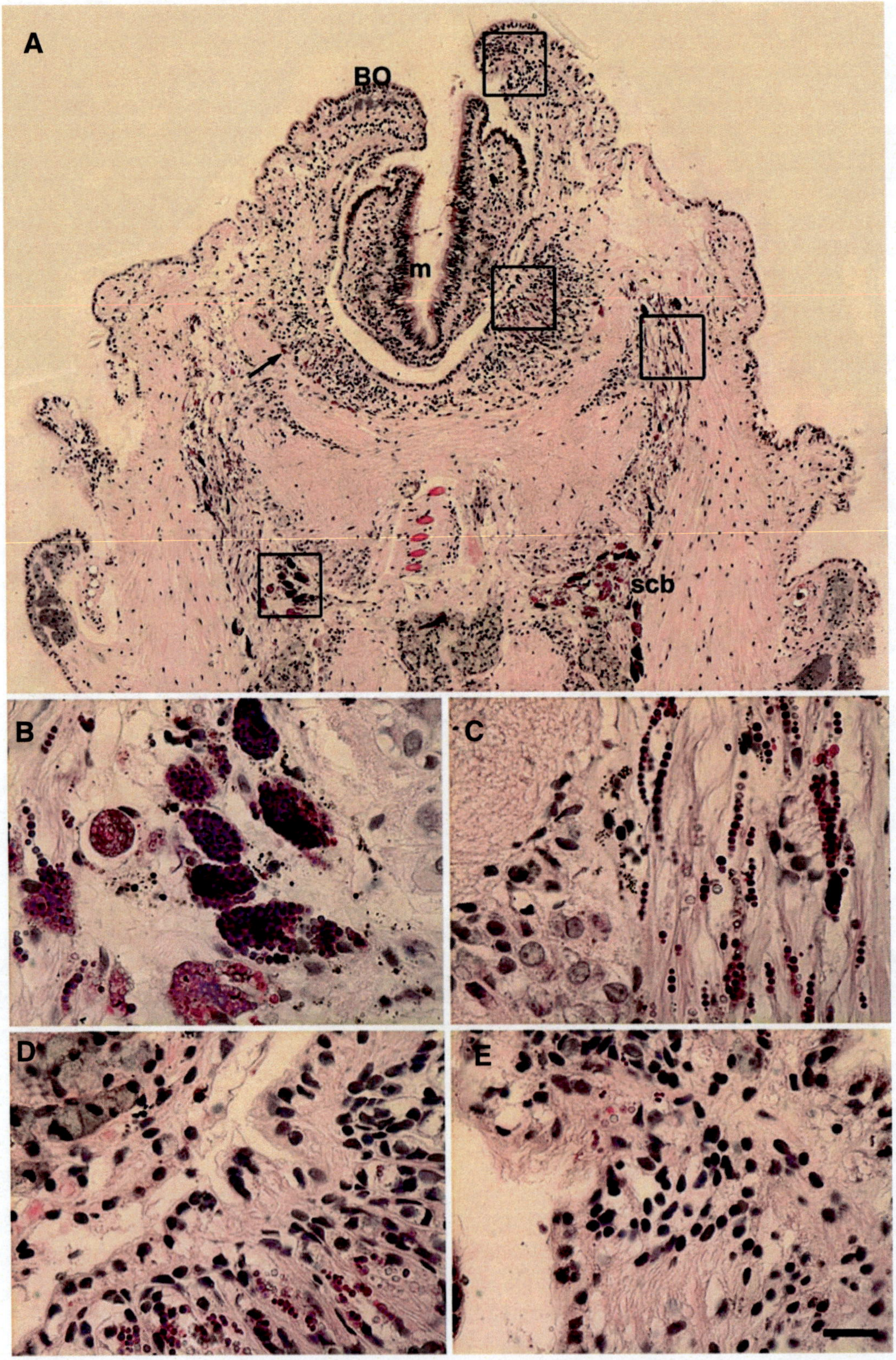

Fig. 10.6 Coronal sections stained with hematoxylin and eosin showing various regions of the secretory system. (A and B) Densely packed secretory cells containing homogeneous (*dark*) or heterogeneous granules (*light*) in the lower left boxed region; (C) Boxed channel region. Granules are segregated in single file channels; (D and E) Two regions containing granules from the periphery of the building organ. Scale bar in E applies to B–D; (B–E) 10 μm

tion. A second possibility is that the worms were fixed with paraformaldehyde which has been shown to create fluorescence in dopamine containing brain tissue (Lofstrom et al., 1976). While the fluorescence may have been enhanced by paraformaldehyde fixation this was not the sole mechanism because isolated unfixed granules are strongly fluorescent with the same broad spectrum (not shown). Other potential structures responsible for the autofluorescence, though speculative, may be DOPA-metal complexes in the adhesive precursors, or the cofactors of redox enzymes, such as tyrosinase, involved in modifying the glue proteins. The autofluorescence merits further investigation because it could reveal details of the adhesive structure and the curing mechanism.

Coronal sections stained with H&E (Fig. 10.6) reveal two prominent secretory cell types in the adhesive gland loaded with either heterogeneous that stain light reddish purple or homogeneous granules that stain dark purple. The granules appear to leave the secretory cells in single file ranks and enter a broad secretory highway with separate lanes for each secretory cell (Fig. 10.6C). The homogeneous and heterogeneous granules remain segregated within their ranks as they move along their lanes to the building organ where they are positioned for rapid deployment (Fig. 10.6D). More dorsal sections reveal similar secretion channels connected to the most posterior regions of the adhesive gland. The structure of the channels is not well understood. Vovelle described them as cellular extensions (Vovelle, 1979). Likewise, the mechanisms

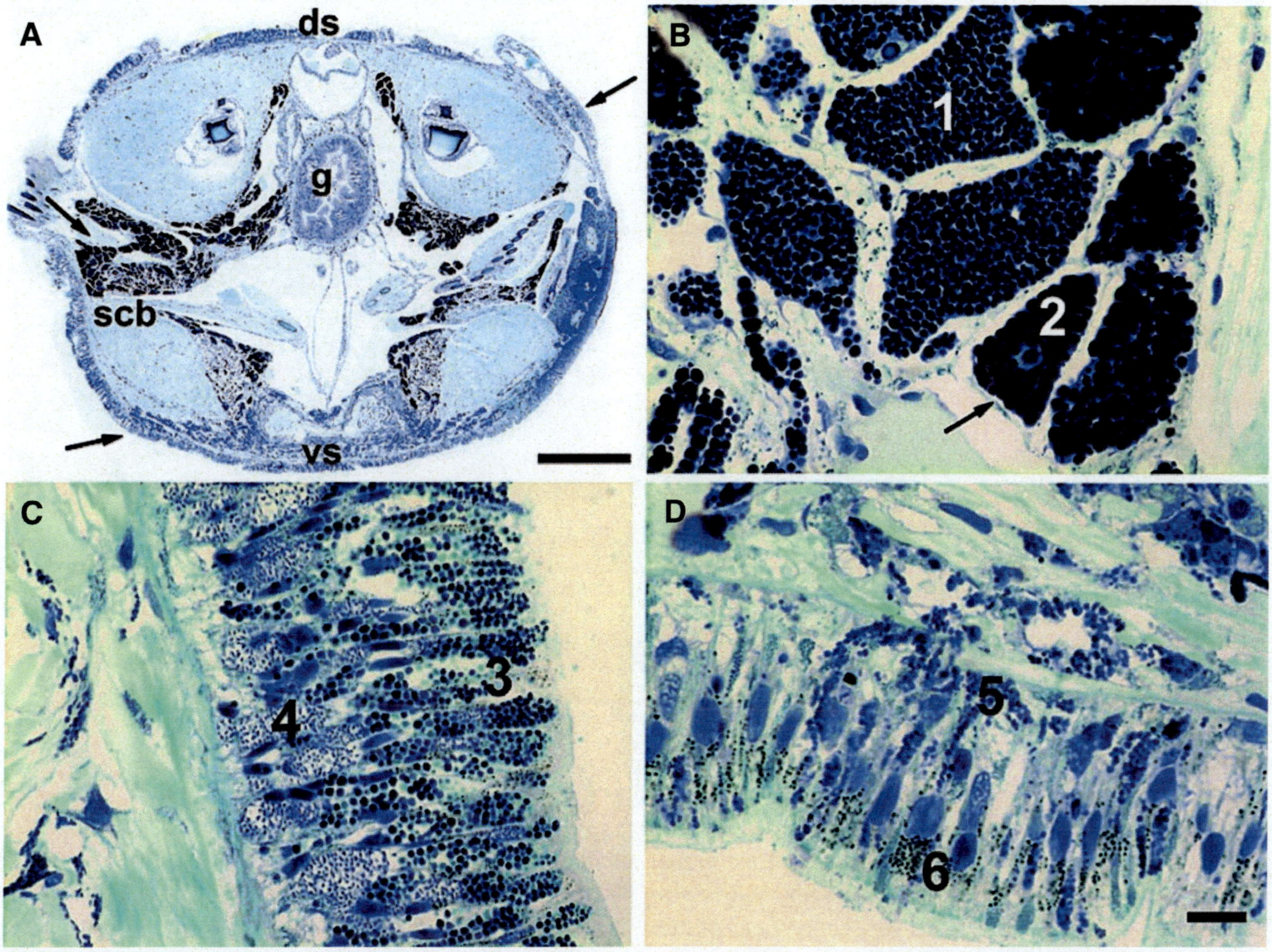

Fig. 10.7 Cross section through the center of the adhesive gland stained with toluidine blue. At least six distinct granule types are apparent (*1–6*); (A) Abbreviations are *ds* dorsal region; *vs* ventral shield; *g* gut; and *scb* secretory cell body; (B) Region of adhesive gland indicated by top-left arrow. Homogeneous granules (*1*), heterogeneous (*2*); (C) Region indicated by top right arrow; (D) Region of the ventral shield. Scale bar in (A) 500 μm; (D) 10 μm

that power granule transport through the channels are unknown. The bundled channels are interspersed with what appears to be loose connective tissue and are surrounded by bands of smooth muscle tissue.

The most prominent secretory cells in the adhesive gland produce the homogeneous and heterogeneous granules, but several other types of secretory granules are also present. Small cells with small bright red granules (<1 μm) line the outer edges of the building organ (Fig. 10.6A, arrow). In addition to the size differences, the contents of these granules are not autofluorescent. The small granules were often observed in the proximity of the heterogeneous and homogeneous granules in the building organ. We cannot rule out that the contents of the small granules are components of the adhesive.

In a cross section through the second thoracic segment stained with Toluidine Blue (TB) the adhesive gland clings to the outside wall of the body cavity (Fig. 10.7). The heterogeneous and homogeneous granules stained distinctly, the heterogeneous granules are dark blue, homogeneous lighter blue (Fig. 10.7B). Several types of columnar epithelial cells with distinct granules are apparent around the entire circumference of the worm (Fig. 10.7C, D). The products of these cells are not known but at least some are likely to produce mucus secretions as described by Vovelle (1965) in *Sabellaria alveolata*. Other products could include proteins identified by sequencing random clones from a cDNA library made with RNA isolated from tissues that included the epithelial secretory cells (Endrizzi and Stewart, 2009). Over 20 cDNAs encoding proteins with secretion signal peptides, most with characteristics of structural proteins, were identified during the expression survey. Some of these secreted proteins are likely components of the sandcastle glue but others could be components of the thin, reddish-brown, fabric-like sheath that lines the inside of the tubes. Neither the composition nor origin of the lining has been carefully studied but it could be secreted from the ventral shield. The punctate Arnow staining (Fig. 10.4A) suggests secretions originating from the ventral shield contain DOPA, which could account for the reddish-brown color of the lining. Other candidate products of the epithelial secretory cells identified in the expression survey include a small collagen with a secretion signal peptide that could be a component of the lining and a secreted laccase enzyme that may be the phenoloxidase activity reported by Vovelle (1965) in the ventral shield. The laccase may be responsible for quinonic tanning of the lining.

10.2.2.1 Granule Composition

A thin section of an epon embedded Sandcastle worm was examined by SEM and energy dispersive X-ray spectroscopy to investigate the composition of the homogeneous and heterogeneous granules (Fig. 10.8). Heterogeneous granules contain high-contrast sub-structures of variable size while no sub-structure is apparent in the homogeneous granules. Spectra acquired in a line scan through both granule types revealed abundant phosphorus and magnesium colocalized with the bright sub-granules in the heterogeneous granules. The relative height of the peaks in Fig. 10.8B does not represent the molar ratios

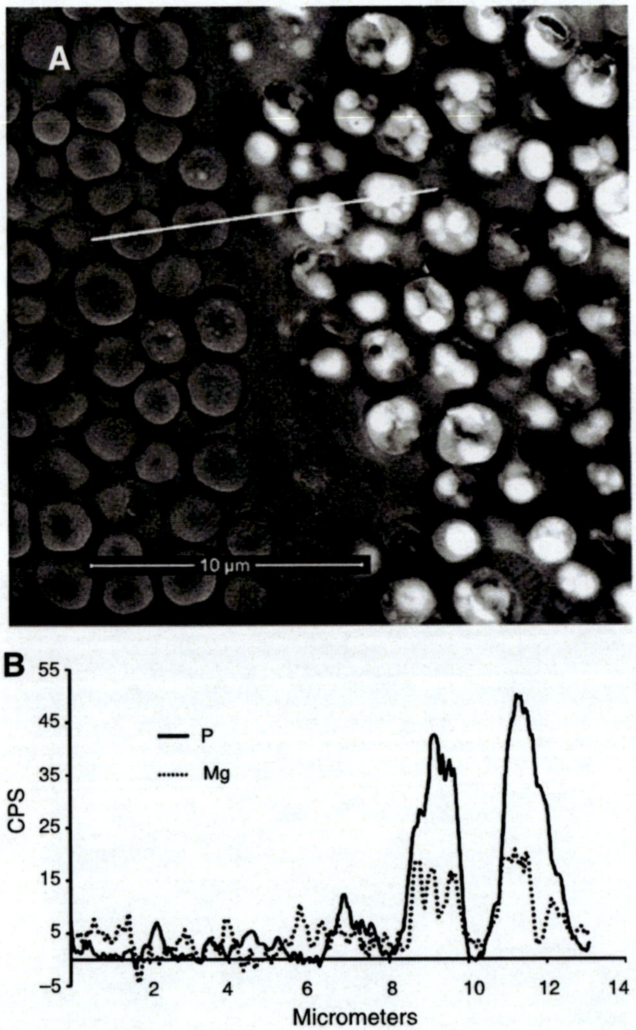

Fig. 10.8 Energy dispersive X-ray spectroscopy of granules within secretory cells. (A) Heterogeneous granules contain bright irregularly sized inclusions; (B) Phosphorus and magnesium along the line crossing through homogeneous and heterogeneous granules in panel A. Phosphorus and magnesium were colocalized in the heterogeneous sub-granules

of P and Mg. The homogeneous granules contained only background levels of phosphorus and magnesium. The results demonstrate the highly phosphorylated Pc3 proteins (Zhao et al., 2005) are located exclusively in the heterogeneous granules. The colocalized Mg may play a role in condensing the highly charged Pc3 proteins into the heterogeneous substructures. Efforts to identify other protein components of each granule type are in progress using antibodies raised against synthetic peptides from Pc1 and 2, which were identified biochemically (Waite et al., 1992), and Pc4 and 5, which were identified in the adhesive gland cDNA library (Endrizzi and Stewart, 2009). In another approach, isolated granules of both types are being analyzed by tandem mass spectrometry to identify peptides for comparison to predicted peptides in the expressed sequence database.

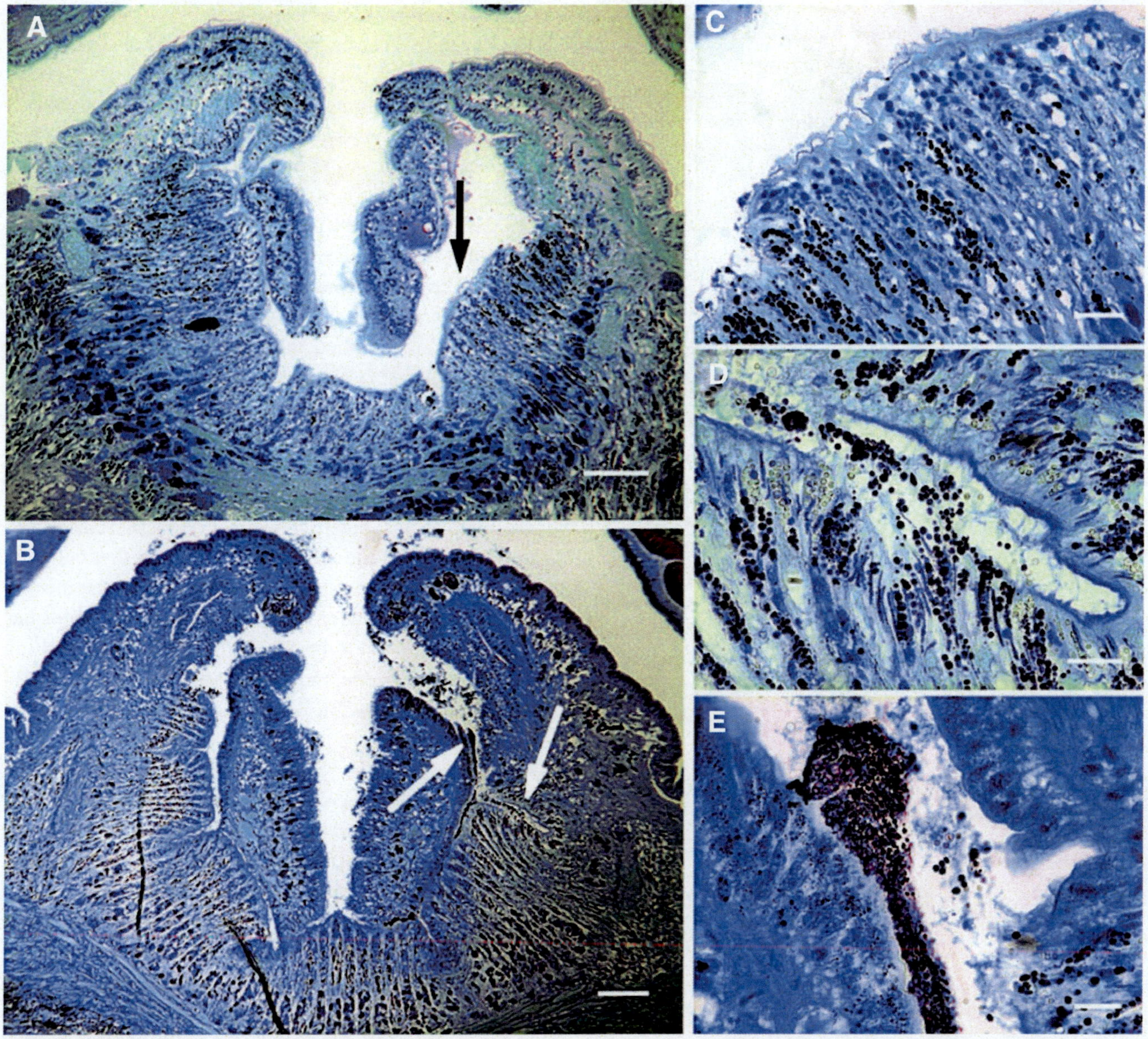

Fig. 10.9 Stimulated secretion. (A) Coronal section through the center of the building organ of an unstimulated worm; (B) Building organ after exposure to dopamine and electric shock. Glue appeared at several points around the building organ; (C) Region of building organ indicated by black arrow in panel A; (D) Secretory granules liberated from channels in the building organ flood into an exterior wrinkle. Granules begin to agglomerate; (E) Adhesive composed of mostly intact secretory granules along the outer portion of the mouth. Scale bar in (A and B) 100 μm; (C) 20 μm; (D and E) 10 μm

10.2.2.2 Stimulated Secretion

The Sandcastle worm glue is a multi-part adhesive that appears to be mixed just after secretion and before application to its gathered building materials. To uncover the biochemical and physiological details of the process it will be helpful to be able to stimulate secretion during observation. We found that one method to artificially trigger secretion is to externally expose the worms to 20 mM dopamine for 20 min followed by electric shocks in the vicinity of the building organ. Secretion did not occur when either stimulus was applied separately. Viscous white glue oozed from several areas around the building organ. As mentioned above, during naturally triggered secretion glue issues only from regions of the building organ in the vicinity of contact points between particles. Worms were immediately fixed during secretion events for histological analysis.

Coronal sections of secreting worms stained with TB revealed a dramatic widening of channels within the building organ compared to unstimulated worms (Fig. 10.9). The channels mostly terminated at the surface of the building organ but some terminate in the folded clefts. The widened channels may be due to muscular contractions resulting from electric shocks, are not the exit pathways for adhesive granules, and their role if any in the natural secretion process is unknown. At the site of secretion in the example shown in Fig. 10.9D, three distinct granule types, dark blue homogeneous, purplish heterogeneous, and unstained, were secreted intact into a cleft before exiting the building organ. Shortly after secretion the granules agglomerate into a viscous mass and the color of the TB stain changed from blue to a purple-violet. The color change, known as metachromasy, is a characteristic of some dyes, such as TB, when bound to particular substances (Bergeron and Singer, 1958). Polyphosphates are known to cause a metachromatic shift with basic dyes (Harold, 1966). The TB color change immediately after secretion is evidence of an unknown chemical change in the adhesive. Perhaps the phosphate sidechains of the Pc3x proteins may become more exposed as a result of heterogeneous granules rupturing.

10.3 Adhesive Models

Our first depictions of the complex coacervate model (Stewart et al., 2004; Stevens et al., 2007) to describe the assembly, secretion, and underwater bonding mechanisms of the Sandcastle worm glue did not take detailed anatomy of the adhesive gland and secretory pathways into full consideration. From the observations reported here and Vovelle's earlier studies it is clear that the adhesive components are packaged into granules in separate cell types. The distinct granules, as they move from the secretory cells to the point of exit at the building organ, remain segregated from one another as they travel by an unknown mechanism in channels of unknown structure. The granules exit intact from points distributed evenly around the entire building organ surface and only then do the contents of the distinct granules mix and become a viscous, presumably adhesive mass. If complex coacervation through electrostatic association of the oppositely charged proteins is part of the process of adhesive formation it must occur immediately after secretion rather than in a compartment of the secretory system. If so, this implies coacervation would have to occur in the high ionic strength of seawater.

In the original sandcastle glue model the foam structure of the set adhesive was attributed to phase separation of the complex coacervate from water with the water phase coalescing into vacuoles to become the cells of the foam. Vovelle described the heterogeneous granules as dilating during secretion with the dense inclusions becoming the pores of the foamy glue, and the contents of the homogeneous granules forming a matrix around the porous heterogenous granules. The observations reported here are consistent with Vovelle's obervations; the dense inclusions of the heterogeneous granules seem to become the pores of the foamy adhesive. Whether complex coacervation as commonly defined is part of the mechanism or not the Sandcastle Worm-inspired concept has nevertheless proven to be a useful paradigm for creating synthetic water-borne underwater adhesives (Shao et al., 2009; Shao and Stewart, 2010).

10.4 Materials and Methods

10.4.1 Animal Preparation

Phragmatoma Californica colony were collected near Santa Barbara, California and maintained in a laboratory aquarium with circulating artificial salt water at 14°C. Worms were carefully removed from their tubes and rinsed with filtrated seawater 3 times before fixation. Fixation was done with 4% paraformaldehyde in PBS at 4°C overnight. After fixation, the worms were dehydrated in a series of ethanol solutions (50, 70, 95, and 100%) at room temperature. The worms were then embedded with ImmunoBed (PolyScience, CA, USA) according to manufacturer's instruction.

10.4.2 Scanning Electron Microscopy

Worms were fixed with 4% paraformaldehyde and 2% glutaraldehyde in 0.1 M sodium cacodylate buffer at 4°C overnight, post-fixed with 2% osmimium tetroxide, dehydrated with serial ethanol at room temperature for several hours, and then mounted on carbon tape. The specimens were examined with a FEI Quanta 600 SEM in low vacuum at the University of Utah Surface and Nanoimaging Facility. An EDAX X-ray detector was used for elemental analysis. The fixed worms were embedded in epon plastic medium according to the manufacturer's instructions. Semi-thin (0.5 μm) sections were cut with an ultramicrotome and immediately mounted on a superfrost slide (VWR).

10.4.3 Fluorescent Microscopy

ImmunoBed embedded worm blocks were cut with a microtome to 2 μm and mounted on a superfrost slide (VWR) and covered with a No. 1 coverslip using antifade embedding medium. Longitudinal sections were imaged with either 20× or 100× objectives using a green filter set. ImageJ was used for image processing and montages were created with Photoshop.

10.4.4 Histological Staining

Two or four micrometer ImmunoBed embedded sections were rehydrated with serial ethanol solutions (95%, 70%, and 50%, 5 min each), rinsed with DI water at RT for 5 min, and then stained with Toluidine Blue according to routine histological protocols.

References

Arnow LE (1937) Colorimetric determination of the components of 3,4-dihydroxyphenylalanine-tyrosine mixtures. Journal of Biological Chemistry 118: 531–537.

Bergeron JA and Singer M (1958) Metachromasy: an experimental and theoretical reevaluation. Journal of Biophysical and Biochemical Cytology 4(4): 433–457.

Bungenberg de Jong HG (1949) Crystallization – coacervation – flocculation. In: Kruyt HR (ed) Colloid Science, vol. II. Elsevier Publishing Company, New York: pp 232–255.

Endrizzi BJ and Stewart RJ (2009) Glueomics: an expression survey of the adhesive gland of the sandcastle worm. The Journal of Adhesion 85: 546–559.

Gruet Y, Vovelle J, and Grasset M (1987) Composante biominerale du ciment du tube ches *Sabellaria alveolata* (L.) Annelide Polychete. Canadian Journal of Zoology 65: 837–842.

Harold FM (1966) Inorganic polyphosphates in biology: structure, metabolism, and function. Bacteriological Reviews 30(4): 772–794.

Jensen RA and Morse DE (1988) The bioadhesive of *Phragmatopoma californica* tubes: a silk-like cement containing L-DOPA. Journal of Comparative Physiology B 158: 317–324.

Lofstrom A, Jonsson G, Wiesel FA, and Fuxe K (1976) Microfluorimetric quantitation of catecholamine fluorescence in rat median eminence. II. Turnover changes in hormonal states. Journal of Histochemistry and Cytochemistry 24(2): 430–442.

Shao H and Stewart RJ (2010) Biomimetic underwater adhesives with environmentally triggered setting mechanisms. Advanced Materials 22(6): 729–733.

Shao H, Bachus KN, and Stewart RJ (2009) A water-borne adhesive modeled after the sandcastle glue of *P. californica*. Macromolecular Bioscience 9(5): 464–471.

Stevens MJ, Steren RE, Hlady V, and Stewart RJ (2007) Multiscale structure of the underwater adhesive of *Phragmatopoma californica*: a nanostructured latex with a steep microporosity gradient. Langmuir 23(9): 5045–5049.

Stewart RJ, Weaver JC, Morse DE, and Waite JH (2004) The tube cement of *Phragmatopoma californica*: a solid foam. Journal of Experimental Biology 207(Pt 26): 4727–4734.

Truchet M and Vovelle J (1977) Study of the element secretory glands of a tubicolous polychaete (*Pectinaria* (= *Lagis*) *koreni*) with the help of electron microprobe and ion microanalyzer. Calcified Tissue Research 24(3): 231–238.

Vovelle J (1965) Le tube de *Sabellaria alveolata* (L.) Annelide Polychete Hermellidae et son ciment. Etude ecologique, experimentale, histologique et histochimique. Archives de Zoologie Experimentale & Generale 106: 1–187.

Vovelle J (1979) Les Glandes Cementaires de Petta Pusilla Malmgren, Polychete Tubicole Amphictenidae, et leur Secretion Organo-minerale. Archives de Zoologie Experimentale & Generale 120: 219–246.

Vovelle J and Grasset M (1976) Les Pectinaires et leur ciment. Bulletin de la Société Zoologique de France 101(5): 1022–1023.

Waite JH, Jensen RA, and Morse DE (1992) Cement precursor proteins of the reef-building polychaete *Phragmatopoma californica* (Fewkes). Biochemistry 31(25): 5733–5738.

Waite JH and Tanzer ML (1981) Polyphenolic Substance of *Mytilus edulis*: Novel Adhesive Containing L-Dopa and Hydroxyproline. Science 212(4498): 1038–1040.

Zhao H, Sun C, Stewart RJ, and Waite JH (2005) Cement proteins of the tube-building polychaete *Phragmatopoma californica*. Journal of Biological Chemistry 280(52): 42938–42944.

11 Adhesive Dermal Secretions of the Amphibia, with Particular Reference to the Australian Limnodynastid Genus *Notaden*

Michael J. Tyler

Contents

11.1 Introduction

The class Amphibia includes three orders: the Caudata or Urodela, commonly known as salamanders and newts and characterized by possessing four limbs and a tail, the Gymnophiona or Caecilians, which are worm-like, have narrow bodies and lack limbs, and the Anura, which are the frogs and toads characterized by the possession of limbs but lack of a tail in the adult.

Adhesive secretions from dermal glands are known in all orders of amphibians. Evans and Brodie (1994) examined the adhesive strength of six species of salamanders and one gymnophionan (*Dermophis mexicanus*). How extensive adhesive secretions occur within these two orders is not known.

Currently, more than 4000 species of anurans are recognized. Of this total, knowledge of species with adhesive glands is extremely limited. Inger (1954) reports the external appearance of a "belly gland" in males of the Asian microhylid genus *Kaloula* and reports that in *K. picta* "The male [is] stuck to the female by a 'gelatinous' substance secreted by the male's belly gland", during amplexus.

Possibly the first species recognized to have an adhesive compound in the skin is the American Bullfrog *Rana catesbeiana* of North America. In 1937 the American Frog Canning Company produced a handbook on raising frogs, and stated that one of the by-products was Frog Glue which was suitable for mending crockery and glassware (Fig. 11.1).

An undated bulletin on bullfrog farming produced by the Department of Agriculture in Florida states "The skins of frogs are used for glue .... One large skin is said to make three ounces of the finest glue, which is used to

Fig. 11.1 Advertisement for frog glue appearing in "Frog Rearing" published in 1937 by the American Frog Canning Company, New Orleans, Louisiana, USA

repair crockery and the like". Further information about the source and preparation of the glue have not been located in the literature.

By far the best known example of adhesive secretions in amphibians is a powerful glue in the species of the brevicepitid genus *Breviceps* in southern Africa (Channing, 2001; Channing and Howell, 2006). Commonly called Rain Frogs or Rubber Frogs their glue functions to adhere males to females during amplexus.

Finally, the most recently studied group with adhesive secretions are the four species of north Australian frogs of the limnodynastid genus *Notaden* (Tyler and Knight, 2009).

11.2 Anuran Dermal Structure

Frog skin is more complex than that of any other vertebrate. Structurally it is composed of three layers. The most superficial is a layer of protective, flattened, horny, dead cells of keratin forming the stratum corneum. Beneath this layer is the stratum spongiosum and, deeper still, the stratum compactum. Between the deeper layers are three types of glands, communicating to the outside *via* narrow ducts.

There is a variety of dermal glands. There are the mucous and seromucous glands which release mucus to produce evaporative cooling, granular (commonly and often inappropriately termed poison glands), which contain a wide variety of compounds, of which peptides and alkaloids are the most significant (Williams et al., 2000; Apponyi et al., 2004). These compounds have diverse roles including anti-microbial protection and insect repellancy. Finally, many species have lipid glands which they secrete and apply with their hands to the dorsal body surface, and so halt or substantially inhibit water loss when they are basking in the sun.

11.2.1 *Breviceps* Species

Species of the southern African brevicepitid (formerly microhylid) genus *Breviceps* and their close relatives within the subfamily Brevicepitinae are extremely rotund. As a consequence, for the process of fertilization, the males are unable to grasp the females in either of the customary amplectic positions: around the waist (inguinal) or the axilla (axillary) grasp. Thus they need an alternative method. They possess adhesive glands: the female upon the dorsal surface and the male upon the ventral. Accordingly each sex contributes to the process of adhesion. Previously there has been debate about which partner provides the glue (Jurgens, 1978; Visser et al., 1982). Alan Channing (pers. comm.) advises that 16 species of *Breviceps* are currently known, but that several others are yet to be described. To date no chemical studies of the secretions have been undertaken.

11.2.2 *Notaden* Species

The only frogs in Australia with adhesive glands are the four members of the genus *Notaden*. It comprises *N. bennetti*, The Holy Cross Toad of Queensland and northern central New South Wales; *N. melanoscaphus*, The Northern Spadefoot Toad of northern Western Australia, the Northern Territory and northern Queensland; *N. weigeli*, Weigel's Spadefoot Toad of the extreme west of the Kimberley in Western Australia, and *N. nichollsi*, the Desert Spadefoot Toad of central and Western Australia.

All *Notaden* species share a globular body form (Fig. 11.2), they also share a vocal sac unique amongst Australian frogs in being bilobular (Tyler, 1972) and an advertisement call of "whoop...whoop...whoop..." repeated for several minutes.

As demonstrated in Fig. 11.3, the principal histological difference is in the size and number of the vesicles – those of the *Notaden* species are much larger and consequently less numerous than those of other species.

As is the case with other secretions, the adhesive secretions of *Notaden* are produced by the walls of the granular glands. Because of the unique nature of the secretions amongst Australian frogs, histological sections of *Notaden bennetti* and *N. melanoscaphus* skin were compared with those of 18 Australian species of Limnodynastidae, Myobatrachidae, and Hylidae (listed in Appendix 1).

Fig. 11.2 *Notaden* species. Copyright F. Knight. Page 97 in "Field Guide to the Frogs of Australia" by Tyler and Knight (2009)

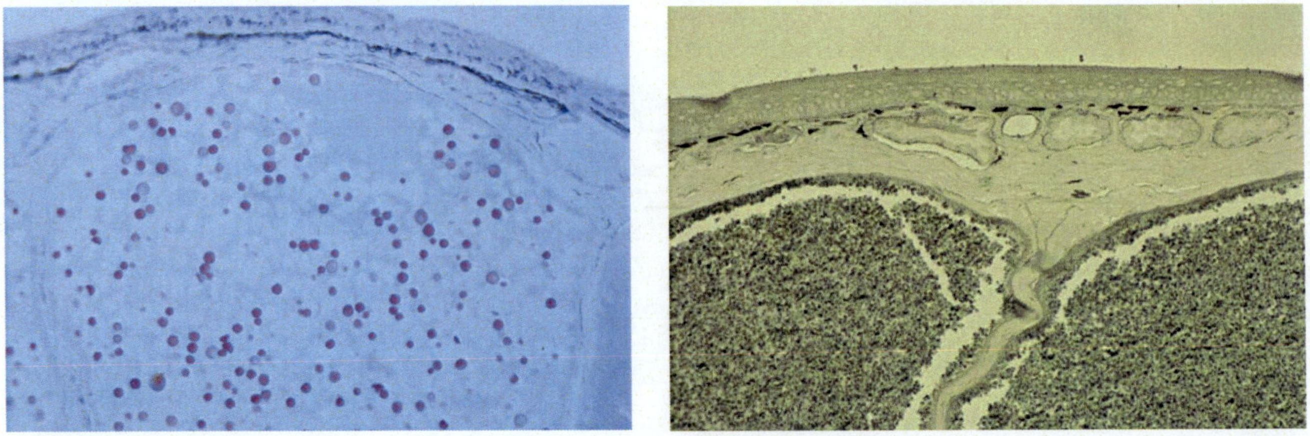

Fig. 11.3 Section of dorsal skin of *Notaden melanoscaphus* (*left*) and *Litoria caerulea* (*right*)

11.3 Range of Adhesive Activity of *Notaden* Secretions

11.3.1 Inorganic Material

Whilst undertaking field studies in the Kimberley Division of north-western Australia I had access to numerous specimens of *N. melanoscaphus*, and the opportunity to test the adhesive power of the dermal secretions upon a variety of materials that happened to be available. The first material tested was cardboard in the form of unopened containers of 35 mm film cassettes. As shown in Fig. 11.4, these stuck together. Adhesion was so strong that separation after two minutes stripped the outer layer of paper covering the boxes at the point of contact.

The strength of the glue also was demonstrated by sticking together two full, wet, cold cans of beer (Fig. 11.5).

Fig. 11.4 Cardboard cartons attached with *Notaden* glue
Fig. 11.5 Full cans of beer attached with *Notaden* glue

Fig. 11.6 Area of surface contact of cans shown in Fig. 11.5
Fig. 11.7 Lateral attachment of two full cans of beer with *Notaden* glue

Subsequent measurements revealed that the average weight of a full can is 408 g. The area of contact between the cans is a narrow peripheral rim with a total surface area of no more than two cm (Fig. 11.6). Full cans also could be attached laterally (Fig. 11.7).

Glass (in the form of microscope slides), wood, paper and silicon tubing were all found to adhere to similar material, whether it was hot or cold, wet or dry. In their patent application. Tyler and Ramshaw (2002) reported that the *Notaden* glue would adhere tightly to leather, wood and plastics including polypropylene, polystyrene and even the so-called non-stick Teflon polymer poly-(tetrafluorethylene).

11.3.2 Surgical Adhesives

Early tests on isolated pharmacological preparations of the guinea pig ileum, perfused isolated rabbit ear artery, and toad rectus abdominis muscle indicated that the glue was not toxic. As a consequence of these encouraging signs I approached Professor George Murrell of the Orthopaedic Research Unit at St George Hospital in Sydney. Professor Murrell's principal research interest was the knee. As a consequence of our discussions a programme was developed to examine the potential of the glue for the repair of damage to the cartilaginous meniscus of the knee.

Damage to the meniscus is common in athletes and in the elderly. Repair commonly involves the stitching or stapling of the separated portions of the meniscus.

Currently three surgical adhesives are in use: gelatin, which is weak; fibrin which, being a blood product bears the risk of disease contamination, and the synthetic polyacrylates.

The meniscal repair studies involved the creation of a longitudinal tear along the periphery. Sheep were selected for study "because of their anatomical and biomechanical similarities to the human meniscus" (Joshi et al., 1995; Allen et al., 1998). The frog glue was obtained by the electrical stimulation technique (Fig. 11.8) developed by Tyler et al. (1992).

Szomor et al. (2001, 2008) compared the results of repairs involving each of the above compounds. They found that the *Notaden* glue was five times stronger than the fibrin glue and 2.5 times stronger than the gelatine glue. Using the tear test it was found that the cyanoacrylate glue proved the strongest bond but it is toxic and possibly carcinogenic; its brittle nature may inhibit healing (Ellman et al., 2005). The glue of *Notaden* has the additional advantage of remaining pliable. The results of the study on the meniscus were that the frog glue has great potential for further investigations and could be considered for meniscal repair in the future.

Fig. 11.8 Glue emerging from dermal glands of *Notaden melanoscaphus* during electrical stimulation

Following a period of surgery at St George Hospital a wooden applicator, used for introducing the glue into the split meniscus, was glued to the ceiling of the operating theatre. There it oscillated in front of an air conditioner outlet vent for several years. Effectively it moved literally millions of times indicating how the flexible nature of the glue can be sustained.

Millar et al. (2009) explored the use of the *Notaden* glue to enhance rotator cuff repair in sheep cadavers. Noting that rotator cuff tears are a "significant cause of morbidity in the adult shoulder" (Neviaser, 1987), they explored the manner in which the use of the frog glue could enhance repair. Because of its unique properties they concluded that it "may ultimately lead to its use as an adjunct to rotator cuff repair in humans".

The study on rotator cuff repairs by Millar et al. (2009) increased the load to failure. Three techniques were used to repair 42 fresh frozen sheep infraspinatus tendons with a mattress stitch configuration; trans-osseous sutures; two traditional metallic suture anchors, with one suture per anchor, and two knotless metallic anchors with one suture per anchor. When stress was applied it was found that addition of the frog glue doubled the load that was required to cause the various repairs to fail.

Currently a large sequence of the molecular structure of the *Notaden* glue is known (Graham et al., 2005, 2006, 2010). Although the potential of the glue is recognized, further development hinges upon the collaboration of a major drug company with the resources to produce a synthetic product.

The functional significance of the secretions of *Notaden* remains uncertain but it can be deduced from the normal process of ecdysis whereby frogs shed their entire skin approximately once each week and eat it. If the *Notaden* species are stressed to release their dermal secretions, perhaps in response to attack by groups of insect predators, such as ants and termites, the insects will adhere to the surface of the skin. Subsequent ecdysis and ingestion will lead to the insects being consumed with the sloughed skin. The possible evolution of adhesiveness in amphibians has been explored by Williams and Anthony (1994).

Appendix 1

Species of Australian native frogs whose skin was compared histologically with that of *Notaden bennetti* and *N. melanoscaphus*

Family Hylidae	*Cyclorana alboguttata*
	Cyclorana cultripes
	Cyclorana maculosus
	Cyclorana novaehollandiae
	Cyclorana platycephala
	Litoria caerulea
	Litoria dahlii
	Litoria eucnemis
	Litoria raniformis
	Litoria splendida
Family Limnodynastidae	*Limnodynastes dorsalis*
	Limnodynastes dumerilii
	Limnodynastes tasmaniensis
	Neobatrachus pictus
	Rheobatrachus silus
Family Myobatrachidae	*Crinia signifera*
	Pseudophryne bibroni

Acknowledgments

I thank my many colleagues who have collaborated with me on this study. Dr. Alan Channing very kindly provided me with information on the genus *Breviceps*.

References

Allen MJ, Houlton JEF, Adams SB, and Rushton N (1988) The surgical anatomy of the stifle joint in sheep. Veterinary Surgery 27(6): 596–605.

Apponyi MA, Pukala TL, Brinkworth CS, Maselli VM, Bowie JH, Tyler MJ, Booker GW, Wallace JC, Carter JA, Separovic F, Doyle J, and Llewellyn JE (2004) Host defence peptides of Australian anurans: structure, mechanisms of action and evolutionary significance. Peptides 25: 1035–1054.

Channing AE (2001) Amphibians of Central and Southern Africa. Cornell University Press, New York.

Channing AE and Howell KM (2006) Amphibians of East Africa. Cornell University Press, New York.

Ellman PI, Brett RT, Maxey TS, Tache-Leon C, Taylor JL, Spinosa DJ, Pineros-Fernandez AC, Rodeheaver GT, and Kern JA (2005) Evaluation of an absorbable cyanoacrylate adhesive as a suture line sealant. Journal of Surgical Research 125(2): 161–167.

Evans CM and Brodie ED (1994) Adhesive strength of amphibian skin secretions. Journal of Herpetology 28(4): 499–502.

Graham LD, Glattauer V, Huson MG, Maxwell JM, Knott RB, White JW, Vaughan PR, Peng Y, Tyler MJ, Werkmeister JA, and Ramshaw JA (2005) Characterization of a protein-based adhesive elastomer secreted by the Australian frog *Notaden bennetti*. Biomacromolecules 6(6): 3300–3312.

Graham LD, Glattauer V, Peng YY, Vaughan PR, Werkmeister JA, Tyler MJ, and Ramshaw JAM (2006) An adhesive secreted by Australian frogs of the genus *Notaden*. In: Smith AM and Callow JA (eds) Biological Adhesives. Springer-Verlag, Heidelberg: pp 207–223.

Graham LD, Danon SJ, Johnson J, Braybrook C, Hart NK, Varley RJ, Evands MDM, McFarland GA, Tyler MJ, Werkmeister JA, and Ramshaw JA (2010) Biocompatibility and modification of the protein-based adhesive secreted by the Australian frog *Notaden bennetti*. Journal of Biomedical Materials Research Part A 93: 421–441.

Inger RF (1954) Systematics and zoogeography of Philippine Amphibia. Fieldiana: Zoology 33(4): 183–531.

Joshi MD, Suh JK, Marui T, and Woo SL (1995) Interspecies variation of compressive biomechanical properties of the meniscus. Journal of Biomedical Materials Research 29(7): 823–828.

Jurgens JD (1978) Amplexus in *Breviceps adspersus* – who's the sticky partner? Journal of the Herpetological Association of Africa 16: 6.

Millar NL, Bradley TA, Walsh NA, Appleyard RC, Tyler MJ, and Murrell GAC (2009) Frog glue enhances rotator cuff repair in a laboratory cadaveric model. Journal of Shoulder and Elbow Surgery 18: 639–645.

Neviaser R (1987) Ruptures of the rotator cuff. Orthopedic Clinics of North America 18: 387–394.

Szomor ZL, Appleyard RC, Tyler MJ, and Murrell GAC (2001) Meniscal repair with a new biologic glue. 47th Annual Meeting, Orthopaedic Research Society, San Francisco.

Szomor ZL, Murrell GAC, Appleyard RC, and Tyler MJ (2008) Meniscal repair with a new biological glue: an ex vivo study. Techniques in Knee Surgery 7(4): 261–265.

Tyler MJ (1972) Superficial mandibular musculature, vocal sacs and the phylogeny of Australo-Papuan Leptodactylid frogs. Records of the South Australian Museum 16(9): 1–20.

Tyler MJ and Knight F (2009) Field Guide to the Frogs of Australia. CSIRO Publishing, Melbourne.

Tyler MJ and Ramshaw JA (2002) An adhesive derived from amphibian skin secretions. Australia Patent No. WO2002/022756.

Tyler MJ, Stone DJ, and Bowie JH (1992) A novel method for the release and collection of dermal glandular secretions from the skin of frogs. Journal of Pharmacological and Toxicological Methods 28: 199–200.

Visser J, Cei JM, and Guterrez LS (1982) The histology of dermal glands of mating *Breviceps* with comments on their possible functional value in microhylids (Amphibia: Anura). South African Journal of Zoology 17(1): 24–27.

Williams CR, Brodie ED, Tyler MJ, and Walker SJ (2000) Antipredator mechanisms of Australian frogs. Journal of Herpetology 34(3): 431–443.

Williams TA and Anthony CD (1994) Technique to isolate salamander gland granular gland products with a comment on the evolution of adhesiveness. Copeia 1994(2): 540–541.

Part B

Part A of this book has described and compared the adhesive systems of some plant and animal species. The chapters focused mainly on "production machinery" such as glands, secretion ducts, their morphology, and the mechanisms involved. However, we also need to understand the "materials" that are produced, in our case the adhesives. The aim is to identify the building blocks or compounds of, for example, biopolymeric adhesives, to characterize and analyze the underlying curing, adhesion, and cohesion mechanisms, and to transfer this knowledge to new adhesive systems. This knowledge will enable us to use biological materials directly as adhesives, to use components of them, to generate hybrid materials, or to synthesize wholly biomimetic adhesives for medical or technical applications.

Part B of this book will focus on technical and medical applications of biological adhesives and biomimetic systems. Industrial adhesives of biological origin or components of them are used in various sectors of industry. Chapter 12 describes some applications of renewable raw materials, for example based on proteins, sugars, and fatty acids. Chapter 13 introduces polyphenolic adhesives from plants and animals and discusses various technical and medical applications. Polyphenolic adhesives are well known and include marine mussel adhesives (Mef-proteins) and adhesives from synthetic phenolic resins (used in brake shoes and clutches) and from tannins (used in the tannery industry). Some possible medical applications of biological polyphenolic adhesives are discussed in Chapter 13, while some of the medical issues raised there are discussed in more detail in Chapter 14. The latter also highlights the adhesion of medical adhesives to different tissues/organs and wound healing. In all cases where adhesives are used for medical treatments, the biocompatibility and biodegradability of the materials play an important role. Doctors are already using a natural biological adhesive – the fibrin system – for tissue repair and other applications. Chapter 15 gives a historical overview of the fibrin adhesive system and in Chapter 16, new scientific data on the fibrin adhesive are presented. In general, many synthetic adhesives are unable to cope with moist conditions and/or have weak adhesion in the presence of water or underwater. Chapter 17 presents a new biodegradable (meth) acrylate based adhesive for surgical applications. In contrast, biological adhesives such as those from the mussel or other (marine) species have evolved to function very effectively in the presence of water as described in Chapter 18. Finally, Chapter 19 reviews some new biomimetic fibrillar adhesives with enhanced adhesive performance under wet conditions.

12 Renewable (Biological) Compounds in Adhesives for Industrial Applications

Hermann Onusseit

Contents

12.1 Introduction

In the summer of 1991, a German couple hiking in the Alps came upon the newly revealed body of a man sticking out of the ice. He was 5000 years old and his possessions included a quiver of arrows. The arrow feathers were glued to the shafts with birch sap, an adhesive from a natural, renewable raw material. Furniture buried with the great kings of Egypt was beautifully made with fine wood veneers. The veneers were glued to a core with adhesives made by boiling hooves and hides of animals.

Glues – adhesives – have been a part of daily life for a very long time. In fact, without adhesive bonding, the industrial manufacture of many common things would not be imaginable today. In their current form and cost, the packaging, books, CDs, cars, and airplanes that we know today would not be possible without adhesive technology (Onusseit, 2008b). The usage of adhesives in manufacturing has a long tradition. Gutenberg made his printed matter by using adhesives: casein glue and fish glue stuck together the paper and leather of his Bible.

Industrial adhesives efficiently join materials in a fast and safe way, to provide for trouble-free, inexpensive production in existing manufacture lines (using automatic machinery as far as possible). These adhesives are subject to a number of requirements: they must resist the conditions to which the finished product will be exposed during use; they should be ecologically friendly; and of course their price is an important aspect that must be considered when selecting an adhesive. Ecological aspects play an increasingly important role today. For example, more and more care has been taken that adhesives can be processed so that as little waste and wastewater as possible accrues from their production and use. In addition to that, the impact of adhesives on the reuse (e.g., returnable products) or recycling of bonded products is a growing matter of interest.

Bonding by adhesives is a well-used technique in nature, where both physical-setting and chemical-curing systems exist. Among the former, either the substances used as adhesive are dissolved in organic solvents, such as the abietic acid of balsam resin dissolved in turpentine, or they are present as dispersions, as is the case with polyisoprene in rubber latex from trees. Carnivorous plants of the genus *Drosera* (sundew) obtain their prey by creating sticky tentacles on the external surfaces of their leaves to catch animal prey. Nature's chemical-curing adhesive systems may be one-component systems, in which a "monomer" is synthesized by the animal and reacts, for example in air with oxygen, to form a high-molecular-weight polymer, or they may be multicomponent systems, such as those that mussels employ to adhere to rocks.

Owing to these examples in nature, it is not surprising that humans began using bonding as a joining technology thousands of years ago. Those early users of adhesives, like the makers of arrows and furniture, learned from examples in nature as they developed bonding as a joining technology. In search of materials adaptable to adhesives, new nature-based substances like plant resin, casein, and gelatin were tested over millennia. By this continuous search, a significant adhesive technology has been developed since medieval times, especially for the bonding of paper and wood, and by the 17th and 18th century an adhesive industry had emerged from these beginnings. Until the 20th century, this adhesive industry was based solely on renewable raw materials. In 1909, with the development of phenolic resin, the century of synthetic polymers began, and in 1914 the manufacture of polyvinyl acetate was patented. In the following decades, more and more synthetic raw materials came into the market, with specific, customized characteristics for specialized uses. Nevertheless, renewable raw materials from biological sources are still an interesting option, and can provide advantages that are increasingly important. Compared to synthetic raw materials, renewables have particular physicochemical characteristics, generally have better biodegradability, and enable us to reduce usage of non-renewable raw materials such as oil. This discussion will focus on adhesive applications for biological raw materials, but it should be noted that this is a relatively small market for natural products compared to other industrial uses.

12.2 Renewable Biobased "Chemical Raw Materials"

Because a lot of conventional adhesive raw materials have the limitations discussed above, it is not surprising that renewable raw materials in the chemical industry have gained importance. But what exactly do we mean by "renewable raw material"? A common but limited definition is "agricultural and forestry manufactured products, which are in usage in the non-food range". Today, millions of tons of specialty and fine chemicals are manufactured by biotechnological processes, and even some basic chemicals like ethanol or lactic acid are produced by this means. Biotech processes create roughly 5% of the products of the chemical industry. In 2006 in Germany renewable raw materials grew on 1.56 million hectares, which is 13% of German arable land. The range of applications for renewable materials has increased greatly in recent years, for both economic and ecological reasons: even agricultural residual materials like whey, molasses or lignocelluloses are in heavy industrial use. One hopes that more and more renewable raw materials will replace fossil raw materials and thus conserve the depleting inventory for future generations. In comparison to fossil raw materials, they are as far as possible CO_2-neutral. Products, produced from plants that release only the amount of CO_2 that they are taking from the atmosphere during their growing, could have a big advantage because they mitigate the greenhouse-effect and thus counter climate change. Their true ecological influence, however, can only be determined by considering an eco-balance over the lifecycle of the product. An eco-balance can be defined as an exploitable comparison of the environmental influences of two or more different products, groups of products, systems, procedures or procedure controls. Production of renewable, industrial agricultural crops as an alternative to food stuff production also offers direct advantages when it enables the usage of marginally-fertile land or when processing plants increase local employment. Staying local reduces the cost of transport, fuel usage, and pollution, and further adds to the eco-balance.

Because the chemical industry needs products with special properties, raising the right crops is not just a matter of agriculture. Innovative research is also needed. By means of plant breeding, biotechnology, and gene technique, the desired raw material qualities are evolved and their yields optimized. Thus, agriculture provides high-quality precursors that can eliminate the need for some complex conversions of fossil fuels. One major beneficiary of this is the adhesives industry.

12.2.1 Renewable Biobased Raw Materials in Industrial Adhesives

Although rapid growth of synthetic raw materials over recent decades has led to important improvements in adhesion, cohesion, and processability, there are still many applications for biological, renewable raw materials (Otten et al., 2007). Some of these applications take advantage of specific physico-chemical properties imparted by such raw materials, such as viscosity response under shear. Often the desired characteristic can be achieved only through chemical modification of the renewable raw material. Cellulosics like starch are an example of this; as they are initially non-water-soluble yet develop useful industrial properties in aqueous solution or colloidal form when modified by chemicals or natural enzymes.

12.2.2 Requirements for Adhesive Raw Materials

The requirements for industrial adhesive raw materials are defined both by what the adhesive must do during the life of the product and by how it is applied. In that context, many important trends have appeared in the last decades, such as new materials to be adhered, faster production speeds, and enhanced safety and exposure restrictions. All of these put demands on a material, the adhesive, which is usually a complex compound made of several ingredients. The key ingredient is a polymer, which in a physical-setting system will be of high molecular weight, and in a chemical-setting system will start out with a low molecular weight that greatly increases during cure. For physical-setting systems, all the key properties that one needs in the adhesive to meet the requirements of the end use must be incorporated in the polymer. Specific characteristics can be adjusted with additives, but without the right polymer, proper performance in service and during recycling is impossible.

The requirements of an adhesive during its application are met by the combination of polymer and the other compounding ingredients in the adhesive, which together give its wet tack, surface-active, or setting speed character, for example. Bonding operations today demand adhesives with a very high quality – and a very broad set – of physical and performance properties, at ever-faster production speeds, and this in turn generates the need for new raw materials.

12.2.3 Requirements for Renewable Raw Materials

Industrial processing of natural ingredients has set completely new requirements on plant and animal products. For example, to extract starch from potato for industrial purpose, it is irrelevant whether the potato is small or big, mealy or waxy – very important factors in food production. Industry cares that the potato has a high starch content, an ideal starch composition, and that the crop is of uniform quality. The latter is of special importance for industrial processes because they run at ever-faster speeds: the process window gets smaller and smaller continuously and raw materials must have tighter tolerances.

12.3 Polymers

The bases for all physical-setting adhesives are polymers, which are either distributed in liquid form as a solution or dispersion, or melted to be liquid during their application. The liquid phase is necessary to achieve good wetting of the substrates and to insure that individual molecules can get close enough (in atomic distances) to the surface of each part to form the required adhesion. Thus, polymers as adhesive raw materials need to be integrated into a liquid matrix or be meltable. The liquid state must have a viscosity that allows easy, precise application at high machine velocities. Many synthetic polymers achieve those requirements. Parameters can be adjusted by varying polarity, branching, chain length, and molecular weight distribution of the polymer.

12.3.1 Renewable Biopolymers as Adhesive Raw Materials

Renewable biopolymers have been used in adhesive formulations for a long time. One can differentiate between plant-based and animal-based biopolymers. Plant-based polymers include, for example, starch, cellulose, and natural latex. Among the animal-based products, proteins like casein and gelatin are of special importance. With regard to quantity, the most important biopolymers are polysaccharides, proteins, and polyisoprene (natural rubber). The most important polysaccharides are homo polysaccharide cellulose and starch, respectively amylose and amylopectin. Starch-based adhesives are made from roots, tubers, and seeds of plants such as maize, potatoes, wheat, rice, and tapioca based on the performance, machining, and cost requirements of the application. Starch-based adhesives are either cold or warm water-soluble depending on the application requirements. Starch adhesives are made by dispersing starch granules in water and then by heating the mixture until the granules burst and absorb water.

As the granules swell, they increase in viscosity and form a paste or adhesive. Gum arabic, which is a polysaccharide, is a copolymer containing carboxyl groups.

Proteins are macromolecules built from amino acids. Proteins are used in nature to bond organs and cells and contain, besides the 20 genetically coded proteinogenic amino acids, post-translational modified non-genetically-coded amino acid groups that are generated after the biosynthesis of the protein through a specific enzymatic reaction.

Although polysaccharides and proteins have long been used in adhesives, only since the 19th century natural rubber has been used as an adhesive raw material, e.g., for band-aids and adhesive tape production. Natural rubber comes from the latex of tropical trees and is a high-molecular-weight carbohydrate polymer with average MW approximately 350,000. The properties of many native biopolymers nevertheless limit their application areas in adhesive technologies. Many of those polymers, such as starch and cellulose, have very high molecular weights and in their native form are insoluble in water in the concentrations needed for technical adhesive applications. Some are not water-soluble in the native form at all or result in colloidal solutions with a viscosity too high to process the material. Those specific solubility and rheological properties can in some cases be exploited, as is done with casein-based adhesives, whose disproportional increase in viscosity while cooling down can be used in the field of bottle labeling, or they can be added to low viscosity systems based on synthetic raw materials. Melting is the other way to arrive at the liquid state, but most biopolymers are not thermoplastic and cannot be liquefied to stable melts. This non-thermoplastic behavior can be used to advantage, however. Starch adhesives, for example, do not show the disruptive inclination to build so-called "stickies" during the paper recycling process as many synthetic thermoplastic polymers do. With regard to biological degradation, the behavior of natural polymers is in general more favorable than synthetic polymers. Those special properties of natural polymers can be emphasized with physical and chemical modifications.

12.4 Application of Adhesives Based on Renewable Biopolymers

12.4.1 Production of Corrugated Board

On a quantity basis, the most important application of biopolymers for adhesives is the use of starch for the production of corrugated board, an important material for the production of shipping containers (Onusseit, 2008a). The corrugated board industry was started in 1874 when a patent was issued in England claiming that bonding a flat linerboard with a fluted corrugated sheet generated a material with an excellent cushioning effect that was pre-eminently well suited as packaging material. Owing to the large amount of adhesive needed for this application, its commodity price has played an important role besides its technical properties. The first adhesives used for making corrugated board were conventional paper glues, for instance a mixture of cereal flours and water. In 1939, the Stein Hall process was invented, an adhesive system based on starch which made it possible to considerably increase the efficiency of the corrugated board producing machinery. Stein Hall corrugating adhesives are a mixture of fully gelatinized starch, available as a colloidal solution, with native, ungelatinized starch dispersed in water. This combination makes it possible to obtain a high-solid adhesive formula without increasing the viscosity to such an extent that the adhesive cannot be applied from the roller application system. However, in this state, the adhesive does not have satisfactory adhesive property for bonding. Following application, during the compression of the corrugated board the adhesive layer is heated quickly to such an extent that the dispersed native starch gelatinizes, resulting in a steeply rising viscosity which allows the freshly bonded corrugated board to resist the forces occurring in the further production process. With starch-based adhesives, excellent technical properties can be combined with a low price. The production of corrugated board has therefore become the most important field of application for biopolymers in adhesives. Corrugating adhesives are under continuous development to cope with ever-advancing manufacture technologies. The very fast speeds of modern corrugators require a constant adaptation of the technological parameters of the adhesives, such as viscosity, gelatinization temperature, and initial "tack".

12.4.2 Labeling of Glass Bottles

Labeling with water-borne adhesives has a long tradition (Onusseit, 2008a). The first adhesives used for labeling were water-borne products based on biopolymers such as dextrin, starch, or casein, which is one of the most ancient raw materials for adhesives. To date, casein-based adhesives are at the forefront of glass labeling. They are water-borne systems that can be applied on wet or dry,

Fig. 12.1 Glass-bottle labeling with casein adhesives

cold or hot glass surfaces and yield excellent bonds. With casein-based adhesives, special needs can be achieved, like ice-water resistance. Water-borne systems are usually applied by means of applicator rollers and segments. Furthermore, they have excellent running performance on labeling machines, even at the highest speeds (up to 80,000 bottles per hour or more) (Fig. 12.1). However, as the availability of casein is not consistent, their commodity price is not consistent either and is frequently expensive relative to other adhesives. In recent years, the development of new (synthetic) raw materials has made it possible to achieve running performance nearly as good as casein-based adhesives.

12.4.3 Making of Books

Whereas protein-based adhesives today have only minor importance in softcover, unsewn bookbinding, they still play an important role in the making of cases for hardcover books. Glutine (animal) glues have proven to be particularly well suited here. Glutine adhesives are derived from skin or animal bone collagen and rank among the most ancient adhesives ever used by humankind. Their sol–gel transition is characterized by a very rapid gelatinization, i.e., change from a liquid to a glassy state, induced by a decrease in temperature by only few degrees below the gelatinization point. This fast setting feature in a bonding process, coupled with excellent "tack", is taken advantage of the turning-in, that is the folding-over, of cover cloths around the edges of the two boards of book covers. To this end, the gelatins are heated in hot-water jacketed tubs and applied *via* rollers at a temperature of 50–60°C. The tubs are filled *via* premelters provided with a water quench, a stirring unit, and a water-circulating pump. After application to the cover stock, these glues cool down to room temperature and very rapidly build up a high tack enabling them to safely overcome the restoring forces during the turning-in process. High-speed case manufacturing machines produce up to 100 cases per minute. Glutine glues are very popular in the traditional bookbinding sector, because they do not affect the paper, can be removed without leaving any residues when rebinding during book restoration, and can easily be removed from the machinery and tools (Onusseit, 2008a).

12.4.4 Lamination (Ply Adhesion) of Tissue Products

Cellulose as raw material is used in the lamination process (ply adhesion) of tissue products. Water-based lamination adhesives join two or three (or even more) tissue plies together. There are several ways in which the plies of embossed products are brought together. One way is to apply the adhesive to the tips of the raised elements of one of the two layers, and those tips come into contact with and are bonded to the recessed areas or floors of the other layer (nested laminating). In pin-to-pin, tip-to-tip or point-to-point laminating, the raised elements of each layer contact each other at the tips, which is where the bonding occurs. The paper typically has a weight of 15–25 g/m^2 and can be either bleached or unbleached. A typical machine speed is 600–800 m/min, but there is even faster equipment running at 1000 m/min. The key characteristics of these adhesives are appropriate wet bonding capacity, no or very little foaming, and good processability. A short setting period enables the lamination to perform well in the machinery, and makes it possible to avoid staining of the tissue paper. Laminating adhesives have to be flexible enough not to impair the softness of the finished product, that is, have a "soft grip". In the past, and today, laminating adhesives were derived from dissolved cellulose. There are challenges in the field. These products release the majority of their water content into the paper during lamination, risking the destruction of the embossed structure of the tissue laminate. Especially on tissue papers that are more difficult to bond due to added wet strength, or on tissue papers impregnated with lotions, cellulose-based adhesives often do not adhere well enough. To improve their properties, some synthetic polymers, particularly polyvinyl alcohol, can be added.

Fig. 12.2 Core winding with dextrin adhesives

Fig. 12.3 Production of envelopes with dextrin adhesives

12.4.5 Core Winding of Tubes

Water-borne adhesives derived from dextrin are used for the production of cores on medium-to-fast winders. Dextrin is a commonly used dry roasted starch because it can be developed into a variety of gums and pastes and the solubility and viscosity can be controlled based on acidity, moisture content, and roasting temperature. In core winding, they are usually applied at a temperature of 50°C to increase their wet tack and setting speed (Fig. 12.2). When using dextrin for core winding, it is important to know that during the recycling of the paper, the pulp will contain adhesives plus other compounding ingredients such as borates that are commonly added to dextrin adhesives to increase their tack and speed. Borates can react with polyvinyl alcohol from other sources, such as lamination, and cause flocculation that may give rise to problems during recycling (e.g., so-called "fisheyes").

12.4.6 Production of Envelopes

Water-based dextrin adhesives are also important for the production of envelopes and advertising mail, which are manufactured from a wide variety of materials and in many different types and sizes. Production speeds have been ever increasing, with today's state of the art reaching 1500 units per minute. Although some plastic films are used, paper is still the most important raw material for the production of envelopes. A special feature of the typical envelope is an adhesive application used only by the consumer for closing the envelope. The widely-known water-reactivated gummed envelopes are manufactured by applying dextrin-based, dextrin/synthetic resin-based, or synthetic resin-based adhesives and force drying them (Fig. 12.3). To provide for a safe and durable closure of the envelope, particularly in auto-insertion machines, the gummed flap must be quickly reactivable to a high tack level, to ensure a quick and secure seal. The dried, gummed adhesive needs to have anti-blocking properties to prevent premature bonding even under high humidity conditions.

Another kind of envelope closure utilizes natural rubber latex, which has an unusual property – dried films of natural rubber latex adhere strongly to themselves, but not to other surfaces. In these so-called "self-seal" envelopes, both sides of the envelope closure are coated and dried. The envelope is sealed by compressing the two layers together (diffusion bonding).

12.4.7 Tapes and Plaster

The first pressure-sensitive adhesives were based on natural rubber and rosin. They were applied as patches and tapes for surgical use (Johnston, 2000). In the 19th century, some early technical tapes were derived from there, but the application range was very narrow. Natural

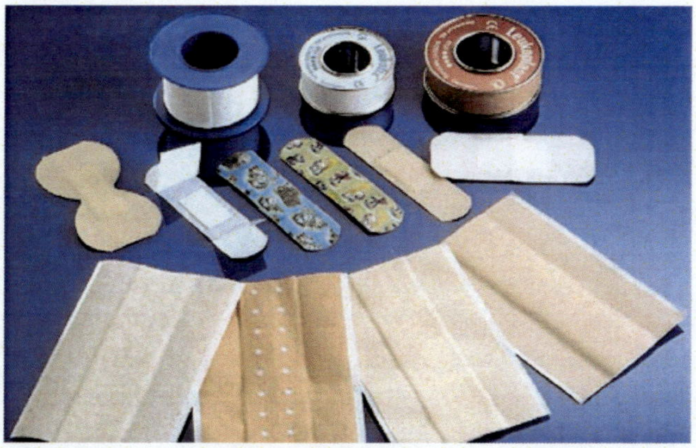

Fig. 12.4 Plaster with natural rubber adhesives

rubber, like several synthetic polymers, is still used for the manufacturing of tapes and plaster (Fig. 12.4). It is generally dissolved in organic solvents and then coated on the relevant substrates. Pressure sensitive adhesive (PSA) for medical tapes and plaster must serve different purposes in different situations. Some applications require high tack, good moisture resistance, and 24–48 h durability. Others, such as cosmetic or facial applications, require lower tack and softer removal. Adhesives for skin contact have proven high performance properties, as well as the breathability and moisture vapor transmission properties required for longer term applications. In addition, medical tapes and plaster shall be removed easily and painlessly.

12.4.8 Cigarette Manufacturing/Packaging

Today cigarettes are being produced at a velocity of up to 700 m/min. Among other polymers, starches are used for the straight seam bonding of cigarette papers. These starches are applied with the aid of discs or nozzles. To adapt this raw material to ever-increasing production velocity, an optimum rheological behavior is especially important. One option to optimize this is the variation of molecular weight, for example using enzymatically degraded starches. These systems enable the development of faster setting adhesives with higher solids content, which in turn allow higher production speeds.

Cigarettes are wrapped either in hard boxes or in soft packages. For soft-package wraps, dextrin adhesives are used because their high strength while still wet – their "tack" – can overcome the restoring forces during the bonding process and thereby permit machine speeds up to 500 packages and more per minute.

12.5 Polymer from Renewable Biobased Building Blocks

The production of polymers for adhesives is based on the classical methods of organic chemistry: ionic or free-radical polymerization, polyaddition and polycondensation. Synthetic monomer units predominate; however, in recent years a trend to use ingredients from plants can be seen. This is not just for economic reasons, because other properties like biological degradability also get attention. Thus, certain ingredients from plants and animals are used as building blocks for the synthesis of adhesive raw materials.

Polyamides are an important example. Polyamides are polymers whose repeating units have a characteristic feature of the amide group, resulting from the reaction of dicarboxylic acids and diamines. The dicarboxylic acid component of polyamides is often derived from plants. Polyamides from these natural components are very flexible, are ideal for bonding difficult surfaces, and in principle are compostable. Dimerized fatty acids used for the production of polyamide resins have a marked influence on the structure and morphology of polyamides, and are obtained by dimerization of unsaturated, long-chain fatty acids derived from natural raw materials. After dimerization, they still contain a number of byproducts, of which mono- and trifunctional carboxylic acids in particular have a quality-reducing effect on the polyamides

Fig. 12.5 Laminating with polyurethane adhesives

to be produced therefrom. For this reason the dimerizate must be purified before use. Highly branched dimeric fatty acids can also be subjected to further chemical modification: in a process which is very time- and money-consuming, the carboxyl groups can be converted into amino groups. The dimeric fatty acid diamines so obtained are particularly valuable monomers for the production of polyamide resins.

Another polymer that is built from natural biobased monomers and is used as a material for adhesives is polyester. In polyester synthesis, where dicarboxylic acids and dialcohols are combined in a polycondensation reaction, the diol component often comes from a natural raw material such as castor oil. Castor oil is extracted from the seeds of the castor plant, in which the main substance is ricinoleic acid, 1,2-hydroxy-octadecane-9-acid.

12.6 Application of Adhesives Based on Polymers from Renewable Biobased Building Blocks

12.6.1 Laminating Adhesives

Packaging, such as bags, pouches, sacks, and wraps is made of materials such as film, foil, or paper (flexible packaging). To optimize the properties of the various materials, often two or more materials are joined together (Fig. 12.5). Through this process packaging materials can be optimized for low weight and material use. Lamination of individual layers is usually done with the help of adhesives, for plastic films polyurethane adhesives are frequently used. In addition to polyurethane adhesives based on synthetic raw materials, biobased components such as special polyols, are also used. The previously mentioned castor oil derivatives are useful here, too. Formulations of polyurethane adhesives with a high content of polyols based on renewable raw materials, which can be cured preferably by aliphatic polyisocyanates have been optimized toward a majority of these renewable raw materials and product performance. Formulated polyurethane adhesives for laminating can have 65–80% content of raw materials based on biomass.

12.6.2 Two-component Polyurethane Adhesives

Two-component polyurethane adhesives are used in a wide variation of industrial applications such as the production of skis and snowboards, industrial filter systems, panels for the furniture industry, the building industry, and in shipbuilding. Biobased products are often preferred because of the hydrophobic properties of these polyols. Castor oil is the most important polyol in the production of two-component polyurethane adhesives.

12.6.3 Thermoplastic Polyamide Adhesives

Polyamide adhesives based on biobased dimerized fatty acids already have a long tradition, for example as hot melt adhesives in the shoe industry (Wichelhaus, 1988). In the past few years, new powerful types have been developed, by means of which especially durable and high-strength bonds can be obtained between most different substrates. An example is the manufacturing of cables. One of the most feared causes of failure of buried power cables is the destruction of the insulation around the conductors by moisture having entered the cable core. Water does not only make its way into the cable through mechanical damage to the cable sheathing, but also can enter in the form of water vapor diffusing through the jacket. The use of polyamide hot melts based on biobased dimerized fatty acids enables a power cable assembly with longitudinally bonded aluminum and copper tapes. This assembly not only prevents the further penetration of moisture in the core, but also has an electrical shielding function. In addition to high bond strengths, the polyamide hot melts based on biobased dimerized fatty acids specially developed for this application excel in their trouble-free integration into a fully automated manufacturing process and in their stability to the permanent strain exerted on the bond when the cables are in operation.

The latest development in telecommunications, fiber optic cable, requires protection against environmental influences as well. Both during cable laying and later on in service, extreme demands are made on the mechanical stability of the cable assembly and on the protection of sensitive glass fiber against the effects of moisture. In order to seal against penetrating moisture, the cable core, which is surrounded by high-strength armoring, is connected durably with the outer jacket. High mechanical stability is obtained by a closed linkage between the reinforcing layer and the cable jacket. Here, tailor-made polyamide hot melts based on biobased dimerized fatty acids are in use.

Polyamide adhesives based on biobased dimerized fatty acids are also used for the production of heat shrinkable components for the cable industry (Fig. 12.6). These components consist of cross-linked polyolefins and are

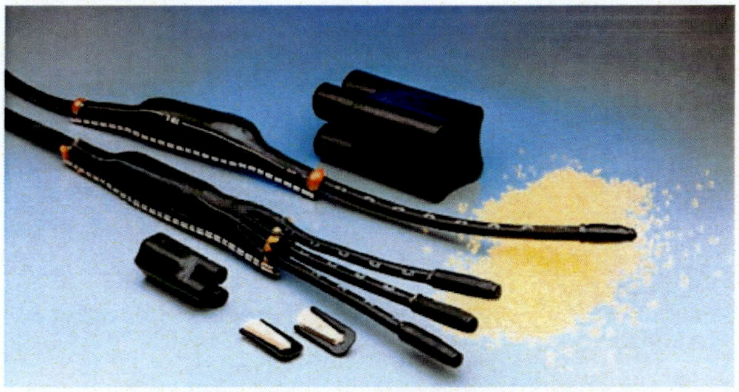

Fig. 12.6 Cable shrink articles with polyamide adhesive

manufactured in expanded form as extruded and molded components such as end caps, shrink sleeves, and wrap arounds. They serve as sealing compounds for cables of all types and diameters to make them pressure- and moisture tight. Inside they are coated with polyamide hot melt adhesives based on biobased dimerized fatty acids. When heated with an open gas flame, these components shrink down onto the cable, and the hot melt presses tightly around all uneven areas of the cable covering. These heat-shrinkable components have been in use all over the world for many years. They excel in easy, fast, and safe assembly, while guaranteeing maximum reliability in service. In addition, heat-shrinkable components are increasingly being used for connecting power cables and for protecting gas- and water pipes against corrosion.

12.7 Reactive Adhesives

The polymer in reactive adhesives, which ultimately provides the high cohesive strength typical of this adhesive type, starts out with very little cohesive strength – that strength is developed in the glue joint after the bonding process. It is therefore necessary to include components in the formulation that promote this reaction, and for that there are different possibilities.

12.7.1 Reactive System from Renewable Biobased Raw Materials

Reactive adhesive systems exist in nature. Often these are one-component systems where the monomer is produced by an animal and reacts in contact with air to form a polymeric material. An insect, the green lacewing, is an example for such a chemically hardening system. The female green lacewing lays her eggs on posts, which she creates with the help of reactive substances. A low-molecular-weight liquid is extruded from a gland and is drawn to a short fiber on the end of which the eggs stick. The liquid is a single-component mix of short protein strings that reacts upon air contact and hardens within seconds. Marine mollusks provide another example of chemically-hardening systems, this time with more than one component. Mussels attach themselves to rocks using byssal filaments, three-component underwater adhesives which consist of proteins such as 3,4-dihydroxy-phenyl-L-alanin (DOPA). Research on how those systems or raw materials can be used in technical applications is actively underway and recent discoveries have greatly increased our understanding of these biological reactive materials. Technical applications are not far in the future, as we learn how to replicate nature, perhaps using nature herself: with the help of bacteria, a precursor gene of the mussel adhesive protein has been incorporated into the tobacco gene. New raw materials for reactive adhesives for industrial applications made from genetically engineered species are on the horizon.

12.8 Additives in Adhesives

Every adhesive formulation contains additives to adjust its properties. Some additives are process aids during manufacture, such as antifoams and stabilizers, some affect application properties like flowability or setting speed, and others impact water resistance, heat resistance, or other end use property important to the service life of the adhesive.

12.8.1 Additives on Renewable Biobased Raw Materials

Adhesive formulators use biopolymers not only as the basic component, as described above, but also as an additive to synthetic-based adhesives to impart special properties. Many hot melt adhesives based on synthetic polymers incorporate natural resins or their derivatives – rosin types are the best known – to improve the adhesion of these adhesives. For many water-based products based on synthetic polymers, colloidally-dissolved, natural biobased polymers are used to control rheological properties or to improve cleaning behavior. In the emulsion polymerization of many synthetic polymers, natural source raw materials have an important role as protective colloids.

Besides its use as a reaction component, ricinus oil or castor oil is also used as a softener in adhesives.

12.9 Application of Adhesives with Biobased Additives

12.9.1 Resins for Hot Melt Adhesives

Hot melt adhesive technology was one of the fastest-growing adhesive technologies of the last few decades (Onusseit, 2008a). Typical hot melt adhesives consist of a base polymer, which determines cohesion, plus wax, which mainly determines the viscosity and setting behavior, and also resins, which influence adhesion. Resins may be either synthetic or biobased. The most important natural resins are those derived from coniferous trees and sold as rosin. There is a distinction between gum rosin and tall oil rosin depending on the extraction process. The traditional method for the extraction of resin from trees is to cut the bark and obtain resin solely from the resin ducts (gum rosin). The other way is to extract it from wood fiber during pulp production by sulfate process, in which case the rosin is removed as a soap known as a crude tall oil which contains, depending upon its source, resin acids, fatty acids, and unsaponifiable material. The tall oil is then distilled at temperatures between 200 and 285°C to tall oil rosin. In addition to native rosin, resins modified by esterification or dimerization are also used in hot melts. Hot melts formulated with rosins are used for packaging applications, in the production of books and catalogs, or in the processing of wood.

12.9.2 Hot Melt Adhesives for Packaging Applications

The most important applications of hot melts in the packaging industry are carton closing (Fig. 12.7), where they are expected to adhere securely to different surfaces and to harden in an extremely short time. Biobased resins are often used to improve the adhesion properties of the hot melt.

Another application in the packaging industry is the use of hot melts in the field of labeling, especially in wrap-around labeling. It is a very popular technique for labeling polyethylene teraphthalate (PET) bottles because of the relatively low surface energy of plastic materials.

Fig. 12.7 Carton closing with hot melt adhesives

Hot melts used here must have excellent adhesion, which is often achieved using biobased resins.

12.9.3 Hot Melt Adhesives for Bookbinding Applications

The most important application of hot melts in the field of bookbinding is spine gluing in perfect binding. Production speeds have been ever increasing, with today's state of the art reaching more than 10,000 units per hour (Fig. 12.8). To meet the challenge from difficult-to-bond papers used in this industry, biobased resins are incorporated into the hot melt to increase adhesion properties.

Fig. 12.8 Book binding with hot melt adhesives

12.9.4 Hot Melt Adhesives for Woodworking Applications

In the manufacture and processing of products from wood, hot melts are used for assembly, profile-wrapping, and laminating applications. Here also, difficult-to-adhere materials must be bonded at very high production speeds, and again the biobased resins in the hot melts are helpful.

12.9.5 Plasticizer for Dispersion Adhesives

When absorbent materials such as paper or wood must be bonded, dispersion adhesives are still very important. Suitable adhesives can easily be made from available synthetic polymer dispersions, and these building blocks may be easily improved with additives such as plasticizers. In addition to synthetic softeners, softeners based on natural raw materials are also in use, such as triacetin, a triester of glycerol, and acetic acid (1,2,3-triacetoxypropane), which both softens the polymer and improves its dispersibility for faster cleanup of dried glue.

12.10 Summary

For thousands of years, people have used glue to create complex objects. The raw materials for these first adhesives came from nature and were adapted from empirical observations. Gradually, many adhesive substances were found and employed, but without an understanding of the principles behind adhesion. With the emergence of the chemical industry, synthetic materials began to replace biologically based materials, but the special properties of biologically based materials means they are still present in the marketplace, often with their performance enhanced through chemical modification. Traditional processing methods were changed and new ones developed. Environmental benefits and interesting new markets for renewable biobased raw materials are additional rewards. The production of natural raw materials that are precisely suited to the adhesives industry requires not only the growth of specialized crops but also intensive research in plant breeding and biotechnology.

References

Johnston J (2000) Pressure Sensitive Adhesive Tapes. Pressure Sensitive Tape Council, Northbrook: pp 3–12.

Onusseit H (2008a) Paper and Packaging Industry. In: Brockmann W, Geiß PL, Klingen J, and Schröder KB (eds) Adhesive Bonding – Materials – Applications and Technology, 1st Ed. Wiley-VCH, Weinheim: pp 257–293.

Onusseit H (2008b) Praxiswissen Klebtechnik. Band 1: Grundlagen, Verlagsgruppe Hüthig, Jehle, Rehm, Heidelberg.

Otten A, Eipel D, and Ermatschenko N (2007) Adhesives from renewable raw materials. Market overview and potentials. Coating International 8: 28–31.

Wichelhaus J (1988) Thermoplastic polyamid adhesives. European Adhesives and Sealants 6: 26–29.

13 Bio-inspired Polyphenolic Adhesives for Medical and Technical Applications

Klaus Rischka, Katharina Richter, Andreas Hartwig, Maria Kozielec, Klaus Slenzka, Robert Sader and Ingo Grunwald

Contents

13.1 Introduction
(*Katharina Richter and Ingo Grunwald*)

Nature has been developing adhesives for millions of years, mankind for just a few thousands of years. For this reason it is worth having a closer look at what nature does and how we can develop bio-inspired adhesives for technical and medical applications. Some examples of natural materials which have already been used for technical adhesives are casein, latex rubber, tree gum, and adhesives derived from natural sources used for the waterproofing of natural textiles, the production of paper, and the sealing of jars (Papov et al., 1995; Creton and Papon, 2003). Bio-inspired adhesives can be found in all areas of the natural world. Because of their origin, those adhesives are also called biological adhesives or bioadhesives and they fulfill several different functions (Smith and Callow, 2006; Carrington, 2008; Antonietti and Fratzl, 2010). Plants use adhesives, for example, for self-healing and for protecting themselves against wood defects, while animals use sticky materials for protecting themselves against predators and for hunting prey (Keckes et al., 2003; Schreiber et al., 2005; Flammang, 2006; Voigt and Gorb, 2008; Plaza et al., 2009). Microorganisms use adhesive material for settlement, surface attachment, and colonization (Melzer et al., 2008; Flammang et al., 2009; Santos et al., 2009; Scholz et al., 2009). Higher organisms, such as humans, rely on an inducible adhesive system: the wound healing promoter fibrinogen ((Berlind et al., 2010), which is discussed in detail in Chapter 15, p. 225 of this book).

One important advantage of nearly all kinds of bioadhesives is independence from the need for surface preparation and special curing conditions, for example 150°C and more for some epoxy resins. The substrates do not need to be freed from humidity and do not require special activation, as often do substrates for technical adhesives.

In addition, something which is difficult to realize for a technical adhesive but which can be easily done in nature is bonding in the presence of water or even underwater (Holten-Andersen and Waite, 2008). The advantages of these bioadhesives are due to their special chemical composition. One class of chemical compound which one frequently finds in bioadhesives are polyphenolic molecules. These molecules are not only used in the animal and plant kingdoms but are also very common in technical applications such as those discussed in Sect. 13.3.

Nature impressively demonstrates the use of polyphenolic compounds in different materials. Polyphenolic compounds are found in many organisms that do not have any evolutionary relationship, clearly emphasizing its functional significance (Rios et al., 2003; Vatten et al., 2005; Arranz et al., 2009). On the one hand these compounds are used as an adhesive by different marine organisms (e.g. algae, mussels), on the other hand they are used as a building material in plants (lignin) or as a part of the defense system of plants against predators (tannins). The functions of tannins are explained in more detail in Sect. 13.4.

The reactive moiety of interest is frequently a special phenolic-based structure, which has the ability to react in different ways with itself and/or other functional groups (Yu et al., 1999; Burzio and Waite, 2000). For example, in plants the secondary metabolites can be subdivided into phenolic and isoprenoide compounds. Phenolic compounds lead in the next hierarchy to antioxidants, polyphenols, and polyhydroxyphenols. As the name already suggests, these molecules contain hydroxyl groups. In particular, hydroxyl groups in the ortho position or so-called cathechols are useful if chemical linkage or chelation is required. For example, tannins in wine refine the taste during crosslinking and precipitation. Furthermore, phenolic acids such as ellagic acid (green tea) or caffeic acid show, in addition to their radical-scavenging features, good adhesion to the cups drinks are served in (Gao et al., 2006). As already mentioned above, cathechols are also found as a building block in marine adhesives. This will be discussed in more detail in Sect. 13.2.

Polyphenolic adhesives, like those based on mussel adhesives, have been discussed for a number of years (Heiss et al., 2006) as potentially new medical adhesives. Sect. 13.5 will give a brief overview of the use of bio-inspired adhesives in medical treatments. In the future there will also be a need for less toxic adhesive materials in space exploration, with astronauts living for longer periods in closed habitats. The advantage and possible use of biomimetic adhesive will be discussed in Sect. 13.6.

13.2 Phenolic Adhesives in *Mytilus edulis* (*Klaus Rischka and Katharina Richter*)

A very well-known polyphenolic compound is the protein-based adhesive of the blue marine mussel *Mytilus edulis*. These mussels show the ability to adhere to a wide variety of surfaces in seawater, for example glass, plastic, metal, wood, and even PTFE (Young and Crisp, 1982).

The adhesive is composed of different proteins called Mefp (*Mytilus edulis* foot protein), which have different functions. Mefp-1 is the essential part of the byssus and plaque coating (Lin et al., 2007). The composition of this protein is remarkable. The basic unit has been identified as a specific repetitive decapeptide (Fig. 13.1) with a molecular weight of about 108 kDa (Waite, 1983).

The proteins responsible for the adhesion are Mefp-3 and Mefp-5. These proteins are located in the boundary layer close to the surface and have a molecular weight of between 5 and 10 kDa (Papov et al., 1995; Vreeland et al., 1998). All these proteins contain a significant amount of the post-translational modified amino acid DOPA (3,4-dihydroxyphenylalanine), namely 10–15 mol% (Mefp-1), 21 mol% (Mefp-3), and 27 mol% (Mefp-5) (Papov et al., 1995; Waite and Qin, 2001). Mefp-2 and Mefp-4 are structural elements of this natural adhesive. Mefp-2 has a significantly lower DOPA content than Mefp-1. It contains a high amount of the amino acid cysteine (up to 6–7 mol%), and is responsible for the foamy morphology and structural stabilization of the byssus cement (Rzepecki et al., 1992). Mefp-4 acts like a linker protein in the junction region between the collagen fibers of the distal byssus thread and the plaque proteins (Zhao and Waite, 2006).

DOPA is a key amino acid for adhesion and cohesion due to the various possible chemical reactions (Yu et al., 1999). Many physical and chemical interactions have been observed with DOPA or other substances containing catechol moieties as shown in Fig. 13.2. The catechol moiety of DOPA has the ability to form extremely stable metal complexes, promoting the adhesion to metal surfaces. The dissociation force between DOPA and TiO_2 surfaces was measured by single molecule AFM (Lee et al., 2006). This is remarkably high at 805 pN, and this compares with the strong dissociation force of the carbon–silicon covalent bond (2000 pN) and the comparatively weak forces applied for unzipping the hydrogen bonds of DNA strands (10–20 pN) (Hamming et al., 2008). The hydroxyl groups of the catechol group (hydrochinon) is a potential hydrogen bond donor. In oxi-

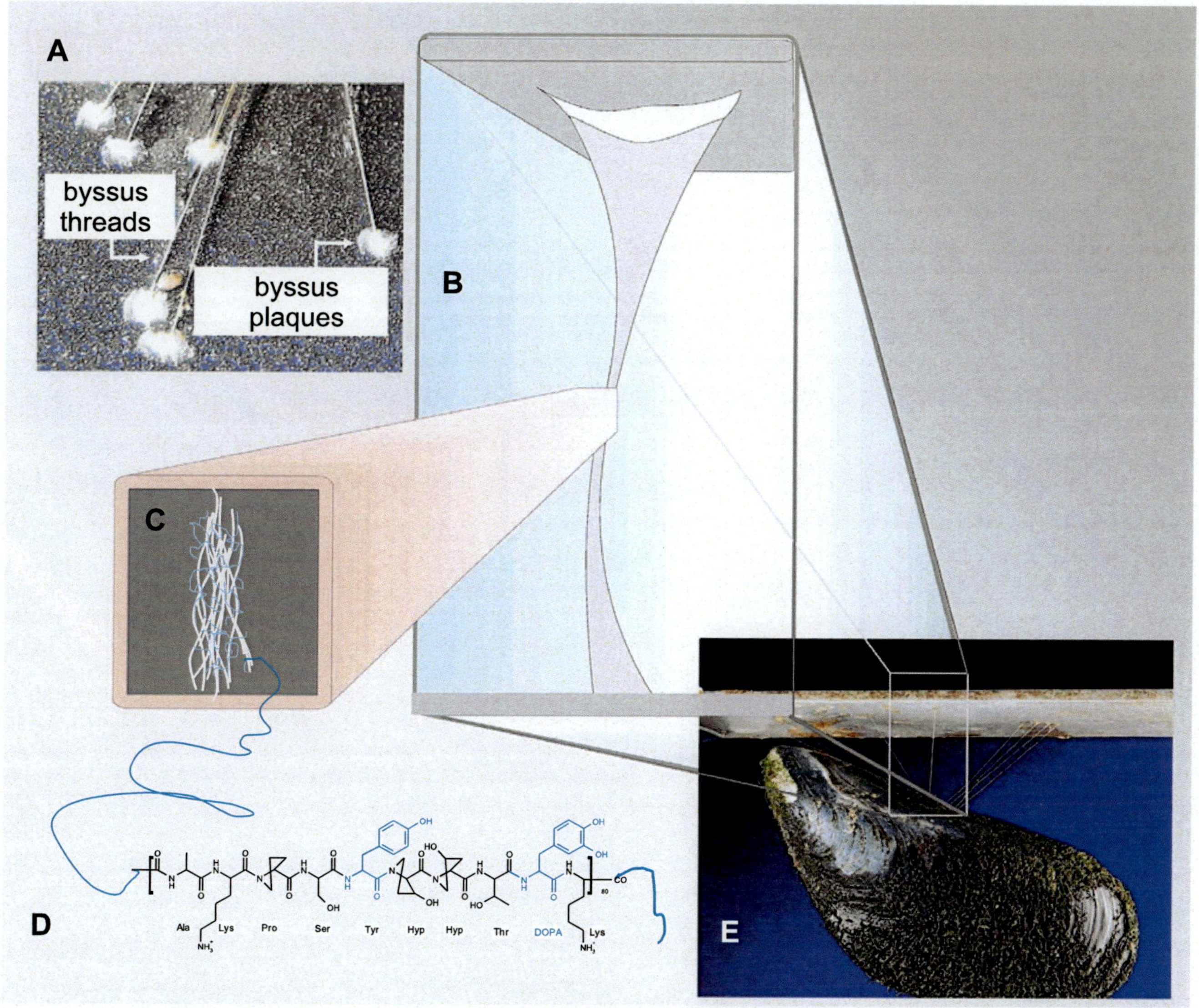

Fig. 13.1 *Mytilus edulis* adhesive system. The blue mussel is attached to an ordinary Teflon® pan (A) and to a metal plate (E). A byssus thread and a byssus plaque are shown schematically magnified in (B). Further magnification shows the byssus thread seen in the amino acid sequence as repeating unit of the adhesive peptide Mefp-1 (C and D). The phenolic elements are emphasized blue. Threads and plaques are produced together by an organ of the mussel called the byssus foot

dized mode (o-chinon) this group behaves as an acceptor. Covalent crosslinks are mandatory for the cohesion. They are formed in the mussel adhesive proteins by radical dimerization of two catechol moieties to form diDOPA (Burzio et al., 2000) and by 1,4-Michael additions of amino groups of the lysine, histidine, and cysteine residues with o-chinon (Liu et al., 2006). The crosslinking process is induced by oxidation of the catechol moiety to the o-quinone (Burzio et al., 2000) either by agents such as $NaIO_4$ or H_2O_2 (Yu et al., 1999) or enzymatically by, for example, mushroom tyrosinase. Furthermore, tyrosinase has the ability to modify tyrosine residues of peptides in DOPA by post-translational hydroxylation (Kitamura et al., 1999). Alternatively, crosslinking can be induced by the addition of Fe(III) ions (Sever et al., 2004). The byssus threads are composed of an outer shell which is five times stronger than the inner area. This effect is said to depend on iron ions complexometric bounds in this area (Harrington et al., 2010). These ions form stable complexes of three DOPA residues in the plaque. Due to the presence of air the protein–iron complex is oxidized and forms radical species. The mussel is thereby a very good example of where and how polyphenolic substances are used for adhesion and cohesion.

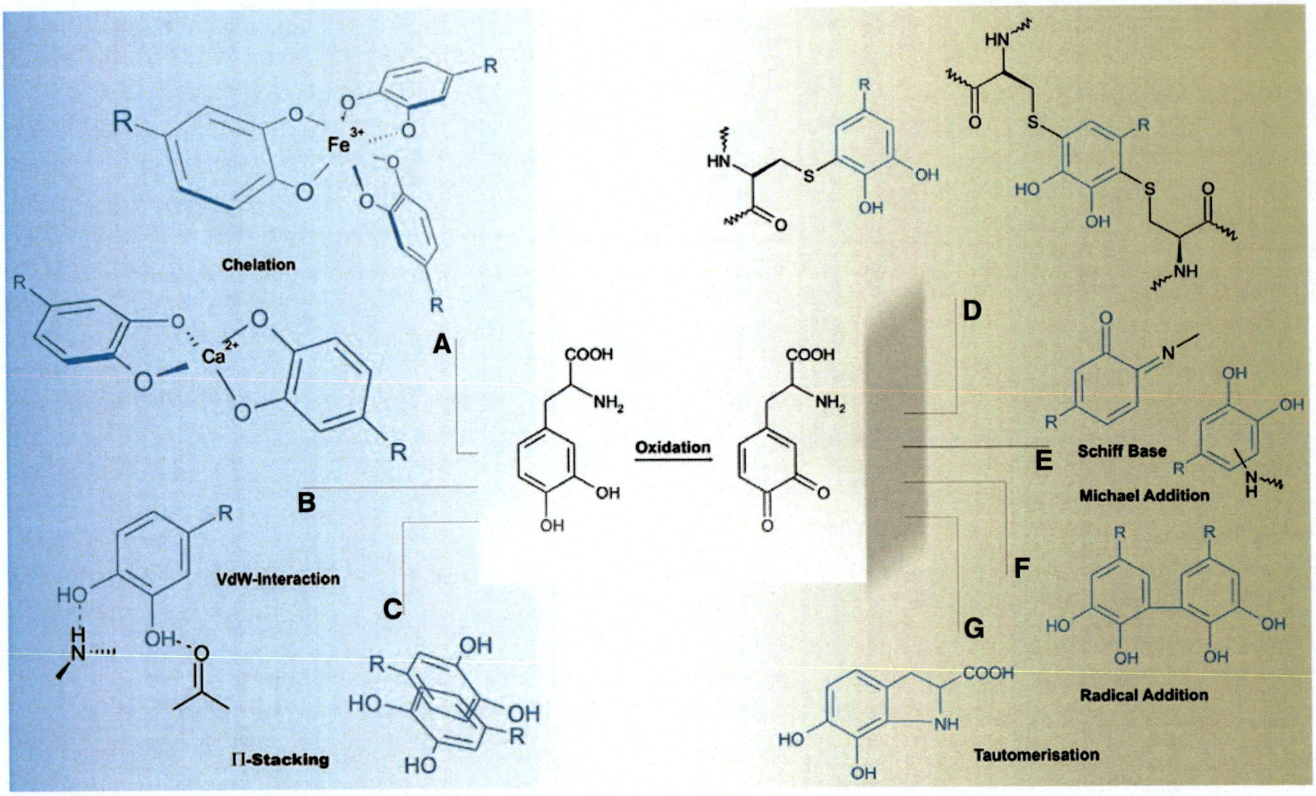

Fig. 13.2 Possible interactions of the DOPA-hydrochinon shows (A) complexation of metalions (M^{n+}), (B) physical interactions with organic components (collagen, keratin), (C) π-stacking due to the aromatic system, and DOPA-Chinon reacts in presence of (D) Cystein or Glutathione to Cysteinyldopa, (E) primary amines to Schiff base or undergoes a Michael reaction, (F) another DOPA in radical addition to diDopa or DOPA-Chinon tautomerize to 5,6-Dihydroxyindol-2-carboxylat

13.3 Synthetic Phenolic Resins and Their Applications

(*Andreas Hartwig*)

What have phenolic resins and bio-inspired adhesives such as those from the blue mussel in common? At closer glance, strong similarities in structure and properties can be observed. Natural adhesives of, for example, the blue mussel have extraordinary properties and phenolic resins are still among the artificial adhesives with the best ageing behavior, and especially so in wet environments. This relationship is caused by structural similarities, although there are also significant differences.

Phenolic resins were developed by Leo Baekeland and were the first synthetic polymers produced on a large scale. They are prepared via a condensation reaction between phenol, or a phenol derivative, and formaldehyde. The condensation reaction is carried out in either a basic environment, forming so-called resols, or at acidic pH, forming a novolac (Wegler and Herlinger, 1963; Ratna, 2009). The main structural elements (including adhesive interaction with a metal atom on a surface) are summarized in Fig. 13.3 and are compared to the typical catechol structure of the blue mussel adhesive. Novolacs need a separate curing agent for the formation of a crosslinked polymer, usually hexamethylene tetramine as a latent formaldehyde source, whereas resols undergo self-curing in a condensation reaction as they are produced with an excess of formaldehyde. Mainly resols are used in adhesives.

During the self-curing reactions the methylol groups attack the aromatic rings, forming methylene bridges, and water is split off. Typical processing temperatures are in the region of 150°C. In order to prevent foaming of the resol due to evaporation of the water, curing at elevated pressure is required. This is one of the main reasons why resols are not widely used in general industry even though they have extremely interesting properties.

In general, phenolic resins have very good mechanical properties and long-term properties even at elevated

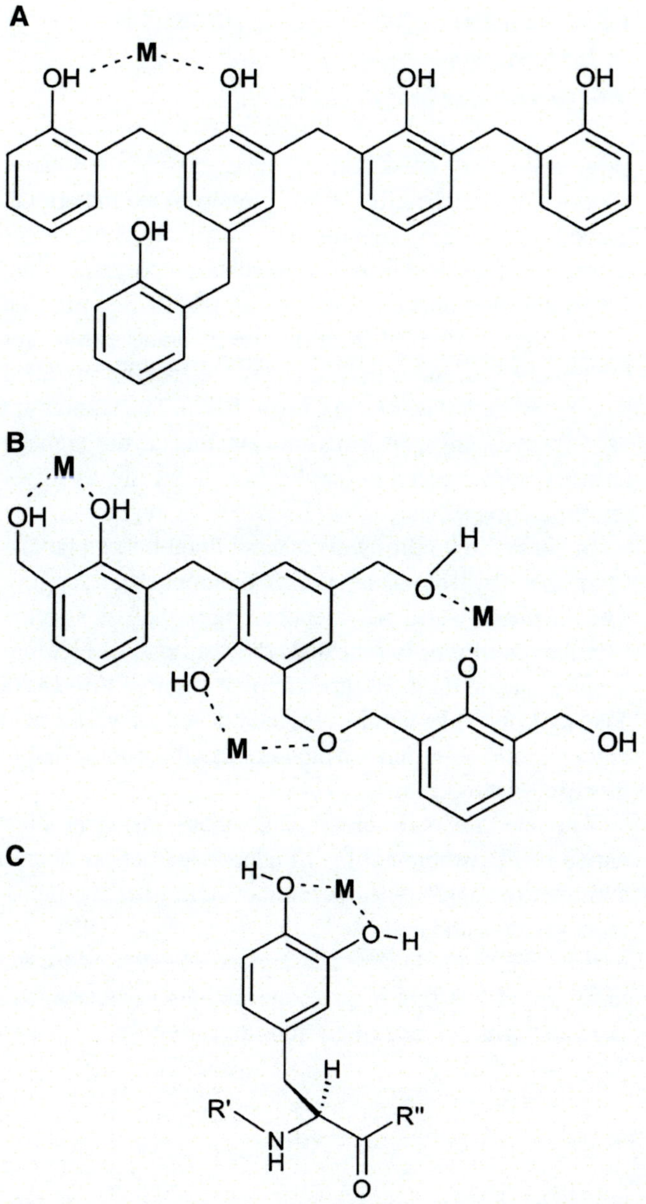

Fig. 13.3 Typical structural elements of a novolac (A) and a resol (B) compared to the DOPA unit of the Mefp protein of the blue mussel (C). Possible chelation reactions are shown with metal atoms M, which are the proposed adhesion mechanisms for these compounds on metallic surfaces

temperatures and high humidity. Of all synthetic organic polymers they have almost the best thermal stability and have excellent fire and smoke characteristics. Furthermore, they are cheap compared to most other thermosetting polymers. Brittleness is a disadvantage of phenolic resins and therefore toughening is required. This is, for example, carried out by the addition of different kinds of rubbers or core-shell rubber particles and alternatively by integration of longer hydrocarbon chains into the polymer network (Gietl et al., 2006; Iyim, 2007). This is again a similarity: The adhesive of the blue mussel is also brittle, and water is required for flexibilization.

Phenolic resins are mainly used as a binder for wood composites, for adhesive bonding of wood and for insulation materials (Habenicht, 1987). The main reason for this is their good durability in moist environments. This is also the main reason why phenolic resin based adhesives are used to bond the veneer plies of exterior-grade plywood panels, the flakes of oriented strandboard (OSB), and particle board panels. By modifying the resin with starch and tannin (Sect. 13.4) a polyphenolic compound – property improvements were demonstrated (Moubarik et al., 2009). Other typical applications are binders for foundry materials, abrasives, and friction linings. For the latter in particular, automotive brake shoes are an example where phenolic resins are used as binders as well as adhesives (Fig. 13.4). They are suitable for the high thermal stability required for this application. The wear behavior of the friction materials strongly depends on the composition of the phenolic resin binder. It was shown that special phosphorous and boron containing

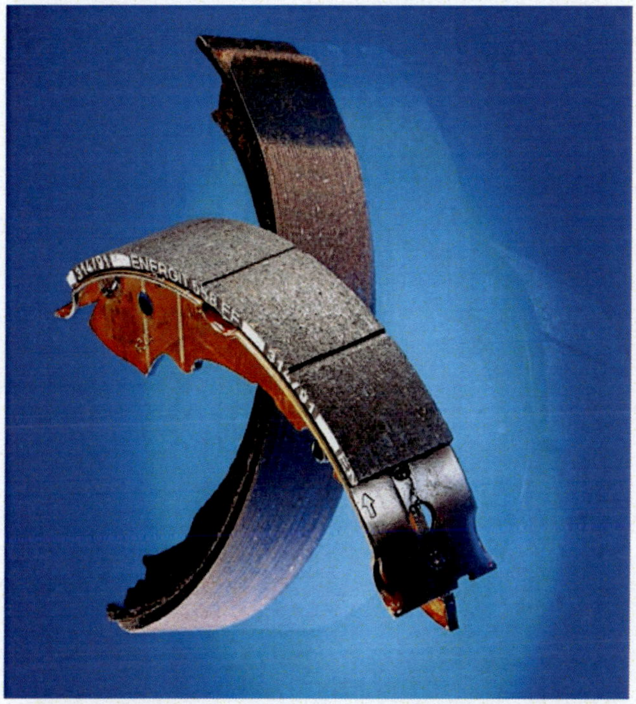

Fig. 13.4 Brake shoes prepared with a phenolic resin as a binder and adhesive

phenolic resins give improved wear resistance and friction stability (Hong et al., 2009). Due to their inherent good fire resistance, the paper of honeycomb structures is impregnated with phenolic resin and these composites are used in lightweight structures, especially for the aerospace industry. Resols are usually used in these applications, as well as in phenolic adhesives.

Besides the application of phenolic resins as adhesives in the wood industry, special resol based materials are used as adhesives for aluminum in the aerospace industry. This is due to the excellent long-term stability of the bonds. The drawback, however, is the required curing in an autoclave in order to obtain bubble and defect free bonds. Usually these adhesives are applied as a dry film during melting wets the surface, and crosslinks during thermal curing. In these film adhesives usually poly(vinyl formal) is applied as a toughening agent. The adhesive Redux 775 from Hexcel is probably the workhorse in this area and has been in use since the 1940s (Higgins, 2000). The bond strength and durability strongly depend on the conditions used for the resol preparation which is carried out with a basic catalyst. Kollek et al. (1986) showed that the resol should have a slightly acidic pH during application because at the original basic pH the resulting peel strength is low. Kollek (1985) also examined the adhesion mechanism of resols on aluminum surfaces. He showed by IR spectroscopy that adhesion mainly occurs via the acidic protons of the phenol and the hydroxymethyl groups forming a chelate complex with the aluminum atoms (Fig. 13.3) (Kollek, 1985).

Additionally the ethylene bridged diphenol units appear to form chelate complexes. This has high similarity to the adhesion mechanisms proposed for the blue mussel adhesives containing large amounts of DOPA. Durability is essential for a lot of applications, namely in the aerospace industry with product lifetimes above 30 years. Sargent showed that aluminum samples bonded with aerospace grade phenolic resins did not show any reduction of peel strength after storage in warm water for over 7 years (Sargent, 2005). Such excellent durability is not found for other adhesive systems. Brockmann also examined the durability of anodized aluminum/phenolic resin adhesive bonds. He concluded that mechanical interlocking is important for the adhesion and that durability stabilization of the oxide layer by the phenolics is an important effect (Brockmann, 1989). Not only for aluminum but also for other metallic substrates such as titanium it was shown that phenolic resins are excellent adhesives (Molitor and Young, 2002).

13.4 Tannins and Their Application in Adhesives

(Maria Kozielec and Klaus Rischka)

Despite the very good bonding properties of the mussel adhesive, the application of this protein-based material is limited by its price. Tannins are a relatively cheap natural source of a catechol moiety-containing polymer. They are natural polyphenols found in the plant kingdom. The word tannin is derived from the use of plants extract for the treatment of hides in leather production, the so-called tanning process (Haard and Chism, 1996). The tannins are used in dyes, inks, pharmaceuticals, and in the tannage industry. Other potential application fields are as finger jointing wood adhesives and for paper impregnation.

In general, two different classes of tannins can be distinguished: hydrolyzable tannins and condensed tannins. The flavonoid-based polymers belong to the latter group. The flavonoid unit is repeated, for example, 2–11 times in mimosa and up to 30 times in pine tannins (Fig. 13.5). The hydrolyzable tannins are composed of gallic acid and epigallic acid units, which are usually condensed to glucose (Natori, 1983).

The need for plant-based panel adhesives as an alternative to petrochemical-based adhesives such as phenol and urea formaldehydes became evident after the petroleum crisis of the early seventies (Singleton, 1992). Due to the crisis, tannins have subsequently been used for exterior wood bonding of products such as particle board, plywood, glulam, and finger jointing.

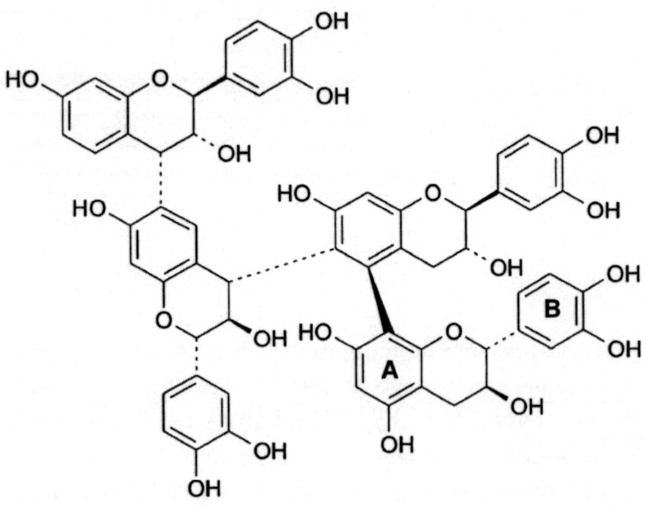

Fig. 13.5 Condensed tannin with the flavone subunit. The resorcinol ring (A ring) is more reactive than the catechol ring (B ring) (Pizzi, 2003)

The extraction of tannins from natural sources such as chestnut wood, sumach leaves, wood and bark of various kinds of oaks, wattles (acacias), and pines is necessary for their use in adhesive systems (Connolly, 1993). Depending on the source of the tannin extract, the tannins have different characteristics. For example, tannins extracted from wattle are less reactive than the extracts derived from pine trees. Pine tannins are especially suitable as wood adhesives for plywood and particle boards. Wattle tannins are generally of lower molecular weight and lower viscosity.

The traditional method of tannin extraction is performed with boiling water. Some 80% of tannin extract is composed of phenolic ingredients, simple sugars, and high molecular weight hydrocolloid gums. When using tannin for adhesives, however, the presence of sugars reduces the strength and water resistance in direct proportion to the amount added. The hydrocolloid gums, in contrast, show a positive effect on both the original strength and water resistance of the adhesive. For the preparation of adhesives and resins, condensed tannins are of greater interest. In contrast to plywood adhesives, the viscosity of the tannin extract is crucial for particle board adhesives and affects the strength and water resistance. The high viscosity of the tannin extracts can be manipulated by dilution. The treatment of the tannin extract with acid and alkaline media is the most commonly used process for eliminating these disadvantages (Pizzi, 1997).

Tannin formaldehyde adhesives are obtained by the reaction of condensed tannins with formaldehyde via polycondensation. Methylene bridge linkages are formed between the reactive positions on the flavonoid molecules, mainly the A ring, by the reaction of formaldehyde with tannins. Compared to the reaction of phenol with formaldehyde, the reaction of tannins with formaldehyde is between 10 and 50 times faster (Pizzi, 2003). Due to the high requirements put on many wood adhesives the reinforcement of the tannin with phenolic or amino plastic resins is necessary.

13.5 Phenolic Adhesives for Medical Applications

(Robert Sader and Ingo Grunwald)

Industrial adhesives, such as the phenolic resins introduced in Sect. 13.3, have replaced traditional joining techniques such as screws, nails, brazing, and welding. The main advantage of adhesive bonding, besides its speed and relative simplicity, is the ability to join different materials. When adhesives are used, the bonded joint leads to homogenous stress distribution over the surface area. In addition, the adhesive material can compensate shortcomings in the surface topology such as defects, irregularities in thickness, or roughness. Furthermore, it can also function as a vibration damper (Schraven, 2007). Another advantage of adhesive bonding is that only the areas to be bonded have to be prepared.

In medicine, and in particular in surgery, adhesive bonding techniques have not become so well established that they have replaced traditional methods such as suturing wounds or stabilizing bone fractures with screws or plates (Bitton et al., 2009). These fixation systems have the drawback that they periodically lead to adverse effects such as, for example, circulatory disorders or inflammatory reactions caused by the use of screws/plates for bone fractures. Because of the greater surgical work required for bone fixation, the surrounding soft tissue is frequently damaged. In addition, the fixation materials have to be removed in a second operation, and the metal of the implant can negatively react with the tissue (Schraven, 2007).

Overall, it can be stated that for many surgical techniques the use of adhesives opens up opportunities for minimum invasive treatments. In principle, most known adhesives can be differentiated into synthetic adhesives and biological adhesives. Biological adhesives, some of them containing polyphenolic compounds, are produced for example in various organisms such as barnacles, mussels, starfish, algae, and many other species (Smith and Callow, 2006). In theory those adhesives are ideal for medical and technical applications: they cure in the presence of water, cure at moderate temperatures, are biodegradable, and have relatively high adhesion forces. However, their use in medicine will only be feasible if no immunological reactions occur and if they can (ideally) be produced synthetically to avoid contamination with other biological material.

An ideal medical adhesive has to fulfill a number of requirements. It should at least:

- be biocompatible and biodegradable
- be easy to handle and the curing should be relatively fast
- have sufficient adhesion and cohesion strength
- be free of contamination with other biological material
- be, as a prerequisite, sterilizable (heat, radiation, or chemical)
- be able to be produced at an affordable price

- have a curing temperature that does not destroy body tissue
- be suitable for different tissues/bones
- be able to be cured in the presence of water (blood) or in moist environments
- not hinder the healing process.

In the late 1960s first experiments with aldehydes and/or polyphenol containing adhesives were performed to create a medical adhesive material. These adhesives systems, composed of gelantine, resorcinol, and aldehyde, were used to glue skin, vessels, lung, and liver wounds. This medical adhesive also showed adverse effects such as tissue necrosis. After the formaldehyde was replaced by less toxic dialdehydes, it was possible to use this adhesive clinically. Other resorcinol/glutaraldehyde-based adhesives were used for the repair of osteochondral defects in the knees of rabbits (Sakihama et al., 2006). They found that this adhesive is not suitable for fixation of cartilage but is suitable for bone repair.

The protein-aldehyde adhesives have only relatively low binding forces and hence are mainly used for their sealant properties. These sealant adhesives are used for sealing applications involving aortic valve replacement, vascular prostheses, and blood-tight surgical sutures. As a further development of protein-aldehyde adhesives, the surgical adhesive BioGlue® has been commercially available since 1998 (CryoLife Inc., USA). This adhesive is used, for example, as a support for conventional surgical sutures of aortic dissections (Raanani et al., 2001).

Since the 1970s fibrin adhesives have also been used as medical adhesives (for detailed information see Chapter 15, p. 225 of this book). One advantage of fibrin adhesives is that they are produced from autologous substances (the same as used in our body for blood-clotting). This ensures the biocompatibility of the adhesive material. The material is decomposed quite fast – within a few days – by the body. The fibrin adhesives have, like the protein-aldehyde adhesives, only weak adhesion strength. They are mainly used for supporting surgical sutures or for less mechanically loaded bonds, for example the closure of small disruptions in organs like the liver and spleen. When adhesive materials of human origin are used there is a theoretical risk of infection by hepatitis or HIV. The use of adhesive materials from non-human sources, like for example the marine mussel adhesives, at least avoids the infection risk.

Ninan et al. (2003) have compared the adhesive strength of marine mussel extracts with other adhesives like, for example, fibrin adhesives in a skin model (porcine). In their experiments the mussel adhesive gave a relatively strong (~1 MPa) tissue joint if the skin was glued in the end-to-end configuration. The bond strength here was better than with fibrin adhesives, but with much longer curing times (Ninan et al., 2003) and so this will not be suitable in an emergency case. The adhesive protein from the blue mussel *Mytilus edulis* was also tested in an *in vivo* study for chondrocyte transplantation (Pitman et al., 1989). The authors of this study report highly promising results for the use of the mussel adhesive in medical applications for chondrocyte transplantation (Pitman et al., 1989). In an *in vitro* system Benthien et al. (2004) tested the effects of bone cement, marine mussel, and cyanoacrylate adhesives on human osteoblasts and fibroblasts. They found that the mussel and cyanoacrylate adhesives had negative effects on the cell cultures that were used.

The above-mentioned examples of the use of mussel adhesives for medical applications have shown that different results are obtained depending on the experimental set-up, the used tissue/cells, and the applied mussel adhesive. These results show there is a strong need for in-depth testing of mussel extracts before any *in vivo* research is performed.

In a recent study Brubaker et al. (2010) reported the use of a mussel-inspired adhesive in islet transplantation. In contrast to other experiments with mussel extracts or purified mussel proteins they followed a biomimetic approach by using a fully synthetic branched poly(ethylene glycol) core modified at the end groups with catechols (functional group in the mussel adhesive proteins). This biomimetic adhesive cured in less than 60 sec and displayed minimal chronic and acute inflammatory reactions. Furthermore, this adhesive was able to glue the transplanted islets to the liver surface (Brubaker et al., 2010).

In another biomimetic approach Bitton et al. (2009) described a phloroglucinol-based adhesive material. The phloroglucinols belong to the family of phenols and play a role in algae adhesion mechanisms. The adhesive is composed of phloroglucinol (the synthetic monomer of the polyphenol in the algae), alginate, and calcium ions. This adhesive displayed similar tensile strength values to fibrin adhesives when tested with porcine muscle tissue (Bitton et al., 2009). In addition, and important for medical applications, the non-oxidized adhesive formulation does not induce cell cytotoxicity (Bitton et al., 2009).

In summary, there is no universal adhesive which can be used for multiple medical applications. Each adhesive has to be developed for a specific application. It will be a challenge for scientists and companies to meet all the aforementioned requirements for an ideal medical adhesive. Some of the above-mentioned adhesives, for example the protein-aldehyde and fibrin adhesives, are already commercially available. However, at present they do not meet all the cited requirements. The polyphenolic adhesives from marine organisms, such as mussels and algae, are some distance away from being medically approved. It maybe worth considering systems which contain biomimetic combinations of elements from different biological adhesives and biocompatible polymer systems with synthetically produced compounds (Grunwald et al., 2009).

13.6 Special Applications: Space Exploration

(*Klaus Slenzka*)

The use of adhesives and related strategies in closed environments including Life Support Systems used in spacecrafts, submarines, etc. is mainly limited to liquids and gels not containing any organic solvents (Truong et al., 2000). Regarding future long-term strategies, the development of strategies for adhesives having the same physical features as, for example, epoxy adhesives will become of utmost importance for the above-mentioned applications (Brazill, 1985; Niebuhr, 2007).

Nature demonstrates very impressively what adhesives may become available in the future. Clams and other marine species are able to use adhesive even when submersed in water, something which is actually not possible with the products currently available in the marketplace.

In addition, alternative adhesive strategies based on biomimicry must be developed. These are of utmost importance for so-called "green chemistry" and are based on the development of sustainable products. For applications in closed environments/habitats including Ecological Life Support Systems during space travel and in extraterrestrial habitats, and also for example in deep sea stations or in submarines, etc., the adhesives must have well-defined properties (Bhowmik et al., 2006, 2009). Due to the isolated environmental conditions, the adhesives must not emit monomers into the air and should have good biocompatibility. For the long-term durability and to minimize the risk of infection, the bonded joints should be more resistant to bacterial attacks than, for example, silicone-based adhesive systems (Morscher, 2003).

Adhesives are already widely used in the design, development, and manufacture of spaceflight hardware for manned spaceflight programs. In particular, specific hardware for several experiments flown on the US shuttle, the space station MIR, or to be flown aboard ISS uses adhesives extensively. One example is the aquatic habitat C.E.B.A.S. (Closed Equilibrated Biological Aquatic System) which has been flown in space 3 times. For these experiments several water tanks were designed with a volume of 8.6 L and, based on safety requirements, were designed to withstand 2.5 bar overpressure. The tanks were manufactured from polycarbonate and were bonded with special adhesives.

The future goal to have human missions to the moon again and later to Mars and beyond requires sustainable technologies, because of the lack of intensive re-supply and rapid return missions. Bio-inspired adhesives are thus an important step toward human exploration of the solar system.

The use of biomimetics is the opportunity to solve a lot of these issues and in addition an opportunity for applications on Earth.

13.7 Conclusion

Polyphenolic substances are widely found in living organisms and fulfill various functions. Using these materials, for example, in adhesives allows organisms to cope with complex living conditions. Polyphenolic substances are already used in technical applications, similar to the natural applications, but still lack some of the advantages of bioadhesives such as curing underwater at moderate temperatures. These technical adhesives, for example the phenolic resins, are far away from having the required biocompatibility for medical applications.

Nature presents a wide variety of adhesives in the same way a DIY superstore offers a large range of adhesives for different substrates and applications. Common technical adhesives are available for wide-ranging applications, but in some special areas there is a great need for better and improved products, for example for underwater adhesion or improved biocompatibility for medical adhesives. A possible route is to develop hybrid materials based on natural and technical components, thereby generating synergistic effects. Certainly, more work has to be undertaken to investigate the chemical and physical mechanisms of bioadhesives resulting in new customized applications for adhesives.

Acknowledgment

The authors would like to thank Stuart Fegan for proofreading the manuscript and for helpful comments.

References

Antonietti M and Fratzl P (2010) Biomimetiv principles in polymer and material science. Macromolecula Chemisty and Physics 211: 166–170.

Arranz S, Saura-Calixto F, Shaha S, and Kroon PA (2009) High contents of nonextractable polyphenols in fruits suggest that polyphenol contents of plant foods have been underestimated. Journal of Agricultural and Food Chemistry 57(16): 7298–7303.

Benthien JP, Russlies M, and Behrens P (2004) Investigating the effects of bone cement, cyanoacrylate glue and marine mussel adhesive protein from Mytilus edulis on human osteoblasts and fibroblasts *in vitro*. Annals of Anatomy – Anatomischer Anzeiger 186: 561–566.

Berlind T, Poksinski M, Tengvall P, and Arwin H (2010) Formation and cross-linking of fibrinogen layers monitored with in situ spectroscopic ellipsometry. Colloids and Surfaces B: Biointerfaces 75: 410–417.

Bhowmik S, Bonin HW, Bui VT, and Weir RD (2006) Durability of adhesive bonding of titanium in radiation and aerospace environments. International Journal of Adhesion and Adhesives 26(6): 400–405.

Bhowmik S, Benedictus R, Poulis JA, Bonin HW, and Bui VT (2009) High-performance nanoadhesive bonding of titanium for aerospace and space applications. International Journal of Adhesion and Adhesives 29(3): 259–267.

Bitton R, Josef E, Shimshelashvilli I, Shapira K, Seliktar D, and Bianco-Peled H (2009) Phloroglucinol-based biomimetic adhesives for medical applications. Acta Biomaterialia 5: 1582–1587.

Brazill D (1985) The physical effects of different stress situations on graphite/epoxy compounds. NASA Ames Summer High School Apprenticeship Research Program: pp 21–28.

Brockmann W (1989) Durability of adhesion between metals and polymers. Journal of Adhesion 29: 53–61.

Brubaker C, Kissler H, Wang LJ, Kaufman DB, and Messersmith PB (2010) Biological performance of mussel-inspired adhesive in extrahepatic islet transplantation. Biomaterials 31: 420–427.

Burzio LA and Waite JH (2000) Cross-linking in adhesive quinoproteins: studies with model decapeptides. Biochemistry 39: 11147–11153.

Burzio LA, Burzio VA, Pardo J, and Burzio LO (2000) In vitro polymerization of mussel polyphenolic proteins catalyzed by mushroom tyrosinase. Comparative Biochemistry and Physiology Series B Biochemistry & Molecular Biology 126: 383–389.

Carrington E (2008) Along the silk road, spiders make way for mussels. Biochemistry 39: 11147–11153.

Connolly DL (1993) Tannin Extraction. USA Patent No. US5238680.

Creton C and Papon E (2003) How to bond things together. MRS Bulletin 28: 419–421.

Flammang P (2006) Adhesive secretions in echinoderms: an overview. In: Smith AM and Callow JA (eds) Biological Adhesives. Springer-Verlag, Heidelberg: pp 183–206.

Flammang P, Lambert A, Bailly P, and Hennebert E (2009) Polyphosphoprotein-containing marine adhesives. The Journal of Adhesion 85: 447–464.

Gao K, Xu A, Krul C, Venema K, Liu Y, Niu Y, Lu J, Bensoussan L, Seeram NP, Heber D, and Henning SM (2006) Of the major phenolic acids formed during human microbial fermentation of tea, citrus, and soy flavonoid supplements, only 3,4-dihydroxyphenylacetic acid has antiproliferative activity. Journal of Nutrition 136(1): 52–57.

Gietl H, Lengsfeld V, and Altstädt J (2006) The efficiency of various toughening agents in novel phenolic type thermoset resin systems. Material Science 41: 8226–8243.

Grunwald I, Rischka K, Kast SM, Scheibel T, and Bargel H (2009) Mimicking biopolymers on a molecular scale: nano(bio)technology based on engineered proteins. Philosophical Transactions of the Royal Society: Series A, Mathematical, Physical and Engineering Sciences 367(1894): 1727–1747.

Haard NF and Chism GW (1996) Characteristics of edible plant tissues. In: Fennema OR and Fennema F (eds) Food Chemistry, 3rd Ed. Marcel Dekker Inc., New York: pp 943–1012.

Habenicht G (1987) Kleben Grundlagen, Technologie, Anwendungen, 3rd Ed. Springer-Verlag, Berlin.

Hamming LM, Fan XW, Messersmith PB, and Brinson LC (2008) Mimicking mussel adhesion to improve interfacial properties in composites. Compost Science and Technology 68(9): 2042–2048.

Harrington MJ, Masic A, Holten-Andersen N, Waite JH, and Fratzl P (2010) Iron-clad fibers: a metal-based biological strategy for hard flexible coatings. Science 328(5975): 216–220.

Heiss C, Kraus R, Schluckebier D, Stiller AC, Wenisch S, and Schnettler R (2006) Bone adhesives in trauma and orthopedic surgery. European Journal of Trauma 2: 141–148.

Higgins A (2000) Adhesive bonding of aircraft structures. International Journal of Adhesion and Adhesives 20: 367–376.

Holten-Andersen N and Waite JH (2008) Mussel-designed protective coatings for compliant substrates. Journal of Dental Research 87: 701–709.

Hong US, Jung SL, Cho MH, Kim SJ, and Jang H (2009) Wear mechanism of multiphase friction materials with different phenolic resin matrices. Wear 266: 739–744.

Iyim TB (2007) Modification of high ortho novolac resin with diacids to improve its mechanical properties. Journal of Applied Polymer Science 106: 46.

Keckes J, Burgert I, Frühmann K, Müller M, Kölln K, Hamilton M, Burghammer M, Roth SV, Stanzl-Tschegg S, and Fratzl P (2003) Cell-wall recovery after irreversible deformation of wood. Nature Materials 2: 810–814.

Kitamura M, Kawakami K, Nakamura N, Tsumoto K, Uchiyama H, Ueda Y, Kumagai I, and Nakaya T (1999) Expression of a model peptide of a marine mussel adhesive protein in *Escherichia coli* and characterization of its structural and functional properties. Journal of Polymer Science: Part A Polymer Chemistry 37: 729–736.

Kollek H (1985) Some aspects of chemistry in adhesion and adhesives. International Journal of Adhesion and Adhesives 5: 75–80.

Kollek H, Brockmann H, and Mueller von der Hagen H (1986) Chemistry of curing and adhesion properties of phenolic resins. International Journal of Adhesion and Adhesives 6: 37–41.

Lee H, Scherer NF, and Messersmith PB (2006) Single-molecule mechanics of mussel adhesion. Proceedings of the National Academy of Sciences of the United States of America 103(35): 12999–13003.

Lin Q, Gourdan D, Sun C, Anderson NH, and Waite JH (2007) Adhesion mechanisms of the mussel foot proteins mfp-1 and mfp-3. Proceedings of the National Academy of Sciences of the United States of America 104(10): 3782–3786.

Liu B, Burdine L, and Kodadek T (2006) Chemistry of periodate-mediated cross-linking of 3,4-dihydroxylphenylalanine-containing molecules to proteins. Journal of the American Chemical Society 128(47): 15228–15235.

Melzer B, Steinbrecher T, Kraft O, Schwaiger R, and Speck T (2008) The attachment of English ivy (*Hedera helix* L.): Biomechanical aspects. Comparative Biochemistry and Physiology: Series A Molecular & Integrative Physiology 150: S89.

Molitor P and Young T (2002) Adhesives bonding of a titanium alloy to a glass fibre reinforced composite material. International Journal of Adhesion and Adhesives 22: 101–107.

Morscher GN (2003) SiC/SiC life modeling at intermediate temperatures under NASA UEET. Conference held in Coco Beach Florida, USA 28 Jan. 2003, NASA Technical Reports Server (NTRS), Coco Beach: pp 1–9.

Moubarik A, Pizzi A, Allal A, Charrier F, and Charrier F (2009) Cornstarch and tannin in phenol-formaldehyde resins for plywood production. Industrial Crops and Products 30: 188–193.

Natori S (1983) Carboaromatic and releated compounds. In: Nakonishi K, Goto T, Natori S, and Nozoe S (eds) Natural products chemistry, Vol. 3. Ed. Kodansha Scientific Books, Tokyo: pp 427–524.

Niebuhr D (2007) Friction and wear behavior of engineering materials in a simulated Martian (CO_2) environment, a preliminary study. Wear 263: 88–92.

Ninan L, Monahan J, Stroshine RL, Wilker JJ, and Shi RY (2003) Adhesive strength of marine mussel extracts on porcine skin. Biomaterials 24: 4091–4099.

Papov VV, Diamond TV, Biemann K, and Waite JH (1995) Hydroxyarginine-containing polyphenolic proteins in the adhesive plaques of the marine mussel *Mytilus edulis*. Journal of Biological Chemistry 270(34): 20183–20192.

Pitman MI, Menche D, Song EK, Benyishay A, Gilbert D, and Grande DA (1989) The use of adhesives in chondrocyte transplantation surgery – In vivo Studies. Bulletin of the Hospital for Joint Diseases Orthopaedic Institute 49: 213–220.

Pizzi A (1997) Wood/bark extracts as adhesives and preservatives. In: Bruce A and Palfreyman J (eds) Forest Products Biotechnology. Taylor and Francis, London: pp 167–182.

Pizzi A (2003) Natural phenolic adhesives I: tannins. In: Pizzi A and Mittal KL (eds) Handbook of Adhesive Technology, 2nd Ed. Marcel Dekker, New York: pp 573–587.

Plaza GR, Corsini P, Marsano E, Perez-Rigueiro J, Biancotto L, Elices M, Riekel C, Agullo-Rueda F, Gallardo E, Calleja JM, and Guinea GV (2009) Old silks endowed with new properties. Macromolecules 42: 8977–8982.

Raanani E, Latter DA, Errett LE, Bonneau DB, Leclerc Y, and Salasidis GC (2001) Use of "BioGlue" in aortic surgical repair. The Annals of Thoracic Surgery 72: 638–640.

Ratna D (2009) Handbook of Thermoset Resins. Smithers Rapra Technology, Shawbury.

Rios LY, Gonthier MP, Rémésy C, Mila I, Lapierre C, Lazarus SA, Williamson G, and Scalbert A (2003) Chocolate intake increases urinary excretion of polyphenol-derived phenolic acids in healthy human subjects. American Journal of Clinical Nutrition 77(4): 912–918.

Rzepecki LM, Hansen KM, and Waite JH (1992) Characterization of a cystine-rich polyphenolic protein family from the blue mussel *Mytilus edulis*. Biological Bulletin 183(1): 123–137.

Sakihama AK, Fagundes DJ, Inouye CM, Aydos RD, Juliano Y, Arakaki JC, Odashiro M, and de Tarso Camilo de Carvalo P (2006) The use of resorcine adhesive in repairing osteochondral's defect in knees of rabbits. Acta Cirurgica Brasileira 21: 45–50.

Santos R, da Costa G, Franco C, Gomes-Alves P, Flammang P, and Coelho AV (2009) First insights into the biochemistry of tube foot adhesive from the sea urchin *Paracentrotus lividus* (Echinoidea, Echinodermata). Marine Biotechnology 11: 686–698.

Sargent JP (2005) Durability studies for aerospace applications using peel and wedge tests. International Journal of Adhesion and Adhesives 25: 247–256.

Scholz I, Barnes WJP, Smith JM, and Baumgartner W (2009) Ultrastructure and physical properties of an adhesive surface, the toe pad epithelium of the tree frog, *Litoria caerulea* White. Journal of Experimental Biology 212: 155–162.

Schraven J (2007) Biocompatibility of a new bone glue in sheep model. Dissertation Thesis. Justus-Liebig-University Giessen, Germany.

Schreiber L, Rochus F, and Hartmann K (2005) Wax and suberin development of native and wound periderm of potato (*Solanum tuberosum* L.) and its relation to peridermal transpiration. Planta 220: 520–530.

Sever MJ, Weisser JT, Monahan J, Srinivasan S, and Wilker JJ (2004) Metal-mediated cross linking in the generation of a marine-mussel adhesive. Angewandte Chemie International Edition 116: 454–456.

Singleton VL (1992) Tannins and the qualities of wines. In: Hemingway RW and Laks PE (eds) Plant Polyphenols: Synthesis, Properties, Significance. Basic Life Sciences, Vol. 59. Ed. Plenum Press, New York: pp 859–880.

Smith AM and Callow JA (2006) Biological adhesives. Springer-Verlag, Berlin.

Truong VT, Olsson-Jacques CL, Russo MS, and Muscat RF (2000) A sensor for water detection in aircraft adhesive bondlines. DSTO Aeronautical and Maritime Research Laboratory RR-0172: 32.

Vatten D, Ghaedian R, and Shetty K (2005) Enhancing health benefits of berries through phenolic antioxidant enrichment: focus on cranberry. Asia Pacific Journal of Clinical Nutrition 4(2): 120–130.

Voigt D and Gorb SN (2008) An insect trap as habitat: cohesion-failure mechanism prevents adhesion of *Pameridea roridulae* bugs to the sticky surfae of the plant *Roridula gorgonias*. Journal of Experimental Biology 211: 2647–2657.

Vreeland V, Waite JH, and Epstein L (1998) Polyphenols and oxidase in substratum adhesion by marine algae and mussel. Journal of Phycology 34: 1–8.

Waite JH (1983) Evidence for a repeating 3,4-dihydroxyphenylalanine- and hydroxyproline-containing decapeptide in the adhesive protein of the mussel, *Mytilus edulis* L. Journal of Biological Chemistry 258(5): 2911–2915.

Waite JH and Qin X (2001) Polyphosphoprotein from the adhesive pads of *Mytilus edulis*. Biochemistry 40(9): 2887–2893.

Wegler R and Herlinger H (1963) Polykondensations- bzw. Polyadditionsprodukte von Carbonyl- und Thiocarbonylverbindungen. In: Müller E, Bayer O, Meerwein H, and Ziegler K (eds) Houben-Weyl: Methoden der Organischen Chemie, Band 14/2 Makromolekulare Stoffe, 4th Ed. Georg Thieme Verlag, Stuttgart: pp 193–424.

Young GA and Crisp DJ (1982) Marine animals and adhesion. In: Allen KW (ed.) Adhesion, Vol. 6. Ed. Applied Science Publishers, London: pp 19–39.

Yu M, Hwang J, and Deming TJ (1999) Role of 3,4-dihydroxyphenylalanine in mussel adhesive proteins. Journal of American Chemical Society 121: 5825–5826.

Zhao H and Waite JH (2006) Proteins in load-bearing junctions: the histidine-rich metal-binding protein of mussel byssus. Biochemistry 45(47): 14223–14231.

14 Medical Products and Their Application Range

Jessica Blume and Willi Schwotzer

Contents

14.1 Objectives, Application, and Sources of Medicinal Adhesives

14.1.1 Objectives

Of all dysfunctions of the human body physical injuries are intuitively the easiest to comprehend: if a bone is broken or a tissue ruptured, there treatment will in all likelihood lead to a full recovery of health. It is therefore hardly surprising that wound treatment and healing of broken bones are among the oldest disciplines of medicine. Physical injuries were and are part of daily life because accidents do occur. In addition, ancient medicine was strongly influenced by warfare. In those ages the strength of an army depended very much on the number and on the condition of its soldiers. For this reason military surgeons and their science were held in high esteem. The treatises on wound healing by well-known ancient physicians such as Hippocrates of Cos (*ca.* 460 BC–*ca.* 370 BC) or, e.g., the suggestions of Aristotle (384 BC–322 BC) to stock up on certain medicinal plants prior to going to war are witnesses to the history of medicine (Schlathölter, 2005).

For the ancient physicians the choices of methods of wound and bone fracture healing were very limited. Surgical screws or even sutures were essentially unknown because their successful application requires advanced enabling technologies. Adhesive bonding was therefore the method of choice.

Thus, while methods and materials improved a lot over the historical time frame, the essence and objectives of medicinal adhesives remained very similar. They include

- fixation of the structures and surfaces which are to be bonded;
- mechanical protection of the wound;

- microbial protection of the wound in order to avoid sepsis;
- filling of cavities;
- regulation of the moisture content;
- supporting – or at least not negatively influencing – the healing process.

14.1.2 The State of the Art

Adhesives play an important role in all segments of modern industrial life. Their range of application reaches from the transportation industry (cars, trains, and aircrafts) to electronics and from food packaging to safety devices. New composite materials exhibit properties which were unknown and sometimes even unthinkable only a few decades ago. They consist of combinations of materials held together by tailor-made adhesives. Adhesives are also found in construction applications. Glue-laminated timber is widely used for load-bearing structures (Brockmann et al., 2009).

In light of the ubiquitous use of the adhesive technology in most areas of life it may come to a surprise that, in medical science, it is mainly restricted to external applications (Wheat and Wolf, 2009). External, in a medical context, refers to organs and tissue exposed to the outside world like the skin or the teeth. These areas actually account for close to 90% of all commercial applications. The literature reflects this trend in that also most of the patents point in the same direction.

It is the objective of this chapter to collect and to compare the teachings on the issue of medicinal adhesives.

14.1.3 Historical Sources and Applications of Medicinal Adhesives

Plants were the predominant sources of early medical adhesives and this for several reasons. Firstly, the adhesive material can easily be harvested by either directly collecting the plant secretions or extracting them with water or alcohol. Secondly, herbal adhesives are less susceptible to microbial contamination or, on the contrary even show antimicrobial activity. Herbal adhesives are predominantly based on polysaccharides, resins, or on natural rubber. The polysaccharide-based products – gum arabic as an example – are notoriously water-soluble and therefore of limited use in the presence of body fluids. Adhesives based on natural rubber can be used for skin applications. However, they became replaced by petrochemical products because of the high allergenic potential of natural rubber (Fig. 14.1).

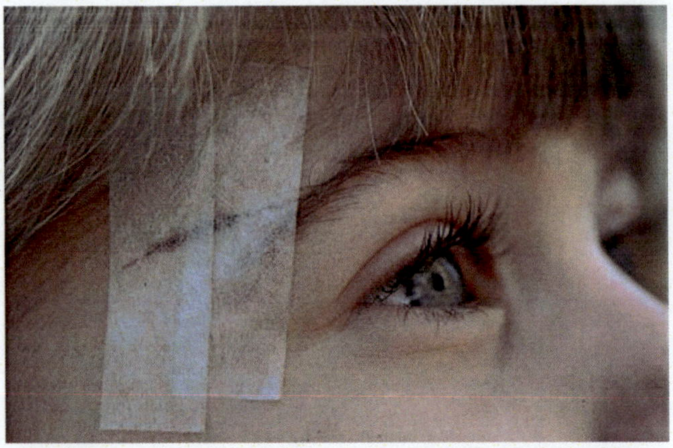

Fig. 14.1 The primary objective of wound treatment is mechanical and microbial protection

Animal glues played an important role in the history of adhesives. A patent of glue made from fish is reported as early as 1754. Adhesives with remarkable adhesive profiles can be produced from animal hide and bones. They are protein-based and can be cross-linked by suitable reagents such as formaldehyde or isocyanates.

Early attempts to use biogenous materials for the bonding of animal tissue date back to 1891. Extracts from chicken blood and mixtures of rosin, gypsum, and pumice stone were tested. These protein containing substances were highly allergenic and never reached importance. The only commercially sizable biogenous adhesives are fibrinogen and fibrinogen/thrombin systems. These systems are widely used in various fields of surgery, e.g., internal organs, skin transplants, ophthalmology, cosmetic surgery and in the area of ear, nose, and throat. They can fully or in part replace sutures, which is very welcomed in connection with sensitive tissues. One of the limiting factors of fibrin adhesives is their high costs. The products are isolated from human blood in processes of very high quality in order to exclude complications. Also, their bond strength is limited which excludes them from load-bearing applications.

Other approaches to biological adhesives include gelatine (Preiss-Bloom and Attar, 2008) as a backbone polymer. Gelatine is biodegradable and contains a large number of functional groups, which can be employed both for adhesion and cohesion. Early gelatine-based adhesives included resorcinol and aldehydes in the cross-linking process, which often entailed histotoxic properties. Necrotization of the bonded tissue will consequently inhibit or even totally prevent the natural healing process.

Table 14.1: Types of medicinal adhesives

Type of adhesive	Remarks
Man-made adhesives	
Alkyl-cyanoacrylates	Surgical applications. Rapid closing of wounds and blood vessels. Excellent adhesion. Bonds are not stable with respect to water; hard bond line; only used in small amounts because of questionable degradation products.
Pressure sensitive adhesives based on thermoplastic synthetic rubber or on acrylates	Widely used for skin adhesives, e.g., adhesive plasters (band-aid®), trans-dermal patches, incision films, etc.
Poly-(methylmethacrylates)	Widely used in dental applications and for the fixation of artificial joints. Problems with a substantial heat of polymerization, toxic components and/or degradation products.
Epoxies	Early attempts in orthopaedic surgery; problems with heat development during the curing process, toxicity of degradation products.
Polyurethanes	Improvement with respect to epoxies but similar problems with heat. Inhibition of healing processes.
Photochemical Tissue Bonding (PTB)	This method employs electromagnetic energy (light) and photosensitizers to produce a tissue seal.
Biological tissue adhesives	
Natural rubber based Pressure Sensitive Adhesives (PSAs)	For external use in wound care; no longer in use because of allergenic potential.
Lithocolle-ossocol	Early attempts of gluing bones with protein-based adhesives; highly allergenic.
Fibrin based	Widely used; comparatively low bond-strength, high-cost.
Gelatine-resorcinol	Problems with bio-toxicity.
Protein-aldehyde	Used in conjunction with sutures.
Polylactone	Used as a sealant in conjunction with sutures.
Peptide based (marine organisms, mussels, etc.)	Promising new systems, still very expensive.
Mussel mimetic catechol-functionalized polyethylene glycol (cPEG)	Synthetic hydrogel-type tissue adhesive mimicking mussel adhesives.

In modern times, adhesives from biological sources have, interestingly, almost disappeared from the range of products used in the fields of wound healing or surgery. Adhesives presently employed in the medicinal field stem both from man-made and biological sources (Belykh et al., 1991; Mobley et al., 2002; Flock and Marchitto, 2005; Heckroth et al., 2007; Bilic et al., 2010). There are also methods on record which bond tissues by means of electromagnetic energy (light) (Kochevar et al., 2008). Table 14.1 summarizes the commonly used systems. It has to be noted that commercial "biological adhesives" are manufactured under strict industrial conditions but are biological by origin (Figs. 14.2, 14.3).

In summary, it is fair to say that there is only a very limited choice of adhesives for applications inside the body. The medical society is therefore in search of adhesives which swiftly and strongly bond body tissue, which will replace mechanical fasteners, which are biocompatible and biodegradable and which, above all, do not negatively interfere with the natural healing processes.

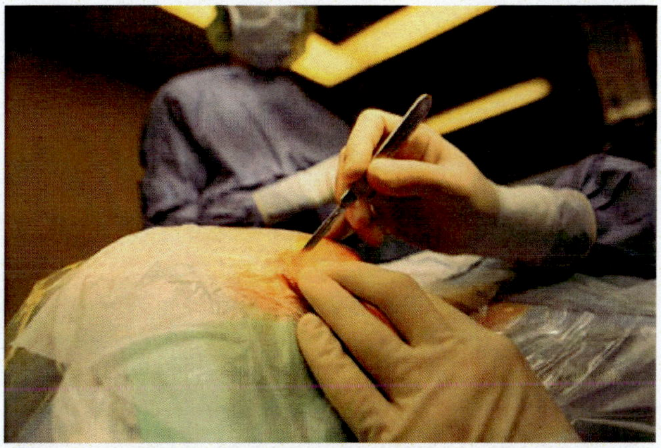

Fig. 14.2 Incision film with anti-microbial properties

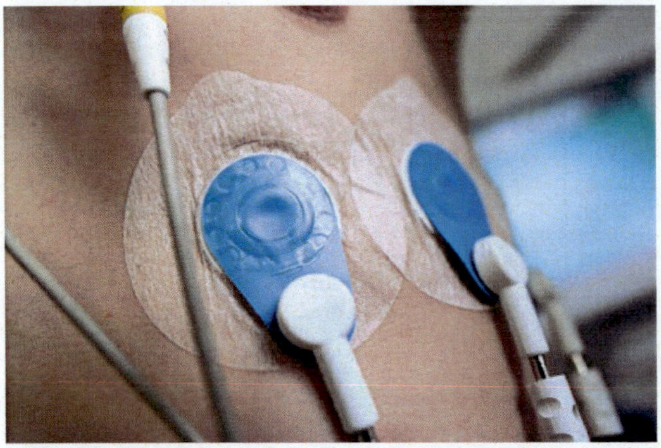

Fig. 14.3 Electrically conductive hydrogel; adhesive with auxiliary function

14.2 Adhesion in Medical Systems

14.2.1 Definitions

For the sake of clarity the terms "adhesive" and "adhesion" will refer to the definitions of DIN EN 923. It specifies an adhesive as *a non-metallic substance capable of joining materials (also called substrata) by surface bonding (adhesion), and the resulting bond possesses adequate internal strength (cohesion).* The adhesive bonding can be permanent or removable. Both, the adhesive and cohesive properties are often deliberately chosen according to the requirements of the bond. With the above definition there is no clear borderline between adhesives and sealants if the sealant binds adhesively to the substratum.

14.2.2 Cohesive Properties

The cohesive strength of an adhesive, i.e., its "inner" strength, is a material property of the bonding substance and independent of the site of application. Consequently, it is characterized by the traditional methods and parameters used in polymer material sciences. Materials for the bond line can be chosen from a vast variety of polymeric substances of biological or man-made nature. The only limitation with respect to medical applications is the required biocompatibility, i.e., the total absence of toxic, allergic, or irritant properties. These requirements apply not only to the original adhesive but also to its degradation products.

For medical adhesives, the same standards as for medical devices (e.g., implants) apply, because adhesives are considered to be medical devices in their own right. The ISO standard ISO 10993-1 provides for a multitude of tests. The choice of test(s) relevant for a specific application depends on the type of tissue contact and contact duration. A summary of the most important tests is given in Table 14.2.

The cohesive strength of the bond line does not have to exceed the properties of the substrata because in any composite material the overall strength is determined by the weakest component. Body tissues differ strongly depending on their functionality. Tendons can withstand loads of up to 100 N/mm^2 (100 MPa) (Sharifi et al., 2009) while typical values for skin are in the range of 0.01–8 MPa.

14.2.3 Adhesion Properties

On a molecular level there is no indication that adhesion in living organisms is any different from adhesion phenomena observed in other areas of applications. Adhesion describes the interaction of surfaces of materials which differ in their chemical nature. Although there is no limit with respect to the topology of the adhering surfaces, all adhesion mechanisms are two-dimensional by definition. One might object that there is ample evidence for mechanical "adhesion" by means of three-dimensional interlocking structures. However, this type of bonding should probably be classified as a macroscopic phenomenon. It is best described in terms of the microscopic mechanic properties of adhesive and substratum, e.g., elastic moduli, angle of interacting planes, etc., rather than by molecular parameters. Perhaps instead of entertaining semantics, more extensive consideration should be given to the diversity of molecular interactions that can occur between surfaces.

The classical adhesion theories consider the interaction between surfaces to be of physical–chemical nature. Such bonds are of varying bond strength and can be classified as follows in approximate order of their strength:

- *Van-der-Waals or dispersion forces* arise from fluctuating dipoles created by the movement of the electrons within the molecular orbital.
- *Dipolar interactions* can be subdivided into two classes: permanent dipoles and induced dipoles. Permanent dipoles exist in bonds between atoms of large differences in electro-negativity. Alternatively, dipoles can be induced when delocalized electron systems are placed in the vicinity of a polarizing dipole.

Table 14.2: Biocompatibility tests for medical adhesives and medical devices (according to ISO 10993-1)

Adhesive categorization			Biological effect											
Nature of body contact		Contact duration: A – limited, <24 h B – prolonged, 24 h – 30 days C – permanent, >30 days	Cytotoxicity	Sensitization	Irritation	Acute toxicity	Subchronic toxicity	Genotoxicity	Implantation	Hemocompatibility	Chronic toxicity	Carcinogenicity	Reproductive toxicity	Biodegradation
Adhesive applied to the surface	Skin	A	X	X	X									
		B	X	X	X									
		C	X	X	X									
	Mucosal membrane	A	X	X	X									
		B	X	X	X	O	O		O					
		C	X	X	X	O	X	X	O		O			
	Breached or compromized surface	A	X	X	X	O								
		B	X	X	X	O	O		O					
		C	X	X	X	O	X	X	O		O			
External adhesive communicating with internal tissues	Blood path indirect	A	X	X	X	X				X				
		B	X	X	X	X	O			X				
		C	X	X	O	X	X	X	O	X	X	X		
	Tissue, bone, dentine	A	X	X	X	O								
		B	X	X	X	X	X	X	X					
		C	X	X	X	X	X	X	X		X	X		
	Circulating blood	A	X	X	X	X		O		X				
		B	X	X	X	X	X	X	X	X				
		C	X	X	X	X	X	X	X	X	X	X		
Adhesives used for implants	Tissue, bone	A	X	X	X	O								
		B	X	X	X	X	X	X	X					
		C	X	X	X	X	X	X	X		X	X		
	Blood	A	X	X	X	X	X		X	X				
		B	X	X	X	X	X	X	X	X				
		C	X	X	X	X	X	X	X	X	X	X		

X tests to be executed according to ISO 10993, *O* tests to be executed according to FDA "Blue Book Memorandum G95-1".

- *Hydrogen bonds* are formed by a hydrogen atom sharing electron pairs from different molecules. They are directional with defined in bond length and bond angle.
- *Covalent bonds* involve direct chemical bonding between the adhesive and the substrata. Although this mechanism is invoked for many reactive adhesives the question whether or not covalent bonds play an important role is controversial.
- *Ionic bonds* are potentially the strongest interactions between molecules and exist between sites of opposite electrical charge. Unlike covalent bonds they do not require a strict geometrical regime. Their bond strength will only depend on the distance between the charges.

Adhesive bonds of strength relevant for technical applications can only exist if the atoms and molecules involved are brought to very close approximation (0.2–0.5 nm). A comparison of the results of theoretical calculations and experimentally determined values of bond strengths reveal an interesting result. It shows that less than 10% of the theoretically possible bond strength is practically achieved. This suggests that only a fraction of the bonding interactions can be realized in a given set-up, possibly because of geometrical constraints.

In the following subsections the types of substrata are described. Despite significant differences there is a common denominator in that all relevant biological surfaces are highly polar and always coated with a layer of water or even immersed in an aqueous medium. Consequently, substances for use as medical adhesives have to exhibit hydrophilic bonding sites with affinity for polar areas.

A first adhesive profile for adhesives suitable for medical application will include the following parameters:

- matrix and its potential degradation products without toxic, allergenic, or irritant properties;
- cohesive strength 0.01–6 MPa (substantially higher for tendons);
- adhesive strength 0.01–6 MPa (substantially higher for tendons).

14.2.4 Medical Bonding Sites

For medical adhesives one of the substrata will always be a body tissue. Tissues are ensembles of cells which are not necessarily identical but which all stem from a common origin. In general, tissues are categorized as epithelial, connective, muscle, and nervous tissues.

Another, and for adhesion purpose perhaps more important way of classification refers to the location of the tissue in the body. It is conducive to discriminate between topical substrata (i.e., on the outside of the body) such as skin, and substrata inside the body. While substrata on the outside of the body are exposed to air, substrata within the body are always submerged in body fluids. The latter add an additional level of difficulty. It is not by coincidence that in the technical world there is a multitude of adhesive application on record for normal environmental conditions while there are very few systems on the market which will operate under water. Biological systems are different in that respect and will hopefully teach us the art of underwater gluing.

14.2.5 Topical Tissues/Organs (Tissues/Organs Exposed to the Outside)

Almost 90% of all medicinal adhesives are used in topical applications in the form of patches, other wound dressings or as structural dental adhesives and fillings. Topical tissues, by definition, are exposed to the outside world. They are built as to withstand many of the environmental influences and they are "dry" in the sense that their moisture content correlates with the ambient humidity. The cells of topical tissues often are "dead" in the sense that they no longer have a metabolism. Consequently, adhesives which bond to topical tissues are mostly man-made substances which are also known and used in other areas of adhesive bonding. Pressure sensitive adhesives are widely used for the bonding to skin tissues. Synthetic petrochemical raw materials such as thermoplastic rubbers and hydrocarbon tackifiers, because of their inertness, are preferred over bio-based products such as natural rubber.

The art of developing adhesives for topical tissues is to control the adhesive and cohesive forces such as to match the purpose. They have to cope with body fluids such as sweat or other materials such as sebum or electrolytes surfacing at the topical layer. Obviously, they must not emit chemicals which could negatively influence the tissue or even the entire body.

14.2.5.1 Skin

In patches and wound dressings, the adhesive bonding occurs on the healthy skin surrounding the wound bed. Skin consists of three layers: the epidermis, dermis, and subcutis. While the epidermis is an epithelial tissue, dermis and subcutis are part of the connective tissue. The epidermis is characterized by a stratified, keratinized squamous epithelium with a thickness varying between 30 µm and 4 mm. The stratum corneum, the outer layer of the epidermis, is composed by dead corneocytes without any cell organelles but filled intracellularly with keratin and multilamellar, epidermal lipids placed in the cell interstitium. The membrane of the corneocytes is characterized by its inflexible cornified envelope and is made up of cross-linked proteins like loricrin, involukrin, and SRPs1-3. The corneocytes are attached to each other *via* desmosomes. Aged skin is thinner and dryer, caused by reduced cell division and equally reduced activity of sebaceus glands. The epidermis is penetrated by hair shafts and skin gland ducts, while hair roots and terminal secretory parts of skin glands are placed in the dermis. Without the multilamellar epidermal lipids the human body would loose 20 liters water per day (Melnik, 1990). The human stratum corneum contains mainly ceramids (40%), cholesterol (25%), free fatty acids (30%), and small amounts of phospholipids, glycosylceramide, cholesterolsulfate, sterole, triacylglycerole, sterolester, squalene, and *n*-alkane in different percentages depending on age, gender, and season (Lampe et al., 1983). The cross linked proteins of the cornified envelope in turn are connected with the multilamellar epidermal lipids.

Sweat consists of 99% water and 1% of a mixture of minerals, urea, lactate, lipids, amino acids, and also some sugars.

Sebaceus glands, which are placed next to the hair follicle in the dermis, secrete sebum along the hair shaft. Sebum is made of 41% triglycerides, 25% wax monoesters, 16% free fatty acids, and 12% squalene (Cheng and Russell, 2004).

Adhesives for skin application therefore have to be able to attach to substrata with a permanently changing surface containing mainly lipids as well as water. The adhesive bonding to the epidermis as substratum is strongly influenced by the biochemical composition of the stratum corneum and by the presence (or absence) of sweat and sebum.

14.2.5.2 Teeth

Adhesives in dental applications are mainly used as structural adhesives, for fillings and inlays or for gingival reconstitutions. Most of the bonding processes can also occur in more or less dry situations as the saliva is carefully removed by the dentist.

Teeth are developed during fetal development from layers of specialized epithelial tissue with exoskeleton structure and connective tissue. The mature tooth consists of enamel, dentine, pulp, and tooth cementum, surrounded with gingiva. According to the depth of the cavity in the tooth, adhesion occurs on more than one substratum with different biochemical compositions. However, most of the dental adhesives bind to hydrophilic dentine or in the case of very small cavities to the enamel. Enamel, the tooth outer layer, is the hardest "tissue" in the human body and is composed to 95% of crystalline carbonated hydroxylapatite covered by a thin protein layer. Underneath the enamel, the living dentine is laying, which is mineralized connective tissue. It contains collagen fibers embedded in 70% carbonated hydroxylapatite (Ten Cate, 1998). Nerves, blood vessels, connective tissue, and odontoblasts, which generate the dentine, are part of the pulp. Adhesives which are applied on dentine close to the pulp, must tolerate fluids during the bonding process as dentine liquor is leaking out from the pulp almost permanently (Söderholm, 1995). The tooth cementum is softer than the enamel and dentine, because it contains only 45% carbonated hydroxylapatite and a higher fraction of collagen fibers.

Dental adhesives must be very durable and resistant to saliva. Saliva mainly consist water, but also polysaccharides, enzymes for first digestion of carbohydrates, electrolytes, or urea. Another problem for adhesive bonding can be the slightly acidic pH range between 6.0 and 7.4 during and after meals.

14.2.5.3 Gingiva

The gingiva, as a part of the mouth mucosa, is an epithelial tissue related to the skin, although the stratum corneum is much thinner and the subcutis is missing. The gingival stratum corneum contains less keratin and the surface is at all times exposed to the saliva. Therefore, gingiva can only partially be referred as an external tissue. In the mouth mucosa, no sebaceus glands can be found, thus the gingiva as a substratum is less rich in lipids compared to the skin.

14.2.6 Internal Tissues/Organs

14.2.6.1 Eye

In human eyes, medicinal adhesives are mainly used for treatments of injured cornea as well as retina. The eye ball is composed by all four tissue types: epithelial, connective, muscle, and nervous tissue. The connective tissue is important for all backing functions. The epithelium always constitutes the outer layer, including the stratum pigmentosum of the inner retina, which is in contact with the vitreous body. Transmission of images occurs at the interface between the epithelial part of the retina and the underlying nervous tissue. The outer skin in front of the lens consists of the lucent cornea, composed of epithelial cells without stratum corneum and a stroma with collagen fibers and glycoproteins as filler. Ciliary muscles play an important role, whenever the image needs to be focused.

Consequently, adhesive bonding in ophthalmology is mainly applied to epithelial tissue components.

14.2.6.2 Connective Tissues: Bones, Cartilages, and Ligaments

Although bones, cartilage, and ligaments look different, they are also part of the connective tissue as they are all derived from the mesenchym. While epithelial, muscle, and nerve tissues are characterized by their cellular structure, connective tissue contains, in addition to the cells, an extracellular substance called "extracellular matrix" (ECM). The ECM can be a fluid, a semi fluid, or a solid. It determines the mechanical properties of bones, cartilage, and ligaments by means of its varying composition and degree of mineralization. Adhesives used in orthopaedic applications must therefore cope with fluids during the bonding process.

The characteristics of cartilage depend on the composition of its ECM. Cartilage tissues that show a high compressive strength are visco-elasticly deformable and resistant to high shear forces. Unlike bones, cartilage does not contain any blood vessels. The cells of cartilage, the chondrocytes, secrete the proteins of the surrounding ECM. They are composed of collagen types 1 and 2, elastin, proteoglycans, and glycosaminoglycans.

Up to 50% of the bones consist of the mineral "calcium hydroxylapatite". The degree of mineralization is therefore lower as compared to the enamel of teeth or the dentine. Osteoblasts, the cells which are instrumental for the bone formation, secrete the proteins and minerals for the bone ECM (Osteoide). Mineralization occurs when the Osteoide layer is growing thicker: Binding of calcium ions to the various ECM proteins is triggered by the activity of the alcaline phosphatase.

Adhesives for orthopaedic applications, especially for the restoration of bone fractures, should preferably be injectable and fast curing in order to rapidly resume mechanical loads. They must not develop extensive heat upon curing to avoid necrosis. They need to be absorbable similar to other wound healing systems in order not to interfere with the biological processes.

14.2.6.3 Cardiovascular System (Blood Vessels)

Man-made medicinal adhesives like cyanoacrylates are classically used on injured blood vessels to stop uncontrolled loss of blood. They mimic the process of blood coagulation, where platelets, fibronectin, and the von Willebrandt factor (vWf) form blood clots and occlude the disrupted vessel wall.

Composition of arteries and veins can be divided into three layers as follows:

- the inner layer: endothelial cells, which are in direct contact with the blood and a thin basement membrane derived from connective tissue;
- the middle layer: smooth muscle cells with elastic fibers in between (on top of the basement membrane);
- the outer layer: cells and ECM of the connective tissue.

In artery walls the three layers are more prominent as compared to veins. Capillary walls have no smooth muscle cells but only endothelial cells.

In comparison with the natural occlusion and the following regeneration, the bio-degradation of cyanoacrylates is all but optimal. Application of cyanoacrylates leads to enhanced acute inflammation as macrophages react on the foreign body. After 60 days, the cyanoacrylate is still present in all tested blood vessels (Kaplan et al., 2004) and more scar tissue formation can be detected in comparison to normal wound healing. Nevertheless, cyanoacrylates are widely used for lack of alternatives.

14.2.6.4 Muscles

In skeletal muscles, muscular cells (also called "fibers") and connective tissue are in close functional association. The muscle is surrounded by a thick layer of collagen-rich connective tissue (muscle fascia) and a subjacent more irregular connective tissue (epimysium). Furthermore, the fibers are arranged in bundles, which are separated by connective tissue. The muscle is bonded by means of tendons to the bone. Tendons are connective tissues in their own right: dry tendons are composed of 86% collagen, 2% elastin, 1–5% proteoglycans, and 0.2% inorganic components such as copper, manganese, and calcium (Lin et al., 2004).

Table 14.3: Specifics of tissue surfaces

Tissue type	Functional groups on surface	Specific requirements for applied adhesive
Skin	Ceramides: –OH, $-NH_2$ Fatty acids: –COOH Proteins: –OH, –COOH, $-NH_2$, $-CONH_2$, –S–S– Triglycerides: –COOR	Resistant to water (sweat) and hydrophobic components ($>C_{24}$)
Teeth	Proteins: –OH, –COOH, $-NH_2$, $-CONH_2$, –S–S– Carbonated hydroxylapatite: –OH, PO_4	Resistant to water (saliva)
Bones	Glycoproteins: –OH, CHO, –S–S– Proteins: –OH, –COOH, $-NH_2$, $-CONH_2$, –S–S– Calcium hydroxylapatite: –OH, PO_4, Ca^{2+} Phospholipids: –OH, –COOH, $-NH_2$, PO_4	"Under-water" bonding process, mineralization possible, injectable, resorbable
Blood vessels	Phospholipids: –OH, –COOH, $-NH_2$, PO_4 Proteins: –OH, –COOH, $-NH_2$, $-CONH_2$, –S–S–	"Under-water" bonding process, injectable, resorbable

Adhesive applications can be foreseen between muscle and tendons or tendons and bone. Application of adhesives between muscle cells would appear to be more critical because of the risk of an obstruction of the impulse transmission and thus the activity of the muscle.

14.2.7 Summary of Parameters for Adhesive Bonding on Human Tissues

As described above, there is a multitude of biological surfaces to which medicinal adhesives can bond. It is to be expected that the biochemical nature of these surfaces will strongly influence the parameters required for successful bonding and healing. The functional groups on the surfaces of the tissue do play an important role with respect to the bonding mechanism. Table 14.3 summarizes the predominant functional groups on the surface of tissues and the resulting requirements for successful adhesion.

Recent publications show that tissue adhesion to biomaterials (adhesives) can strongly be improved by combining different functional groups to yield hybrid adhesive materials and/or by adjusting the hydrophilic–hydrophobic balance (Nakabayashi et al., 1991). Furthermore, the mobility of those functional groups within the adhesive surface may play an important role as cells rearrange the connection between their cytoskeleton and plasma membrane proteins (e.g., integrins) during the adhesion process (Dogic et al., 1998). Biomaterial or adhesives consisting of more than one polymer (and therefore carry a blend of different functional groups) reflect in a better way the natural situation described by the fluid mosaic model for cell membranes or the complex models for the extracellular matrix.

14.3 The Healing Process

Treat the whole (patient) and not the hole (in his body).

The ultimate goal of medicinal adhesive bonding is not the closing of but the healing of the wound. The latter can only be performed by the body and only if all the necessary conditions are fulfilled. Primary fixation of the tissue or bone fragments, avoidance of infection and regulation of the moisture content are necessary but not sufficient conditions for recovery. Above all, the bond line must not interfere with the healing process. Ideally, the adhesive will be supportive by assuming a scaffolding function, delivering drugs that expedite healing and acting as an antimicrobial shield.

14.3.1 Wound Healing

Upon injury to a tissue, a highly complex biochemical process is initiated which leads to the healing of the wound. In a simplified model of the healing of skin injuries three phases can be distinguished. They are schematically shown in Fig. 14.4.

The first phase which immediately follows the injury of the tissue is referred to as coagulation and irritation step. Its main aim is to stop the bleeding, which is achieved by aggregation of blood platelets (thrombocytes). During a second phase, also called the inflammatory phase, secretion of exudate takes place which cleans the wound from pathogenic germs and flushes out foreign matter and cell debris. The action of the exudate is supported by immuno cells migrating into the wound area. They produce and secrete proteases for disintegrating the ECM. During this phase of high activity which is accompanied by swelling

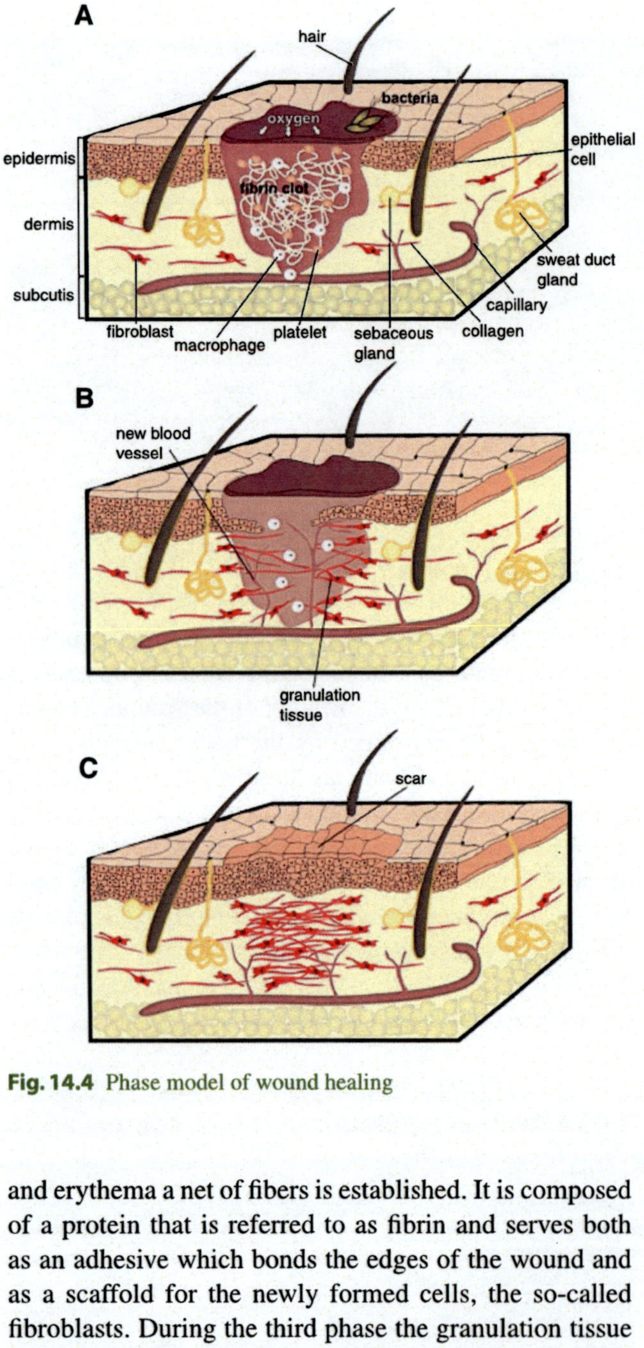

Fig. 14.4 Phase model of wound healing

and erythema a net of fibers is established. It is composed of a protein that is referred to as fibrin and serves both as an adhesive which bonds the edges of the wound and as a scaffold for the newly formed cells, the so-called fibroblasts. During the third phase the granulation tissue is formed, followed by the growth of new blood vessels and contraction of the wound.

The reconstitution of the mechanical tensile strength of rat skin wounds was investigated by Hudak et al. (2004). Over a period of 7 days measurements were performed at intervals of 24 h. The largest increase of the tensile strength was observed within the first 24 h. This corresponds to the inflammatory and the beginning of the proliferative phase, i.e., the phase during which the fibrin network is formed.

The healing of bone fractures follows to a greater or lesser extent the same pattern. The degradation of the injured tissue occurs simultaneously with the formation of the new bone at the so-called cutting cones. In secondary fracture healing, a soft callus is formed from unmineralized connective tissue in order to stabilize the fracture zone. Consequently, bridges are formed between the fragments of the fracture (Schenk and Willenegger, 1977; Hahn, 2007). The callus serves as a scaffold for the ingrowing fibroblasts, which then differentiate themselves to become osteoblasts.

14.3.2 A Critical View on Existing Medicinal Adhesives

In the technical world adhesive bonds are an alternative to and often in competition with mechanical fasteners, and vice versa. Medical applications are no exception to this rule. With respect to surgical sutures the existing medicinal adhesive systems seem to exhibit one or more of the following disadvantages:

- insufficient initial and/or final bond strength;
- inhibition of the natural wound healing processes;
- high costs.

The lack of adhesive strength comes to no surprise since the adhesive forces have to be established toward a difficult substrate (tissue) under difficult conditions (surrounded by liquid). Even in the area of technical adhesives there are not many systems on record which would operate satisfyingly under such conditions. In addition, most of these systems are not suitable for medical applications because of toxicological or allergenic reasons.

The remaining problems, i.e., the inhibition of wound healing and the costs of the existing systems, may be interrelated. According to the prevalent practice adhesives are applied such as to coat the entire surfaces of the substrates. This is preferred because mechanical forces are best dissipated by a large surface. For medical applications such a complete coating of the injured tissue will be totally counterproductive to healing because the breeding ground for new cells will be completely clogged. In order to leave ample space to grow for the new cells (fibroblasts or osteoblasts), the adhesive zones must be reduced to small spots like the fibrin or collagen fibers in the natural healing process. This requirement entails the use of tissue adhesives with very high adhesive properties since the spot-like surfaces have to carry the entire load. Some of the biological "super glues" that were presented earlier in this book appear to be suitable for this purpose. And although their costs by unit weight may be high, their spotty application will keep the amounts and thereby the costs within reasonable limits.

14.3.3 A Blueprint for Medicinal Adhesives

Summarizing the above findings a requirement profile of a medicinal adhesive could be:

- cohesive strength in the range of a few MPa;
- adhesive strength in the range of 0.01–6 MPa;
- minimal interference with the tissue;
- non-toxic, non-irritant, non-allergenic;
- scaffolding properties;
- preferably antimicrobial.

One possibility to satisfy the above conditions is the use of a double-sided adhesive tape consisting of a layer of bio-absorbable foam equipped with a spotty pattern of biocompatible adhesive on both sides. The foam is characterized by an open-cell structure such as to permit blood vessels to interpenetrate. Additionally, its surface properties must foster the formation and growth of fibroblast and assist in their proliferation by providing a scaffold. Topical foams of that kind are known (Blume et al., 2010a).

The adhesive must both bond to the foam and to the tissue surface (Fig. 14.5). It must perform under the natural conditions, viz., immersed in body fluids. Such a compound will, in all likelihood be of highly polar nature with abundant sites for hydrogen bonding. The strength of the bonds to the body tissue and to the foam should be comparable to the cohesive strength of the foam. Printing methods to apply such an adhesive to the surface of a foam in a spotty pattern are known in the industry (Blume et al., 2010b).

Adhesives suitable for the purpose can be man-made or from biological sources. The biological adhesives are of special interest since they were designed by nature to adhere to the medicinal surfaces described above. Moreover, most of them seem to perform superbly in environments encountered in wounds. By means of a combination between man-made bio-absorbable foam and very small quantities of high performance biological adhesives a reasonable economic structure of such a system can be expected.

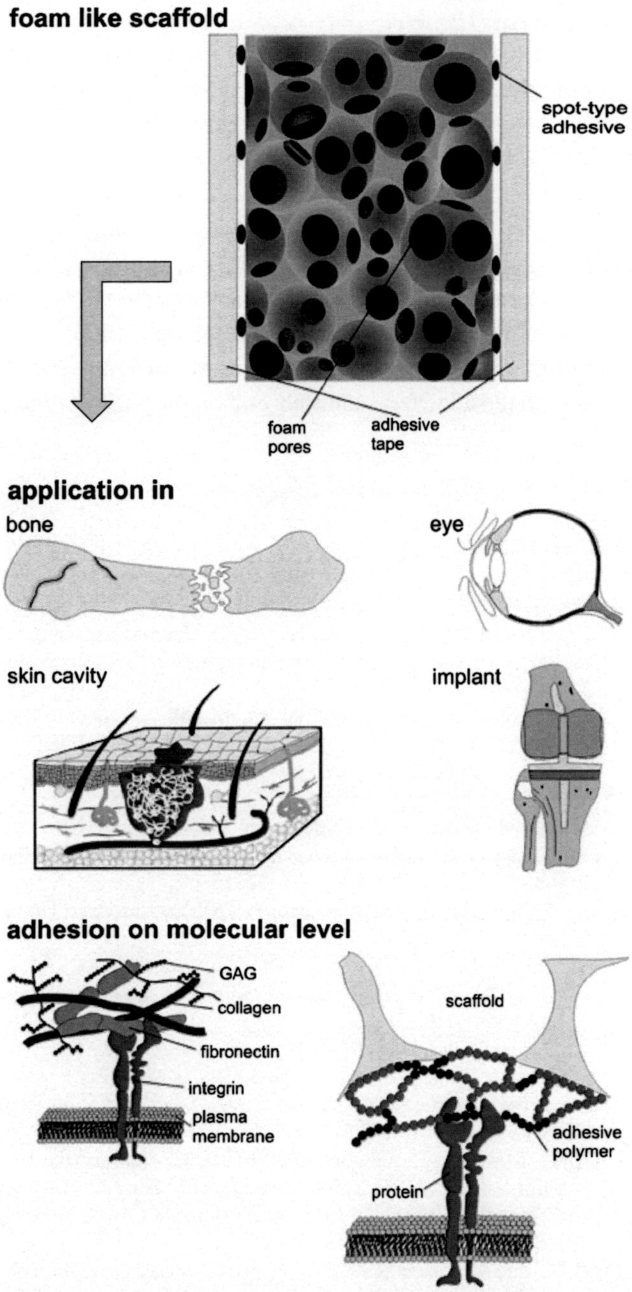

Fig. 14.5 Scaffold with double-sided adhesive tape for internal medical applications

As an alternative to using combination of foam equipped with an adhesive such a foam can be created *in situ* from an adhesive formulation, provided the formulation in the unreacted form is bio-compatible, that the foaming process will not negatively influence the wound tissue and that the foam created in situ will support the healing process.

14.4 Conclusion

Medicinal adhesives are widely used for topical applications. However, their use inside of the body is still very limited despite of the prospective advantages they could offer. Several reasons contribute to this situation in which the state of the art of medicinal adhesive technology is not yet up to the expectations of the potential users, i.e., the surgeons and the physicians. It is suggested to let nature influence the design strategies for medicinal adhesives more extensively than it did in the past, both with respect to the adhesive materials and the way they are put to action.

References

Belykh SI, Gigauri VS, Davydov AB, Keshelava VV, Mlynchik VE, Utyamyshev RI, and Firsova EV (1991) Material for connecting members for inner soft tissue and organs. USA Patent No. 5,011,493.

Bilic G, Brubaker C, Messersmith PB, Mallik AS, Quinn TM, Haller C, Done E, Gucciardo L, Zeisberger SM, Zimmermann R, Deprest J, and Zisch AH (2010) Injectable candidate sealants for fetal membrane repair: bonding and toxicity in vitro. American Journal of Obstetrics and Gynecology 202(1): 85–89.

Blume J, Buser S, Dobmann A, and Zimmermann E (2010a) European Union Patent No. 2009012613.

Blume J, Dobmann A, and Schwotzer W (2010b) European Union Patent No. 10152528.

Brockmann W, Geiss PL, Klingen J, and Schröders B (2009) Adhesive Bonding – Materials, Applications and Technology. Wiley-VCH Verlag GmbH, Weinheim.

Cheng JB and Russell DW (2004) Mammalian wax biosynthesis. II. Expression cloning of wax synthase cDNAs encoding a member of the acyltransferase enzyme family. Journal of Biological Chemistry 279(36): 37798–37807.

Dogic D, Rousselle P, and Aumailley M (1998) Cell adhesion to laminin 1 or 5 induces isoform-specific clustering of integrins and other focal adhesion components. Journal of Cell Science 111(Pt 6): 793–802.

Flock ST and Marchitto KS (2005) Progress towards seamless tissue fusion for wound closure. Otolaryngologic Clinics of North America 38(2): 295–305.

Hahn NE (2007) Properties and degradation of a new bioresorbable bone glue: results of an animal experimental analysis on the rabbit femur condyle (Original German). Medizinisches Zentrum für Chirurgie, Justus-Liebig-Universität Gießen: 132 pp.

Heckroth H, Köhler B, and Dörr S (2007) Medizinische Klebstoffe für die Chirurgie. European Union Patent No. EP 2011 808 A1.

Hudak R, Gal P, Toporcer T, Kilik R, and Sabo J (2004) Mechanical tensile strength tests of rat skin wounds during the first seven days of the healing process. Journal of Metals, Materials and Minerals 14(1): 11–19.

Kaplan M, Bozkurt S, Kut MS, Kullu S, and Demirtas MM (2004) Histopathological effects of ethyl 2-cyanoacrylate tissue adhesive following surgical application: an experimental study. European Journal of Cardio-Thoracic Surgery 25(2): 167–172.

Kochevar IE, Redmond RW, and Azar D (2008) Photochemical tissue bonding. USA Patent No. US 7,331,350 B2.

Lampe MA, Burlingame AL, Whitney J, Williams ML, Brown BE, Roitman E, and Elias PM (1983) Human stratum corneum lipids: characterization and regional variations. Journal of Lipid Research 24(2): 120–130.

Lin TW, Cardenas L, and Soslowsky LJ (2004) Biomechanics of tendon injury and repair. Journal of Biomechanics 37(6): 865–877.

Melnik BC (1990) Biochemie und Pathobiochemie des epidermalen Lipidstoffwechsels. Georg Thieme Verlag, Stuttgart

Mobley SR, Hilinski J, and Toriumi DM (2002) Surgical tissue adhesives. Facial Plastic Surgery Clinics of North America 10: 147–154.

Nakabayashi N, Nakamura M, and Yasuda N (1991) Hybrid layer as a dentin-bonding mechanism. European Journal of Esthetic Dentistry 3(4): 133–138.

Preiss-Bloom O and Attar I (2008) Hemostatic materials and dressing. USA Patent No. US 2008/0213243 A1.

Schenk RK and Willenegger HR (1977) Zur Histologie der primären Knochenheilung. Modifikation und Grenzen der Spaltheilung in Abhängigkeit der Defektgröße. Unfallheilkunde 80: 155–160.

Schlathölter M (2005) Geschichte der Theorie und Praxis der Wundheilung unter besonderer Berücksichtigung des 19. und 20. Jahrhunderts. Universität Münster, Fachbereich Medizinische Fakultät.

Sharifi D, Kazemi D, and Latifi H (2009) Evaluation of tensile strength of the superficial digital flexor tendon in horses subjected to transcutaneous electrical neural stimulation therapeutic regimen. American Journals of Applied Sciences 6(5): 816–819.

Söderholm KJ (1995) Does resin based dentine bonding work? International Dental Journal 45(6): 371–381.

Ten Cate AR (1998) Oral Histology: Development, Structure, and Function, 5th Ed. Mosby-Year Book, St. Louis.

Wheat JC and Wolf JS (2009) Advances in bioadhesives, tissue sealants, and hemostatic agents. Urologic Clinics of North America 36: 265–275.

15 Fibrin: The Very First Biomimetic Glue – Still a Great Tool

James Ferguson, Sylvia Nürnberger and Heinz Redl

Contents

15.1 Introduction

The phenomenon of blood clotting fascinated even the very earliest of philosophers and scientists. They marveled at the fact that the body could repair itself, stopping blood flow and healing its own defect. The idea of using the clotting mechanism to heal and close wounds also seems to have been a dream of the first medical doctors. Galen (129–199 AD) observed the clotting process in more detail and described "fibrae" or "threads" in both circulating blood and in clots and in 1666 Malpighi found "long tough threads" and "a nerve-like network of threads" in washed clots. These "threads" or fibers would later be identified as fibrin, the substance that is the end product of the coagulation cascade and that forms the basis of the blood clot. In 1905, the modern coagulation cascade as we know it was consolidated by Morawitz.

Shortly, after 1909, Bergerl was the first to demonstrate the practical use of fibrin powder. Grey (1915) and Harvey (1918) used fibrin to control bleeding from par-

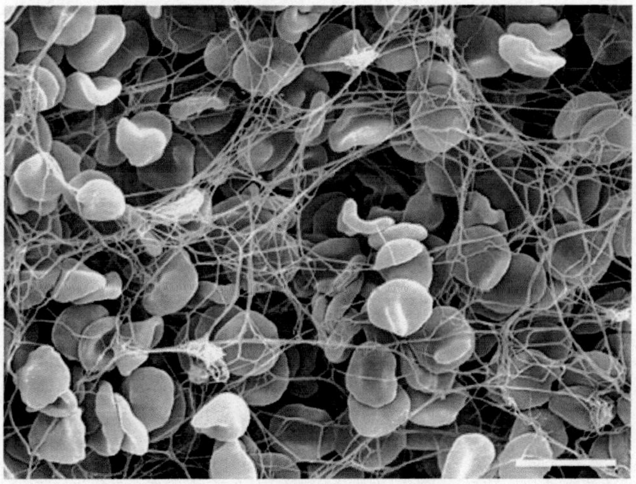

Fig. 15.1 Fresh blood clot in electron microscope showing erythrocytes braided by fibrin fibers. Scale bar: 10 μm

enchymatous organs. Decades later, in 1944 Tidrick and Warner as well as Cronkite et al. reported on plasma fibrinogen clotted with thrombin used for skin grafts in burn injuries in soldiers.

These early attempts already showed how promising fibrin glues and sealants could be in wound closure and hemostasis and highlighted the basic advantages of fibrin as a hemostat, including excellent tissue tolerance, improved wound healing and complete *in vivo* resorption. However, the inadequate strength and stability of early glues ensured that fibrin sealants would not be widely accepted until decades later.

15.2 Mechanisms

The body reacts to blood loss after injury with the clotting of blood, a process that results from the interaction of blood platelets, blood plasma, and damaged tissue (Fig. 15.1). This interaction triggers a cascade of enzymatic reactions, activating a string of so-called coagulation factors that are identified with the Roman numerals I to XIII. The final stage of the coagulation cascade is the activation of the glycoprotein fibrinogen (factor I) by the enzyme thrombin (factor II). This activated protein, fibrin, is essentially what makes up the fascinating "threads" seen by early doctors and forms the basis of the blood clot.

The fibrinogen molecule is composed of two outer D domains and a central E domain. Each D domain has three polypeptide chains; an alpha chain, a beta chain and a gamma chain (Fig. 15.2). When thrombin reacts with fibrinogen it causes the proteolytic cleavage of peptides from these polypeptide chains and initiates the polymerization of the resulting fibrin molecules to strands of fibrin. These strands continue to polymerize and form a network of fibrin fibers that comprise the blood

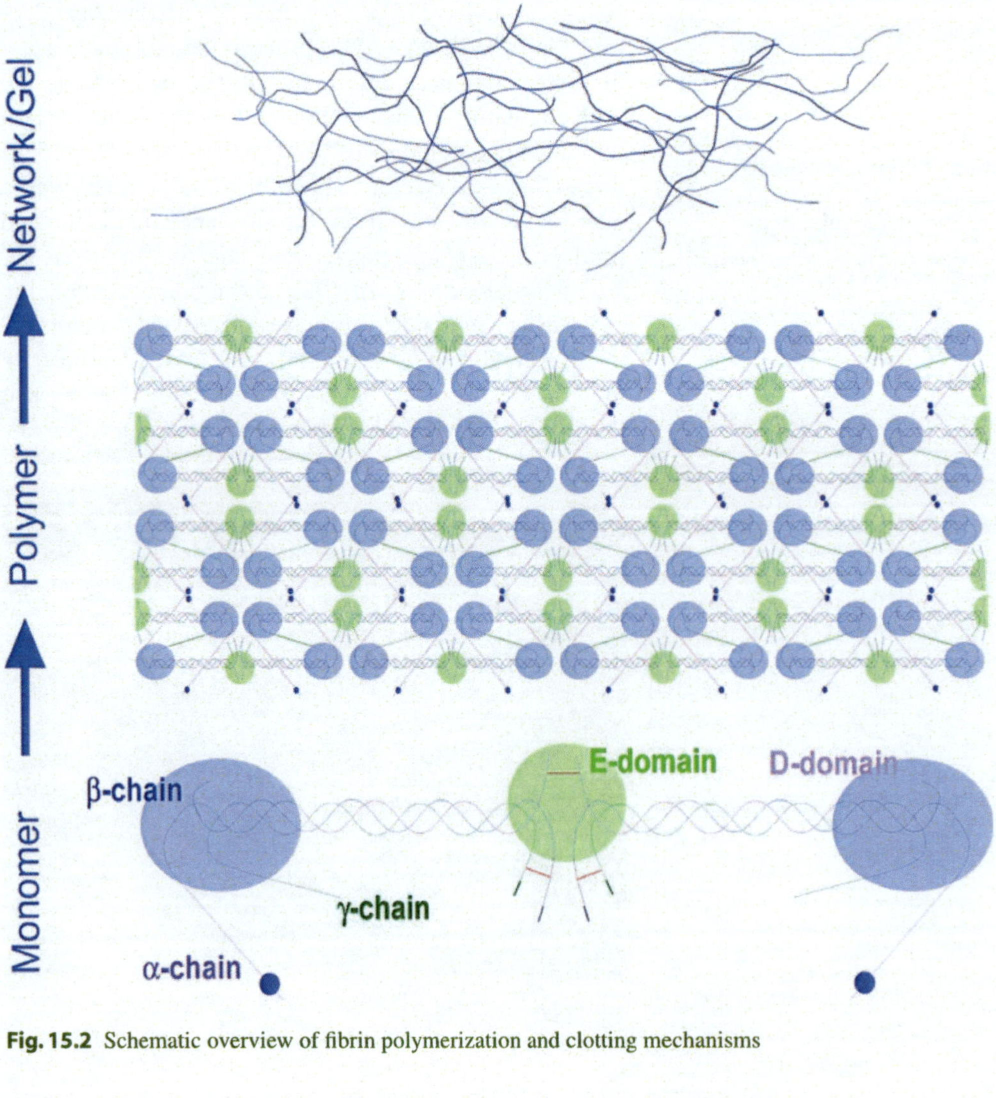

Fig. 15.2 Schematic overview of fibrin polymerization and clotting mechanisms

clot (Fig. 15.1). In the further course of the process, the activated factor XIII causes a crosslinking of the fibrin monomers (Lorand, 1972; Schwartz et al., 1973). Fibrin-γ-chains are completely cross-linked within a few minutes after fibrin polymerization begins. The process of fibrin α-chain cross-linking takes up to two hours with about 70% of cross-linking taking place within 10 min. Calcium is necessary for the reactions and, in order to achieve an optimum rate of cross-linking, $CaCl_2$ concentration must be at least 5 mmol/l (Seelich and Redl, 1979). With time, an increasingly rigid clot is formed, which is later dissolved and absorbed during the wound healing process.

The importance of fibrin cross-linking lies in the resulting increase in clot rigidity (Shen et al., 1974). The elastic modulus, tensile strength, and adhesiveness are functions of fibrinogen concentration (Sierra, 1993; Khare et al., 1998). Investigations by Duckert and Nyman (1978) show that the presence of FXIII also causes collagen to crosslink with fibrin, thus providing adhesive strength when attaching to tissues.

Factor XIII also cross-links fibrin to plasma-derived fibronectin, a glycoprotein which is important for cell adhesion (Danen and Yamada, 2001) and growth, an all important factor for later healing and tissue regeneration. Furthermore, α_2-antiplasmin and plasmin activator inhibitor 2 (PAI-2) which both help to control subsequent fibrinolysis of the clot (Mosesson et al., 2001) are also cross-linked to fibrin by FXIIIa.

15.2.1 Role of Fibrin in Wound Healing

The fibrinopeptides released when fibrinogen is polymerized to fibrin as well as the fibrin degradation products that are released during fibrinolysis act to attract and activate leukocytes. Fibrin degradation products are also essential for the chemotaxis (Kay et al., 1973) and recruitment and activation of macrophages. Macrophages are an important part of the cell population in the wound because they contribute growth factors as well as help with fibrin degradation.

The replacement of the wound with new tissue growth relies on both a viable blood supply and cells that produce new tissue. In 1960, Banerjee and Glynn first demonstrated that implanted fibrin clots are invaded by new capillaries and tissue producing cells, or fibroblasts.

The fibrin network produced under physiological conditions serves as a matrix for the ingrowth of fibroblasts and the formation of collagen fibers (Beck et al., 1961; Ross, 1968). Dinges et al. (1986) determined in a granulation tissue model that fibrin sealant significantly stimulated fibroblast migration. According to Bruhn and Pohl (1981), fibroblast proliferation is stimulated by the presence of factor XIII, while there is an inhibitory effect on epidermal cell proliferation (Hashimoto and Marks, 1984). According to Kasai et al. (1983), crosslinked fibrin rather than factor XIII as such is essential for cell proliferation.

Thrombin remains bound to fibrin within the clot. The effects of thrombin on wound healing are manifold and are a vital part of wound repair aside from mere platelet activation. Thrombin itself has mitogenic properties in cell cultures, such as long-lasting hormone-like influence on fibroblast proliferation (Pohl et al., 1979; Gugerell et al., 2010). Furthermore thrombin is involved in transformation of factor XIII to XIIIa, conversion of fibrinogen to fibrin, prostaglandin production, and activation of protein C (Gustafson, 1984). Moreover, partial structures of thrombin act to improve wound healing (Carney et al., 1992; Stiernberg et al., 1993).

15.2.2 Degradation

A major advantage of fibrin sealant as a tissue adhesive, especially compared to synthetic glues, is its total biocompatibility and the relatively swift degradation that occurs in the body. The rate of sealant degradation depends on the fibrinolytic activity in the area of application, the thickness of the sealant layer, which in most cases should be as thin as possible as excessively long sealant persistence may not be desirable (Schlag and Redl, 1986), and the amount of plasminogen, the proenzyme of plasmin, that degrades fibrin, in the clot.

The rate of degradation of a fibrin sealant *in vivo* can be tailor-made to fit the specific wound healing situation by adding an antifibrinolytic agent (Pfluger and Redl, 1982). Studies have demonstrated that aprotinin, a natural protease inhibitor, is superior to synthetic antifibrinolytic agents (Redl et al., 1982b; Stemberger et al., 1982). For example tranexamic acid (tAMCA) (Seelich and Redl, 1979) was shown to cause convulsive seizures following subdural application in a rat model (Schlag et al., 2000) and is contraindicated in neurosurgical operations. Another approach to prevent premature fibrinolysis is to remove plasminogen from the fibrinogen component (Tse et al., 1993). Since the proteolytic activity in wounds is not only related to plasminogen/plasminogen activator, but also dependent on leukocyte proteases, elastase/cathepsin G inhibitors may also be used to prevent premature clot lysis (Redl et al., 1997).

15.3 Clinical Use of Fibrin Sealants

The clinical use of fibrin sealants can be divided into three main groups. They can be used as:

- hemostats, which stop bleeding
- sealants, for example to close gas leaks from the lung
- adhesives/glues, which allow tissue surfaces to be attached for example in skin transplantation

The tissue sealant/adhesive material should also:

- support wound healing, or at least not interfere with the process
- allow convenient application
- be safe
- be biodegradable

The first commercial fibrin sealant from human plasma was produced in 1978 (Immuno AG, now Baxter Healthcare) and has proven to be effective in hemostasis, sealing of tissue, and wound healing. It is a two-component system. The first component, the sealer protein, contains highly concentrated fibrinogen and factor XIII (FXIII) in addition to other plasma proteins such as albumin and cold-insoluble globulin (fibronectin). The second component, the hardener, is a solution of thrombin and calcium chloride. A fibrinolysis inhibitor is generally added to one of the two components.

This product may well be regarded as the conceptual prototype of fibrin sealant products now approved or under development. More than 2000 papers have been published and an early overview of work with fibrin sealants was provided by two multi-volume monographies by Schlag and Redl (1986) and Schlag et al. (1986 as well as 1994).

There are now several commercially available fibrin sealants: Tisseel® (Baxter Healthcare, Illinois), Artiss® (Baxter Healthcare), Beriplast® (CSL Behring, Pennsylvania), Evicel® (Omrix Biopharmaceuticals Ltd., NY & Ethicon Inc., New Jersey) and Quickseal® (Ethicon Inc., New Jersey). Several other fibrin sealants have been or are currently being developed.

Apart from the highly concentrated fibrinogen obtained by the cryoprecipitation of pooled plasma, fibrin based sealants can also be derived from single donor or autologous sources (Saltz et al., 1991). These "homemade" sealants have typically less favorable sealing properties (Hamm and Beer, 1985; Redl and Schlag, 1986b), with a low fibrinogen (clottable protein) concentration, only moderate α-chain cross-linking and therefore, limited tensile strength (Redl and Schlag, 1986b; Buchta et al., 2005).

While commercial fibrin sealants were used together with bovine thrombin only until around the early 90s, bovine thrombin is still used with "home brews" or as free thrombin together with nonsealant-based hemostats like gelatine sponges. Large amounts of free bovine thrombin are considered a risk factor due to the potential formation of antibodies against bovine thrombin and factor V. These antibodies may cross-react with human thrombin and factor V, which in a worst case scenario can even lead to disseminated intravascular coagulation (DIC) (Schoenecker et al., 2000).

A variation of the standard fibrinogen-thrombin system was introduced in 1997 with the use of the snake venom enzyme batroxobin (Dascombe et al., 1997), in which acid-soluble fibrin I is formed first and then aggregated to a fibrin clot by an increase in pH (alkaline buffer as the reaction partner in a two component system).

In addition to plasma-derived products, fibrin sealant systems were also put together using recombinant components produced by transgenic sheep (fibrinogen – milk) and cell cultures (prothrombin, FXIII) (Butler et al., 1997).

15.3.1 Hemostasis

Fibrin sealants have been used to stop bleeding in many difficult indications. In a multicenter clinical study in 1989 Rousou et al. were able to demonstrate that fibrin sealant could achieve hemostasis in a significantly shorter period of time in cardiac surgery than with any of the conventional methods.

Other randomized studies followed soon after and confirmed the excellent hemostatic effect of fibrin sealant in urology (Gasser et al., 1983) and in spleen and liver surgical patients (Kram et al., 1988). These studies were the clinical basis for Immuno AG, Austria, later Baxter Healthcare, to introduce the first commercial fibrin sealant as Tisseel to the US market in 1998.

Fibrin sealants are now routinely used in cardiovascular, thoracic, plastic and reconstructive surgery as well as dental surgery and neurosurgery and have positive effects on surgical outcomes, including reduced blood loss and reduced complications (Jackson, 2001).

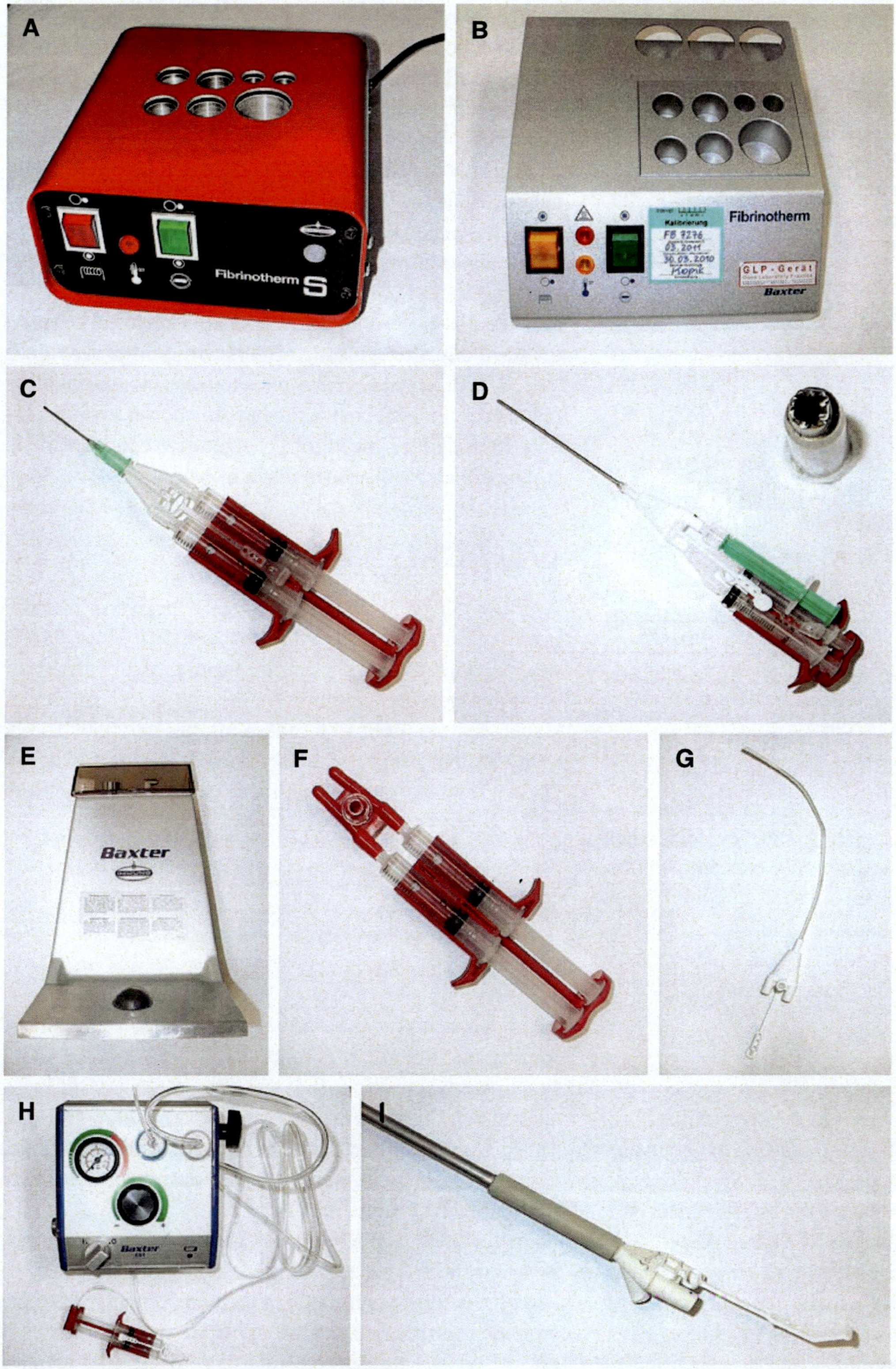

Fig. 15.3 Example of tissue engineering sealant devices (history, present and future). (A) Heating and stirring devices for fast dissolution of the fibrinogen components (1st generation) and (B) second generation. (C) Double syringe system for synchronous delivery of fibrin sealant component with mixing needle. (D) Liver biopsy needle (tip of the needle shown in insert) with biopsy channel (green syringe) and associated delivery channel for fibrin sealant to seal the biopsy lesion. (E) Foot pedal gas controller for double syringe spray system and (F) spray head (1st generation). (G) Bendable spray catheter. (H) Automatic gas controller (thumb activated) and spray head 2nd generation. (I) Prototype laparoscopy spray catheter, e.g., for hernia mesh application

15.3.1.1 Combination of Fibrin with Collagen and Other Carriers

Fibrin sealants have frequently been used for hemostasis in combination with collagen fleeces. A fibrinogen/thrombin precoated collagen fleece (Tachocomb) was introduced for hemostasis (Lippert and Wolff, 1990) and is now in the market as Tachosil. Fibrin was also used on a polylactide dressing (fibrin bandage) to achieve hemostasis (Jackson et al., 1998) and with demineralized spongiosa for sealing of fistulas (Flicker et al., 1986).

15.3.2 Sealing

As far as the adhesive effect of fibrin is concerned, critics have repeatedly pointed out that it is limited compared with synthetics, e.g., cyanoacrylate. This limited adhesive effect is, however, compensated by high elasticity (Redl and Schlag, 1986b), which makes the material especially useful for tissue that must remain flexible as in skin grafts, lung injuries and e.g superior for fixation of hernia meshes (Gruber-Blum et al., 2010). In addition Fibrin has a superior biocompatibility.

15.3.2.1 Nerves

The use of fibrin sealants for sealing began in 1972, when Matras et al. successfully used a highly concentrated human plasma-derived fibrinogen solution in combination with FXIII, thrombin, and calcium chloride to seal severed nerves in rabbit experiments. In 1976, Kuderna used fibrin sealant clinically for nerve anastomosis.

15.3.2.2 Skin Grafts

Skin grafting was one of the first indications in which fibrin sealant was tested and modern skin grafting has been dramatically improved with its use. In 1973, Spängler et al. described the use of fibrin sealant for full thickness skin grafts in rats, and in 1977 and 1985 Staindl reported his clinical experiences using fibrin glue to seal skin grafts. Fisseler-Eckhoff et al. (1988) found an acceleration of the wound healing process in burns. In 1979, Frey et al. and in 1980 Rendl and Staindl showed that fibrin-based sealing and grafting could even be achieved in infected wounds. In 1988, Henrich et al. compared the usefulness of three sealing systems in full thickness skin grafts following burns and he concluded that two fibrin sealants were superior to cyanoacrylate, a synthetic tissue glue.

Normally a fibrin sealant with a lower concentration of thrombin and therefore a slower polymerization time will be used in skin grafting, allowing a longer period for the surgeon to find the optimal positioning of the graft (Mittermayr et al., 2006).

A graft must be nourished by diffusion before full vascularization can take place (Converse et al., 1969). The three-dimensional, net-like structure of physiological fibrin sealants allows diffusion into and nourishment of the graft in the first postoperative days. Furthermore, migration of cells and vascularization into the fibrin net starts on the third day after grafting. The resulting improved graft "take" with all the effects of supported wound healing that fibrin sealant provides results in shorter hospitalization times and less need for postoperative care. Cosmetic results are also superior in fibrin fixed grafts.

15.3.2.3 Hernia

A novel use of fibrin sealant as an adhesive is the gluing of a mesh with or without sutures, to close hernia defects. This can take place in normal, open surgery or using laparoscopic surgery. As the amount of sutures or staples that are used to fix these meshes can be reduced, many patients experience less pain or discomfort when fibrin glues are used (Katkhouda et al., 2001; Schwab et al., 2006; Fortelny et al., 2008).

15.4 Preparation and Application of Fibrin Sealant

Although a high fibrinogen concentration is important for sealant strength, it is more difficult to dissolve when lyophilized or freeze-dried. To overcome this problem, a special heating and stirring device was developed (Redl, 1980; Fig. 15.3A, B) and preassembled, dual syringe, deep frozen systems were also introduced for more user convenience by some manufacturers (Fig. 15.3C).

In the past, application methods involved the use of high thrombin concentrations (several 100 NIH units/ml) and consecutive application of the two components to the wound or tissue. Consecutive application results in a high rate of coagulation on the interface of the two components, but also causes poor component mixing and poor rigidity of the resulting clot. Use of low thrombin concentrations (4–6 UI/ml) results in slow coagulation and allows premixing, which was originally introduced for nerve sealing (Kuderna, 1979).

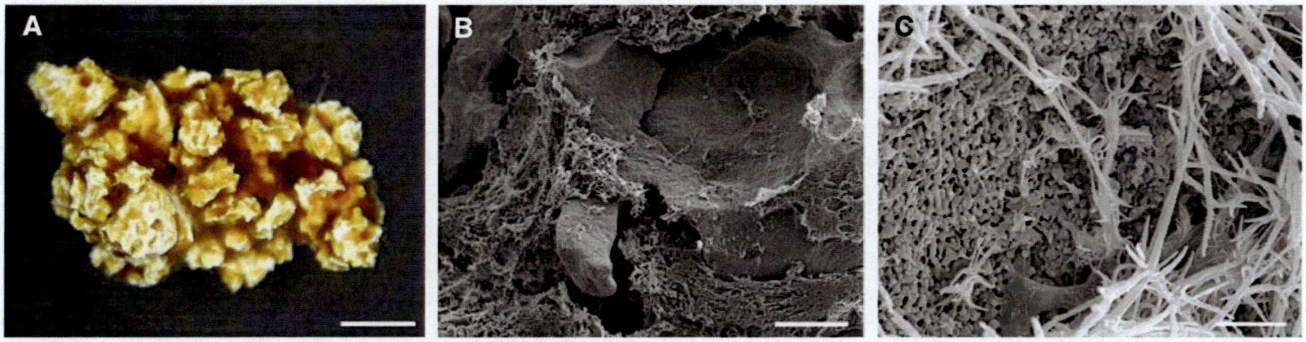

Fig. 15.4 Calcium phosphate combined with fibrin glue in (A) macroscopic view and (B, C) SEM. (A) The chemically fixed, brown fibrin fills the interspaces of the porous calcium phosphate, which appears white. (B, C) The SEM images show the fibrous structure of the fibrin and the rough surface of the Tricos granule. Scale bars: (A) 2 mm; (B) 200 μm; and (C) 5 μm

For these reasons, an application device consisting of a double syringe and two detachable heads was developed (Redl et al., 1982a; Fig. 15.3C). An alternative is to mix the two components on a piece of aluminum foil and apply the premixed sealant with a spatula (Redl and Schlag, 1984). Furthermore fibrin sealant has been used with a special biopsy and closure needle (Zatloukal et al., 1988; Fig. 15.3D). In the double syringe systems the two components can be simultaneously mixed and applied, either through a needle or by spraying.

High and low thrombin concentrations can be employed and the two components mix well. Spraying in combination with a high thrombin concentration (500 NIH units/ml) has proven particularly effective in establishing hemostasis while a low thrombin concentration (4–6 IU/ml) has proven beneficial in all applications where the parts to be sealed require subsequent adaptation, e.g., in skin grafting (Mittermayr et al., 2006) and in some microsurgical operations for skin transplantation (Redl and Schlag, 1986a). Especially in this indication a thin layer of fibrin sealant is the method of choice (O'Grady et al., 2000).

In general, fibrin sealant can be sprayed with the help of gas spray systems or with syringe pressure only. When no gas spraying system is used, fibrin is applied in a more streaming or sprinkling manner than spraying. The advantage is in the non-tethered device and the lack of need of gas source.

Spray systems with gas are connected to a conventional pressurized gas source (Redl et al., 1982a) or to specialized units, which allow automatic activation without foot pedals (e.g., EasySpray developed in our lab; Fig. 15.3H). An additional advantage offered by spraying with gas is that the operating site can be cleaned and dried before application of the sealant. The sealant is thus applied to a "dry" surface, which improves its efficacy. Moreover, no clogging in the device occurs when the sealing procedure is interrupted. A spray head is especially useful for covering large areas, e.g., for fixation of skin grafts and coating the donor area. Specific spray systems were developed for endoscopic application of fibrin (Fortelny et al., 2010; Fig. 15.3I).

15.5 Fibrin as a Biomatrix

A physiologically formed fibrin clot acts as a biomatrix that allows the growth factors to be slowly released and cells to settle and proliferate in the area of the wound and at the same time degrade at the optimum rate to allow space for new tissue growth. This makes it an ideal candidate for a biomatrix or scaffold for medication delivery or tissue engineering.

15.5.1 Fibrin as a Delivery System for Substances (Medication)

As early as 1950, a patent was described in the USA in which the combined application of fibrin and antibiotics was used (Ferry and Morrison, 1950). Substances like antibiotics are released from fibrin by simple diffusion, which depends to a large extent on the concentration gradient between the clot and its environment. Although antibiotics incorporated into fibrin clots are retained for a longer period compared to direct instillation into body cavities, drug retention is much lower than with bone cement-antibiotic mixtures and is insufficient to maintain adequate local drug concentrations for more than a few days. Some antibiotics substantially inhibit the crosslinking reaction of fibrin. However, the rate of fibrin crosslinking in the presence of antibiotics can be normalized by incorporating additional factor XIII (20 units/ml) into the reaction mixture (Redl et al., 1982a). The mixing of freeze-dried fibrin with antibiotics may slow the initial drug release and lengthen the period of adequate drug concentration for up to 18 days (Itokazu et al., 1997).

Fibrin biomatrix has also been used in the delivery of chemotherapeutics such as 5-fluorourazil, taxol or carboplatin (MacPhee et al., 1996; Yoshida et al., 2000).

15.5.2 Fibrin as a Delivery System for Growth Factors

A multitude of studies have been carried out on growth factor delivery with fibrin. Fibrin naturally binds some growth factors including basic fibroblast growth (bFGF), and vascular endothelial growth factor VEGF (isoform 165) (Sahni and Francis, 2000; Mosesson et al., 2001). This allows the use of fibrin as a biomatrix for growth factor delivery (Hashimoto et al., 1992).

Research groups, however, have developed methods of tethering growth factors or other peptides to the fibrin matrix, allowing a slower release of bioactive molecules that occurs parallel to the degradation of the scaffold and thereby links the rates of degradation and wound remodeling with tissue regeneration (Zisch et al., 2001; Morton et al., 2009).

Improved angiogenesis and revascularization have been proven in fibrin scaffolds that include bFGF or VEGF (Mittermayr et al., 2008) and was further enhanced when a combination of growth factors (VEGF isoform 165 and 121 or VEGF isoform 165 and bFGF) were used (Wong et al., 2003).

Kawamura and Urist (1988) used fibrin sealant to deliver bone morphogenetic protein (BMP) and reported a synergistic effect on bone regeneration that has also been extensively studied by other groups (Schmoekel et al., 2004; Jung et al., 2005). In a similar way to growth factor/fibrin constructs, a fragment of parathyroid hormone linked to fibrin has been shown to induce local bone regeneration (Arrighi et al., 2009).

The effect of nerve growth factor (NGF) and neurotrophin 3 (NT-3) on nerve regeneration has also been studied (Lee et al., 2003; Taylor et al., 2006).

15.5.3 Fibrin as a Matrix for Cells

The combined use of keratinocytes and a fibrin matrix for skin regeneration was first described by Hunyadi et al. in 1988. Isolated keratinocytes were cultivated on a fibrin matrix and the relative percentage of epidermal stem cells found *in vivo* was found to be maintained.

In the mean time, research on the delivery of cells with fibrin scaffolds has thrived. Fibrin matrices have been shown to be suitable for corneal cells, fat cells, cardiovascular, urothelial, bone, nerve, muscle, tracheal and meniscal cells. An autologous keratinocyte/fibrin approach to support skin healing has been commercialized as Bioseed S in Europe. Further products combine fibroblasts with fibrin (Flasza et al., 2007) or fibrin microbeads (Gorodetsky et al., 1999) and are used for skin tissue regeneration.

Other cells including human myofibroblasts (Ye et al., 2000) and osteoprogenitor cells (Tholpady et al., 1999), have also been successfully grown in fibrin gels. Fibrin sealant has also been used to deliver tracheal epithelial cells and preadipocytes (Wechselberger et al., 2002), for cell-based urothelial tissue regeneration and hepatocyte transplantation (Bach et al., 2001; Gwak et al., 2004). Fibrin can also be used to deliver stem cells such as bone marrow mononuclear cells and mesenchymal stem cells.

15.5.4 Fibrin as a Carrier for Osteoconductive Materials

Schwarz and colleagues implanted allogenic demineralized bone powder and bone matrix gelatine together with and without fibrin sealant in heterotopic and orthotopic sites in rats (Schwarz et al., 1989). It has also been shown that a fibrin matrix supports handling and osteogenic properties of the ceramic materials including osteointegration, bone remodeling, and new bone formation (Bagot d'Arc et al., 1994; Daculsi et al., 1994). More intensive bone regeneration or more rapid fracture healing was achieved in a number of experiments by using fibrin sealant (Bohler et al., 1977; Bosch et al., 1977, 1980; Wischhoefer et al., 1982; Palacios-Carvajal and Moina, 1986). Synthetic calcium phosphate bone substitutes such as hydroxyapatite (HA), beta-tricalcium phosphate (beta-TCP) or mixtures can be mixed with fibrin (Castellani et al., 2009) and are also provided as kits, e.g., Tricos T (Fig. 15.4).

15.6 Conclusion

The wondrous qualities of fibrin as a biological adhesive and sealant have changed modern surgery. Fibrin sealants not only have a long history, but a bright future as well, with new uses as biomatrices and scaffolds for stem cells and growth factors.

Acknowledgment

This chapter is dedicated to Helene Matras and Thomas Seelich – the pioneers in fibrin sealant research and development.

References

Arrighi I, Mark S, Alvisi M, von Rechenberg B, Hubbell JA, and Schense JC (2009) Bone healing induced by local delivery of an engineered parathyroid hormone prodrug. Biomaterials 30(9): 1763–1771.

Bach AD, Bannasch H, Galla TJ, Bittner KM, and Stark GB (2001) Fibrin glue as matrix for cultured autologous urothelial cells in urethral reconstruction. Tissue Engineering 7(1): 45–53.

Bagot d'Arc M, Corlieu P, and Daculsi G (1994) Mastoid cavity filling for bone reconstruction with a mixture of fibrin glue and ceramic granules. In: Schlag G and Redl H (eds) Fibrin Sealant in Surgical and Nonsurgical Fields. Neurosurgery, Ophthalmic Surgery, Vol. 5. ENT Ed. Springer-Verlag, Berlin: pp 174–179.

Banerjee SK and Glynn LE (1960) Reactions to homologous and heterologous fibrin implants in experimental animals. Annals of the New York Academy of Sciences 86: 1064–1074.

Beck E, Duckert F, and Ernst M (1961) The influence of fibrin stabilizing factor on the growth of fibroblasts in vitro and wound healing. Thrombosis et Diathesis Haemorrhagica Journal 6: 485–491.

Bergerl S (1909) About the effects of fibrin. Deutsche Medizinische Wochenschrift 35(633): 665.

Bohler N, Bosch P, Sandbach G, Schlag G, Eschberger J, and Schmid L (1977) The influence of homologous fibrinogen on osteotomy healing in rabbits. Unfallheilkunde 80(12): 501–508.

Bosch P, Braun F, Eschberger J, Kovac W, and Spangler HP (1977) The action of high-concentrated fibria on bone healing. Archiv für orthopädische und Unfall-Chirurgie 89(3): 259–273.

Bosch P, Lintner F, Arbes H, and Brand G (1980) Experimental investigations of the effect of the fibrin adhesive on the Kiel heterologous bone graft. Archives of Orthopaedic and Trauma Surgery 96(3): 177–185.

Bruhn HD and Pohl J (1981) Growth regulation of fibroblasts by thrombin, factor XIII and fibronectin. Wiener Klinische Wochenschrift 59(3): 145–146.

Buchta C, Hedrich HC, Macher M, Hocker P, and Redl H (2005) Biochemical characterization of autologous fibrin sealants produced by CryoSeal and Vivostat in comparison to the homologous fibrin sealant product Tissucol/Tisseel. Biomaterials 26(31): 6233–6241.

Butler SP, van Cott K, Subrumanian A, Gwazduaskas FC, and Velander WH (1997) Current progress in the production of recombinant human fibrinogen in the milk of transgenic animals. Journal of Thrombosis and Haemostasis 78(1): 537–542.

Carney DH, Mann R, Redin WR, Pernia SD, Berry D, Heggers JP, Hayward PG, Robson MC, Christie J, and Annable C (1992) Enhancement of incisional wound healing and neovascularization in normal rats by thrombin and synthetic thrombin receptor-activating peptides. Journal of Clinical Investigation 89(5): 1469–1477.

Castellani C, Zanoni G, Tangl S, van Griensven M, and Redl H (2009) Biphasic calcium phosphate ceramics in small bone defects: potential influence of carrier substances and bone marrow on bone regeneration. Clinical Oral Implants Research 20(12): 1367–1374.

Converse JM, Uhlschmid GK, and Ballantyne DL (1969) "Plasmatic circulation" in skin grafts. The phase of serum imbibition. Plastic and Reconstructive Surgery 43(5): 495–499.

Cronkite EP, Lozner EL, and Deaver JM (1944) Use of thrombin and fibrinogen in skin grafting. Journal of the American Medical Association 124(14): 976–978.

Daculsi G, Bagot d'Arc M, and Cottreal M (1994) Cacium phosphate ceramics (Triosite®) in otology: evaluation in animals and application in humans. In: Schlag G and Redl H (eds) Fibrin Sealing in Surgical and Nonsurgical Fields, Vol. 1. Wound Healing Ed. Springer-Verlag, Berlin: pp 121–126.

Danen EH and Yamada KM (2001) Fibronectin, integrins, and growth control. Journal of Cellular Physiology 189(1): 1–13.

Dascombe WH, Dumanian G, Hong C, Heil BV, Labadie K, Hessel B, Blomback B, and Johnson PC (1997) Application of thrombin based fibrin glue and non-thrombin based batroxobin glue on intact human blood vessels: evidence for transmural thrombin activity. Journal of Thrombosis and Haemostasis 78(2): 947–951.

Dinges HP, Redl H, Thurnher M, Schiesser A, and Schlag G (1986) Morphometric studies on wound healing after systemic administration of adriamycin and local application of fibrin sealant. Application of a new wound healing model using spongiosa implants. Pathology, Research & Practice 181(6): 746–754.

Duckert F and Nyman D (1978) Factor XIII, fibrin and collagen. Journal of Thrombosis and Haemostasis 63: 391–396.

Ferry JD and Morrison PR (1950) Fibrin clots and methods for preparing the same. USA Patent No. US 2533004(A).

Fisseler-Eckhoff A, Grabosch A, Donati I, and Müller KM (1988) Histological results after skin transplantation with and without fibrin sealant. Die Ellipse 15: 180–186.

Flasza M, Kemp P, Shering D, Qiao J, Marshall D, Bokta A, and Johnson PA (2007) Development and manufacture of an investigational human living dermal equivalent (ICX-SKN). Regenerative Medicine 2(6): 903–918.

Flicker M, Redl H, and Zwick H (1986) Closure of an acquired esophagobronchial fistula with fibrin glue. Praxis und Klinik der Pneumologie 40(10): 419–421.

Fortelny RH, Schwab R, Glaser KS, Puchner KU, May C, Konig F, Redl H, and Petter-Puchner AH (2008) The assessment of quality of life in a trial on lightweight mesh fixation with fibrin sealant in transabdominal preperitoneal hernia repair. Hernia 12(5): 499–505.

Fortelny RH, Petter Puchner AH, Ferguson J, Gruber Blum S, Brand J, Mika K, and Redl H (2010) Spray application of fibrin sealant with an angled spray tip device in laparoscopic inguinal hernia repair. European Surgery: in press.

Frey M, Holle J, Mandl H, and Freilinger G (1979) Die Vorteile der aufgeschobenen Spalthauttransplantation und die Erweiterung ihres Anwendungsbereiches durch die Verwendung des Fibrinklebers – Advantages of delayed split skin transplantation and widened usage due to use of fibrin sealant. European Surgery 11(5): 97–100.

Gasser G, Mossig H, Fischer M, Eidler R, Klaring W, and Lurf H (1983) Modification of suprapubic prostatectomy using a biological gluing technic. Wiener Klinische Wochenschrift 95(12): 399–403.

Gorodetsky R, Clark RA, An J, Gailit J, Levdansky L, Vexler A, Berman E, and Marx G (1999) Fibrin microbeads (FMB) as biodegradable carriers for culturing cells and for accelerating wound healing. Journal of Investigative Dermatology 112(6): 866–872.

Grey EG (1915) Fibrin as a haemostatic in cerebral surgery. Surgery Gynecology & Obstetrics 21: 452–454.

Gruber-Blum S, Petter-Puchner AH, Mika K, Brand J, Redl H, Ohlinger W, Benesch T, and Fortelny RH (2010) A comparison of a bovine albumin/glutaraldehyde glue versus fibrin sealant for hernia mesh fixation in experimental onlay and IPOM repair in rats. Surgical Endoscopy. doi:10.1007/s00464-010-1094-y

Gugerell A, Schossleitner K, Attakpah E, Wolbank S, Bittner M, and Pasteiner W (2010) Compatibility of Fibrin Sealants with Primary Keratinocytes is influenced by Fibrin Bound Thrombin. Conference Proceeding: Tissue Engineering and Regenerative Medicine International Society-EU, 13–17 June, Galway, Ireland: Poster Nr. 568.

Gustafson GT (1984) Ecology of wound healing in the oral cavity. Scandinavian Journal of Haematology, Supplement 40: 393–409.

Gwak SJ, Choi D, Paik SS, Lee EY, Lees KS, and Kim BS (2004) Stable hepatocyte transplantation using fibrin matrix. Biotechnology Letters 26(6): 505–508.

Hamm KD and Beer R (1985) Studies on the adhesive strength of biological tissue sealants. Zeitschrift für experimentelle Chirurgie, Transplantation, und künstliche Organe 18(5): 281–288.

Harvey SC (1918) Fibrin paper as an hemostatic agent. Annals of Surgery 68: 66–70.

Hashimoto T and Marks R (1984) Factor XIII inhibits epidermal cell migration in vitro. Journal of Investigative Dermatology 83(6): 441–444.

Hashimoto J, Kurosaka M, Yoshiya S, and Hirohata K (1992) Meniscal repair using fibrin sealant and endothelial cell growth factor. An experimental study in dogs. American Journal of Sports Medicine 20(5): 537–541.

Henrich HA, Baeumer F, Oberschelp J, and Kilian K (1988) Einfluss von drei verschiedenen Gewebeklebern auf Vollhauttransplantate bei Verbrennungen – Influence of three different tissue sealants of full thickness skin transplants in burns. Die medizinische Welt 39: 701–706.

Hunyadi J, Farkas B, Bertenyi C, Olah J, and Dobozy A (1988) Keratinocyte grafting: a new means of transplantation for full-thickness wounds. Journal of Dermatologic Surgery and Oncology 14(1): 75–78.

Itokazu M, Yamamoto K, Yang WY, Aoki T, Kato N, and Watanabe K (1997) The sustained release of antibiotic from freeze-dried fibrin-antibiotic compound and efficacies in a rat model of osteomyelitis. Infection 25(6): 359–363.

Jackson MR (2001) Fibrin sealants in surgical practice: an overview. American Journal of Surgery 182(2 Suppl.): 1S–7S.

Jackson MR, Taher MM, Burge JR, Krishnamurti C, Reid TJ, and Alving BM (1998) Hemostatic efficacy of a fibrin sealant dressing in an animal model of kidney injury. The Journal of Trauma 45(4): 662–665.

Jung RE, Schmoekel HG, Zwahlen R, Kokovic V, Hammerle CH, and Weber FE (2005) Platelet-rich plasma and fibrin as delivery systems for recombinant human bone morphogenetic protein – 2. Clinical Oral Implants Research 16(6): 676–682.

Kasai S, Kunimoto T, and Nitta J (1983) Cross-linked of fibrin by activated factor XIII stimulated attachment, morphological changes and proliferation of fibroblasts. Biomedical Research 4: 155–160.

Katkhouda N, Mavor E, Friedlander MH, Mason RJ, Kiyabu M, Grant SW, Achanta K, Kirkman EL, Narayanan K, and Essani R (2001) Use of fibrin sealant for prosthetic mesh fixation in laparoscopic extraperitoneal inguinal hernia repair. Annals of Surgery 233(1): 18–25.

Kawamura M and Urist MR (1988) Human fibrin is a physiologic delivery system for bone morphogenetic protein. Clinical Orthopaedics and Related Research 235: 302–310.

Kay AB, Pepper DS, and Ewart MR (1973) Generation of chemotactic activity for leukocytes by the action of thrombin on human fibrinogen. Nature: New Biology 243(123): 56–57.

Khare A, Woo L, McLean A, Stewart JE, DiOrio JP, Amrani DL, and Helgerson S (1998) Mechanical characterization of fibrin gels. Blood Coagulation & Fibrinolysis 9(7 – Session 11. Wound healing): 105.

Kram HB, Reuben BI, Fleming AW, and Shoemaker WC (1988) Use of fibrin glue in hepatic trauma. The Journal of Trauma 28(8): 1195–1201.

Kuderna H (1976) Clinical application of nerve-anastomoses adhesion using fibrinogen. Fortschritte der Kiefer- und Gesichts-Chirurgie 21: 135.

Kuderna H (1979) Das Fibrinklebesystem. Nervenklebung – The fibrin sealant system. Sealing of nerves. Deutsche Zeitschrift für Mund-, Kiefer- und Gesichts-Chirurgie 3: 32–35.

Lee AC, Yu VM, Lowe JB, Brenner MJ, Hunter DA, Mackinnon SE, and Sakiyama-Elbert SE (2003) Controlled release of nerve growth factor enhances sciatic nerve regeneration. Experimental Neurology 184(1): 295–303.

Lippert H and Wolff H (1990) Experiences with fibrin glue-coated collagen fleece. Zentralblatt für Chirurgie 115(18): 1175–1180.

Lorand L (1972) Fibrinoligase: the fibrin-stabilizing factor system of blood plasma. Annals of the New York Academy of Sciences 202: 6–30.

MacPhee MJ, Singh MP, Brady R, Akhyani N, Liau G, Lasa C, Hue C, Best A, and Drohan W (1996) Fibrin sealant: a versatile delivery vehicle for drugs and biologics. In: Sierra DH and Saltz R (eds) Surgical Adhesives and Sealants: Current Technology and Applications. Technomic Publishing AG, Basel: pp 109–120.

Matras H, Dinges HP, Lassmann H, and Mamoli B (1972) Suture-free interfascicular nerve transplantation in animal experiments. Wiener Medizinische Wochenschrift 122(37): 517–523.

Mittermayr R, Wassermann E, Thurnher M, Simunek M, and Redl H (2006) Skin graft fixation by slow clotting fibrin sealant applied as a thin layer. Burns 32(3): 305–311.

Mittermayr R, Morton T, Hofmann M, Helgerson S, van Griensven M, and Redl H (2008) Sustained (rh)VEGF(165) release from a sprayed fibrin biomatrix induces angiogenesis, up-regulation of endogenous VEGF-R2, and reduces ischemic flap necrosis. Wound Repair and Regeneration 16(4): 542–550.

Morawitz P (1905) Die Chemie der Blutgerinnung. Ergebnisse der Physiologie, biologischen Chemie und experimentellen Pharmakologie 4(1): 307–422.

Morton TJ, Furst W, van Griensven M, and Redl H (2009) Controlled release of substances bound to fibrin-anchors or of DNA. Drug Delivery 16(2): 102–107.

Mosesson MW, Siebenlist KR, and Meh DA (2001) The structure and biological features of fibrinogen and fibrin. Annals of the New York Academy of Sciences 936: 11–30.

O'Grady KM, Agrawal A, Bhattacharyya TK, Shah A, and Toriumi DM (2000) An evaluation of fibrin tissue adhesive concentration and application thickness on skin graft survival. Laryngoscope 110(11): 1931–1935.

Palacios-Carvajal J and Moina E (1986) The mixture of fibrin sealant and a porous ceramic as osteoconductor: an experimental study. In: Schlag G and Redl H (eds) Fibrin Sealant in Operative Medicine, Vol. 7. Traumatology – Orthopaedics Ed. Springer-Verlag, Berlin: pp 189–197.

Pfluger H and Redl H (1982) *In vivo* and *in vitro* degradation of fibrin adhesives (studies in rats). Zeitschrift für Urologie und Nephrologie 75(1): 25–30.

Pohl J, Bruhn HD, and Christophers E (1979) Thrombin and fibrin-induced growth of fibroblasts: role in wound repair and thrombus organization. Wiener Klinische Wochenschrift 57(6): 273–277.

Redl H (1980) Verfahren und Vorrichtung zur Beschleunigung des Auflösens von schwerlöslichen lyophilisierten Arzneimitteln – Method and apparatus for the accelerated solubilization of a difficult to dissolve lyophilized medication. Austria Patent No. A 4289/80.

Redl H and Schlag G (1984) Background and methods of fibrin sealing. In: Skjoldborg H (ed) Scientific Workshop 82: Tisseel/Tissucol Symposium – Areas of Application, Problems and Perspectives in Current Surgery Immuno Denmark ApS, Aarhus: pp 11–19.

Redl H and Schlag G (1986a) Fibrin sealant and its modes of application. In: Schlag G and Redl H (eds) Fibrin Sealant in Operative Medicine, Vol. 3. Gynaecology and Obstetrics – Urology Ed. Springer-Verlag, Berlin: pp 13–26.

Redl H and Schlag G (1986b) Properties of different tissue sealants with special emphasis on fibrinogen based preparations. In: Schlag G and Redl H (eds) Fibrin Sealant in Operative Medicine, Vol. 3. Gynaecology and Obstetrics – Urology Ed. Springer-Verlag, Berlin: pp 27–38.

Redl H, Schlag G, and Dinges HP (1982a) Methods of fibrin seal application. Journal of Thoracic and Cardiovascular Surgery 30(4): 223–227.

Redl H, Schlag G, Dinges HP, Kuderna H, and von Seelich T (1982b) Background and methods of fibrin sealing. In: Winter D, Gibbons DF, and Plenk H (eds) Biomaterials. John Wiley and Sons Ltd., New York: pp 669–679.

Redl H, Schlag G, and Eibl J (1997) Fibrinogen-based tissue adhesive. Austria Patent No. AT406120B.

Rendl KH and Staindl O (1980) Hauttransplantation mit Fibrinkleber beim Ulcus cruris – Skin transplantation with fibrin sealant. Aktuelle Dermatologie 6: 199–203.

Ross R (1968) The fibroblast and wound repair. Biological Reviews of the Cambridge Philosophical Society 43(1): 51–96.

Rousou J, Levitsky S, Gonzalez-Lavin L, Cosgrove D, Magilligan D, Weldon C, Hiebert C, Hess P, Joyce L, and Bergsland J (1989) Randomized clinical trial of fibrin sealant in patients undergoing resternotomy or reoperation after cardiac operations. A multicenter study. Journal of Thoracic and Cardiovascular Surgery 97(2): 194–203.

Sahni A and Francis CW (2000) Vascular endothelial growth factor binds to fibrinogen and fibrin and stimulates endothelial cell proliferation. Blood 96(12): 3772–3778.

Saltz R, Sierra D, Feldman D, Saltz MB, Dimick A, and Vasconez LO (1991) Experimental and clinical applications of fibrin glue. Plastic and Reconstructive Surgery 88(6): 1005–1015.

Schlag G and Redl H (1986) Fibrin Sealant in Operative Medicine. Springer-Verlag, Berlin.

Schlag G and Redl H (1994) Fibrin Sealing in Surgical and Nonsurgical Fields. Vol. 1: Wound healing & Vol. 4: Orthopedic Surgery, Maxillofacial Surgery. Springer-Verlag, Berlin.

Schlag G, Redl H, Turnher M, and Dinges HP (1986) The importance of fibrin in wound repair. In: Schlag G and Redl H (eds) Principles of Fibrin Sealant, Vol. 1. Otorhinolaryngology Ed. Springer-Verlag, Berlin: pp 3–12.

Schlag MG, Hopf R, and Redl H (2000) Convulsive seizures following subdural application of fibrin sealant containing tranexamic acid in a rat model. Neurosurgery 47(6): 1463–1467.

Schmoekel H, Schense JC, Weber FE, Gratz KW, Gnagi D, Muller R, and Hubbell JA (2004) Bone healing in the rat and dog with nonglycosylated BMP-2 demonstrating low solubility in fibrin matrices. Journal of Orthopaedic Research 22(2): 376–381.

Schoenecker JG, Hauck RK, Mercer MC, Parker W, and Lawson JH (2000) Exposure to topical bovine thrombin during surgery elicits a response against the xenogeneic carbohydrate galactose alpha1-3galactose. Journal of Clinical Immunology 20(6): 434–444.

Schwab R, Willms A, Kroger A, and Becker HP (2006) Less chronic pain following mesh fixation using a fibrin sealant in TEP inguinal hernia repair. Hernia 10(3): 272–277.

Schwartz ML, Pizzo SV, Hill RL, and McKee PA (1973) Human Factor XIII from plasma and platelets. Molecular weights, subunit structures, proteolytic activation, and cross-linking of fibrinogen and fibrin. Journal of Biological Chemistry 248(4): 1395–1407.

Schwarz N, Redl H, Schlag G, Schiesser A, Lintner F, Dinges HP, and Thurnher M (1989) The influence of fibrin sealant on demineralized bone matrix-dependent osteoinduction. A quantitative and qualitative study in rats. Clinical Orthopaedics and Related Research 238: 282–287.

Seelich T and Redl H (1979) Das Fibrinklebesystem. Biochemische Grundlagen der Klebemethode – The fibrin sealant system. Biochemical background of the sealing method. Deutsche Zeitschrift für Mund-, Kiefer- und Gesichts-Chirurgie 3: 22–26.

Shen LL, McDonagh RP, McDonagh J, and Hermans J (1974) Fibrin gel structure: influence of calcium and covalent cross-linking on the elasticity. Biochemical and Biophysical Research Communications 56(3): 793–798.

Sierra DH (1993) Fibrin sealant adhesive systems: a review of their chemistry, material properties and clinical applications. Journal of Biomaterials Applications 7(4): 309–352.

Spängler HP, Holle J, and Braun F (1973) Tissue adhesion with fibrin (An experimental study with rat skin grafts). Wiener Klinische Wochenschrift 85(50): 827–829.

Staindl O (1977) Tissue adhesion using highly-concentrated human fibrinogen on the example of free, autologe skin-grafting. Archives of oto-rhino-laryngology 217(2): 219–228.

Staindl O (1985) Indications of the fibrin sealant in facial plastic surgery. Archives of Facial Plastic Surgery 2(323): 340.

Stemberger A, Fritsche HM, Haas S, Spilker G, Wriedt-Lübbe I, and Blümel G (1982) Biochemische und experimentelle Aspekte der Gewebeversiegelung mit Fibrinogen und Kollagen – Biochemical and experimental aspects of tissue sealing with fibrinogen and collagen. In: Cotta H and Braun A (eds) Fibrinkleber in Orthopädie und Traumatologie. Georg Thieme Verlag, Stuttgart: pp 290–293.

Stiernberg J, Redin WR, Warner WS, and Carney DH (1993) The role of thrombin and thrombin receptor activating peptide (TRAP-508) in initiation of tissue repair. Journal of Thrombosis and Haemostasis 70(1): 158–162.

Taylor SJ, Rosenzweig ES, McDonald JW, and Sakiyama-Elbert SE (2006) Delivery of neurotrophin-3 from fibrin enhances neuronal fiber sprouting after spinal cord injury. Journal of Controlled Release 113(3): 226–235.

Tholpady SS, Schlosser R, Spotnitz W, Ogle RC, and Lindsey WH (1999) Repair of an osseous facial critical-size defect using augmented fibrin sealant. Laryngoscope 109(10): 1585–1588.

Tidrick RT and Warner J (1944) Fibrin fixation of skin transplants. Surgery 15: 90–95.

Tse D, Mankarious SS, Lui SL, Thomas WR, Alpern M, Enomoto ST, and Garanchon CM (1993) Topical fibrinogen complex. USA Patent No. WO/1993/005067.

Wechselberger G, Russell RC, Neumeister MW, Schoeller T, Piza-Katzer H, and Rainer C (2002) Successful transplantation of three tissue-engineered cell types using capsule induction technique and fibrin glue as a delivery vehicle. Plastic and Reconstructive Surgery 110(1): 123–129.

Wischhoefer E, Hellerer O, Falter E, and Hertlein H (1982) Verhalten autologer Beckenkammspongiosa nach Transplantation mit und ohne Fibrinklebesystem in der Femurepiphyse des Hundes – The behaviour of autologous spongiosa transplants from the ilial crest with and without fibrinadhesive in the canine femoral epiphysis. Unfallheilkunde 85(6): 250–252.

Wong C, Inman E, Spaethe R, and Helgerson S (2003) Fibrin-based biomaterials to deliver human growth factors. Journal of Thrombosis and Haemostasis 89(3): 573–582.

Ye Q, Zund G, Benedikt P, Jockenhoevel S, Hoerstrup SP, Sakyama S, Hubbell JA, and Turina M (2000) Fibrin gel as a three dimensional matrix in cardiovascular tissue engineering. European Journal of Cardio-Thoracic Surgery 17(5): 587–591.

Yoshida H, Yamaoka Y, Shinoyama M, and Kamiya A (2000) Novel drug delivery system using autologous fibrin glue – release properties of anti-cancer drugs. Biological & Pharmaceutical Bulletin 23(3): 371–374.

Zatloukal K, Dinges HP, Thurnher M, Redl H, and Schlag G (1988) A new technical of liver biopsy with plugging of the needle track using a double channel biopsy device. An experimental study. Zeitschrift für Gastroenterologie 26(11): 699–703.

Zisch AH, Schenk U, Schense JC, Sakiyama-Elbert SE, and Hubbell JA (2001) Covalently conjugated VEGF – fibrin matrices for endothelialization. Journal of Controlled Release 72(1–3): 101–113.

16 Properties and Potential Alternative Applications of Fibrin Glue

Sylvia Nürnberger, Susanne Wolbank, Anja Peterbauer-Scherb, Tatjana J. Morton, Georg A. Feichtinger, Alfred Gugerell, Alexandra Meinl, Krystyna Labuda, Michaela Bittner, Waltraud Pasteiner, Lila Nikkola, Christian Gabriel, Martijn van Griensven and Heinz Redl

Contents

16.1 Characterization of Fibrin as Matrix

Clot formation is an essential mechanism for wound closure and its principle is ubiquitous in the animal kingdom, comprising invertebrates such as arthropods, echinoderms, and cephalopods as well as all classes of vertebrates (Alsberg and Clark, 1908; Xu and Doolittle, 1990; Feral, 2010). The general principle of coagulation is the conversion of proteins to fibrous material by enzyme reaction in the presence of blood cells. Although the reacting partners (proteins, enzymes, and cell types) strongly differ between the animal groups, the final product always consists mainly or partly of fibrous material. Its functionality seems to rely on the formation of a gauze-like cover sealing the lesion.

16.1.1 The Components of Fibrin Gels and Their Influence on Morphology and Function (*Sylvia Nürnberger, Alexandra Meinl, Alfred Gugerell and Heinz Redl*)

Fibrin and its constituents, fibrinogen and thrombin, represent the fibrous components of blood clots in vertebrates (Doolittle et al., 1997). In natural fibrin clots, the filaments form a dense network spinning around erythrocytes, individual blood cells, and remaining platelets (Fig. 16.1A). The individual fibrils have a 20–22 nm cross striation (Aho et al., 1983; Weisel et al., 1993) and a thickness in a range from a few fibrinogen molecules (30 Å) up to 300 nm. Fibrils locally branch and fuse again, and likewise form a multiple interconnected network.

The same fibrous substructure occurring in the coagulum *in vivo* can be found in fibrin materials derived from blood plasma such as fibrin gels or commercially available fibrin sealants (Fig. 16.1B–E), which are used for either

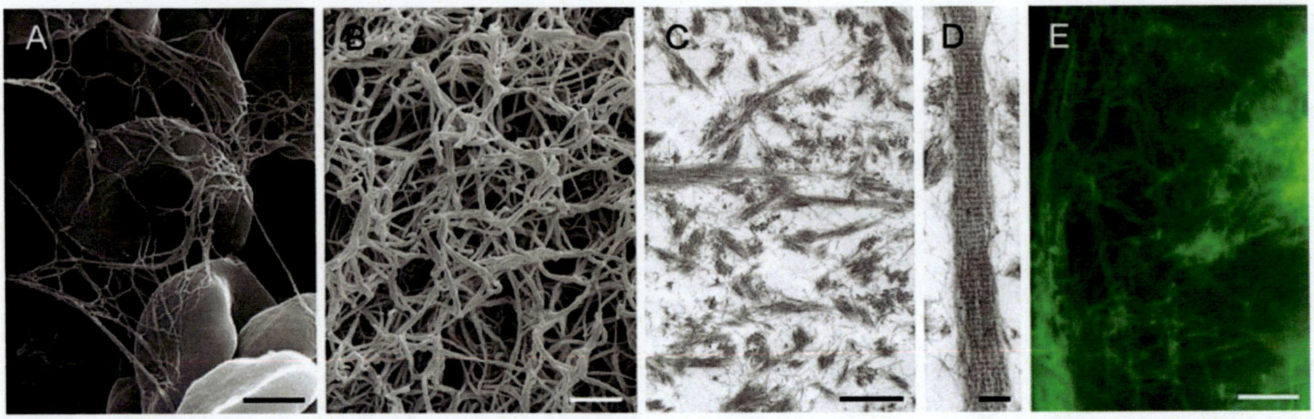

Fig. 16.1 Comparison of (A) a fibrin network spun around erythrocytes in a natural fibrin clot and (B–E) commercially available fibrin in (B) SEM, (C, D) TEM, and (E) confocal microscopy. (D) High magnification of a fibrin fiber showing the periodic striation of the fibrils. Scale bars: (A, B) 2 μm; (C) 500 nm; (D) 100 nm; and (E) 10 μm

clinical application or cell culture. Despite the general fibrous nature of polymerized fibrin, the fine structure has a wide range of possible morphologic properties including fiber size, density, and homogeneity. The morphology of the fiber network in turn is responsible for the mechanical properties of the clots such as stiffness, tensile strength, clotting time, and even the color. Both morphologic and mechanical parameters ultimately influence the behavior of cells in fibrin sealants in clinical as well as in experimental application (see Chapter 15, p. 225).

Parameters affecting the fine structure of the fibrin gels and the individual fibers, respectively, comprise the fibrin formulation (quantity and ratio between fibrinogen and thrombin and additional ingredients), the nature of fibrinogen, the pH, and the temperature (Ferry and Morrison, 1947; Blomback and Okada, 1982; de Maat and Verschuur, 2005).

16.1.1.1 Fibrinogen

Generally, fibrinogen determines the density and stiffness of the clot in a direct correlation implying that more fibrinogen per volume results in a denser clot with decreased porosity and permeation (Blomback et al., 1989; Mooney et al., 2010). Histologically, this effect is visible by the more intense staining of clots with high fibrinogen concentration (Fig. 16.2 upper part). Apart from the density, fibrinogen concentration also determines the homogeneity of the clot. The enormous variability of structural properties becomes obvious even in the range from 0.5 to 50 mg/ml fibrinogen and varying thrombin concentrations of 0.05–250 IU/ml (final clot concentration). Very low fibrinogen concentrations (0.5 mg/ml) cause jelly and translucent gels incapable of keeping their shape. They stain very faintly in histology, except for some regions of higher density. A higher fibrinogen percentage (12.5 mg/ml and more) leads to stiff clots whereas the stain affinity gradually increases with the fibrinogen content.

Increased staining affinity in the histologic images correlates with thick fibers and high fiber density at an ultrastructural level (Fig. 16.2 lower part). Very small fibers are however embedded between the large fibers, concluding that high fibrinogen concentrations not only provoke the increase in fiber size but also the increase in the range of fiber sizes. The combination of two classes of fiber size in one clot supports one of the two dynamic models of polymerization and fiber network formation (Chernysh and Weisel, 2008). This model says that an initial fiber network is constructed in the first period. In a second period, those fibers are enlarged by polymerization and, in addition, new fibers are formed in between until clotting is finished. Therefore, lateral and axial polymerization proceed simultaneously until clotting is completed. According to this model, the large fibers in Fig. 16.2 correspond to the initial polymerized ones which increase in size during the whole clotting process. The small fibers represent the "delayed" network forming after the gel point. On the contrary, in the second model, network construction is completed in an initial period and afterwards fibers become enlarged only by axial polymerization (Kita et al., 2002; Chernysh and Weisel, 2008).

Influence on the fibrin fine structure may further result from different natural subvariants of the Aα, Bβ or γ double chains of fibrinogen where either one pair or one single chain is partially altered. The biological reasons for the variations are genetic polymorphisms, alternative splicing,

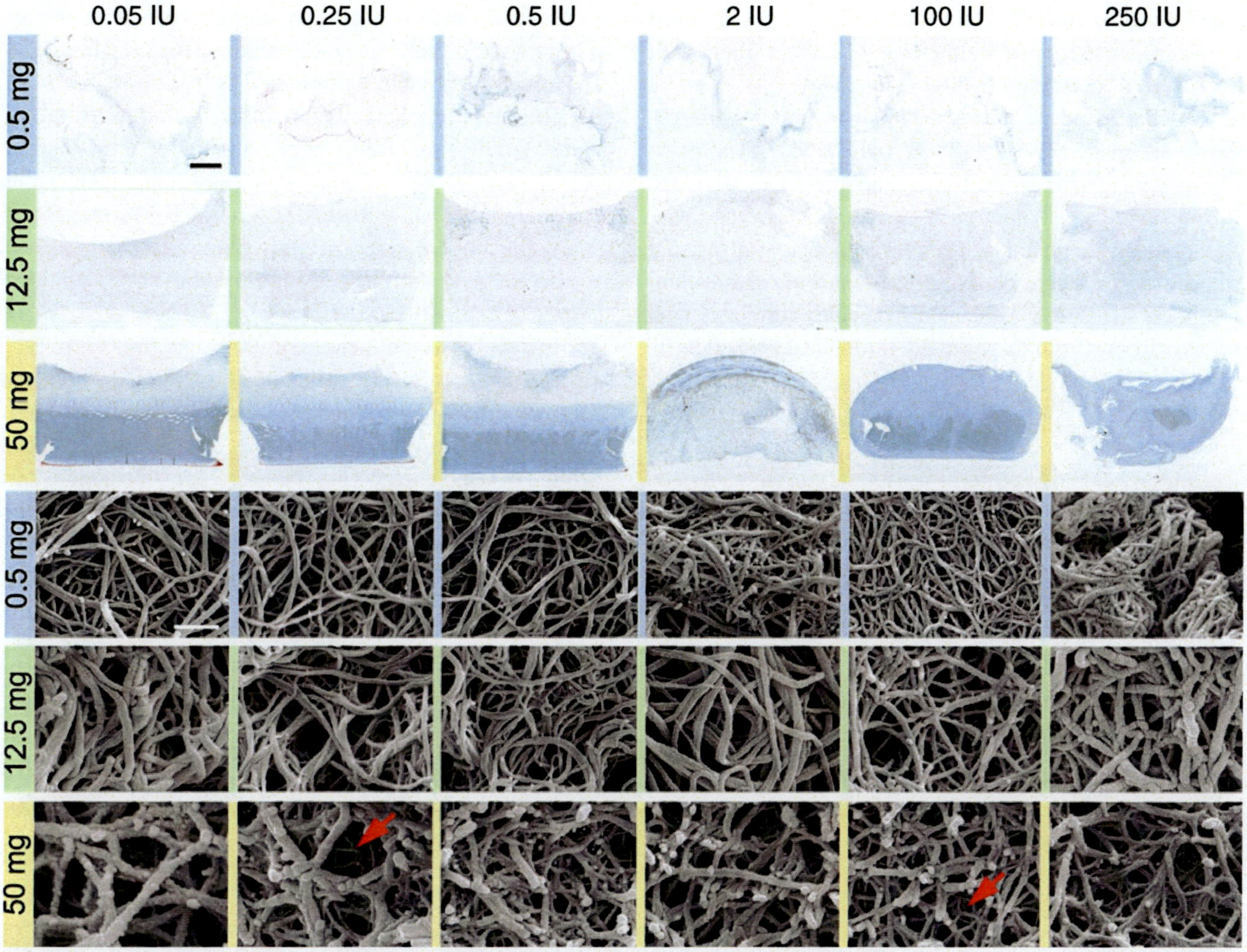

Fig. 16.2 Concentration series of fibrin clots: Thrombin concentrations are horizontally arranged: 0.05–0.25–0.5–2–100–250 IU/ml and the fibrinogen concentrations vertically 0.5–12.5–50 mg/ml. The upper part of the image shows the histologic sections with increasing staining intensity from low to higher fibrinogen concentrations but little changes with increasing thrombin concentration (Trichrom-staining; scale bar = 1 mm). Lower part of the image demonstrates the SEM images of the fiber networks and reveals higher fiber thickness with higher fibrinogen concentrations. With 50 mg/ml a second finer fiber network is visible (arrows). Increasing thrombin concentration does not alter fiber structure in a regular manner. Only the highest thrombin concentration in combination with low fibrinogen concentrations results in very dense clots with thick fibers

posttranslational modifications or proteolytic degradation (de Maat and Verschuur, 2005). Some fibrinogen variants have been found to manifest themselves at the morphologic level *via* alteration of fibrin polymerization but the exact mechanisms have not been clarified in all cases.

The major component in natural, unfractionated fibrinogen is the 340 kD fibrinogen (more than 50% of the total fibrinogen in healthy individuals). Proteolytic degradation is the main reason for the alteration of this molecule and may result in two types of low molecular weight fibrinogens. In those molecules either one (305 kD LMW-forms) or both (270 kD LMW-form) of the Aα chains are partially enzymatically degraded. LMW-variants determine the density of the fiber network, namely denser and more closed fibrin networks are formed than in the main 340 kD variant (Kaijzel et al., 2006).

Elongation of the fibrinogen Aα-chains (isoform α_E) by alternative splicing leads to 420 kD fibrinogen. The corresponding clots contain thinner and more irregular fibers with a higher degree of branching. In addition, nodular structures on the fibers were attributed to globular C-termini and accounted for the reduced rate of polymerization and slow lateral fibril association (Mosesson et al., 2004).

Alternative splicing may further result in the extension of the γ-chain (γ'). Morphologic studies of the influence on the network structures are contradictory and report either smaller fibers, reduced pore size, and increased branch points (Cooper et al., 2003), or thicker fibers with decreased fiber density (Collet et al., 2004).

Bβ-chains and γ-chains contain sialic acid at carbohydrate side chains. The quantity of sialic acid molecules on fibrinogen is posttranscriptionally modified and suggested to modulate protofilament formation and lateral aggregation in an inverse relation. Therefore, an increased amount of sialic acid residues per fibrinogen molecule causes thinner fibers whilst few sialic acid molecules correlates with thicker fibers (Gralnick et al., 1978; Maghzal et al., 2005).

16.1.1.2 Thrombin

Thrombin as a serine protease has a catalytic character and allows the conversion of fibrinogen to fibrin in quantities of several times the thrombin's weight over a high range of concentration (Ferry and Morrison, 1947; Kaminski and McDonagh, 1983). Nevertheless, like fibrinogen, the thrombin concentration influences the fibrin clot structure in a direct correlation, causing increased density, decreased porosity (Blomback et al., 1989; Mooney et al., 2010), and reduced fiber size (Weisel and Nagaswami, 1992; Wolberg et al., 2003). The underlying mechanism was explained by increased fibrinogen cleavage with higher thrombin activity.

Most of those studies were carried out with fibrin in a low range of concentration, below 5 mg/ml fibrinogen and 2 IU/ml thrombin (determined as far as possible from the different nomenclatures in different studies using enzyme activity, molarities or percentage of prothrombin plasma level). With higher concentrations, such as in commercially available sealants (2–250 IU thrombin and 50 mg fibrinogen), a recent study shows that the pore size is also reduced with increasing thrombin concentrations (Mooney et al., 2010).

However, the influence of thrombin on fibrin gel structure is not always evident (Fig. 16.2) suggesting that the effect of thrombin is probably highly variable and depends on other factors (e.g. FXIII, Ca^{++}, plasma proteins) which may vary between the fibrinogen and thrombin batches. Some studies showed that the changes in the fibrin networks depend on the range of thrombin concentration (Weisel and Nagaswami, 1992; Wolberg et al., 2003).

16.1.1.3 Clot Irregularities

Frequently neglected irregularities in fibrin clot morphology are irregularities, which may appear inside the individual fibrin clots of both low and high concentrations. Slight differences in fiber thickness in certain regions are

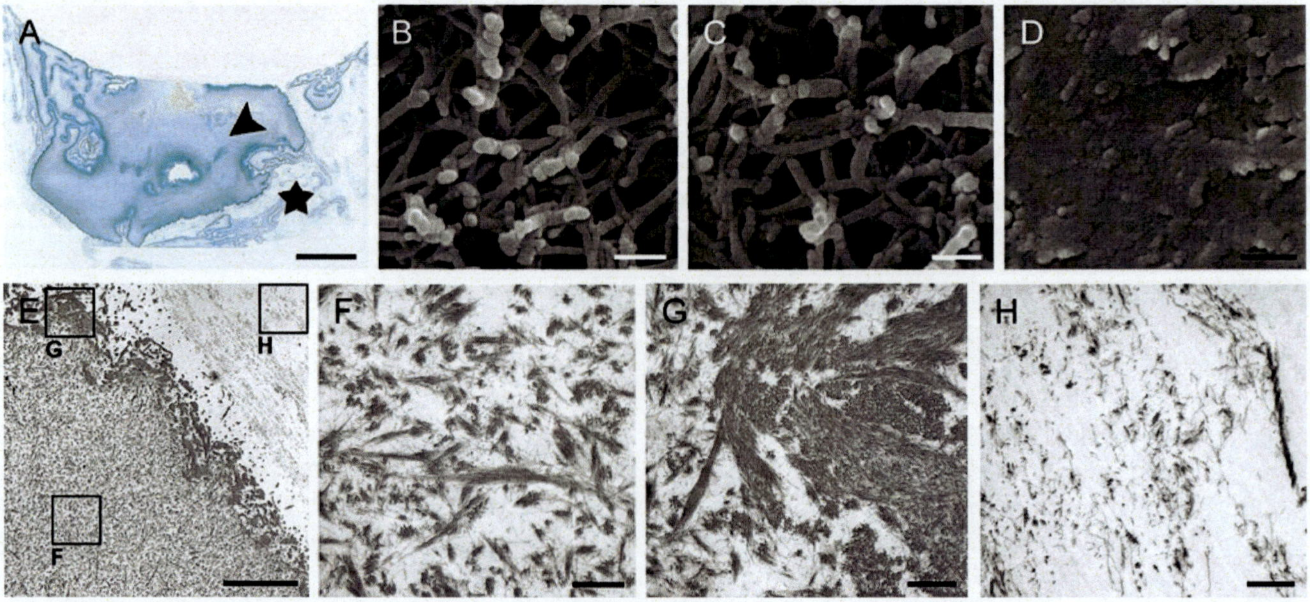

Fig. 16.3 Regions of inhomogeneity inside clots of high fibrinogen and thrombin concentration are visible (A) in histology as looser, fibrous, and brightly staining (asterisk) or denser, homogenous and intensively staining areas (arrowhead) (B, C) in REM as areas of different fiber thickness and density. (D) Extremely dense regions lacking any fibrous substructure refer to locally extremely high content of fibrinogen. (E) TEM image of the different network densities in overview. The squares indicate the regions of higher magnification of (F) with medium and (G) extreme dense regions and (H) very loose arrangement of thin fibers. Scale bars: (A) 1 mm; (B–D) 0.5 μm; (E) 10 μm; and (F–H) 500 nm

frequently visible only at an ultrastructural level, while others are evident using light microscopy (Fig. 16.3) and sometimes even macroscopically. In clots with high concentrations of both fibrinogen and thrombin, the regional differences can be attributed to the increased rate of polymerization with high thrombin concentration and simultaneously the high amount of highly viscous fibrinogen (Kaibara, 1973; Blomback et al., 1989; Weisel and Nagaswami, 1992; Zhao et al., 2008). The consequence is that mixing of the two components remains incomplete until the end of the reaction. Local differences in the thrombin to fibrinogen ratio inside the clot then result in structural differences.

16.1.1.4 Lot Variations

Apart from determinable variations between clots of different fibrin formulations, there remains a certain variability of the clot structure between different batches. The reasons are biological variations of the donors for the pooled fibrinogen and thrombin (Cooper et al., 2003; Wolberg et al., 2003; Kaijzel et al., 2006) and additional components such as growth factors (PDGF, EGF, and bFGF), FXIII (Lim et al., 2003) albumin and ECM components such as fibronectin and vitronectin.

16.1.1.5 Additives – Salts

Fibrin clots are influenced by a number of additional substances, which may be present at the time of polymerization. One of the largest influence is a high concentration of salt, which leads to partial or entire transparency of the clots. This clearing effect has primarily been reported by Ferry and Morrison (1947). By means of spectrometry, they measured the turbidity and suggested that with increasing salt concentration and transparency the clots became increasingly homogenous (termed "fine" *versus* "coarse" or "turbid"). Morphologic analysis later confirmed the structural changes of decreasing fiber size with concomitant increasing salt concentration (Weisel and Nagaswami, 1992). A high amount of salt totally prevents the formation of fibrous elements and causes

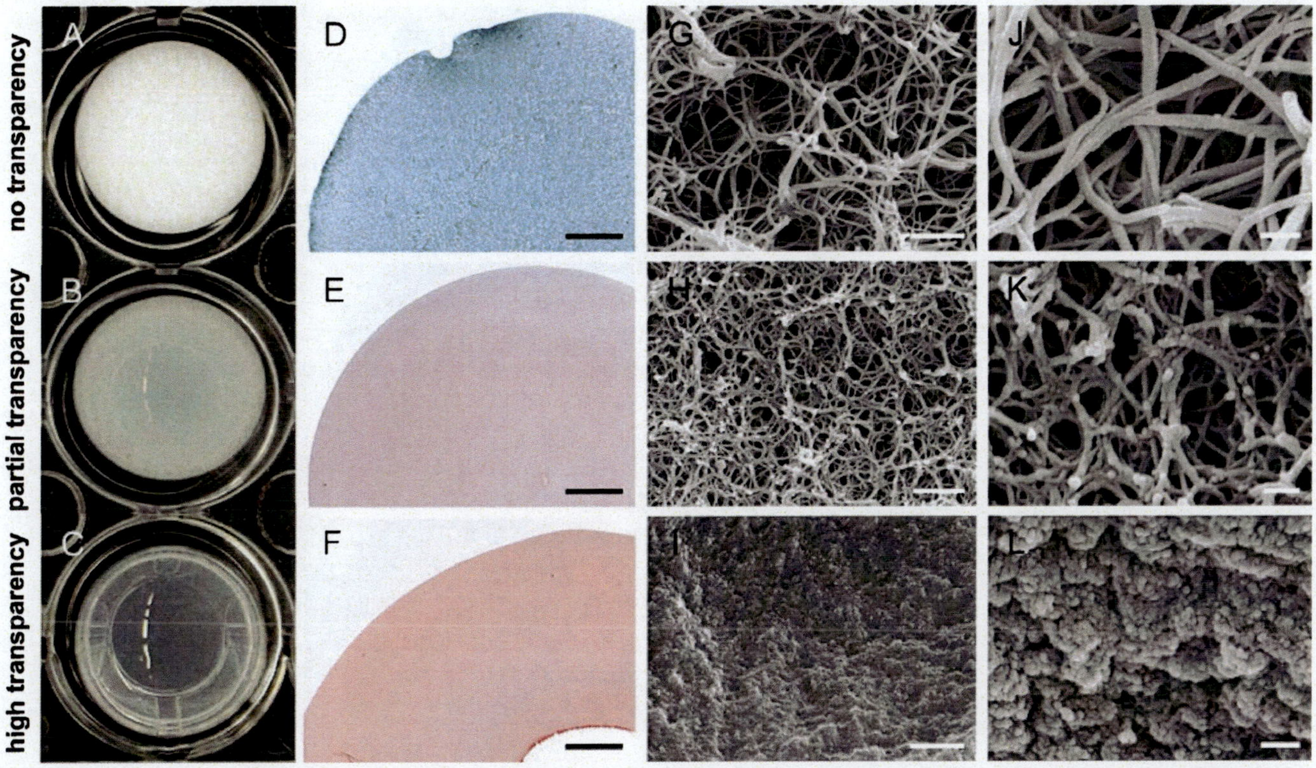

Fig. 16.4 Fibrin clots polymerized (A, D, G, J) without, (B, E, H, K) in the presence of 75 nM NaCl and 37.5 nM arginine and (C, F, I, L) 100 nM NaCl and 50 nM arginine show (A–C) a continuous transparency in macroscopic few, (D–F) a color shift from blue to red with MSB-staining and (G–L) decreasing fiber size until total homogeneity in SEM. The clots contain 50 mg/ml fibrinogen and 2 IU/ml thrombin. Scale bars: (D–F) 1 mm; (G–I) 1 μm; and (J–L) 0.5 μm

an amorphous structure, and a totally transparent appearance (Fig. 16.4; Redl et al., 1985; Goessl and Redl, 2005). This clearance effect is dependent on the pH, temperature, and ionic strength (Ferry and Morrison, 1947; Nair et al., 1986). Previous studies reported that the effect is restricted to divalent cations (Nair et al., 1986). More recently, though, the mechanism causing the structural difference was described as a specific effect of Cl^- which opposes the lateral aggregation of protofibrils (Di Stasio et al., 1998). The structural changes cause further alterations of the mechanical properties. Forfeiting flexibility and tensile strength, the clots become stiff like as highly concentrated gelatine or agarose.

16.1.1.6 Additives – Fibrinolysis Inhibitors

Massive influences on the clot structure as well as physical properties are also exerted by certain fibrinolysis inhibitors, which are used to prevent premature clot lysis of fibrin sealants. Aprotinin and tranexamid acid (t-AMCA) are the most common additives. It was shown that the structure of the fibrin clot massively changes with t-AMCA, in the same way as with Cl^- salts. Instead of the normal fibrous structure, the clot has a very dense network consisting of atypically small fibers or even a homogenous structure lacking any fibrous component. Reduction of the fibrosity leads not only to more transparency of the clot but also to reduction of the tensile strength. Due to the analogy with the behavior of salt clots, it was suggested that t-AMCA affects the fibrin matrix *via* ionic interaction (Furst et al., 2007). Aprotinin in contrast does not influence the typical fibrin structure.

16.1.1.7 Clot Casting

In experimental use it has frequently been observed that a thin layer ("skin") of homogenous material covers the surface of fibrin clots (unpub. obs.). Those layers are dense,

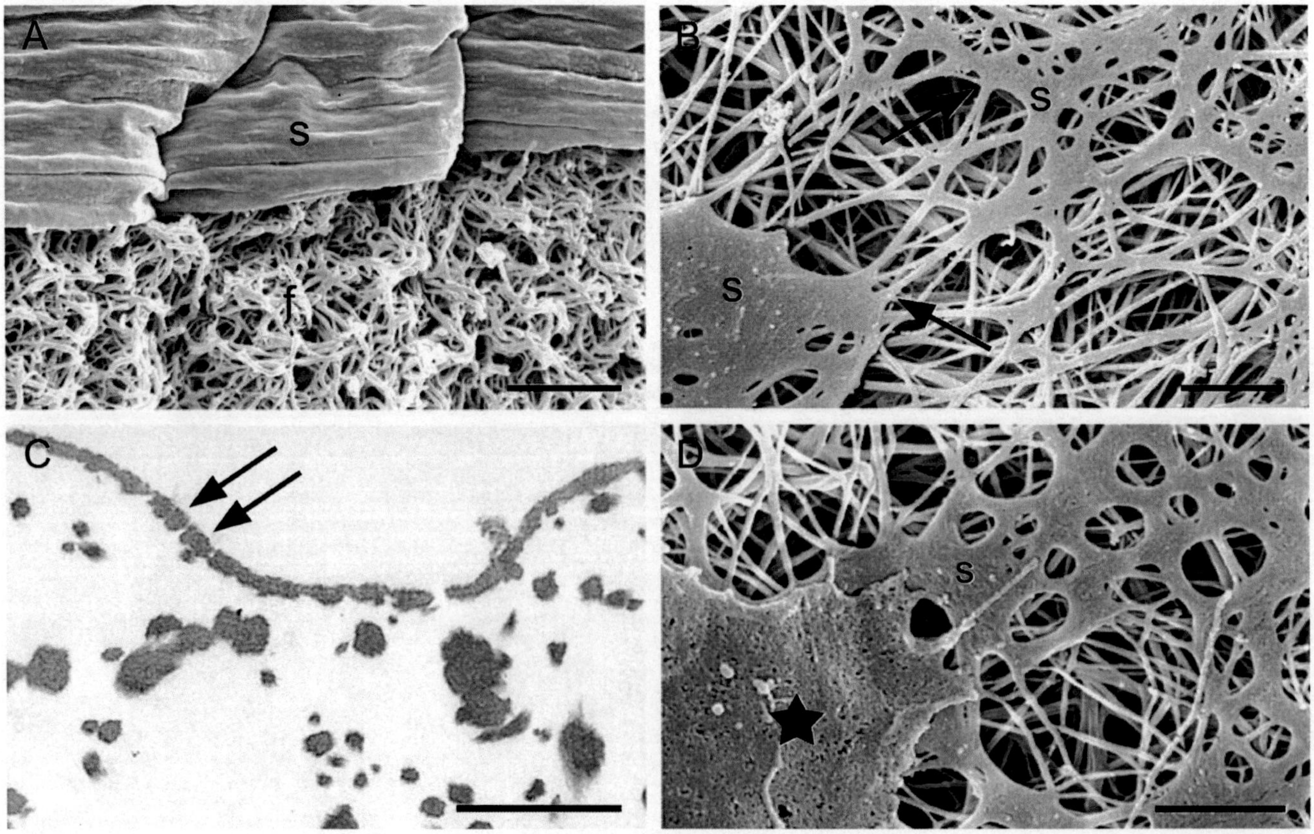

Fig. 16.5 Morphology of the "fibrin-skin" at the air exposed clot surface. (A) SEM image of a dense, smooth layer with an abrupt margin. The "skin" (*s*) is slightly wavy and seems to be detached easily. Note the fibrous (*f*) clot structure in the "skin"-free area. (B) Periphery of a fenestrated region of the "skin" (*s*) with individual fibers (arrows) spreading from the periphery of the "skin". (C) TEM images of a "skin-layer" in cross section with individual fibers identifiable (arrows). (D) A similar situation as in (B) but with a cell (asterisk) growing on the surface. Note the similar appearance between the cell and "fibrin-skin" (*s*). Scale bars: (A) 5 μm; (B) 2 μm; (C) 1 μm; and (D) 2 μm

smooth, have a regular thickness from 40 to 300 nm, and form at the air-exposed surface of the clots (Fig. 16.5). Those sites, which are in contact with the casting mold during polymerization, show the normal fibrous structure instead. At the periphery of the layer, individual fibers merge with it (Fig. 16.5B) suggesting that the membrane develops from individual fibers. Support for this conclusion is given by the thickness of the "skin" which correlates with the thickness of the individual fiber inside the clot. TEM observations further confirm the fibrous composition of the "skin", since individual fibers are sometimes discernible in its cross section (Fig. 16.5C). The reason for the formation of the "skin" has not yet been clarified but is believed to be caused by surface tension during polymerization at the interface with air.

In the presence of cells on the fibrin clot, the "skin" and the cells may reveal very similar morphology and surface structure. Especially if the skin has a network-like appearance and the cells have a polygonal cell shape and a flat and smooth surface, they can easily be mistaken (Fig. 16.5B, D).

16.1.1.8 Fibrinolysis – Clot Dissolution

In vivo, fibrinogen clots are degraded in the course of tissue regeneration mainly driven *via* the plasminogen pathway. It is converted into its enzymatically active form, plasmin, by one of two types of plasminogen activators, the tissue-type or urokinase-type plasminogen activator; tPA and u-PA (Anglés-Cano, 2006). Apart from plasminogen, fibrinolysis can also be realized by metalloproteinases (Hiraoka et al., 1998; Hotary et al., 2002). Both mechanisms are partially mediated by cells, which either synthesize the components or serve as a reaction surface. In addition, cells abolish fibrin by phagocytosis and enzymatic digestion (see Sect. 16.2). However, fibrin also degrades *in vitro* without the presence of any cells. The reasons are that both the precursor of the enzyme (plasminogen) and one of its reactors (t-PA) bind fibrin, are coisolated with the fibrinogen component and are part of the fibrin clot.

Structurally, fibrinolysis manifests itself as increasing inhomogeneities of the clot structure and changing stain affinity in histologic sections (Fig. 16.6). Special stains are apt to support this morphologic feature by a color effect in staining fresh fibrin red and more degraded one bluish (Fig. 16.6). At an ultrastructural level, the fibers become continuously rougher, fuzzy, and decorated with spherical fiber residuals (Fig. 16.7). In advanced stages, the clot is reduced to a small fraction of its original size which mainly consists of granules. Those granules result from cleavage of the fibers which was found to proceed perpendicular to their longitudinal axis rather than by continuous lysis of the fiber from the surface to the center (Veklich et al., 1998; Collet et al., 2000). Lysis does not proceed in all clot regions equally, neither at tissue-level (histology) nor at fiber-level (ultrastructure), but a distinct pattern (periphery to center) could not be determined. *In vitro*, the actual durability and dynamic of clot degradation varies depending on the formulation of the fibrin, its density, and fiber size. Concerning the latter, clots consisting of thin fibers degrade slower due to decreased plasminogen and tPA binding sites and the consecutive plasmin activity (Gabriel et al., 1992; Collet et al., 2000). Apart from the morphologic properties of the fibrin clot itself, the cultivation medium, fibrinolysis inhibitors, and the cells influence fibrinolysis.

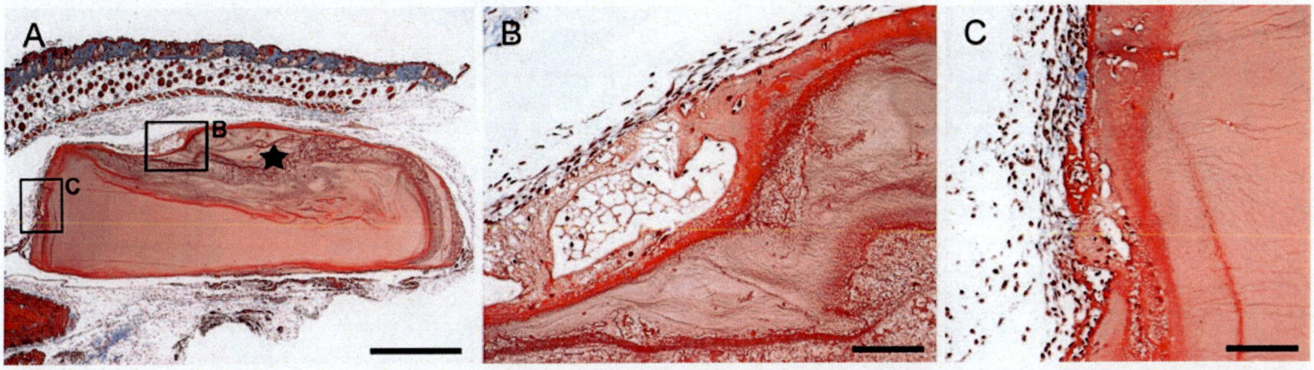

Fig. 16.6 *In vivo* clot degradation: Histologic MSB-staining of a fibrin clot (containing 1500 U/ml Aprotinin) and surrounding tissue three days after subcutaneous application in a mouse. Fresh fibrin and muscle appear bright red while degrading areas of the clot are slightly bluish (asterisk). (A) Overview image with squares indicating the details of (B, C). Both details show areas with cell invasion from the surrounding connective tissue. Scale bars: (A) 1 mm and (B, C) 100 µm

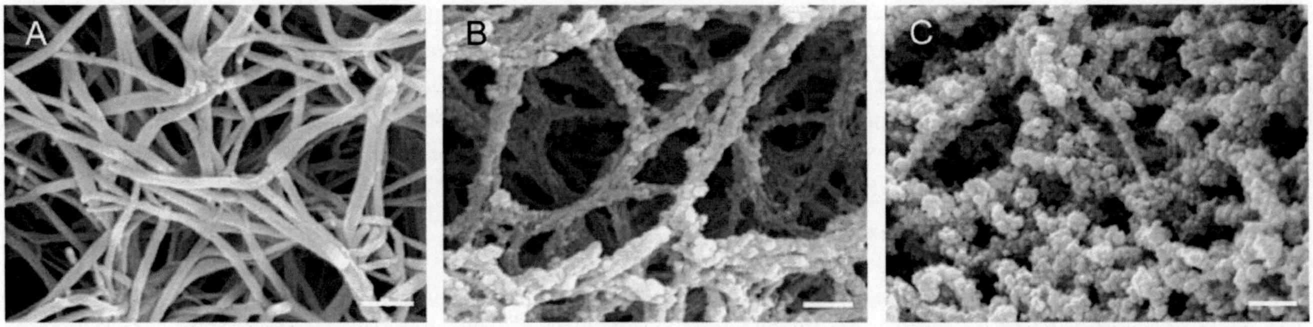

Fig. 16.7 *In vitro* clot degradation: (A) Intact fibrin network of a clot kept for eight days in PBS with the fibrinolysis inhibitor aprotinin (50 mg/ml fibrinogen, 2 IU thrombin). (B) A clot with the same fibrin formulation as in (A) kept in PBS for eight days without fibrinolysis inhibitor. It shows signs of fibrinolysis such as a rough and fuzzy surface decorated with fiber residuals. (C) Advanced fibrinolysis in a clot kept in urokinase. The fibrous material is entirely granular and the original fiber network is hardly discernible (50 mg/ml fibrinogen, 0.5 IU thrombin). Scale bars: (A–C) 0.5 μm

16.2 Fibrin as Matrix for Cells

Apart from their hemostatic function, fibrin clots in wounds serve as temporal scaffolds for cells involved in wound healing. After the initial activity of neutrophils, macrophages and lymphocyte, cells such as fibroblasts, keratinocytes, and endothelial cells prepare and conduct tissue reformation (Martin, 1997; Laurens et al., 2006). Depending on the injured tissue, other tissue-specific cell types are also involved in fibrin clot replacement. Due to its ubiquitous presence in the body, fibrin has a high biocompatibility and broad acceptance by a wide range of cell types (Fig. 16.8) and is biodegradable without any physiological risk. Therefore, fibrin is predestined not only for clinical use as sealant and cell carrier but also for the continuously growing field of tissue engineering. Apart from its basic functionality for tissue regeneration, fibrin gels have further advantages: Because of its main components fibrinogen and thrombin and additional natural ingredients such as fibronectin, vitronectin, and growth factors, it provides a bio-functional environment with beneficial effects on cell activities and tissue regeneration (Buchta et al., 2005). It has already been used for various tissue engineering approaches including skin, cardiovascular, nerve, cartilage, and ocular regeneration (Mann, 2003).

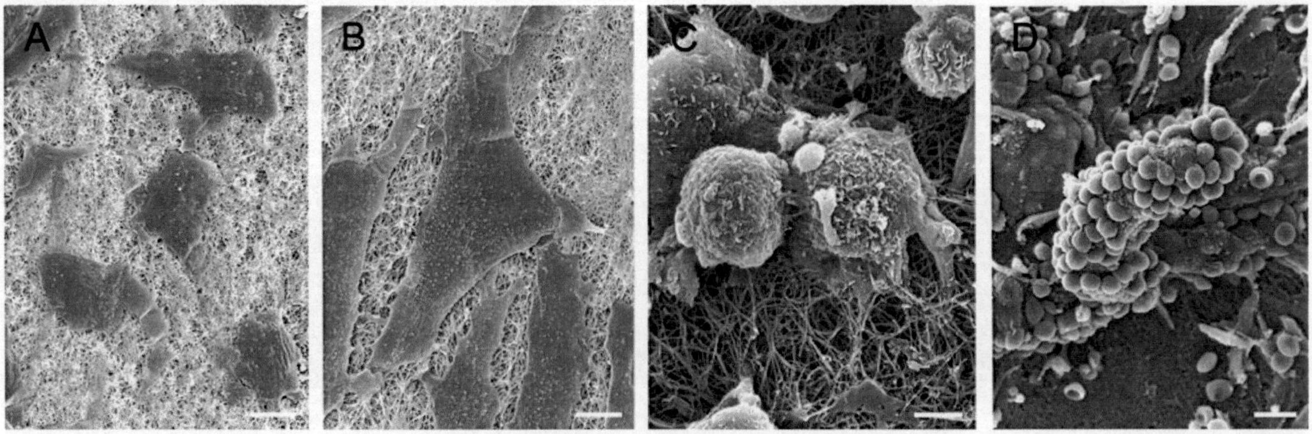

Fig. 16.8 SEM images of 2D cultures on clots (12.5 mg/ml fibrinogen, 2 IU/ml thrombin) diluted from a commercially available fibrin sealant with primary cells of very different origin: (A) Human dermal fibroblasts and (B) human vascular endothelial cells form polygonal cells with a flat surface and single microvilli only. (C) Primary cell cultures of human chondrocytes and (D) canine uterus glands. (A, B and D) 24 h (C) 7 d after seeding all cells are highly vital and retain their typical phenotype. Scale bars: (A) 20 μm; (B) 10 μm; (C) 5 μm; and (D) 10 μm

16.2.1 General Characterization of Cell Culture on and in Fibrin (*Sylvia Nürnberger, Alfred Gugerell, Susanne Wolbank, Michaela Bittner, Waltraud Pasteiner and Heinz Redl*)

In tissue engineering, either commercially available fibrin sealants or alternatively obtained fibrin gels (by cryoprecipitation from blood banks or autologous plasma processing with special devices) are used for clinical or experimental applications (Schlag and Redl, 1986). The conserved nature of the blood coagulation molecules offers the opportunity for cross-species application of fibrin. Fibrin from several species (e.g. salmon, mouse, bovine, and human) can be used for cells of various other species (e.g. mouse, bovine, canine, and human cells; see Fig. 16.8). For cell or tissue culture fibrin gels are used either as cultivation surface (2D cultures) or for encapsulation of the cells in a matrix (3D cultures).

Irrespective of the broad compatibility of the fibrin matrix, there are numerous differences in cellular behavior on and in fibrin clots concerning adhesion, proliferation, morphology, and migration. These differences are due on the one hand, to the nature of the cells and are, on the other hand, closely related to the properties of the fibrin (as discussed above). Strong deviations of the fibrin formulation from the natural fibrin composition may even cause poor adhesion, proliferation, and cell death (Redl et al., 1985; Furst et al., 2007; Macasev et al., 2010).

16.2.1.1 Adhesion

Adequate adhesion is the first critical event in cell-matrix interactions in natural fibrin clots of wounds as well as in fibrin sealants and gels and depends on both counterparts: the cells and the matrix. In fibrin clots adhesion involves several types of receptors on the cell membrane (cell adhesion molecules, CAMs) and different ligands on the fibrin matrix. The main ligand is the fibrin(ogen) molecule, which provides specific binding sites on the fibrin(ogen) polypeptide chains α, β, and γ. Most of the fibrin(ogen) binding cell receptors belong to the class of integrins. One or several integrin types are in turn represented on most cell types. The most important one on fibroblasts and endothelial cells is the $\alpha_v\beta_3$ integrin (Chang et al., 1995; Gailit et al., 1997). Receptors of other classes binding to fibrinogen include ICAM-1 on fibroblast membranes (Farrell and al Mondhiry, 1997) or VE-cadherin on endothelial cells (Bach et al., 1998).

PAR-1 is the major receptor for thrombin and mediates signalling for proliferation and migration. It has been found on plateletes, smooth muscle, and endothelial cells and is the major mediator for vascularization of fibrin (De Cristofaro and De Candia, 2003; Smadja et al., 2008).

Another important adhesion molecule in the fibrin clot is fibronectin. A lot of cell types are known to bind to it by $\alpha_5\beta_1$ integrins and syndecan 4 (Kim et al., 1992; Corbett and Schwarzbauer, 1999; Midwood et al., 2006). One cell type especially dependent on fibronectin is the keratinocytes, which lack fibrinogen receptors and do not bind to fibrinogen (Fig. 16.9) (Kim et al., 1992;

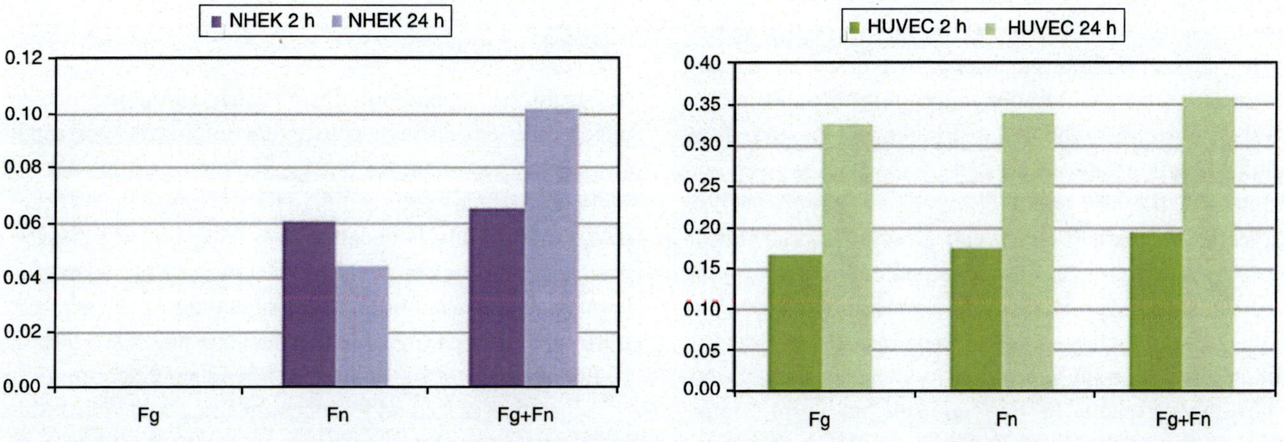

Fig. 16.9 Cell number of human keratinocytes (left) and endothelial cells (right) grown for 2 h and 24 h on 96-well plates coated with fibrinogen (Fg), fibronectin (Fn), and a combination of both matrix proteins (Fg + Fn). The lactate dehydrogenase assay shows that keratinocytes hardly adhere to fibrinogen, but to a combination of fibrinogen and fibronectin, and fibronectin alone. Endothelial cells, in contrary adhere well to all those matrix surfaces. *NHEK* normal human epidermal keratinocytes; *HUVEC* human umbilical vein endothelial cells

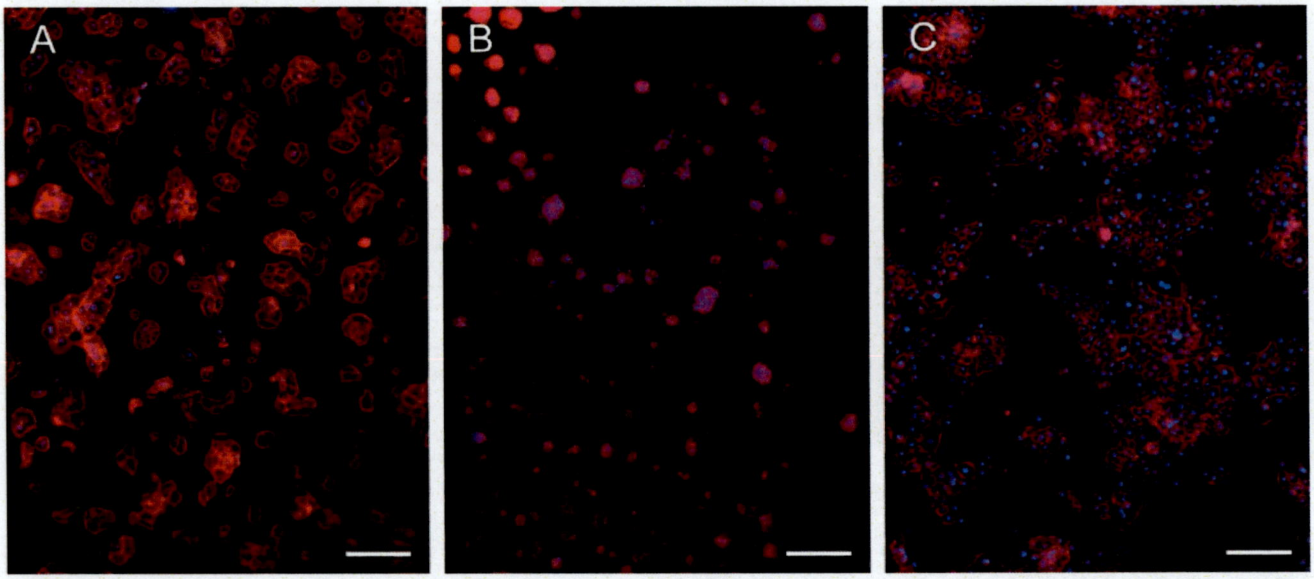

Fig. 16.10 Normal human epithelial keratinocytes seeded on top of fibrin clots and cultivated for 24 h. Cellular actin cytoskeleton was labeled with phalloidin (red) and the nuclei with DAPI (blue). (A) On clots made from a commercial product (50 mg/ml fibrinogen, 2 IU thrombin) keratinocytes showed a normal polygonal morphotype with a well-developed cytoskeleton. (B) Keratinocytes grown on clots deprived of fibronectin (FIB3) show a rounded shape with little amount of cytoskeleton. (C) Spiking of FIB3 clots with 2 μg/ml fibronectin prevents the cell morphology alteration. Scale bars: 200 μm

Gorodetsky et al., 1998; Kubo et al., 2001). Depletion of fibronectin from fibrin matrix causes massive reduction of the actin cytoskeleton and actin-related adhesion as well as alteration of the cell shape (Fig. 16.10). Adding fibronectin during polymerization of fibronectin-free fibrin clots abrogates this effect and keratinocytes adhere in a normal way (Fig. 16.10).

16.2.1.2 Proliferation

Apart from adhesion, cell receptors are involved in signaling for several cellular functions. Therefore, the matrix has an effect on proliferation, migration, differentiation, cellular physiology, and cell death. As in adhesion, those cellular functions are influenced by morphology and formulation of the clot and additional components such as extracellular matrix proteins and growth factors. Formulations with increasing fibrinogen content have a limiting effect on proliferation inside and outside the clot (Cox et al., 2004; Mooney et al., 2010). Inside the clot, this probably relies on high clot-density correlated with higher fibrinogen content. Fibrinogen itself has a proliferative function; the extent of which though differs between the cell types. Thrombin also stimulates cell proliferation, but has a dose-dependent effect with a proliferation peak and subsequent drop and is in addition dependent on the presence of serum (Gorodetsky et al., 1998; Gandossi et al., 2000). Also, fibronectin has a certain proliferative effect (Pankajakshan and Krishnan, 2009) and may in addition stimulate differentiation by conformational changes and related alterations of integrin binding (Garcia et al., 1999). Fibrin gels and sealants are known to contain several growth factors (TGF-β1, FGF, and VEGF) which are isolated with the fibrin components during isolation from the blood plasma and additionally stimulate cell growth.

16.2.1.3 Migration

Migration on, in, and into fibrin is an essential mechanism which proceeds during tissue regeneration in wound repair and is also important for the application of fibrin gels and sealants in tissue engineering. In fibrin clots, if used as scaffold, cell mobility is necessary for adequate cell distribution, either in the course of proliferation or the adaptation to the *in vivo* conditions after implantation. In addition, fibrin clot integration into the defect site and fast vascularization are critical for successful *in vivo* application.

Migration is related to adhesion since focal adhesions of the cells are necessary for local attachment while the cells push forward. On the other hand, permanent adhesion may prevent movement. Receptors responsible for migration are integrins (Guo et al., 1990; Kim et al.,

1992, 1994; Greiling and Clark, 1997; O'Toole, 2001) and syndecan-4 (Lin et al., 2005). The fibrin clot components fibronectin and fibrin(ogen) influence migration *via* those receptors. Even though most cell types bind to fibrinogen, fibronectin is more essential for migration for many cell types such as fibroblasts (Greiling and Clark, 1997) and Schwann cells (Akassoglou et al., 2003). Keratinocytes, which do not bind to fibrinogen, also move *via* fibronectin interaction, mainly through $\alpha_5\beta_1$ integrins (Guo et al., 1990; Kim et al., 1992). A difference in the importance of fibronectin has been found in the type of movement. For example, fibroblasts require fibronectin for 3D migration into the clot, but not for 2D migration on the clot surface (Knox et al., 1986). An inhibitory function of fibronectin was proven for macrophages, which instead move *via* fibrinogen interactions (Lanir et al., 1988). Fibrinogen also varies in its effect on cell migration depending on whether the α- or γ-polypeptide chains are cross-linked. Which of the variants is the better, again depends on the cell types (Ciano et al., 1986; Brown et al., 1993).

One of the most important factors determining the migration of cells is the morphology of the clot. Its influence is congruent in all studies, and rather independent of the cell type. Higher porosity due to either fibrinogen concentration or the composition of the clots (low and high molecular weight fibrinogens; Ciano et al., 1986; Naito et al., 1996; Kaijzel et al., 2006) sustains cell migration. Cox and coauthors (2004) showed that it is the fibrinogen concentration rather than the thrombin which is responsible for this effect. This phenomenon probably relies on the concentration-related increase in density and decrease in pores size in clots with higher fibrinogen concentration. On the basis of this fact the movement behavior may be explained as follows: Movement through the fiber network might be realized by cell shape adaptation

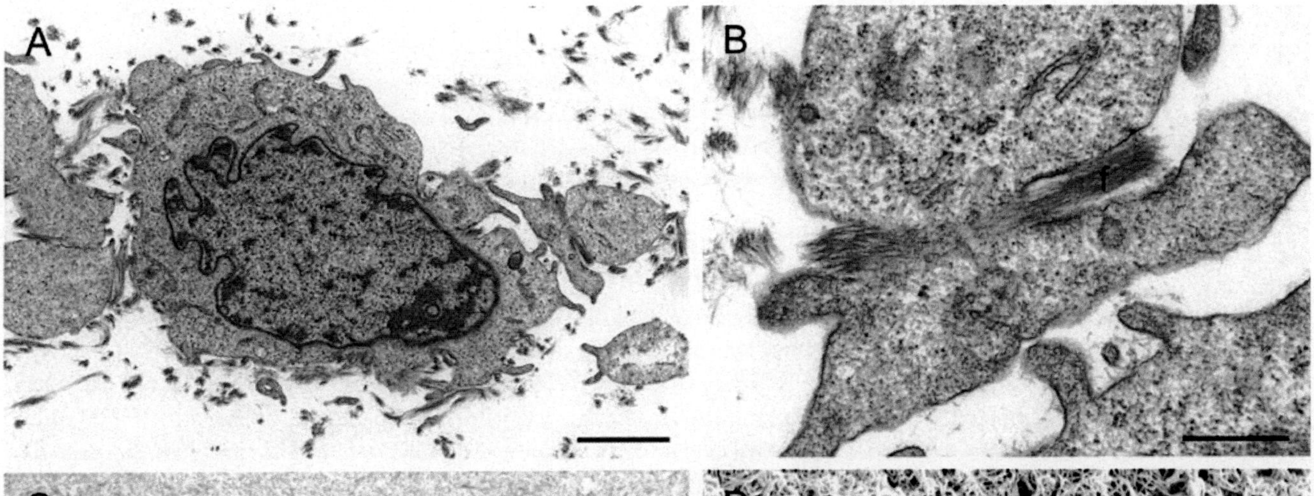

Fig. 16.11 Chondrocytes in fibrin gels of different densities. (A) In gels with high porosity chondrocytes extend several small and large cell processes and (B) form intense contact to the fibers (*f*). (C) Cell process embedded in dense fibrin matrix. The local lumen (asterisk) probably results from proteolytic activity of the chondrocyte. Notice the microvilli (arrow) extending into the free space. (D) SEM image of a cryo-fractured fibrin clot shows a cell densely embedded in a tight fibrin matrix. Scale bars: (A) 2 μm; (B) 500 nm; (C) 2 μm; and (D) 10 μm

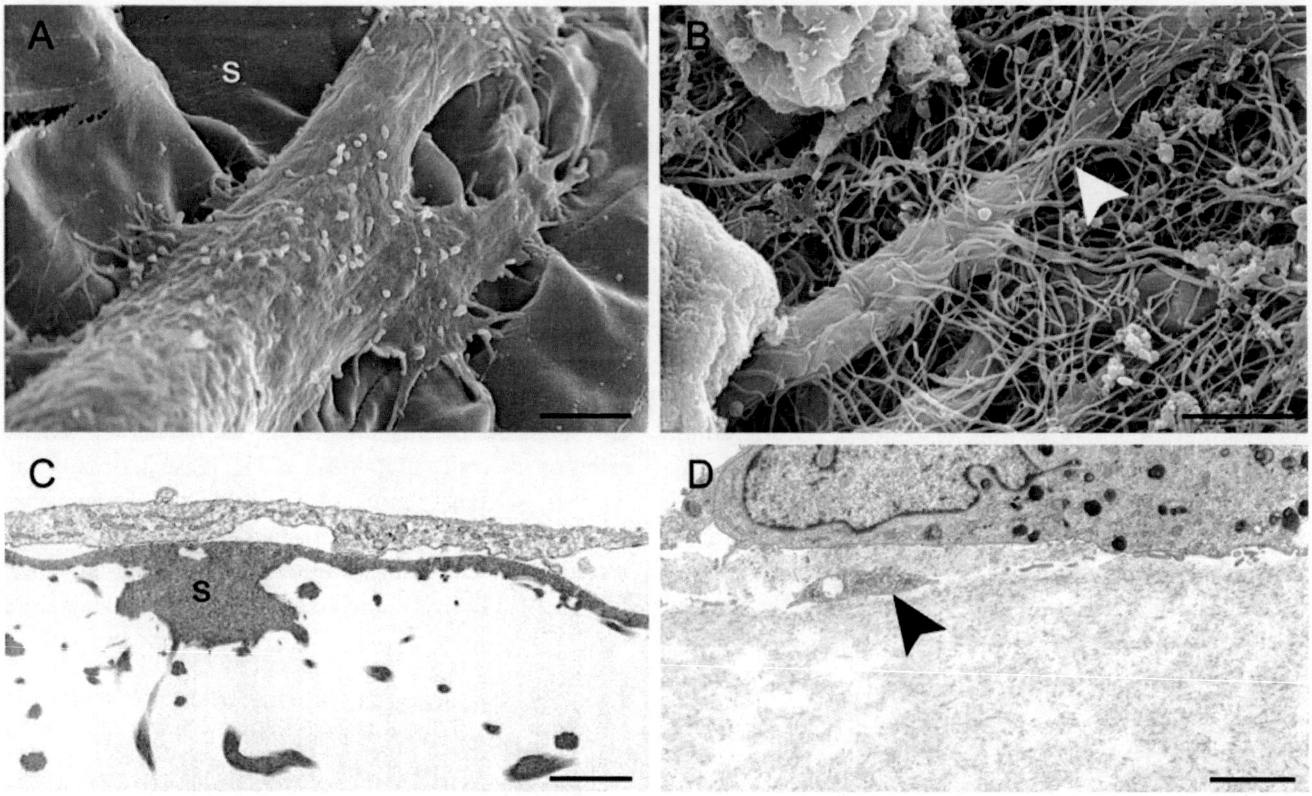

Fig. 16.12 (A, B) SEM and (C, D) TEM images of cells growing on the surface of fibrin gel with (A) or without (B) a "fibrin-skin". In (A) and (C) the "skin" (*s*) forms a dense layer hampering the cells to invade into deeper clot regions. (B) and (D): On the surface without "skin" the cells penetrate the fibrous matrix with cell processes (arrowhead). Scale bars: (A) 3 μm; (B) 5 μm; (C) 1 μm; and (D) 3 μm

and pushing apart fiber strands, as suggested in previous studies (Ciano et al., 1986; Lanir et al., 1988) and not by digestion of the fibrin. In this case, high interfiber spaces support and low porosity hampers migration (Fig. 16.11). Also, if cells digest the matrix, either by cellular plasminogen activation or metalloproteinases (Ronfard and Barrandon, 2001) or phagocytosis (Schlag and Redl, 1986), migration could be somewhat delayed in dense fibrin. Apart from this morphologic aspect, stronger adhesion of the cells by quantitatively increased contact could be another reason for slower migration in a dense fiber network.

Experimental setups for cell migration can be performed in several ways: by inside-out migration of encapsulated cells, by transmigration or sandwich assays, by covering cells with a fibrin layer or by seeding them on the top of a clot. In the latter case, it has to be considered that a "fibrin-skin" may seal the surface of the fibrous clot and prevent ingrowth (Fig. 16.12). In addition, the smooth surface of the "skin" changes the surface roughness, and simultaneously cellular behavior and proliferation (Jiang et al., 2010).

For *in vivo* applications it is probable that "fibrin-skins" do not form because of immediate contact with the body fluid which prevents "skin" formation. In addition, it was found that about 15 min are needed until the "skin" starts to develop in the case of 50 mg/ml fibrinogen and 2 IU (data not shown) which is longer than the period of polymerization during clinical application.

16.2.2 Soft Tissue Engineering Using Adipose-derived Stem Cells in 3D Fibrin Matrix of Low Component Concentration (*Anja Peterbauer-Scherb, Martijn van Griensven, Krystyna Labuda, Christian Gabriel, Heinz Redl and Susanne Wolbank*)

Treatment of soft tissue defects caused by surgical interventions, severe burns, decubital ulcera or complex traumatic injuries with soft tissue deficits provides a clinical challenge. The reconstruction of contour defects by fat transplantation is mostly unsatisfactory as the transplanted adipose tissue is subject to shrinkage leading to unpredictable results. In many cases the transplanted fat is replaced by connective or even calcified tissue (Kral and Crandall, 1999) or liponecrotic cysts are observed (Zheng et al., 2008). Tissue engineering could provide solutions for this clinical need by combining cells with appropriate scaffold materials and stimulating cues such as growth factors. Fibrin has favorable mechanical features comparable to native adipose tissue, allows direct inclusion of cells without sophisticated seeding techniques and application as injectable biomatrix *in vivo*. All of these properties make fibrin a promising biomaterial for delivery of cells in soft tissue engineering. Mature adipocytes would already express the desired phenotype. However, they are fragile and show low expandability (Tanzi and Fare, 2009). Furthermore, mature adipocytes have a significantly higher oxygen consumption than adipose precursor cells making them vulnerable to hypoxic conditions (von Heimburg et al., 2005). The application of stem cells contained in adipose tissue (adipose-derived stem cells; ASC) is therefore a promising approach for overcoming these limitations. ASC are capable of differentiating towards several mature cell types including osteoblasts, chondrocytes, adipocytes, myocytes, hepatocytes, and neuronal cells (Halvorsen et al., 2001; Zuk et al., 2001; Safford et al., 2002; Seo et al., 2005). Furthermore, they exert angiogenic features (Hausman and Richardson, 2004; Rehman et al., 2004; Borges et al., 2006) or even possess endothelial differentiation potential (Miranville et al., 2004; Planat-Benard et al., 2004; Cao et al., 2005), which could facilitate vascularization of the newly formed fat tissue.

The suitability of ASC combined with fibrin matrix of variable composition has been evaluated for adipose tissue-equivalent formation *in vitro* (Peterbauer-Scherb et al., 2010). Therefore, undifferentiated ASC were embedded in fibrin clots composed of 2 IU/ml thrombin and fibrinogen of 6.25, 12.5 and 25 mg/ml (both Tisseel®; Baxter, Austria) and kept under control or adipogenic conditions for up to 28 days. Low component concentrations should support cell viability (Bensaid et al., 2003) and allow homogenous distribution of cells in the clots. The constructs were generated under addition of aprotinin, which prolongs the stability of 3D constructs and limits cellular apoptosis (Gille et al., 2005).

Scanning electron microscopy evaluation showed that the embedded cells were homogenously distributed throughout the clot (data not shown). The applied fibrinogen concentrations did not influence cell viability as demonstrated by calceinAM-DAPI staining. On days 2 and 7, cells appeared elongated in the 6.25 and 12.5 mg/ml fibrinogen clots but remained small and spherical in the 25 mg/ml clots. After 28 days, the cell morphology in adipogenic medium was remarkably enlarged, round, and containing lipid-vacuoles (Figs. 16.13, 16.14).

When quantified by lactate-dehydrogenase (LDH) assay, cell number increased to d7 (>3-fold) and decreased thereafter until day 28 (Fig. 16.15). Proliferation was unaffected by fibrinogen concentration in the control. This is contradictory to previous studies, showing a consistent decrease in bone marrow MSC, or fibroblast proliferation with increasing fibrinogen concentrations (ranging 5–50 mg/ml) (Cox et al., 2004; Catelas et al., 2006; Ho et al., 2006). Adipogenic conditions generally yielded higher cell numbers which were even increased with increasing fibrinogen concentrations. This may partly be attributed to migration of ASC especially from low fibrinogen clots. Increased proliferation in adipogenic medium could potentially be attributed to secreted leptin levels (~50 ng/ml), which have been reported to stimulate rat preadipocyte proliferation (Wagoner et al., 2006).

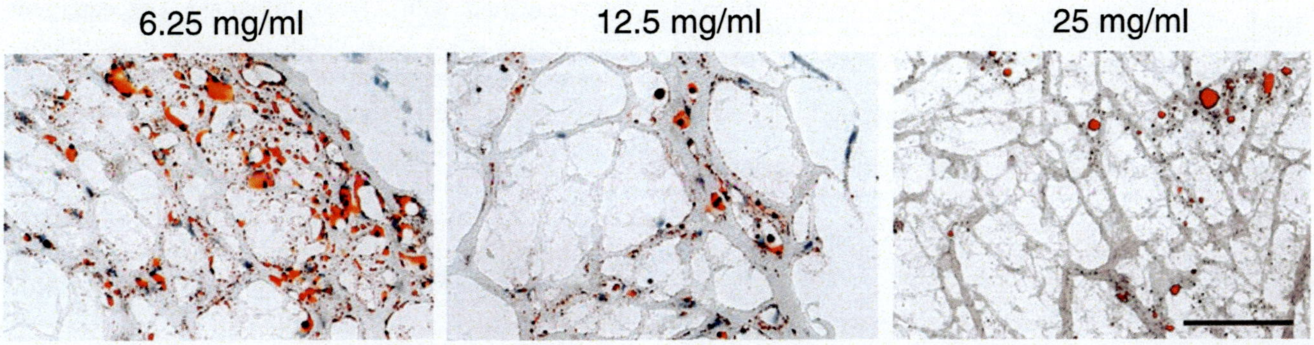

Fig. 16.13 Lipid vesicle formation of human ASC/fibrin constructs composed of 6.25, 12.5, and 25 mg/ml fibrinogen and 2 IU/ml thrombin after 28 days of adipogenic culture as determined by Oilred O (*OO*) staining of cryosections. Representative picture of one donor. Scale bar: 200 µm

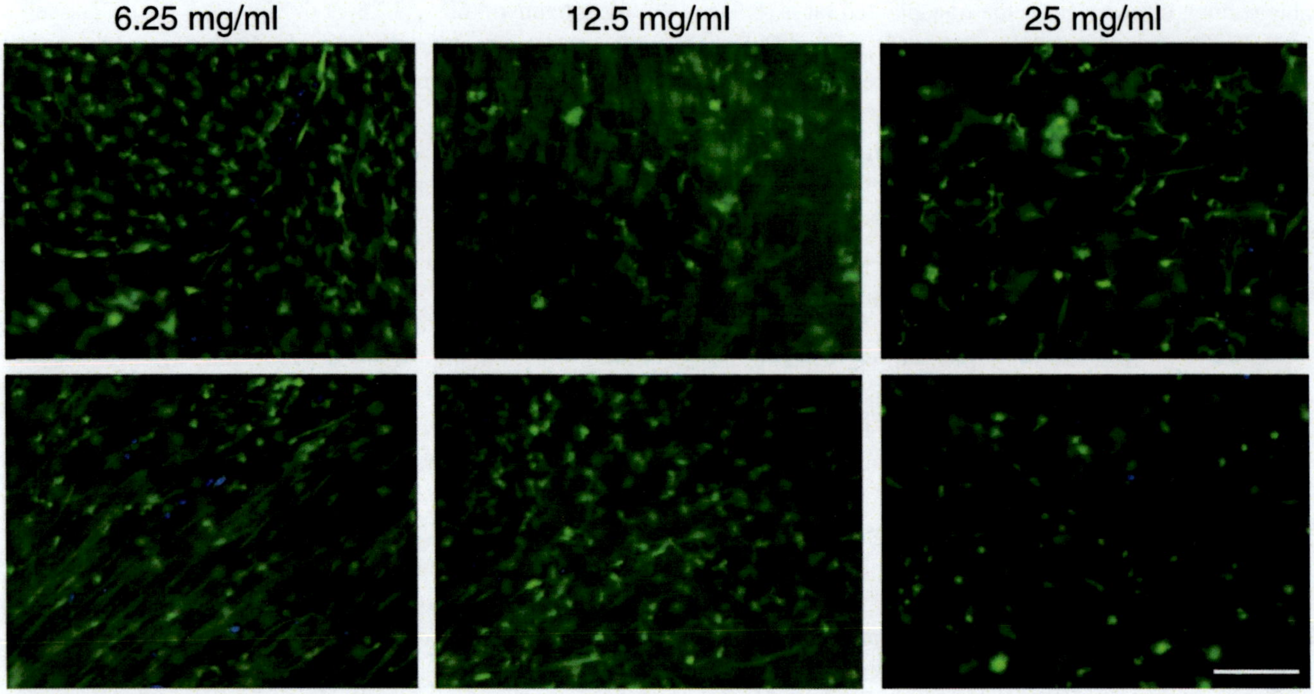

Fig. 16.14 Live dead staining (green/blue) of human ASC/fibrin constructs composed of 6.25, 12.5, and 25 mg/ml fibrinogen and 2 IU/ml thrombin after 28 days of adipogenic (upper panel) and control (lower panel) culture. Scale bar: 200 μm

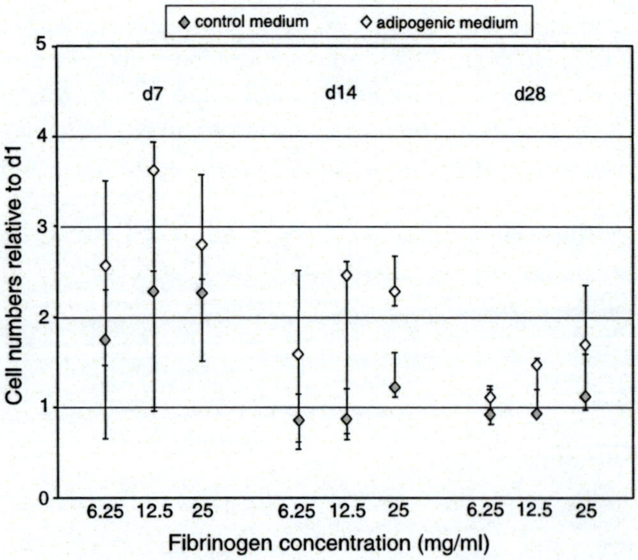

Fig. 16.15 Quantification of human ASC embedded in fibrin using 6.25, 12.5, and 25 mg/ml fibrinogen and 2 IU/ml thrombin, cultured under adipogenic and control medium for 1, 7, 14, and 28 days as determined by LDH assay. Data for 3 donors are shown relative to cell numbers found on d1. Median, Q1, Q3

For the tested adipogenic markers fatty acid binding protein 4 (FABP4) and peroxisome proliferative activated receptor gamma (PPARγ), strong induction was observed under adipogenic conditions demonstrating successful differentiation (Fig. 16.16; qRTPCR; Fig. 16.17 leptin ELISA; Fig. 16.13 Oilred O staining). When embedded in 25 mg/ml fibrinogen clots, ASC showed highest expression levels for FABP4 (up to 629.0-fold), and PPARγ (up to 1.6-fold), corroborated by significantly elevated leptin secretion (33 ng/ml) on day 14. Under these conditions, besides fibrinogen the levels of factors such as fibronectin are more concentrated. This may affect adipogenesis, since fibronectin-coating of growth surfaces can improve adipogenic differentiation of ASC (Kral and Crandall, 1999). Furthermore, enhanced osteogenic differentiation was correlated with higher fibrinogen concentrations using bone marrow MSC in fibrin (Catelas et al., 2006).

Human ASC represent an abundant source of autologously applicable mesenchymal stem cells, which can be isolated with low donor site morbidity (Gimble et al., 2007). Fibrin, on the other hand, originates from human blood and is not associated with immune reactions and could potentially be applied in an autologous manner. Alternatively, approved clinical grade products (Tissucol/Tisseel®, Artiss®, Evicel®, Beriplast®, and Quixil™) are commercially available, but not all of them have been shown to be suitable for cell growth (Fürst et al., 2007). Active growth factors and cell adhesion molecules includ-

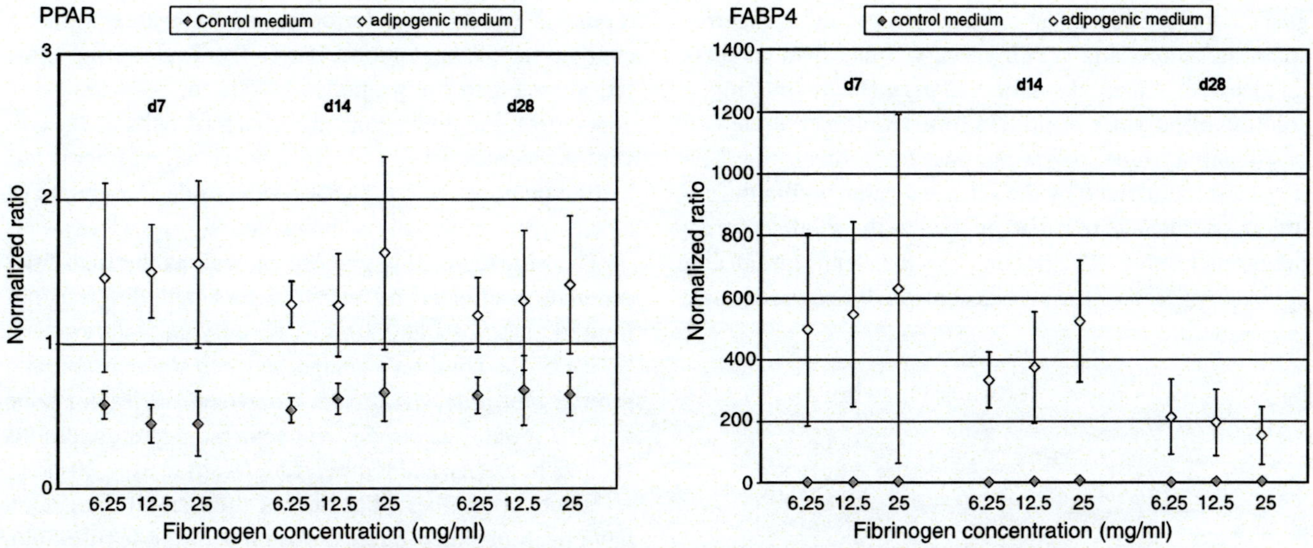

Fig. 16.16 mRNA expression of adipogenic markers PPARγ, FABP4 of human ASC from 4 individual donors embedded in fibrin using 6.25, 12.5, and 25 mg/ml fibrinogen and 2 IU/ml thrombin cultured under adipogenic and control conditions for 7, 14, and 28 days. Ratios normalized to the housekeeping gene hypoxanthine-guanine phosphoribosyltransferase are depicted. mean ± SD

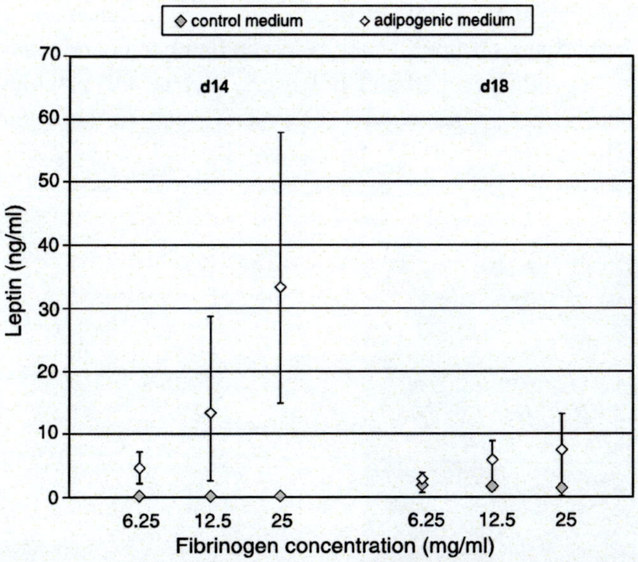

Fig. 16.17 Levels of leptin in the supernatants of human ASC from four individual donors embedded in fibrin using 6.25, 12.5, and 25 mg/ml fibrinogen and 2 IU/ml thrombin cultured under adipogenic and control conditions for 14 and 28 days as determined by ELISA. mean ± SD

ing fibronectin are intrinsically included in the fibrinogen component (Cox et al., 2004). Constructs composed of fibrin matrix of low component concentrations – allowing homogenous cell distribution – with predifferentiated ASC should represent a suitable strategy for adipose tissue formation *in vivo*.

16.2.3 Electrospun Fibrin Nanofiber Matrices (*Tatjana J. Morton, Lila Nikkola, Heinz Redl and Martijn van Griensven*)

As described by Langer and Vacanti in 1993, tissue engineering is "an interdisciplinary field that applies the principles of engineering and life sciences toward the development of biological substitutes that restore, maintain, or improve tissue function" (Langer and Vacanti, 1993). One part of tissue engineering has been the design of scaffolds with biologically and mechanically similarity to native extracellular matrix (ECM). With the technique of electrospinning it is possible to create various structures, shapes, and sizes of fibrin matrices with a high surface/volume ratio and fibers with small diameters of 0.1–1 μm having biological properties similar to the ECM.

Electrostatic spinning, or electrospinning, is a process that utilizes electrostatic forces to create small diameter fibers from the solution of a polymer. The first successful development of electrospinning took place in 1934 by Formhals, who electrospun small fibers of cellulose ester from a solvent of acetone and alcohol. The process can generate generous amounts of fibers at the sub-micron level, smaller in diameter than any standard extrusion process. Electrospinning is based on the electrostatic repulsion within a polymer solution, and the subsequent electrostatic attraction of the polymer solution (or its solute) to a grounded electrode. It is a variant of electrospraying, where droplets of charged solution are attracted to a grounded electrode by an external electric field. Electrospinning succeeds in creating fibers, rather than droplets that leave the charged solution. During electrospinning, a solution of polymer or protein forms a droplet at the end of a capillary tube, or syringe needle, in this setup. The droplet is maintained by the surface tension within the solution, and the electric field applied to the solution forms the droplet into a "Taylor cone", as the electric repulsion begins to overcome the surface tension. When the electric repulsion within the solution reaches a critical value, a charged jet of the solution leaves the Taylor cone. This jet, attracted to the grounded electrode, becomes more concentrated in both solute and charge density as the solvent evaporates. Progressively, a thin charged fiber is left behind that travels to the ground electrode. A diagram of an electrospinning unit is shown below (Fig. 16.18).

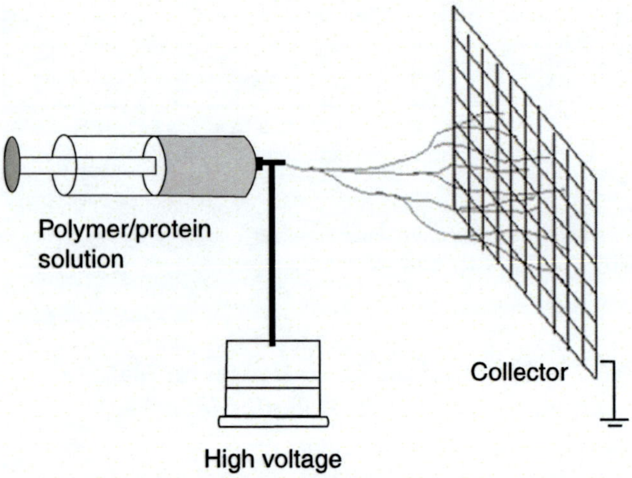

Fig. 16.18 Schematic electrospinning setup

The efficacy of this process, as well as the final fiber product, is affected by many factors including but not limited to the concentration of the polymer solution, viscosity of the solution, voltage between the solution and ground electrode, the distance between the Taylor cone and the ground electrode, and environmental conditions such as humidity and temperature (Katti et al., 2004).

Electrospun fibers, because of their small diameters, have been of much interest not only in the textile field, but also in biomedical research. The small diameter fibers are more attractive for cell attachment, because of their similarity in size to native ECM components, which allow the cell to attach to several fibers in a more natural geometry, rather than the singular flattened orientation that an attached cell would experience on a large diameter fiber. Much research had been conducted with synthetic resorbable polymers, such as polyglycolic acid (PGA), and natural polymers, such as collagen (Boland et al., 2001), fibrinogen (Boland et al., 2001,

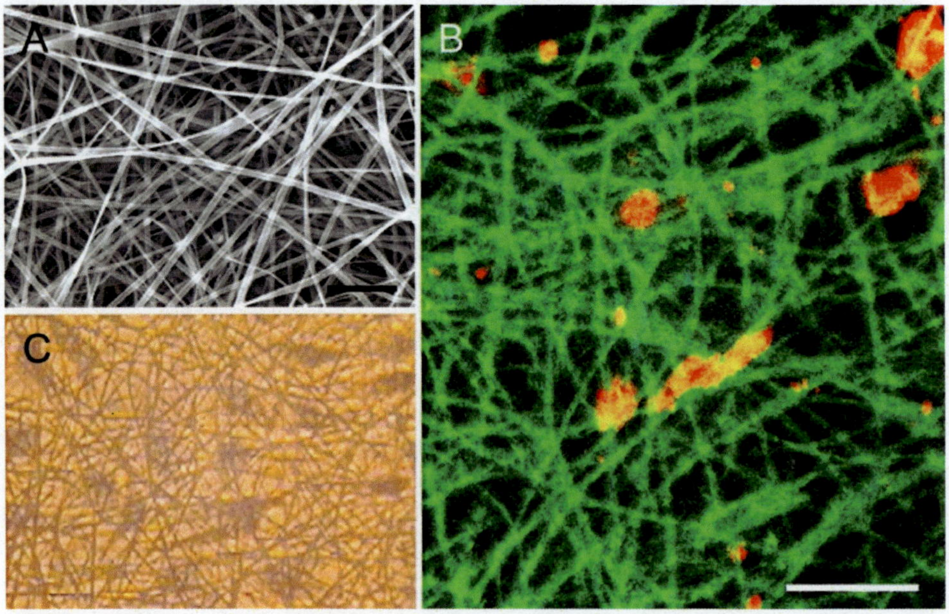

Fig. 16.19 (A) Scanning electron microscope image of electrospun fibrin. (B) Light microscope image of mouse myoblast cells on electrospun fibrin after 5 days of cultivation. (C) Laser scanning microscope image of isolated human adipose-derived mesenchymal stem cells on electrospun fibrin after 5 days of cultivation. Cells migrated between the nanofibers into the scaffold. Scale bars: (A) 20 μm and (B) 30 μm; (C) 200×

2004; Wnek et al., 2003; Morton et al., 2010), and fibrin (Morton et al., 2010).

Cell seeding experiments on electrospun fibrin proved the similarity in size of nanofiber to native extracellular matrix components and the three-dimensional structure allows cells to attach to several fibers in a natural geometry. Both a mouse myoblast cell line and isolated human adipose-derived mesenchymal stem cells showed localization to the scaffold, well-formed morphology, and high viability (Fig. 16.19).

With the function of fibrin as a drug delivery depot (see following Sect. 16.3), electrospun fibrin nanofibers represent a major potential biocompatible and biodegradable scaffold for tissue-engineering applications.

16.3 Fibrin as Matrix for Substances

Natural extracellular matrices (ECMs) of tissues are regarded as depots for growth factors, which affect many physiologic processes in surrounding tissues (Vlodavsky et al., 1991; Richardson et al., 2001; Wong et al., 2003). Similar to ECMs, biomatrix preparations, such as fibrin-based biomaterials, may act as temporary depots for the sustained release of substances, drugs or DNA. Fibrin-based biomaterials are optimally suited as drug depots because of their biocompatibility, advantageous biological properties, established use in hemostasis, tissue sealing, and support of wound healing (MacPhee et al., 1996; Spotnitz, 2001; van Hinsbergh et al., 2001) (see Chapter 15, p. 225).

16.3.1 Release of Substances and Drugs (*Tatjana J. Morton, Martijn van Griensven and Heinz Redl*)

The three-dimensional fibrin clot is ultimately degraded *via* proteolysis by plasmin, and the degradation products are then resorbed by phagocytosis (Martinowitz and Saltz, 1996). The natural degradation of fibrin sealants is a prerequisite for a controlled release drug depot. Substances of interest (i) can be simply mixed into fibrin, (ii) bound *via* naturally affinity to fibrin, or (iii) linked to naturally occurring fibrin-anchors (i.e. thrombin and fibronectin).

Release studies using substances of low molecular weight (i.e. β-galactosidase) or high molecular weight (i.e. cytochrome C) simply mixed into fibrin demonstrated an immediate release after clot formation (Morton et al., 2009). Also, change to a denser structure i.e. by the addition of tranexamic acid (t-AMCA, 4-(aminomethyl) cyclohexane carboxylic acid), an antifibrinolytic agent (Fürst et al., 2007), did not affect the fast release of substances with different molecular weights (Morton et al., 2009). Tranexamic acid is one of the most common fibrinolysis inhibitors and influences the structure and mechanical properties of fibrin (Liu et al., 1979; Mosher and Johnson, 1983; Richardson et al., 2001).

Some growth factors, i.e. vascular endothelial growth factor 165 (VEGF-165), basic fibroblast growth factor (FGF-2), and interleukin-1β (IL-1β), were evaluated as naturally occurring fibrin-anchored growth factors (Sahni et al., 1998, 2004; Wong et al., 2003). Because of their binding affinity to fibrin, studies showed continuous and slow release of these cytokines out of fibrin (Sahni and Francis, 2000).

For pharmaceutically active substance with no binding affinity to fibrin, Schense and Hubbell (1999) developed a methodology for covalent incorporation of exogenous bioactive peptides by transglutaminase activity of factor XIIIa into fibrin during a coagulation process (Schense and Hubbell, 1999). Using that technology VEGF-121 were covalently conjugated in fibrin matrices to demonstrate continuous release out of fibrin and endothelialization in human umbilical vein endothelial cells (Zisch et al., 2001). Also bound bone morphogenetic protein 2 (BMP-2) led to a slow release and new bone formation in an experimental *in vivo* critical size defect model (Schmoekel et al., 2004, 2005).

Substances can also be linked to fibrin-anchors based on naturally occurring proteins with a fibrin binding moiety such as thrombin (TH) and fibronectin (FN) (Morton et al., 2009). Thrombin as a natural byproduct in fibrin formation shows tight binding to fibrin with high binding capacity but without crosslinking. Binding of proteins to thrombin by random chemical crosslinking reactions bears the risk that lysin residues within the fibrin binding exo-loop of thrombin become modified and the fibrin/fibrinogen binding activity is lost. To avoid this effect, a modified form of the irreversible thrombin inhibitor PPACK (D-phenylalanyl-L-propyl-L-arginine chloromethyl ketone) was used to bind proteins to a specific site on thrombin. The modified PPACK is bound easily to a protein of interest and will direct the protein to the active site of thrombin and form a covalent link without affecting the fibrin/fibrinogen binding activity of thrombin (Lyon et al., 1995). Fibronectin, the second natural byproduct in fibrin sealant formulations, is a large molecule and binds to fibrin *via* affinity and FXIII-crosslinking. Linking to fibrin *via* fibronectin binding was carried out by covalent 1-ethyl-3-3-di-

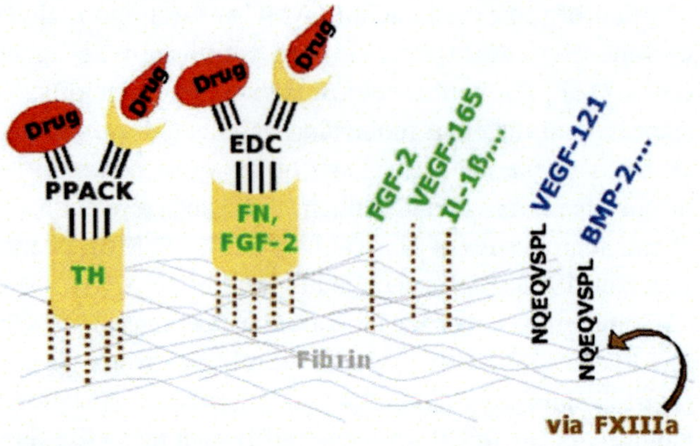

Fig. 16.20 Substances as growth factors (FGF-2, VEGF-165, IL-1β,...) with a high naturally affinity (·····) are bound to fibrin. Substances without naturally affinity to fibrin are linked through fibrin conjugates to fibrin. Fibrin conjugates consist of a fibrin-anchor (FN/FGF-2-EDC or TH-PPACK), which has an affinity binding to fibrin and is covalently (—) bound directly to a pharmaceutically active substance or indirectly *via* a drug binding moiety. Moieties are either affinity bound or covalently bound. Proteins or peptides, which are modified with the factor XIIIa sequence NQEQVSPL, can be covalently linked to fibrin during coagulation process

methyl-aminopropylcarbodiimide (EDC). The fibronectin was bound to a tumor necrosis factor (TNF) antibody for further affinity binding of TNF (Harlow and Lane, 1988; Locksley et al., 2001). Both thrombin and fibronectin are proteins with a high natural binding affinity to fibrin. By using both of these a slower and continuous release of various substances not having a natural affinity to fibrin components could be achieved (Morton et al., 2009). In summary, a pharmaceutically active substance can be modified or the fibrin matrix can be modified to bind the substance in the matrix. A fibrin binding moiety, or fibrin-anchor, can be directly linked to a pharmaceutically active substance or indirectly linked to a drug binding moiety. The resulting structure, whether directly or indirectly bound, are termed fibrin conjugates. As such there are three possibilities for binding substances to fibrin for slow release, namely (1) they either have a natural affinity to fibrin, or (2) are substances with such a natural affinity used as fibrin-anchors, or (3) are substances with an additional FXIIIa reactive moiety linked by FXIIIa to fibrin (Fig. 16.20).

16.3.2 Gene-activated Matrix (*Georg A. Feichtinger, Heinz Redl and Martijn van Griensven*)

Fibrin sealants can be used as biocompatible carrier/drug release system for the delivery of therapeutic nucleic acids. In the past, several biomaterials have been used in tissue engineering applications for the delivery of naked DNA, a method that is generally referred to as non-viral gene activated matrix (GAM) gene delivery (Bonadio, 2000; Bleiziffer et al., 2007; Betz et al., 2008). Such non-viral GAM systems combine biocompatible carriers and DNA with or without transfection reagents (Bonadio, 2000; Andree et al., 2001; Michlits et al., 2007; Betz et al., 2008; Schillinger et al., 2008; des Rieux et al., 2009; Lei et al., 2009) and allow sustained release of the therapeutic DNA locally *in vivo* at a defined target site. The general principle is to transfect either endogenous target cells that are infiltrating the carrier matrix during wound healing, or exogenous cells that are directly incorporated into the matrix prior to implantation (Bonadio, 2000). Tissue regeneration is induced by the GAM-mediated transfer of therapeutic genes that encode growth factor or transcription factor genes, which induce cellular differentiation. The locally transfected target cells produce the growth factors/transcription factors and thus trigger the local induction of tissue regeneration at the implant site. Through application of suitable therapeutic genes (depending on the target tissue) it is therefore possible to induce the *in situ* generation of functional tissue through paracrine and autocrine stimuli that are produced by the transfected cells.

GAMs have been applied in tissue regeneration initially using naked plasmid DNA, on for example, collagen matrices as carriers (Bonadio et al., 1999). More recently fibrin (and other hydrogels) has become an interesting option, because it can be applied to the target site non-invasively and its ease of handling. Furthermore, this treatment option is considered relatively economical (low production cost of plasmid DNA compared to recombinant growth factors) and safe (non-viral gene delivery, *in situ* transfection). DNA, transfection reagents, and cells can easily be incorporated in hydrogels rather than just coated to the matrix (as in, for example collagen GAMs). The charge-based interaction of DNA with fibrin and fibrinogen (Morton et al., 2009) allows a sustained release of naked plasmid DNA from the matrix. It has been demonstrated that fibrin matrices can produce a sustained release of plasmid DNA for a period of over 19 days at the implant site (des Rieux et al., 2009).

In order to enhance the relatively low transfection efficacy of this mode of gene delivery it is possible to apply additional transfection reagents such as liposomes to complex the DNA *prior to* incorporation into the hydrogel. It has been shown that lipoplexes retain their stability within fibrin and can efficiently transfect invading or incorporated cells (Michlits et al., 2007; des Rieux et al., 2009; Lei et al., 2009). Interestingly, there is evidence (Lei et al., 2009) that the toxicity of lipoplexes, an inherent drawback of this transfection reagent, is reduced if applied within fibrin hydrogels. Another approach of incorporating DNA into fibrin hydrogels is to couple the DNA to peptides that are capable of being actively incorporated into the fibrin matrix meshwork, for example through the use of a transglutaminase substrate site (Trentin et al., 2005). Other modalities of DNA condensation that have been used include copolymer protected gene vectors (Schillinger et al., 2008) in which the DNA is complexed to cationic and anionic copolymers to produce a condensed and shielded DNA copolymer complex.

The underlying mechanism of DNA uptake of cells during GAM transfection is still poorly understood. Naked DNA and lipoplexes are taken up by cells during migration into or within the GAMs during the normal wound healing response and thus lead to transfection (Bonadio, 2000). For copolymer protected DNA in fibrin, it has been shown that chondrocytes, for example, mainly incorporate these complexes in a clathrin-independent route of endocytosis *via* phagocytosis and macropinocytosis (Schillinger et al., 2008). The mechanism that governs GAM-mediated transfection of plasmid DNA, however, still has to be elucidated. Nevertheless, several parameters influencing transfection efficacy have already been defined (Lei et al., 2009). DNA-release and the transfection of target cells through fibrin GAMs have been found to be mainly dependent on:

- fibrinogen/fibrin concentration
- plasmid DNA concentration/lipoplex concentration
- cell-mediated fibrin degradation

Examples of non-viral fibrin GAMs that have been experimentally applied in tissue regeneration include cartilage (Schillinger et al., 2008), wound healing (Andree et al., 2001; Michlits et al., 2007; Branski et al., 2010), and myocardial application (Christman et al., 2005). The use of fibrin as a hydrogel for GAM based gene delivery holds great promise for tissue regeneration, but it is necessary to identify the underlying mechanism and the respective bottlenecks of fibrin based transfection in order to further enhance the efficacy and safety of these functionalized biomaterials for future clinical application. The augmentation of the still relatively low transfection efficacy of GAMs in general (either through matrix modification or application of novel transfection reagents) is of great importance to providing a system that can efficiently transfect target cells, and which is therefore capable of producing the strong *in situ* stimulus required for efficient tissue regeneration. Currently applied GAMs that possess limited transfection efficacy can nonetheless achieve an adequate, although not optimal, differentiation stimulus because of the higher bioactivity of endogenously produced growth factors compared to their recombinant counterparts. The limited amounts of growth factor produced in low transfection efficacy GAMs can therefore still be sufficient for improved tissue regeneration.

Despite the mentioned bottlenecks and the lack of knowledge about the mechanisms of this gene therapy approach for tissue regenerative, there is general consensus that fibrin based GAMs and other GAMs hold promising potential for future use in tissue regeneration (Bonadio, 2000).

Acknowledgments

The overview of fibrin research given in this chapter is based on studies partially funded by the Lorenz Böhler Fonds, the European STREP Project Hippocrates (NMP3-CT-2003-505758), and the Marie Curie grant "Alea Jacta EST" (MEST-CT-2004-8104) and carried out within the scope of the European NoE Expertissues (NMP3-CT-2004-500283) and Angioscaff (NMP-2008-214402) programs.

The authors would like to thank the "Cell Imaging and Ultrastructure Research Unit" CIUS and Dr. Guenter Resch from the IMP-IMBA–GMI Electron Microscopy Facility in Vienna as well as the University of Applied Science in Vienna (Fachhochschule Technikum Wien) for providing the equipment for the electron microscope investigations. Furthermore, we would like to thank Cordula Bartl in the Department of Pathology at the University of Veterinary Medicine Vienna for providing cultures of canine uterus glands. We would also like to express thanks to Prim. Dr. Öhlinger, Head of the Institute of Pathology of the Mostviertel Hospital in Amstetten, Austria, for his input on the pathohistological evaluations.

References

Aho HJ, Viljanto J, Raekallio J, and Pelliniemi LJ (1983) Ultrastructural characteristics of cells in human wound collected by Cellstic device. Journal of Surgical Research 35(6): 498–506.

Akassoglou K, Akpinar P, Murray S, and Strickland S (2003) Fibrin is a regulator of Schwann cell migration after sciatic nerve injury in mice. Neuroscience Letters 338(3): 185–188.

Alsberg CL and Clark ED (1908) The blood clot of *Limulus polyphemus*. Journal of Biological Chemistry 5: 323–329.

Andree C, Voigt M, Wenger A, Erichsen T, Bittner K, Schaefer D, Walgenbach KJ, Borges J, Horch RE, Eriksson E, and Stark GB (2001) Plasmid gene delivery to human keratinocytes through a fibrin-mediated transfection system. Tissue Engineering 7(6): 757–766.

Anglés-Cano E (2006) Plasminogen activation, fibrinolysis, and cell proteolytic activity in antiphospholipid syndrome. In: Khamashta MA (ed) Hughes Syndrome, 2nd Ed. Springer-Verlag, London: pp 457–469.

Bach TL, Barsigian C, Yaen CH, and Martinez J (1998) Endothelial cell VE-cadherin functions as a receptor for the beta15–42 sequence of fibrin. Journal of Biological Chemistry 273(46): 30719–30728.

Bensaid W, Triffitt JT, Blanchat C, Oudina K, Sedel L, and Petite H (2003) A biodegradable fibrin scaffold for mesenchymal stem cell transplantation. Biomaterials 24(14): 2497–2502.

Betz VM, Betz OB, Harris MB, Vrahas MS, and Evans CH (2008) Bone tissue engineering and repair by gene therapy. Frontiers in Bioscience 13: 833–841.

Bleiziffer O, Eriksson E, Yao F, Horch RE, and Kneser U (2007) Gene transfer strategies in tissue engineering. Journal of Cellular and Molecular Medicine 11(2): 206–223.

Blomback B and Okada M (1982) Fibrin gel structure and clotting time. Thrombosis Research 25(1–2): 51–70.

Blomback B, Carlsson K, Hessel B, Liljeborg A, Procyk R, and Aslund N (1989) Native fibrin gel networks observed by 3D microscopy, permeation and turbidity. Biochimica et Biophysica Acta 997(1–2): 96–110.

Boland ED, Wnek GE, Simpson DG, Pawlowski KJ, and Bowlin GL (2001) Tailoring tissue engineering scaffolds using electrostatic processing techniques: a study of poly(glycolic acid). Journal of Macromolecular Science: Pure and Applied Chemistry A 38: 1231–1243.

Boland ED, Matthews JA, Pawlowski KJ, Simpson DG, Wnek GE, and Bowlin GL (2004) Electrospinning collagen and elastin: preliminary vascular tissue engineering. Frontiers in Bioscience 9: 1422–1432.

Bonadio J (2000) Tissue engineering via local gene delivery: update and future prospects for enhancing the technology. Advanced Drug Delivery Reviews 44(2–3): 185–194.

Bonadio J, Smiley E, Patil P, and Goldstein S (1999) Localized, direct plasmid gene delivery in vivo: prolonged therapy results in reproducible tissue regeneration. Nature Medicine 5(7): 753–759.

Borges J, Torio-Padron N, Momeni A, Mueller MC, Tegtmeier FT, and Stark BG (2006) Adipose precursor cells (preadipocytes) induce formation of new vessels in fibrin glue on the newly developed cylinder chorioallantoic membrane model (CAM). Minimally Invasive Therapy & Allied Technologies 15(4): 246–252.

Branski LK, Masters OE, Herndon DN, Mittermayr R, Redl H, Traber DL, Cox RA, Kita K, and Jeschke MG (2010) Pre-clinical evaluation of liposomal gene transfer to improve dermal and epidermal regeneration. Gene Therapy 17(6): 770–778.

Brown LF, Lanir N, McDonagh J, Tognazzi K, Dvorak AM, and Dvorak HF (1993) Fibroblast migration in fibrin gel matrices. American Journal of Pathology 142(1): 273–283.

Buchta C, Hedrich HC, Macher M, Hocker P, and Redl H (2005) Biochemical characterization of autologous fibrin sealants produced by CryoSeal and Vivostat in comparison to the homologous fibrin sealant product Tissucol/Tisseel. Biomaterials 26(31): 6233–6241.

Cao Y, Sun Z, Liao L, Meng Y, Han Q, and Zhao RC (2005) Human adipose tissue-derived stem cells differentiate into endothelial cells in vitro and improve postnatal neovascularization in vivo. Biochemical and Biophysical Research Communications 332(2): 370–379.

Catelas I, Sese N, Wu BM, Dunn JC, Helgerson S, and Tawil B (2006) Human mesenchymal stem cell proliferation and osteogenic differentiation in fibrin gels in vitro. Tissue Engineering 12(8): 2385–2396.

Chang MC, Wang BR, and Huang TF (1995) Characterization of endothelial cell differential attachment to fibrin and fibrinogen and its inhibition by Arg-Gly-Asp-containing peptides. Journal of Thrombosis and Haemostasis 74(2): 764–769.

Chernysh IN and Weisel JW (2008) Dynamic imaging of fibrin network formation correlated with other measures of polymerization. Blood 111(10): 4854–4861.

Christman KL, Fang Q, Yee MS, Johnson KR, Sievers RE, and Lee RJ (2005) Enhanced neovasculature formation in ischemic myocardium following delivery of pleiotrophin plasmid in a biopolymer. Biomaterials 26(10): 1139–1144.

Ciano PS, Colvin RB, Dvorak AM, McDonagh J, and Dvorak HF (1986) Macrophage migration in fibrin gel matrices. Laboratory Investigation 54(1): 62–70.

Collet JP, Park D, Lesty C, Soria J, Soria C, Montalescot G, and Weisel JW (2000) Influence of fibrin network conformation and fibrin fiber diameter on fibrinolysis speed: dynamic and structural approaches by confocal microscopy. Arteriosclerosis, Thrombosis, and Vascular Biology 20(5): 1354–1361.

Collet JP, Nagaswami C, Farrell DH, Montalescot G, and Weisel JW (2004) Influence of gamma' fibrinogen splice variant on fibrin physical properties and fibrinolysis rate. Arteriosclerosis, Thrombosis, and Vascular Biology 24(2): 382–386.

Cooper AV, Standeven KF, and Ariens RA (2003) Fibrinogen gamma-chain splice variant gamma' alters fibrin formation and structure. Blood 102(2): 535–540.

Corbett SA and Schwarzbauer JE (1999) Requirements for alpha(5)beta(1) integrin-mediated retraction of fibronectin-fibrin matrices. Journal of Biological Chemistry 274(30): 20943–20948.

Cox S, Cole M, and Tawil B (2004) Behavior of human dermal fibroblasts in three-dimensional fibrin clots: dependence on fibrinogen and thrombin concentration. Tissue Engineering 10(5–6): 942–954.

De Cristofaro R and De Candia E (2003) Thrombin domains: structure, function and interaction with platelet receptors. Journal of Thrombosis and Thrombolysis 15(3): 151–163.

de Maat MP and Verschuur M (2005) Fibrinogen heterogeneity: inherited and noninherited. Current Opinion in Hematology 12(5): 377–383.

des Rieux A, Shikanov A, and Shea LD (2009) Fibrin hydrogels for non-viral vector delivery in vitro. Journal of Controlled Release 136(2): 148–154.

Di Stasio E, Nagaswami C, Weisel JW, and Di Cera E (1998) Cl-regulates the structure of the fibrin clot. Biophysical Journal 75(4): 1973–1979.

Doolittle RF, Spraggon G, and Everse SJ (1997) Evolution of vertebrate fibrin formation and the process of its dissolution. Ciba Foundation Symposium 212: 4–17.

Farrell DH and al Mondhiry HA (1997) Human fibroblast adhesion to fibrinogen. Biochemistry 36(5): 1123–1128.

Feral JP (2010) Wound healing after arm amputation in *Sepia officinalis* (Cephalopoda: Sepioidea). Journal of Invertebrate Pathology 52: 380–388.

Ferry JD and Morrison PR (1947) Preparation and properties of serum and plasma proteins; the conversion of human fibrinogen to fibrin under various conditions. Journal of the American Chemical Society 69(2): 388–400.

Formhals A (1934): Progress and apparatus for preparing artificial threads. USA Patent No. US 1,975,504.

Fürst W, Banerjee A, and Redl H (2007) Comparison of structure, strength and cytocompatibility of a fibrin matrix supplemented either with tranexamic acid or aprotinin. Journal of Biomedical Materials Research Part B: Applied Biomaterials 82(1): 109–114.

Gabriel DA, Muga K, and Boothroyd EM (1992) The effect of fibrin structure on fibrinolysis. Journal of Biological Chemistry 267(34): 24259–24263.

Gailit J, Clarke C, Newman D, Tonnesen MG, Mosesson MW, and Clark RA (1997) Human fibroblasts bind directly to fibrinogen at RGD sites through integrin alpha(v)beta3. Experimental Cell Research 232(1): 118–126.

Gandossi E, Lunven C, and Berry CN (2000) Role of clot-associated (-derived) thrombin in cell proliferation induced by fibrin clots in vitro. British Journal of Pharmacology 129(5): 1021–1027.

Garcia AJ, Vega MD, and Boettiger D (1999) Modulation of cell proliferation and differentiation through substrate-dependent changes in fibronectin conformation. Molecular Biology of the Cell 10(3): 785–798.

Gille J, Meisner U, Ehlers EM, Muller A, Russlies M, and Behrens P (2005) Migration pattern, morphology and viability of cells suspended in or sealed with fibrin glue: a histomorphologic study. Tissue and Cell 37(5): 339–348.

Gimble JM, Katz AJ, and Bunnell BA (2007) Adipose-derived stem cells for regenerative medicine. Circulation Research 100(9): 1249–1260.

Goessl A and Redl H (2005) Optimized thrombin dilution protocol for a slowly setting fibrin sealant in surgery. European Surgery 37(1): 43–51.

Gorodetsky R, Vexler A, An J, Mou X, and Marx G (1998) Haptotactic and growth stimulatory effects of fibrin(ogen) and thrombin on cultured fibroblasts. Journal of Laboratory and Clinical Medicine 131(3): 269–280.

Gralnick HR, Givelber H, and Abrams E (1978) Dysfibrinogenemia associated with hepatoma. Increased carbohydrate content of the fibrinogen molecule. New England Journal of Medicine 299(5): 221–226.

Greiling D and Clark RA (1997) Fibronectin provides a conduit for fibroblast transmigration from collagenous stroma into fibrin clot provisional matrix. Journal of Cell Science 110(Pt 7): 861–870.

Guo M, Toda K, and Grinnell F (1990) Activation of human keratinocyte migration on type I collagen and fibronectin. Journal of Cell Science 96(Pt 2): 197–205.

Halvorsen YD, Franklin D, Bond AL, Hitt DC, Auchter C, Boskey AL, Paschalis EP, Wilkison WO, and Gimble JM (2001) Extracellular matrix mineralization and osteoblast gene expression by human adipose tissue-derived stromal cells. Tissue Engineering 7(6): 729–741.

Harlow E and Lane D (1988) Antibodies: A Laboratory Manual, 1st Ed. Cold Spring Harbor Laboratory Publications Press, New York.

Hausman GJ and Richardson RL (2004) Adipose tissue angiogenesis. Journal of Animal Science 82(3): 925–934.

Hiraoka N, Allen E, Apel IJ, Gyetko MR, and Weiss SJ (1998) Matrix metalloproteinases regulate neovascularization by acting as pericellular fibrinolysins. Cell 95(3): 365–377.

Ho W, Tawil B, Dunn JC, and Wu BM (2006) The behavior of human mesenchymal stem cells in 3D fibrin clots: dependence on fibrinogen concentration and clot structure. Tissue Engineering 12(6): 1587–1595.

Hotary KB, Yana I, Sabeh F, Li XY, Holmbeck K, Birkedal-Hansen H, Allen ED, Hiraoka N, and Weiss SJ (2002) Matrix metalloproteinases (MMPs) regulate fibrin-invasive activity via MT1-MMP-dependent and -independent processes. Journal of Experimental Medicine 195(3): 295–308.

Jiang Y, Ding G, and Lu S (2010) Behavior of dermal fibroblasts on microdot arrays yields insight into wound healing mechanisms. Molecular Biology Reports (in press).

Kaibara M (1973) Dynamic viscoelastic study of the formation of fibrin networks in fibrinogen-thrombin systems and plasma. Biorheology 10(1): 61–73.

Kaijzel EL, Koolwijk P, van Erck MG, van Hinsbergh VW, and de Maat MP (2006) Molecular weight fibrinogen variants determine angiogenesis rate in a fibrin matrix in vitro and in vivo. Journal of Thrombosis and Haemostasis 4(9): 1975–1981.

Kaminski M and McDonagh J (1983) Studies on the mechanism of thrombin. Interaction with fibrin. Journal of Biological Chemistry 258(17): 10530–10535.

Katti DS, Robinson KW, Ko FK, and Laurencin CT (2004) Bioresorbable nanofiber-based systems for wound healing and drug delivery: optimization of fabrication parameters. Journal of Biomedical Materials Research Part B: Applied Biomaterials 70(2): 286–296.

Kim JP, Zhang K, Chen JD, Kramer RH, and Woodley DT (1994) Vitronectin-driven human keratinocyte locomotion is mediated by the alpha v beta 5 integrin receptor. Journal of Biological Chemistry 269(43): 26926–26932.

Kim JP, Zhang K, Chen JD, Wynn KC, Kramer RH, and Woodley DT (1992) Mechanism of human keratinocyte migration on fibronectin: unique roles of RGD site and integrins. Journal of Cellular Physiology 151(3): 443–450.

Kita R, Takahashi A, Kaibara M, and Kubota K (2002) Formation of fibrin gel in fibrinogen-thrombin system: static and dynamic light scattering study. Biomacromolecules 3(5): 1013–1020.

Knox P, Crooks S, and Rimmer CS (1986) Role of fibronectin in the migration of fibroblasts into plasma clots. Journal of Cell Biology 102(6): 2318–2323.

Kral JG and Crandall DL (1999) Development of a human adipocyte synthetic polymer scaffold. Plastic and Reconstructive Surgery 104(6): 1732–1738.

Kubo M, Van de Water L, Plantefaber LC, Mosesson MW, Simon M, Tonnesen MG, Taichman L, and Clark RA (2001) Fibrinogen and fibrin are anti-adhesive for keratinocytes: a mechanism for fibrin eschar slough during wound repair. Journal of Investigative Dermatology 117(6): 1369–1381.

Langer R and Vacanti JP (1993) Tissue engineering. Science 260(5110): 920–926.

Lanir N, Ciano PS, Van de WL, McDonagh J, Dvorak AM, and Dvorak HF (1988) Macrophage migration in fibrin gel matrices. II. Effects of clotting factor XIII, fibronectin, and glycosaminoglycan content on cell migration. Journal of Immunology 140(7): 2340–2349.

Laurens N, Koolwijk P, and de Maat MP (2006) Fibrin structure and wound healing. Journal of Thrombosis and Haemostasis 4(5): 932–939.

Lei P, Padmashali RM, and Andreadis ST (2009) Cell-controlled and spatially arrayed gene delivery from fibrin hydrogels. Biomaterials 30(22): 3790–3799.

Lim BC, Ariens RA, Carter AM, Weisel JW, and Grant PJ (2003) Genetic regulation of fibrin structure and function: complex gene-environment interactions may modulate vascular risk. The Lancet 361(9367): 1424–1431.

Lin F, Ren XD, Doris G, and Clark RA (2005) Three-dimensional migration of human adult dermal fibroblasts from collagen lattices into fibrin/fibronectin gels requires syndecan-4 proteoglycan. Journal of Investigative Dermatology 124(5): 906–913.

Liu CY, Nossel HL, and Kaplan KL (1979) The binding of thrombin by fibrin. Journal of Biological Chemistry 254(20): 10421–10425.

Locksley RM, Killeen N, and Lenardo MJ (2001) The TNF and TNF receptor superfamilies: integrating mammalian biology. Cell 104(4): 487–501.

Lyon ME, Fine JS, Henderson PJ, and Lyon AW (1995) D-phenylalanyl-L-prolyl-L-arginine chloromethyl ketone (PPACK): alternative anticoagulant to heparin salts for blood gas and electrolyte specimens. Clinical Chemistry 41(7): 1038–1041.

Macasev DJ, DiOrio JP, Gugerell A, Goppelt A, Gulle H, and Bittner M (2010) Cell compatibility of fibrin sealants: in vitro study with cells involved in soft tissue engineering. Journal of Biomaterials Applications (in press).

MacPhee MJ, Singh MP, Brady R, Akhyani N, Liau G, Lasa C, Hue C, Best A, and Drohan W (1996) Fibrin sealant: a versatile delivery vehicle for drugs and biologics. In: Sierra DH and Saltz R (eds) Surgical Adhesives and Sealants: Current Technology and Applications. Technomic Publishing AG, Basel: pp 109–120.

Maghzal GJ, Brennan SO, and George PM (2005) The sialic acid content of fibrinogen decreases during pregnancy and increases in response to fibrate therapy. Thrombosis Research 115(4): 293–299.

Mann BK (2003) Biologic gels in tissue engineering. Clinics in Plastic Surgery 30(4): 601–609.

Martin P (1997) Wound healing – aiming for perfect skin regeneration. Science 276: 75–81.

Martinowitz U and Saltz R (1996) Fibrin sealant. Current Opinion in Hematology 3(5): 395–402.

Michlits W, Mittermayr R, Schafer R, Redl H, and Aharinejad S (2007) Fibrin-embedded administration of VEGF plasmid enhances skin flap survival. Wound Repair and Regeneration 15(3): 360–367.

Midwood KS, Mao Y, Hsia HC, Valenick LV, and Schwarzbauer JE (2006) Modulation of cell-fibronectin matrix interactions during tissue repair. Journal of Investigative Dermatology Symposium Proceedings 11(1): 73–78.

Miranville A, Heeschen C, Sengenes C, Curat CA, Busse R, and Bouloumie A (2004) Improvement of postnatal neovascularization by human adipose tissue-derived stem cells. Circulation 110(3): 349–355.

Mooney R, Tawil B, and Mahoney M (2010) Specific fibrinogen and thrombin concentrations promote neuronal rather than glial growth when primary neural cells are seeded within plasma-derived fibrin gels. Tissue Engineering: Parts A 16(5): 1607–1619.

Morton TJ, Furst W, van Griensven M, and Redl H (2009) Controlled release of substances bound to fibrin-anchors or of DNA. Drug Delivery 16(2): 102–107.

Morton TJ, Nikkola L, Grasl C, Wolbank S, Moritz M, Pfeifer S, Redl H, and van Griensven M (2010) Electrospun fibrin nanofibers for the use in tissue engineering. (unpublished).

Mosesson MW, DiOrio JP, Hernandez I, Hainfeld JF, Wall JS, and Grieninger G (2004) The ultrastructure of fibrinogen-420 and the fibrin-420 clot. Biophysical Chemistry 112(2–3): 209–214.

Mosher DF and Johnson RB (1983) Specificity of fibronectin – fibrin cross-linking. Annals of the New York Academy of Sciences 408: 583–594.

Nair CH, Shah GA, and Dhall DP (1986) Effect of temperature, pH and ionic strength and composition on fibrin network structure and its development. Thrombosis Research 42(6): 809–816.

Naito M, Nomura H, and Iguchi A (1996) Migration of cultured vascular smooth muscle cells into non-crosslinked fibrin gels. Thrombosis Research 84(2): 129–136.

O'Toole EA (2001) Extracellular matrix and keratinocyte migration. Clinical and Experimental Dermatology 26(6): 525–530.

Pankajakshan D and Krishnan LK (2009) Design of fibrin matrix composition to enhance endothelial cell growth and extracellular matrix deposition for in vitro tissue engineering. Artificial Organs 33(1): 16–25.

Peterbauer-Scherb A, Danzer M, Gabriel C, van Griensven M, Redl H, and Wolbank S (2010) In vitro adipogenesis of adipose-derived stem cells in 3D fibrin matrix of low component concentration. Journal of Tissue Engineering and Regenerative Medicine (submitted).

Planat-Benard V, Silvestre JS, Cousin B, Andre M, Nibbelink M, Tamarat R, Clergue M, Manneville C, Saillan-Barreau C, Duriez M, Tedgui A, Levy B, Penicaud L, and Casteilla L (2004) Plasticity of human adipose lineage cells toward endothelial cells: physiological and therapeutic perspectives. Circulation 109(5): 656–663.

Redl H, Schlag G, and Dinges HP (1985) Vergleich zweier Fibrinkleber. Medizinische Welt 36: 769–776.

Rehman J, Traktuev D, Li J, Merfeld-Clauss S, Temm-Grove CJ, Bovenkerk JE, Pell CL, Johnstone BH, Considine RV, and March KL (2004) Secretion of angiogenic and antiapoptotic factors by human adipose stromal cells. Circulation 109(10): 1292–1298.

Richardson TP, Murphy WL, and Mooney DJ (2001) Polymeric delivery of proteins and plasmid DNA for tissue engineering and gene therapy. Critical Reviews™ in Eukaryotic Gene Expression 11(1–3): 47–58.

Ronfard V and Barrandon Y (2001) Migration of keratinocytes through tunnels of digested fibrin. Proceedings of the National Academy of Sciences USA 98(8): 4504–4509.

Safford KM, Hicok KC, Safford SD, Halvorsen YD, Wilkison WO, Gimble JM, and Rice HE (2002) Neurogenic differentiation of murine and human adipose-derived stromal cells. Biochemical and Biophysical Research Communications 294(2): 371–379.

Sahni A and Francis CW (2000) Vascular endothelial growth factor binds to fibrinogen and fibrin and stimulates endothelial cell proliferation. Blood 96(12): 3772–3778.

Sahni A, Odrljin T, and Francis CW (1998) Binding of basic fibroblast growth factor to fibrinogen and fibrin. Journal of Biological Chemistry 273(13): 7554–7559.

Sahni A, Guo M, Sahni SK, and Francis CW (2004) Interleukin-1beta but not IL-1alpha binds to fibrinogen and fibrin and has enhanced activity in the bound form. Blood 104(2): 409–414.

Schense JC and Hubbell JA (1999) Cross-linking exogenous bifunctional peptides into fibrin gels with factor XIIIa. Bioconjugate Chemistry 10(1): 75–81.

Schillinger U, Wexel G, Hacker C, Kullmer M, Koch C, Gerg M, Vogt S, Ueblacker P, Tischer T, Hensler D, Wilisch J, Aigner

J, Walch A, Stemberger A, and Plank C (2008) A fibrin glue composition as carrier for nucleic acid vectors. Pharmaceutical Research 25(12): 2946–2962.
Schlag G and Redl H (1986) Fibrin Sealant in Operative Medicine. Springer-Verlag, Berlin.
Schmoekel HG, Weber FE, Seiler G, von Rechenberg B, Schense JC, Schawalder P, and Hubbell J (2004) Treatment of nonunions with nonglycosylated recombinant human bone morphogenetic protein-2 delivered from a fibrin matrix. Veterinary Surgery 33(2): 112–118.
Schmoekel HG, Weber FE, Schense JC, Gratz KW, Schawalder P, and Hubbell JA (2005) Bone repair with a form of BMP-2 engineered for incorporation into fibrin cell ingrowth matrices. Biotechnology and Bioengineering 89(3): 253–262.
Seo MJ, Suh SY, Bae YC, and Jung JS (2005) Differentiation of human adipose stromal cells into hepatic lineage in vitro and in vivo. Biochemical and Biophysical 328(1): 258–264.
Smadja DM, Basire A, Amelot A, Conte A, Bieche I, Le Bonniec BF, Aiach M, and Gaussem P (2008) Thrombin bound to a fibrin clot confers angiogenic and haemostatic properties on endothelial progenitor cells. Journal of Cellular and Molecular Medicine 12(3): 975–986.
Spotnitz WD (2001) Commercial fibrin sealants in surgical care. American Journal of Surgery 182(Suppl 2): 8S–14S.
Tanzi MC and Fare S (2009) Adipose tissue engineering: state of the art, recent advances and innovative approaches. Expert Review of Medical Devices 6(5): 533–551.
Trentin D, Hubbell J, and Hall H (2005) Non-viral gene delivery for local and controlled DNA release. Journal of Controlled Release 102(1): 263–275.
van Hinsbergh VW, Collen A, and Koolwijk P (2001) Role of fibrin matrix in angiogenesis. Annals of the New York Academy of Sciences 936: 426–437.
Veklich Y, Francis CW, White J, and Weisel JW (1998) Structural studies of fibrinolysis by electron microscopy. Blood 92(12): 4721–4729.
Vlodavsky I, Bar-Shavit R, Ishai-Michaeli R, Bashkin P, and Fuks Z (1991) Extracellular sequestration and release of fibroblast growth factor: a regulatory mechanism? Trends in Biochemical Sciences 16(7): 268–271.
von Heimburg D, Hemmrich K, Zachariah S, Staiger H, and Pallua N (2005) Oxygen consumption in undifferentiated versus differentiated adipogenic mesenchymal precursor cells. Respiratory Physiology & Neurobiology 146(2–3): 107–116.
Wagoner B, Hausman DB, and Harris RB (2006) Direct and indirect effects of leptin on preadipocyte proliferation and differentiation. American Journal of Physiology – Regulatory, Integrative and Comparative Physiology 290(6): R1557–R1564.
Weisel JW and Nagaswami C (1992) Computer modeling of fibrin polymerization kinetics correlated with electron microscope and turbidity observations: clot structure and assembly are kinetically controlled. Biophysical Journal 63(1): 111–128.
Weisel JW, Veklich Y, and Gorkun O (1993) The sequence of cleavage of fibrinopeptides from fibrinogen is important for protofibril formation and enhancement of lateral aggregation in fibrin clots. Journal of Molecular Biology 232(1): 285–297.
Wnek GE, Carr ME, Simpson DG, and Bowlin GL (2003) Electrospinning of nanofiber fibrinogen structures. Nano Letters 3: 213–216.
Wolberg AS, Monroe DM, Roberts HR, and Hoffman M (2003) Elevated prothrombin results in clots with an altered fiber structure: a possible mechanism of the increased thrombotic risk. Blood 101(8): 3008–3013.
Wong C, Inman E, Spaethe R, and Helgerson S (2003) Fibrin-based biomaterials to deliver human growth factors. Journal of Thrombosis and Haemostasis 89(3): 573–582.
Xu X and Doolittle RF (1990) Presence of a vertebrate fibrinogen-like sequence in an echinoderm. Proceedings of the National Academy of Sciences USA 87(6): 2097–2101.
Zhao H, Ma L, Zhou J, Mao Z, Gao C, and Shen J (2008) Fabrication and physical and biological properties of fibrin gel derived from human plasma. Journal of Biomedical Materials 3(1): 015001.
Zheng DN, Li QF, Lei H, Zheng SW, Xie YZ, Xu QH, Yun X, and Pu LL (2008) Autologous fat grafting to the breast for cosmetic enhancement: experience in 66 patients with long-term follow up. Journal of Plastic, Reconstructive & Aesthetic Surgery 61(7): 792–798.
Zisch AH, Schenk U, Schense JC, Sakiyama-Elbert SE, and Hubbell JA (2001) Covalently conjugated VEGF – fibrin matrices for endothelialization. Journal of Controlled Release 72(1–3): 101–113.
Zuk PA, Zhu M, Mizuno H, Huang J, Futrell JW, Katz AJ, Benhaim P, Lorenz HP, and Hedrick MH (2001) Multilineage cells from human adipose tissue: implications for cell-based therapies. Tissue Engineering 7(2): 211–228.

17 Biodegradable (Meth)acrylate-based Adhesives for Surgical Applications

Albrecht Berg, Fabian Peters and Matthias Schnabelrauch

Contents

17.1 Introduction

The use of adhesives in surgery is an old but mostly unfulfilled dream (Donkerwolcke et al., 1998). Compared to conventional bonding techniques employed in surgery today like stitching, fixing with screws, pins, and plates, gluing has several advantages because it represents a fast and uncomplicated technique that causes no or only slight injuries of surrounding tissue and enables a homogenous load distribution between bonded materials (Rimpler, 1996). If such an adhesive would be gradually self-degrading in the body, newly formed tissue could replace the adhesive during the healing process and a complete regeneration of the damaged tissue would be possible. A gradual degradation of the adhesive would also maintain the necessary bonding strength within the tissue repair period and finally no foreign material would remain in the body.

Potential applications of those adhesives include bonding of soft tissue, internal organs, and bone fragments or fixation of implants in the body, wound closure, tissue defect filling, and transdermal drug delivery (Reece et al., 2001). Despite this promising potential, the current clinical use of biodegradable adhesives is limited to a few selected indications. A major reason for this situation is the numerous demands a medical adhesive has to fulfill. Among these requirements essentially for a degradable adhesive used in surgery are the following:

- Good adhesion on wet soft and hard tissue in the presence of blood and tissue liquor
- Sufficient adhesion strength (tensile, shear strength, and compression stability), especially for applications in loaded bone
- Easy applicability as injectable paste, suspension or solution of adjustable viscosity
- Rapid curing of the adhesive after application (adjustable preferentially in a range between 2 and 10 min)

- Biocompatibility, namely no or only very slight local or systemic cytotoxicity of the cured adhesive, its single components and formed degradation products and no interference with the natural tissue regeneration process
- No heat generation during curing process leading to necrotic tissue reactions
- Little volume change during curing
- Gradual and complete biodegradability within a predictable time period
- Sterilizability using validated sterilization procedures
- Stability of monomers and other additives under defined storage conditions

Considering all requirements mentioned above, the development of a clinically usable biodegradable adhesive represents a challenging task for biomaterial scientists and engineers (Heiss et al., 2006).

During the last decades numerous classes of adhesives have been tested for their potential use in surgery including fibrin glues (Sierra, 1993; MacGillivray, 2003), gelatin-resorcinol-aldehyde adhesives (Bachet et al., 1997), cyanoacrylates (Vauthier et al., 2003; Shalaby and Shalaby, 2004), protein-dialdehyde systems (Chao and Torchiana, 2003), polysaccharide-derived glues (Oelker and Grinstaff, 2008) or methacrylate and acrylate-based adhesives (Lewis, 1997). The latter class of adhesives has found widespread application in nearly all industrial areas and, based on the pioneering work of Charnley, polymerizable methacrylates have also been introduced into surgery as bone cements to fix endoprosthesis into the bone tissue (Lewis, 1997). Currently, methacrylates and acrylates (= (meth)acrylates) are clinically used not only as bone cements but also as dental adhesives and composites, contact and intraocular lenses, drug release systems and liquid wound closure sprays. In all of these applications, non-biodegradable (meth)acrylate polymers are used mainly because the degradation of the biomaterials is not advantageously in these cases. Nevertheless, numerous efforts have been undertaken to chemically modify (meth)acrylates in order to obtain biodegradable surgical adhesives, implant coatings, or tissue defect-filling materials.

In this chapter a survey will be given about the synthesis of (meth)acrylate monomers and macromers which can be used to create *in situ* curable fully or at least partly degradable polymer and composite biomaterials. Application-relevant properties of these (meth)acrylate-based systems will be reported and their potential as clinically usable adhesives will be discussed with regard to the requirements such materials have to fulfill.

17.2 General Features of (Meth)acrylate Polymerization

(Meth)acrylate monomers used in adhesive systems are conventionally cured by radical polymerization reaction initiated by the addition of radical initiators or the irradiation with light (Fig. 17.1).

The initiation of (meth)acrylate polymerization with common low-molecular weight initiators like peroxides or azo compounds normally requires temperatures above body temperature to promote efficient radical formation. For this reason, in cases where an *in situ* curing of the adhesive within the body is aimed, tertiary amines (e.g., *N,N*-dimethyl-p-toluidine) are added to the adhesive mixture to accelerate radical formation. A (meth) acrylate monomer or mixtures of different (meth)acrylate monomers bearing only one (meth)acrylate function per molecule can be polymerized in the presence of dibenzoyl peroxide and a tertiary amine at room or body temperature with curing times of 2–10 min. Such an adhesive system can be conveniently handled in surgery.

In many technical and also dental adhesive systems bis- and tris-(meth)acrylates are added as co-monomers to produce crosslinks increasing the mechanical stability of the bonding. In contrast to mono-(meth)acrylates which follow a first-order polymerization rate kinetic, (meth)acrylates containing two or more (meth)acrylate functions per molecule polymerize with polymerization

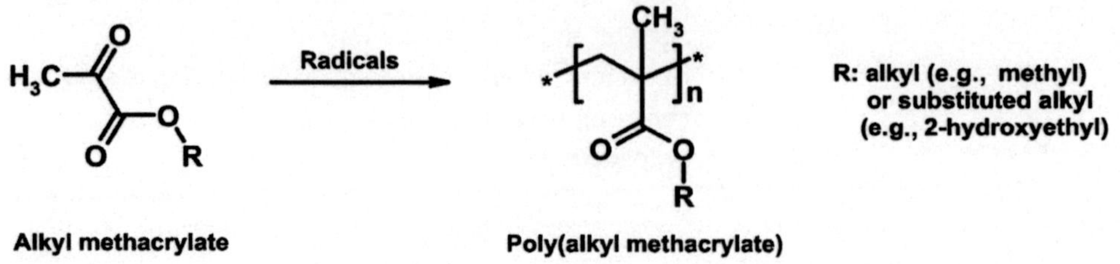

Fig. 17.1 Radical polymerization of alkyl methacrylate

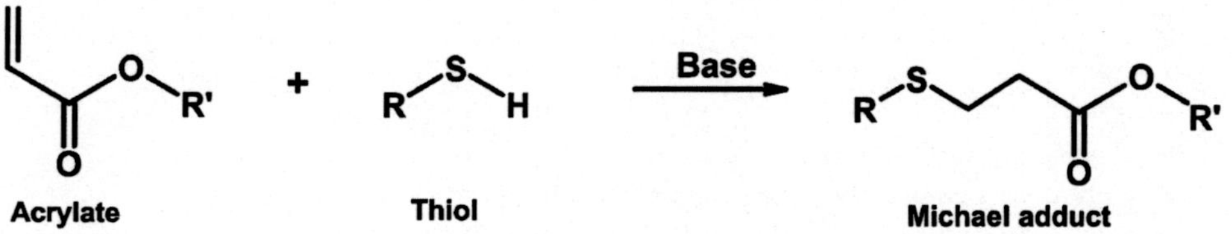

Fig. 17.2 General scheme of Michael-type reaction between an acrylate and a thiol

rates of higher order resulting in very short curing times often within seconds. Furthermore, the use of higher amounts of bis- and tris-(meth)acrylate crosslinkers increases the brittleness of the resulting polymeric network. The design of a (meth)acrylate-based adhesive which can be cured *in situ* in the body therefore needs to be carefully optimized with regard to its composition.

Exposure to UV or UV-near visible light in the presence of suitable radical-generating photoinitiators is another common method to cure (meth)acrylates. Most conventional photoinitiators are based on benzoin, benzophenone or thioxanthone chromophores. An often used photoinitiator in biomedical applications is the α-diketone camphorquinone. Compared to thermal radical polymerization, photopolymerization has the advantage that chain transfer processes terminating chain propagation are minimized. A disadvantage of this method is the limited depth of penetration of light which makes it difficult to cure compact masses. Curing of photo-crosslinkable adhesives in layers is alternatively recommended.

In general, during polymerization (meth)acrylates form non-biodegradable polymers consisting in long hydrocarbon chains not susceptible to hydrolytic or enzymatic cleavage. Therefore, a special material design is necessary to create completely or at least partly biodegradable materials from (meth)acrylate precursors. Two different approaches are known to obtain biodegradable (meth)acrylate systems, one based on the attachment of (meth)acrylate functions at the ends of hydrolytically cleavable or dissoluble synthetic oligomers and the other one on the incorporation of (meth)acrylate functions into degradable biopolymers. Both approaches will be discussed in detail in the following sections.

In addition to radical polymerization, (meth)acrylates are able to undergo a base-catalyzed Michael-type addition with nucleophilic reaction partners like malonates, amines, or thiols. Because a fast curing reaction is needed for adhesives used in surgery, reactive Michael acceptors like acrylates are favored in this case to react with strong nucleophilic partners as thiols. This reaction system is well suited for biomaterials because of its selectivity (acrylates react orders of magnitudes faster with thiols than with other nucleophils) and the absence of toxic by-products. The general reaction scheme of a Michael-type addition between an acrylate and a thiol is shown in Fig. 17.2.

17.3 Oligo- and Polylactone-based (Meth)acrylate Adhesives

Liquid or low-melt oligomeric esters derived from α-hydroxy carboxylic acids represent an interesting class of starting materials for the generation of *in situ* curable adhesives, coatings, or defect-filling materials (Roller and Bezwada, 1996; Wang et al., 1997). These oligoesters can be easily prepared from the lactones and dilactones, respectively, of the corresponding α-hydroxy carboxylic acids by a ring-opening oligomerization in the presence of a suitable hydroxyl or amino group-containing initiator and an esterification catalyst (e.g., stannous octanoate) in the melt of the monomer (Fig. 17.3). The molecular weights of the resulting oligolactones can be controlled by the employed monomer to initiator feed ratio. Based on a set of commercially available lactones (ε-caprolactone, p-dioxanone, and trimethylene carbonate) and dilactones (L- and D,L-lactide, glycolide) and a large number of suitable initiators including monohydric alcohols, di-, tri-, and polyols, a wide variety of structurally different oligoesters can be easily prepared. Amino acid esters have also been successfully used as initiators of the ring-opening oligomerization (Schnabelrauch et al., 2002).

A first attempt to generate adhesive macromers by esterification of the terminal hydroxyl groups of oligomeric α-hydroxy carboxylic acids with reactive acrylic or methacrylic acid derivatives like their anhydrides or chlorides was described by Ritter in 1986. Following this concept, numerous adhesive macromers of varying structures and molecular weights have been synthesized (Fig. 17.3, Storey et al., 1993; Sandner et al., 1997; Vogt et al., 2005). Alternatively, oligolactones can be treated

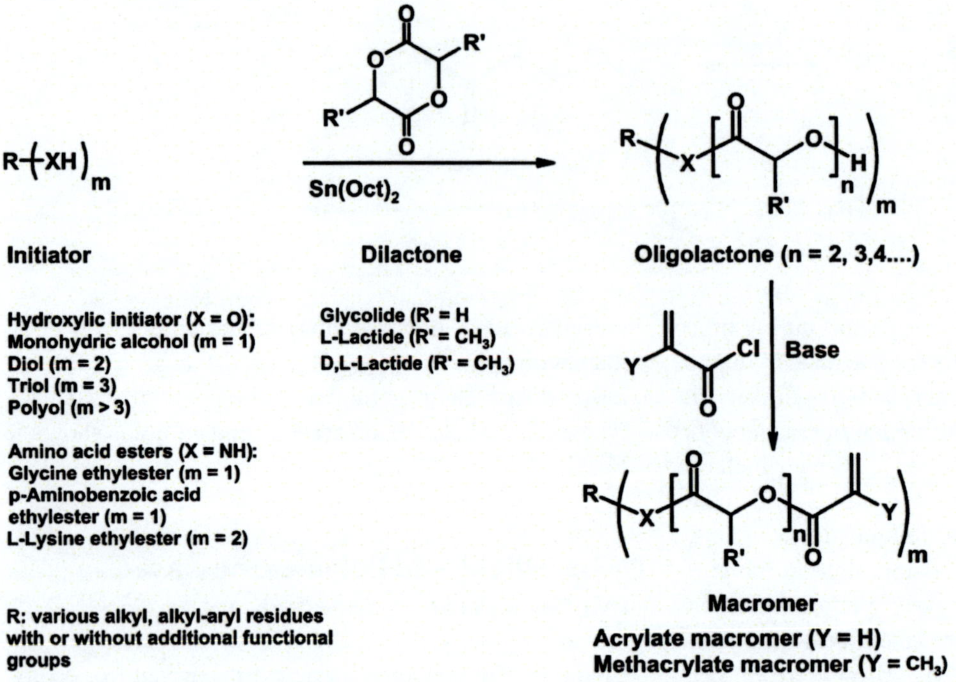

Fig. 17.3 Synthesis of (meth)acrylated oligolactones by ring-opening oligomerization and subsequent end-capping with (methacrylate) functions

with 2-isocyanatoethylmethacrylate in organic solvents to introduce methacrylate groups *via* urethane moieties at the termini of the oligomer chain (Ferreira et al., 2008).

Sawhney et al. (1998) reported a first surgical sealant system based on (meth)acrylated oligolactone macromers. The adhesive was prepared by linking lactide or trimethylene carbonate units on both ends of polyethylene glycol (PEG) and subsequent end-capping with acrylate groups. The water-soluble macromers were photochemically cured with visible light in the presence of an eosin Y/triethanol amine photoinitiation system. At the end of the nineties this system was brought into the clinic under the trade names FocalSeal and Advaseal to seal air leaks associated with lung surgery. FocalSeal-L Synthetic Absorbable Sealant (Focal Inc., USA) became the first surgical adhesive with a crosslinking reagent to be approved by the US Food and Drug Administration (FDA) for clinical use. The product is currently distributed by Genzyme Biosurgery (Lexington, USA). As reported in the literature the adhesive acts as a mechanical leakage barrier in the human body for up to 14 days (Yüksel, 2005).

During the last decade various photo-curable polymers were developed potentially usable as tissue adhesives. In the group of Anseth, a multifunctional macromer was prepared by attaching acrylate groups *via* succinic acid ester units to the hydroxyl groups of poly(vinyl alcohol) (Martens et al., 2002). The resulting macromer (see Fig. 17.7) can be crosslinked *via* photopolymerization with UV light and an Irgacure 2959 photoinitiator forming a hydrophilic biodegradable hydrogel. The ester bonds in the crosslinks are cleaved homogeneously at a rate dictated by the hydrolysis kinetic constant for the ester bond and the number of degradable ester linkages. Another photo-curable sealant system was generated by melt polycondensation of polyethylene glycol of different molecular weights with succinic acid and acylation of the resulting polyester polyol with acryloyl chloride (see Fig. 17.7, Nivasu et al., 2004). The polyesterpolyol acrylate could be easily photo-crosslinked into non-tacky films by long wavelength UV irradiation in the presence of benzophenone/hydroxycyclohexyl acetophenone.

A similar adhesion system containing (meth)acrylated ethylene glycol and glycerol oligolactones was described as sealant in dental applications (Wenz and Nies, 1998). Photo-curing was performed in the presence of campherquinone. Based on the latter photochemically curable system a two-component adhesive for gluing of bone fragments was developed containing ethylene glycol bis(oligolactide methacrylate) as reactive macromer component and the non-methacrylated oligolactone as a second non-reactive component. This second component contains an organoboron compound, namely

9-borabicyclo[3.3.1]nonane (9-BBN) dissolved in PEG as radical initiator initiating polymerization after mixing the components (Wenz, 1998). The adhesive started hardening after 1 min and reached its final stability after 24 h. It was shown that the cured material is slowly degraded in aqueous medium by hydrolytic cleavage forming lactic acid and ethylene glycol. The residual oligomeric methacrylic acids are supposed to be excreted from the body. The developed adhesive exhibited a good biocompatibility and even beginning degradation in a rabbit small animal model using the adhesive in a monocondylar osteotomy of the distal femur after 3 months (Heiss et al., 2005). However, deleterious tissue reactions were found in a long-term study in a sheep model after 6 months (Ignatius et al., 2005; Grossterlinden et al., 2006).

Recently, a poly(propylene glycol-co-lactide) dimethacrylate adhesive with monocalcium phosphate monohydrate (MCPM)/β-tricalcium phosphate (β-TCP) fillers in various levels has been investigated (Zhao et al., 2010).

Curing of this adhesive was performed by treatment with visible light in the presence of camphorquinone/*N*,*N*-dimethylamino-p-toluidine. Water sorption by the photo-cured materials catalyzed varying filler conversion to dicalcium phosphate (DCP). With greater DCP levels, faster release of phosphate and calcium ions and improved buffering of polymer degradation products were observed. Cured films of the adhesive composite implanted into chick embryo femurs became closely apposed to the host tissue and did not appear to induce adverse immunologic reaction.

In our own research at INNOVENT we focused on a dianhydroglucitol bis(oligo-L-lactide) terminated with methacrylate functions (DLM-macromers, Fig. 17.4) as a thermally curable macromer (Schnabelrauch and Vogt, 1999; Vogt et al., 2002). In corresponding adhesive systems we used bis-methacrylated 1,8-octanediol-bis(oligo-L-lactide) and 1,8-octanediol-bis(oligo-L-lactide-co-glycolide) as co-macromers (Fig. 17.4). The macromer synthesis was performed in a two-step procedure by ring-opening oligomerization followed by methacrylate end-capping as described above. Oligo-L-lactones with 2, 4, or 6 lactic acid repeating units per hydroxylic group of the initiating diols were synthesized controlling the molecular weight of the prepared oligomers by the employed molar initiator:lactide ratio. The synthesized methacrylates were obtained as yellowish low to highly viscous liquids at room temperature. Mixtures comprising one of the prepared oligolactone macromers and a conventional radical initiation system (e.g., dibenzoyl peroxide/p-dimethyltoluidine) readily polymerize at room or body

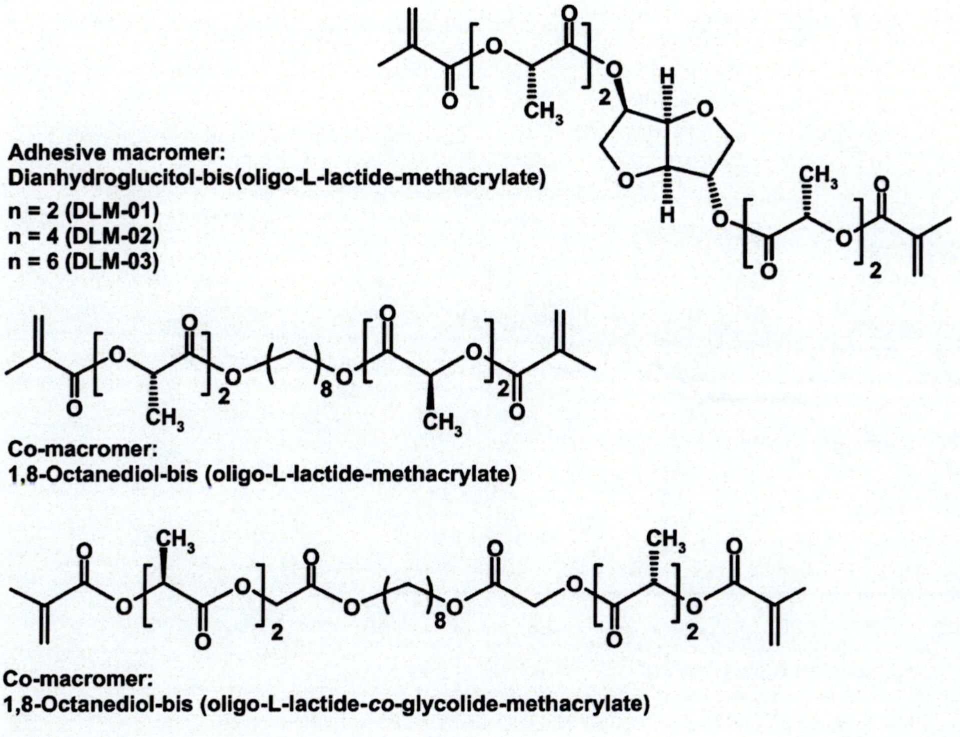

Fig. 17.4 Adhesive macromers and co-macromers

temperature to form solid polymer networks. Curable adhesive mixtures of the macromers were developed varying their composition by adding different organic (sorbitol and glycine) and inorganic (NaCl, $NaHCO_3$, $CaCO_3$, and calcium phosphates) fillers and diluents (Heiss et al., 2009).

Based on these investigations, a three-components adhesive system was established containing a liquid macromer component and two solid components. The macromer component comprises the DLM-macromer, one of the octanediol-based co-macromers, and a reactive diluent (e.g., 2-hydroxyethyl-methacrylate (HEMA)). Both the solid components mainly consist of β-tricalcium phosphate and $NaHCO_3$. In addition, one of the solid components contains the radical initiator and the other one an activator. Mixing the three components thoroughly, the resulting adhesive has a desired curing time of 90–150 s during which the processing of the adhesive is possible (Schnabelrauch et al., 2003).

The compression strength of cured samples of the prepared adhesion compositions were measured using a tensile testing machine (Instron 4467). The addition of low-molecular weight organic compounds like glycine or sorbitol as fillers resulted in values for the compression strength of about 20 MPa. With composite materials containing β-tricalcium phosphate (Cerasorb®, curasan Kleinostheim, Germany) as filler, compression strengths of up to 90 MPa were measured. In a next step defatted dried and wet bone specimens were bonded with the new adhesive compositions. After storing the bonded samples under dry and wet atmosphere, respectively, the tensile shear strength of the adhesive joints were determined. It was found that the bonding strength is strongly influenced by the type and amount of added filler materials. Bone specimens bonded in the dry state showed tensile shear strengths in a range between 2 and 22 MPa. Using the same adhesive compositions for the bonding of wet bone samples led to a remarkable decrease of the resulting tensile shear resistance. Nevertheless, relatively high values for the tensile shear strength of up to 15 MPa were measured for optimized adhesive compositions with Cerasorb® filler under wet conditions mimicking the human body medium.

In vitro degradation of the adhesive system was determined in SBF (simulated body fluid) medium and Sörensen buffer at 37°C using cured cylindrical samples of 10 × 10 mm (diameter × height). Dependant on their composition, the cured adhesive systems showed a different but continuous weight loss over 40 weeks as shown in Fig. 17.5. During longer storage in SBF medium a stagnation in the degradation process was observed.

The *in vitro* cytocompatibility of the adhesive systems was tested using a fluorescein diacetate (FDA)/ ethidium bromide (EtBr) viability assay. Cell viability of osteoblast-like MC3T3-E1 cells on sterilized disks of the cured adhesive systems was assayed after 1 and 4 days. It was found that the percentage of dead cells was less than 5% in each case. Fluorescence micrographs showed that the cells had adhered on the sample surface and formed

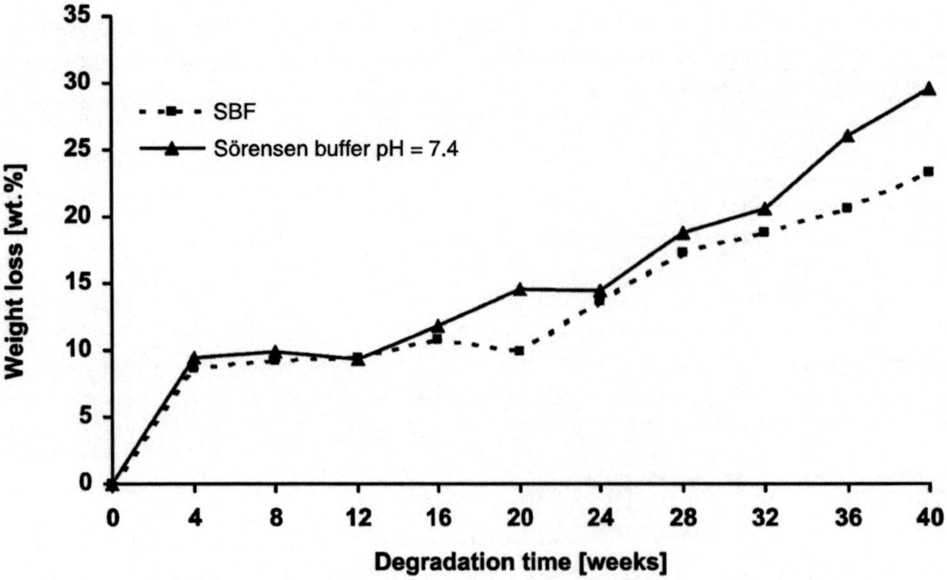

Fig. 17.5 Degradation of a cured adhesive system (macromer DLM-01, filler: $NaHCO_3$/β-TCP (10:90), filler content: 60%) at 37°C in SBF medium and Sörensen buffer

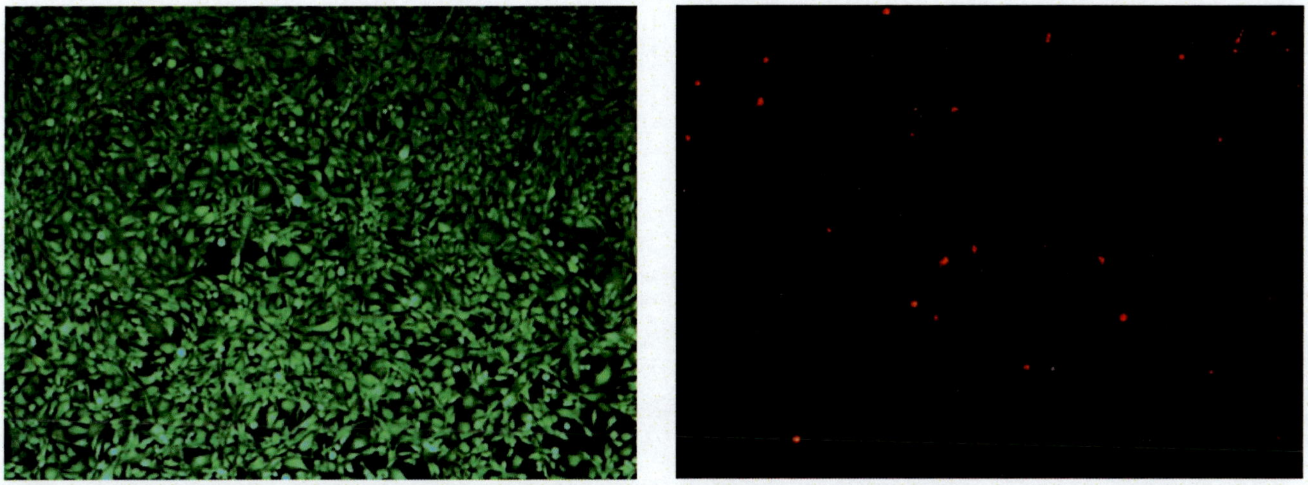

Fig. 17.6 Attachment and viability of pre-osteoblast MC3T3-E1 cells at the surface of the cured adhesive after 4 days (green fluorescence (left): viable cells; red fluorescence (right): dead cells)

a confluent cell layer after 4 days of cultivation illustrating that the cells proliferate (Fig. 17.6). Determination of alkaline phosphatase (AP) activity of MC3T3-E1 cells settled on cured adhesive samples exhibited a considerably lower AP activity compared to the control (tissue culture polystyrene) but a uniform increase in AP activity over a cultivation period of 12 days. A first *in vivo* study performed in a rabbit model over a period of 3 months displayed a good biocompatibility without any signs of local or systemic toxicity for the novel bone adhesive. Further studies are in progress to evaluate the adhesive for surgical applications.

The use of (meth)acrylate-based monomers in combination with other radically polymerizable monomers known to form biodegradable polymer networks has also been described. For example, poly(propylene fumarate) (PPF), a well-known monomer to produce injectable and *in situ* curable materials (Wang et al., 2006) was employed as a blend with phloroglucinol triglycidyl methacrylate to generate partially degradable bone cements (Jayabalan et al., 2001). Recently, PPF-based compositions have been reported as composite adhesives containing hydroxyapatite filler for orthopedic applications (Mitha and Jayabalan, 2009). Although these compositions have shown promising mechanical properties and have also been evaluated to possess a good cytocompatibility, PPF-based adhesives have not been introduced into clinical practice up to now.

A recent development in ophthalmic adhesives is the use of dendritic molecules as macromers or cross-linkers to create synthetic hydrogels (Oelker and Grinstaff, 2008). Dendrimers are highly branched monodisperse molecules that can be functionalized for subsequent crosslinking. Due to their high crosslinking density even at low concentrations, they are attractive macromers for the formation of stable hydrogels. In the group of Grinstaff hybrid linear-dendritic oligomers have been developed composed of components like glycerine, succinic acid and PEG which are known to be biocompatible. The dendritic oligomers are synthesized by esterification reactions to provide biodegradable ester bonds and are capped with methacrylate groups (e.g. $[(G1)PGLSA\text{-}MA]_2$-PEG, Fig. 17.7, Carnahan et al., 2002). The macromers were photo-cured by 514 nm laser irradiation in the presence of 1-vinyl-2-pyrrolidinone, triethanolamine, and eosin Y. These biodendrimers have been used as ophthalmic adhesives for *in vitro* repair of central corneal incisions, stellate lacerations, and *in vivo* central corneal incisions using a chicken model. In the latter study it was found that, compared to suture-treated wounds, those treated with biodendrimer adhesive ones showed a more uniform epithelium and stroma as well as less corneal haze and scarring during healing (Berdahl et al., 2009).

In the group of Hubbell a new crosslinkable system was developed where diacrylates (e.g., PEG-diacrylate) as well as tri- and tetra-acrylates react in a Michael-type addition with thiols like pentaerythritol tetrakis (3-mercaptopropionate). Performing the reaction in aqueous dispersion or reverse emulsion curing times were in a surgically relevant time scale of 5–10 min and the resulting materials had compression strength of up to 7 MPa (Vernon et al., 2003). Those injectable high-modulus materials might have potential for tissue fixation and augmentation.

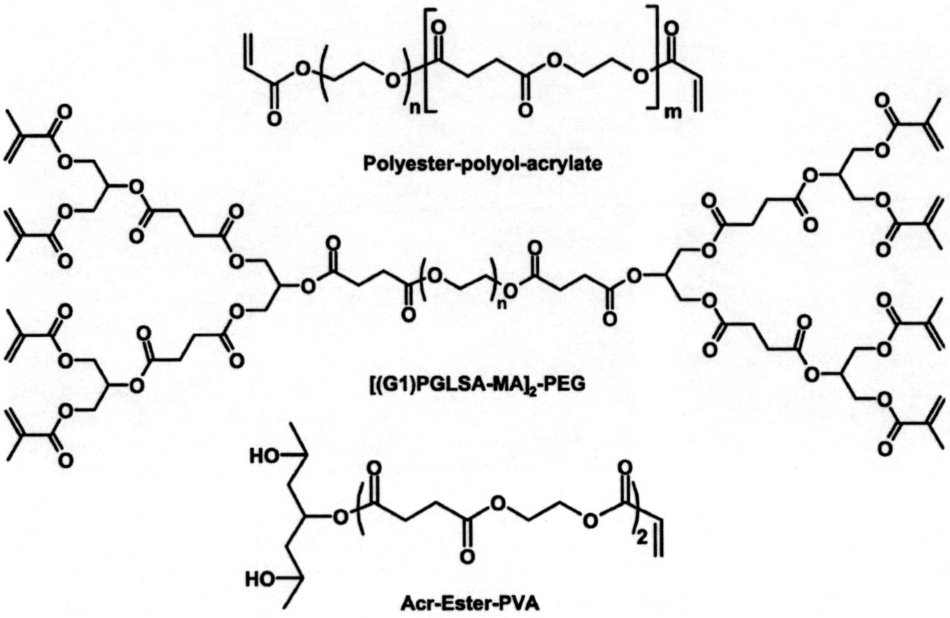

Fig. 17.7 Examples of (meth)acrylate macromers: a polyester-polyol-acrylate (Nivasu et al., 2004), a methacrylated biodendrimer, [(G1)PGLSA-MA]$_2$-PEG (Degoricija et al., 2007), and a poly(vinyl alcohol)-derived acrylate, Acr-Ester-PVA (Martens et al., 2002)

17.4 Biopolymer-based (Meth)acrylate Adhesives

Naturally occurring peptides and proteins offer several advantages as biomaterials including an excellent biocompatibility and an advantageous biodegradability. Considering the successful development of fibrin adhesives, their use as surgical adhesives has already been realized. Further protein-based systems like gelatin-resorcinol-aldehyde or protein-dialdehyde adhesives have found occasional applications in surgery (Ennker et al., 1994; Sung et al., 1999). Mimicking adhesives of marine organisms is a relatively new and promising concept to use peptides or proteins for the design of clinically usable glues (Lee et al., 2006).

17.4.1 Protein-based Systems

Due to the lack of bonding strength of the most of the currently available protein-based glues, there exist several attempts to combine peptides and proteins with synthetic organic monomers like (meth)acrylates to generate mechanically more robust adhesive systems.

A first approach to create peptide-based surgical adhesives was undertaken by Rimpler (1996) who prepared radically curable (meth)acrylated amino acids and peptides derived for example from glycine, lysine, aspartamic acid, or di- and triglycines (for examples see Fig. 17.8). Curing of the so-called "peptoplasts" was performed as a two-compound system with one component comprising the corresponding amino acid or peptide (meth)acrylate and a second component containing the initiator, the reaction product from 9-BBN and a dilinoylated ethylene diamine. At room temperature curing times of about 60 s were determined. Using bovine bone samples, relatively high adhesion strengths were found under dry conditions for the peptoplasts. Bonding under wet conditions led to a considerable decrease in the adhesion strength but even in Ringer solution values for the adhesion strength of about 0.3 MPa were measured (Berndt and Rimpler, 1991a, b). A cardinal problem of these adhesive systems was their slow and incomplete biodegradation *in vivo*.

Among proteins collagen and gelatin, the denatured and degraded form of collagen, have been extensively explored as tissue adhesives for different surgical applications. Free hydroxyl or amino groups present in these molecules can be used to incorporate (meth)acrylate groups into both collagen or gelatin by reaction with reactive (meth)acrylic acid derivatives (Benton et al., 2009). The resulting protein macromers form cross-linked hydrogel structures after radical polymerization initiated either by addition of water-soluble initiation systems (e.g., potassium peroxodisulfate/triethanol amine) or by irradiation with UV-light. Normally cured gelatin (meth)acrylates

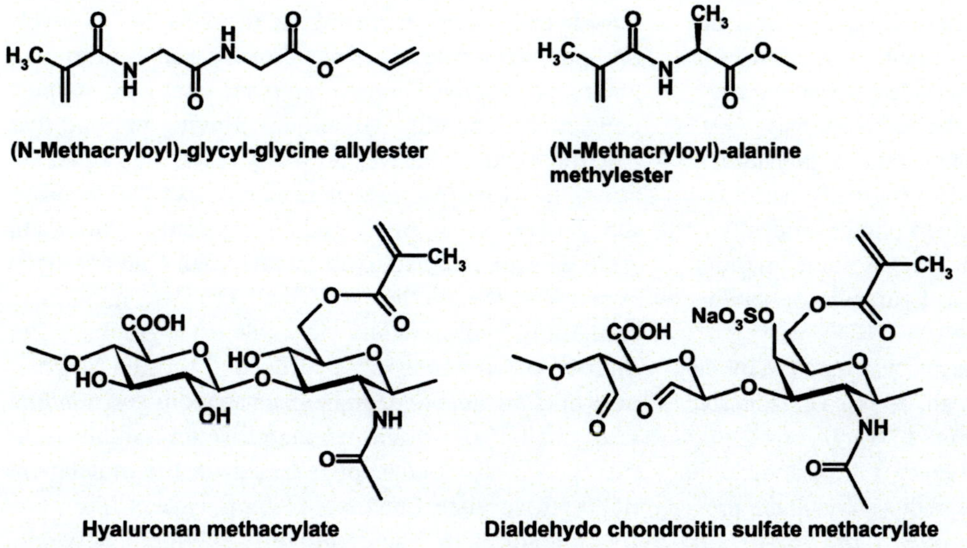

Fig. 17.8 Methacrylate adhesive components based on amino acids and peptides and glycosaminoglycans (hyaluronan and chondroitin sulfate)

show good adhesion on wet tissue, but unfortunately their mechanical stability and adhesion strength in a wet body fluid medium is rather low. In several attempts therefore, curable gelatin derivatives have been used in combination with co-monomers like PEG-di(meth)acrylates. Besides gelatin (meth)acrylates, also styrenated gelatin was used as reactive macromer. In an approach to a photo-curable tissue adhesive glue for artery repair styrenated gelatin was photo-cured with PEG-diacrylate as co-monomer in the presence of a water-soluble carboxylated camphorquinone using visible light at 420–500 nm (Li et al., 2003). The adhesive coated on an incised rat abdominal aorta was immediately converted to a swollen gel upon photocuring and concomitantly hemostasis was completed. As found in histologic examination the formed gel was tightly adhered to the artery shortly after photoirradiation. The cross-linked gel gradually degraded with time and was completely absorbed within 4 weeks after application.

17.4.2 Polysaccharide-based Systems

Among polysaccharides, cellulose, starch, and dextran-based adhesives have found numerous applications in a variety of industrial areas for many years. Some of these polysaccharide derivatives like hydroxypropyl-methyl cellulose are also used in mucoadhesive drug delivery systems (Kharenko et al., 2009) whereas aldehyde derivatives of dextran and hydroxyethyl starch in combination with a star PEG amine and an aminated gelatine, respectively, have been proposed as soft tissue biosealants (Mo et al., 2000; Artzi et al., 2009).

17.4.3 Glycosaminoglycan-based Systems

Recent efforts in the development of tissue adhesives have focused on more complex polysaccharide structures like hyaluronan (HA) or chondroitin sulfate (CS) (Oelker and Grinstaff, 2008). Both polysaccharides belonging to the glycosaminoglycane (GAG) group are constituents of the natural extracellular matrix possessing various biologic functions. HA, a non-sulfated, high-molecular weight biopolymer, is formed from repeating disaccharide units of β-D-*N*-acetylglucosamine (GlcNAc) and β-D-glucuronic acid (GlcA) linked by alternating β-1→3 and β-1→4 glycosidic bonds. It is also found in the synovial fluid, and the vitreous humor in the eye. CS composed of β-1→4-linked repeating disaccharide units of β-D-glucuronic acid β-1→3-linked to β-D-*N*-acetylgalactosamine is an important component of cartilage. Whereas CS is normally provided by isolation from animal tissues (trachea and cartilage), HA can be produced in larger quantities by a biotechnologic fermentation process. In order to obtain curable derivatives, these GAGs have been modified with different types of functional groups including aldehyde, thiol, dihydrazide, and (meth)acrylate moieties (Ifkovits and Burdick, 2007; Prestwich and Kuo, 2008). (Meth)acrylate modification was performed either by acylation of free hydroxyl groups of the GAGs with reactive (meth)acrylic acid derivatives especially anhydrides (Smeds et al., 2001) or by treatment with glycidyl methacrylate (Li et al., 2004; Moller et al., 2007). Adhesives derived from these GAG esters are potentially biodegradable under physiologic conditions. GAG methacrylates are normally water-soluble macromers which can be cured photochemically or thermally by addition of suitable water-soluble initiation systems to form water-insoluble hydrogels of varying cross-linking density.

A HA-based ophthalmic adhesive was reported by the laboratory of Grinstaff (Miki et al., 2002). The adhesive containing HA methacrylate (Fig. 17.8), 1-vinyl-2-pyrrolidinone as co-monomer, and eosin Y/triethanolamine as photoinitiation system can be photo-cured in the presence of low-intensity argon ion laser light ($\lambda = 513\,nm$). The ability of this sealant to repair corneal lacerations was evaluated *in vivo* in four types of full-thickness, 3-mm corneal wounds of rabbit eyes. In 97% of treated eye wounds sealing was observed reforming the anterior chambers and stabilizing the intraocular pressure at a normal level. The adhesive was completely disappeared after 21 days and only mild corneal scarring remained.

Recently a chondroitin sulfate containing both aldehyde and methacrylate groups (dialdehydo chondroitin sulfate methacrylate, Fig. 17.8) was proposed to act as a primer in cartilage defect repair using a synthetic defect-filling biomaterial (Wang et al., 2007). If the chondroitin sulfate primer is deposited at the interface between natural tissue and artificial material, the aldehyde functions can form covalent bonds with the amine groups of the collagen in the host tissue, and methacrylate groups are able to participate in the polymerization of the hydrogel-type biomaterial used to fill the defect. The adhesive primer was evaluated in a model cartilage system where photo-curable PEG-diacrylate (with and without cells) used as defect-filling hydrogel was bonded to an artificial cartilage explant. In different animal models it was found that the new adhesive system not only led to mechanical stability of the defect-filling hydrogel but also to an enhancement of tissue development and repair. Although further studies are required to confirm the clinical relevance of this adhesion system in cartilage repair, the general concept seems to be full of promises.

17.5 Concluding Remarks

Inspired by the successful use of (meth)acrylates in bone cements and dental adhesives considerable efforts have been undertaken during the last 20 years to develop bone and tissue adhesives broadly usable in trauma and reconstructive surgery to replace or at least support conventional bonding techniques. Having in mind that (meth)acrylate adhesives normally reach high adhesion strengths, a main focus of research was dedicated to the generation of biodegradable systems. Two main strategies are pursued to attain this goal. One approach consists in the synthesis of (meth)acrylate monomers containing hydrolytically cleavable ester groups. After thermal or photochemical curing the formed polymer networks are able to degrade in an aqueous surrounding into smaller and often excretable water-soluble fragments. A second concept is based on the attachment of (meth)acrylate functions to biopolymers like gelatin or hyaluronan. The hydrophilic polymeric networks generated after curing are degraded in the body by enzymatic and hydrolytic processes. Both concepts have been shown to lead to partly or completely degradable adhesive systems. Although substrate adhesion and bonding strength of (meth)acrylate adhesives are known to be diminished in wet media, reasonable adhesion strengths were found especially with oligolactone-based (meth)acrylate adhesives. Unfortunately for the latter ones in some cases adverse tissue reactions have been reported *in vivo*. The reasons for these reactions are currently not clear and further studies are necessary to evaluate this adhesive systems. In contrast to the fully synthetic macromers, biopolymer-based (meth)acrylate adhesives exhibit an excellent biocompatibility but because of swelling and lower network density their adhesion strength is often rather low. In combination with the various other requirements a surgical adhesive has to fulfill, this might be the reason that (meth)acrylate adhesives have currently found only very limited clinical applications.

Nevertheless, substantial scientific progress has been made over the last years to design (meth)acrylate adhesives with a promising set of properties. But for their broad application in surgery the key problem which has to be solved is the development of completely degradable adhesives simultaneously conserving their known high adhesion strength over a sufficiently long period of application. A deeper understanding of the relationship between molecular structure, reactivity, and mechanical properties of relevant adhesive systems will be an important key to achieve this goal. Because there is an ongoing need for surgical adhesives, scientists and engineers will be continually encouraged to develop such materials in the future.

References

Artzi N, Shazly T, Baker AB, Bon A, and Edelmann ER (2009) Aldehyde-amine chemistry enables modulated biosealants with tissue-specific adhesion. Advanced Materials 21: 3339–3403.

Bachet J, Goudot B, Dreyfus G, Banfi C, Ayle NA, Aota M, Brodaty D, Dubois C, Delentdecker P, and Guilmet D (1997) The proper use of glue: a 20-year experience with the GRF glue in acute aortic dissection. Journal of Cardiac Surgery 12(Suppl 2): 243–253.

Benton JA, DeForest CA, Vivekanandan V, and Anseth KS (2009) Photocrosslinking of gelatin macromers to synthesize porous hydrogels that promote valvular interstitial cell function. Tissue Engineering Part A 15(11): 3221–3230.

Berdahl JP, Johnson CS, Proia AD, Grinstaff MW, and Kim T (2009) Comparison of sutures and dendritic polymer adhesives for corneal laceration repair in an in vivo chicken model. Archives of Ophthalmology 127(4): 442–447.

Berndt HO and Rimpler M (1991a) Klebstoffe für die Humanmedizin Teil 2: Untersuchung von Peptidklebern auf ihre Verwendbarkeit in biologischen Systemen. Adhäsion 35: 34–37.

Berndt HO and Rimpler M (1991b) Klebstoffe für die Humanmedizin. Teil I. Peptoplaste, Synthese eines neuen resorbierbaren Klebstoffsystems. Adhäsion 35: 23–27.

Carnahan MA, Middleton C, Kim J, Kim T, and Grinstaff MW (2002) Hybrid dendritic-linear polyester-ethers for in situ photopolymerization. Journal of the American Chemical Society 124(19): 5291–5293.

Chao HH and Torchiana DF (2003) BioGlue: albumin/glutaraldehyde sealant in cardiac surgery. Journal of Cardiac Surgery 18(6): 500–503.

Degoricija L, Johnson CS, Wathier M, Kim T, and Grinstaff MW (2007) Photo cross-linkable Biodendrimers as ophthalmic adhesives for central lacerations and penetrating keratoplasties. Investigative Ophthalmology & Visual Science 48(5): 2037–2042.

Donkerwolcke M, Burny F, and Muster D (1998) Tissues and bone adhesives – historical aspects. Biomaterials 19(16): 1461–1466.

Ennker IC, Ennker J, Schoon D, Schoon HA, Rimpler M, and Hetzer R (1994) Formaldehyde-free collagen glue in experimental lung gluing. Annals of Thoracic Surgery 57(6): 1622–1627.

Ferreira P, Coelho JF, and Gil MH (2008) Development of a new photocrosslinkable biodegradable bioadhesive. International Journal of Pharmaceutics 352(1–2): 172–181.

Grossterlinden L, Janssen A, Schmitz N, Priemel M, Pogoda P, Amling M, Rueger JM, and Linhart W (2006) Deleterious tissue reaction to an alkylene bis(dilactoyl)-methacrylate bone adhesive in long-term follow up after screw augmentation in an ovine model. Biomaterials 27(18): 3379–3386.

Heiss C, Hahn N, Wenisch S, Alt V, Pokinskyj P, Horas U, Kilian O, and Schnettler R (2005) The tissue response to an alkylene bis(dilactoyl)-methacrylate bone adhesive. Biomaterials 26(12): 1389–1396.

Heiss C, Kraus R, Schluckebier D, Stiller AC, Wenisch S, and Schnettler R (2006) Bone adhesives in trauma and orthopedic surgery. European Journal of Trauma 2: 141–148.

Heiss C, Kraus R, Peters F, Henn W, Schnabelrauch M, Berg A, Pautzsch T, Weisser J, and Schnettler R (2009) Development of a bioresorbable self-hardening bone adhesive based on a composite consisting of polylactide methacrylates and beta-tricalcium phosphate. Journal of Biomedical Materials Research Part B: Applied Biomaterials 90(1): 55–66.

Ifkovits JL and Burdick JA (2007) Review: photopolymerizable and degradable biomaterials for tissue engineering applications. Tissue Engineering 13(10): 2369–2385.

Ignatius AA, Augat P, Hollstein E, Schorlemmer S, Peraus M, Pokinskyj P, and Claes L (2005) Biocompatibility and functionality of the degradable polymer alkylene bis(dilactoyl)-methacrylate for screw augmentation in vivo. Journal of Biomedical Materials Research Part B: Applied Biomaterials 75(1): 128–136.

Jayabalan M, Thomas V, and Rajesh PN (2001) Polypropylene fumarate/phloroglucinol triglycidyl methacrylate blend for use as partially biodegradable orthopaedic cement. Biomaterials 22(20): 2749–2757.

Kharenko EA, Larionova NI, and Demina NB (2009) Mucoadhesive drug delivery systems (review). Pharmaceutical Chemistry Journal 43(4): 200–208.

Lee BP, Dalsin JL, and Messersmith PB (2006) Biomimetic adhesive polymers based on mussel adhesive proteins. In: Smith AM and Callow JA (eds) Biological Adhesives. Springer-Verlag, Heidelberg: pp 257–278.

Lewis G (1997) Properties of acrylic bone cement: state of the art review. Journal of Biomedical Materials Research 38(2): 155–182.

Li C, Sajiki T, Nakayama Y, Fukui M, and Matsuda T (2003) Novel visible-light-induced photocurable tissue adhesive composed of multiply styrene-derivatized gelatin and poly(ethylene glycol) diacrylate. Journal of Biomedical Materials Research Part B: Applied Biomaterials 66(1): 439–446.

Li Q, Williams CG, Sun DD, Wang J, Leong K, and Elisseeff JH (2004) Photocrosslinkable polysaccharides based on chondroitin sulfate. Journal of Biomedical Materials Research Part A 68(1): 28–33.

MacGillivray TE (2003) Fibrin sealants and glues. Journal of Cardiac Surgery 18(6): 480–485.

Martens P, Holland T, and Anseth KS (2002) Synthesis and characterization of degradable hydrogels formed from acrylate modified poly(vinyl alcohol) macromers. Polymer 43(2): 6093–6100.

Miki D, Dastgheib K, Kim T, Pfister-Serres A, Smeds KA, Inoue M, Hatchell DL, and Grinstaff MW (2002) A photopolymerized sealant for corneal lacerations. Cornea 21(4): 393–399.

Mitha MK and Jayabalan M (2009) Studies on biodegradable and crosslinkable poly(castor oil fumarate)/poly(propylene fumarate) composite adhesive as a potential injectable biomaterial. Journal of Materials Science: Materials in Medicine 20(Suppl 1): S203–S211.

Mo X, Iwata H, Matsuda S, and Ikada Y (2000) Soft tissue adhesive composed of modified gelatin and polysaccharides. Journal of Biomaterials Science, Polymer Edition 11(4): 341–351.

Moller S, Weisser J, Bischoff S, and Schnabelrauch M (2007) Dextran and hyaluronan methacrylate based hydrogels as matrices for soft tissue reconstruction. Biomolecular Engineering 24(5): 496–504.

Nivasu VM, Reddy TT, and Tammishetti S (2004) In situ polymerizable polyethyleneglycol containing polyesterpolyol acrylates for tissue sealant applications. Biomaterials 25(16): 3283–3291.

Oelker AM and Grinstaff MW (2008) Ophthalmic adhesives: a materials chemistry perspective. Journal of Materials Chemistry 18: 2521–2536.

Prestwich GD and Kuo JW (2008) Chemically-modified HA for therapy and regenerative medicine. Current Pharmaceutical Biotechnology 9(4): 242–245.

Reece TB, Maxey TS, and Kron IL (2001) A prospectus on tissue adhesives. American Journal of Surgery 182(Suppl 2): 40S–44S.

Rimpler M (1996) Gluing – a challenge in surgery. International Journal of Adhesion and Adhesives 16: 10–20.

Ritter W (1986) Surgical adhesive system for hard body tissues. USA Patent No. 4,626,310 (DE 3229635).

Roller MB and Bezwada R (1996) Liquid and low melt absorbable copolymers and their blends – synthesis and rheological characterization. T34-Analytical Techniques for Improved Medical Device Design, Vol. III, pp 2848–2851.

Sandner B, Steurich S, and Gopp U (1997) Synthesis and characterization of oligo(lactone) methacrylates. Polymer 38(10): 2515–2522.

Sawhney AS, Poff B, Powell M, Messier K, Doherty E, Yao F, Enscore DJ, and Jarett PK (1998) Bioabsorbable synthetic hydrogel as a surgical lung sealant. Fall Meeting August 23–27, 1998 Polymer Materials: Science and Engineering, Vol. 79. Boston, Massachusettes: p 256.

Schnabelrauch M and Vogt S (1999) Biologisch abbaubares Kompositmaterial (Biologically degradable composite material). Germany Patent No. DE19939403.

Schnabelrauch M, Vogt S, Larcher Y, and Wilke I (2002) Biodegradable polymer networks based on oligolactide macromers: synthesis, properties and biomedical applications. Biomolecular Engineering 19(2–6): 295–298.

Schnabelrauch M, Vogt S, and Reif D (2003) Bioresorbierbares Kompositmaterial. Germany Patent No. 10355992.

Shalaby SW and Shalaby WSW (2004) Cyanoacrylate-based systems as tissue adhesives. In: Shalaby SW and Burg KL (eds) Absorbable and Biodegradable Polymers (Advances in Polymeric Biomaterials Series), 1st Ed. CRC Press, Boca Raton: pp 59–75.

Sierra DH (1993) Fibrin sealant adhesive systems: a review of their chemistry, material properties and clinical applications. Journal of Biomaterials Applications 7(4): 309–352.

Smeds KA, Pfister-Serres A, Miki D, Dastgheib K, Inoue M, Hatchell DL, and Grinstaff MW (2001) Photocrosslinkable polysaccharides for in situ hydrogel formation. Journal of Biomedical Materials Research 54(1): 115–121.

Storey RF, Warren SC, Allison CJ, Wiggins JS, and Puckett AD (1993) Synthesis of bioabsorbable networks from methacrylate-endcapped polyesters. Polymer 34(20): 4365–4372.

Sung HW, Huang DM, Chang WH, Huang RN, and Hsu JC (1999) Evaluation of gelatin hydrogel crosslinked with various crosslinking agents as bioadhesives: in vitro study. Journal of Biomedical Materials Research 46(4): 520–530.

Vauthier C, Dubernet C, Fattal E, Pinto-Alphandary H, and Couvreur P (2003) Poly(alkylcyanoacrylates) as biodegradable materials for biomedical applications. Advanced Drug Delivery Reviews 55(4): 519–548.

Vernon B, Tirelli N, Bachi T, Haldimann D, and Hubbell JA (2003) Water-borne, in situ crosslinked biomaterials from phase-segregated precursors. Journal of Biomedical Materials Research Part A 64(3): 447–456.

Vogt S, Larcher Y, Beer B, Wilke I, and Schnabelrauch M (2002) Fabrication of highly porous scaffold materials based on functionalized oligolactides and preliminary results on their use in bone tissue engineering. European Cells & Materials Journal 4: 30–38.

Vogt S, Berger S, Wilke I, Larcher Y, Weisser J, and Schnabelrauch M (2005) Design of oligolactone-based scaffolds for bone tissue engineering. Bio-Medical Materials and Engineering 15(1–2): 73–85.

Wang DA, Varghese S, Sharma B, Strehin I, Fermanian S, Gorham J, Fairbrother DH, Cascio B, and Elisseeff JH (2007) Multifunctional chondroitin sulphate for cartilage tissue-biomaterial integration. Nature Materials 6(5): 385–392.

Wang N, Wu XS, Lujan-Upton H, Donahue E, and Siddiqui A (1997) Synthesis, characterization, biodegradation, and drug delivery application of biodegradable lactic/glycolic acid oligomers: I. Synthesis and characterization. Journal of Biomaterials Science, Polymer Edition 8(12): 905–917.

Wang S, Lu L, and Yaszemski MJ (2006) Bone-tissue-engineering material poly(propylene fumarate): correlation between molecular weight, chain dimensions, and physical properties. Biomacromolecules 7(6): 1976–1982.

Wenz R (1998) First results with a bioresorbable bone glue. In: Walenkamp GHIM and Bakker FC (eds) Biomaterials in Surgery. Georg Thieme, Stuttgart: pp 132–135.

Wenz R and Nies B (1998) Biologically resorbable polymerization products made of binding agent systems which can be hardened by radiation. Germany Patent No. WO/1998/020839 (DE19646782.9).

Yüksel KÜ (2005) Protein polymers. In: Quinn JV (ed) Tissue Adhesives in Clinical Medicine, 2nd Ed. BC Decker Inc., Hamilton: pp 113–132.

Zhao X, Olsen I, Li H, Gellynck K, Buxton PG, Knowles JC, Salih V, and Young AM (2010) Reactive calcium-phosphate-containing poly(ester-co-ether) methacrylate bone adhesives: chemical, mechanical and biological considerations. Acta Biomaterialia 6(3): 845–855.

18 Byssus Formation in *Mytilus*

Heather G. Silverman and Francisco F. Roberto

Contents

18.1 Introduction

The ability of the *Mytilus* genus of mussels (Phylum Mollusca, Class Bivalvia, and Family Mytilidae) to adhere in marine environments has fascinated researchers from numerous disciplines of science for decades. These relatively small, sessile bivalves attach to a wide range of surfaces present in their natural intertidal and subtidal ocean habitats (rocks, wood, seaweed, other animals, and ship hulls, for example) as well as to surfaces commonly tested in research laboratory settings (glass, plastics [including Teflon®], metals, and biological materials such as teeth, bones, cells, and tissues). No single man-made product on the market to date can claim to possess such a vast application range. An understanding of the unique biological adhesive system in *Mytilus* species (sp.) will undoubtedly aid in the development of biomimetic glues and related products for use in virtually every industry requiring bonding of two materials.

This chapter aims to describe the process in which *Mytilus* sp. attach to surfaces. Details related to the morphology of secretion, biochemistry of the resulting attachment structure, and implications for further research to understand this unique system are presented.

18.2 Overview of Byssogenesis

Mytilus sp. use an exogenous, post-larval attachment structure – the byssus – for permanent or temporary bonding. This process of macroscopic bioadhesion or byssus formation – is termed "byssogenesis". A byssus consists of three primary parts: (1) a single main stem, (2) multiple threads attached to the stem – "byssal threads", and (3) a distal, or terminal "plaque" – "byssal plaque" at the end of each thread (Brown, 1952). Byssi originate from a root which is attached to the anterior and posterior byssal retractor muscles of the bivalve. The foot is a muscular- and glandular-containing organ present in all mussels (Vitellaro-Zuccarello et al., 1983a). Within the *Mytilus* sp. specifically, the small, tongue-shaped foot has a deep groove on the ventral surface which initiates from the byssal root-stem orifice and merges distally with a deeper depression at the tip (Tamarin and Keller 1972; Tamarin, et al. 1976). Byssal adhesive proteins originating from specialized glands within the foot organ are either secreted or released into this distal depression and ventral pedal groove. The proteinaceous mixture consists of possibly eleven (11) distinct mussel-related adhesive proteins that assemble to form the strong, flexible byssal thread and plaque. The concerted actions of the glands and neuromuscular innervations within the foot organ enable the mussel to produce the byssus (Vitellaro-Zuccarello et al., 1983b). Byssogenesis occurs within minutes and can be repeated multiple times with the same substrate (Laursen, 1992). The strength of the attachment is a function of both mechanical and biochemical properties of the mussel's adhesive proteins in concert with the habitat in which the mussel resides (Babarro et al., 2008).

Figure 18.1 depicts *Mytilus edulis*, the "blue mussel" or "bearded mussel" attached to a solid surface. One shell of the bivalve is removed to reveal the relationship of the mussel's anatomy (foot organ and retractor muscle) with the exogenous byssus structure (stem, thread, and plaque).

Figure 18.2 depicts byssogenesis in *Mytilus edulis*: (A) The mussel extends the foot between its two shells. The foot is positioned with the ventral pedal groove towards the surface of interest and the distal tip (distal depression) prepares the surface *via* muscular contractions and movements within the foot. (B) The foot is pressed firmly to the surface – forming an air-tight and water-tight seal. Adhesive proteins are secreted and/or released into the ventral pedal groove and distal depression. Within minutes a single byssal thread and terminal plaque originating from the stem are formed. (C) The foot retracts from the surface. (D) Additional threads are added to the byssus with the mussel extending the foot again to a new location on the rock surface.

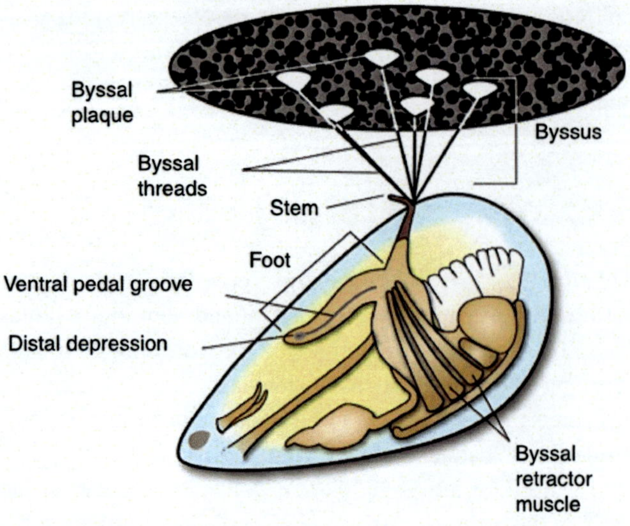

Fig. 18.1 Anatomy of *Mytilus edulis*

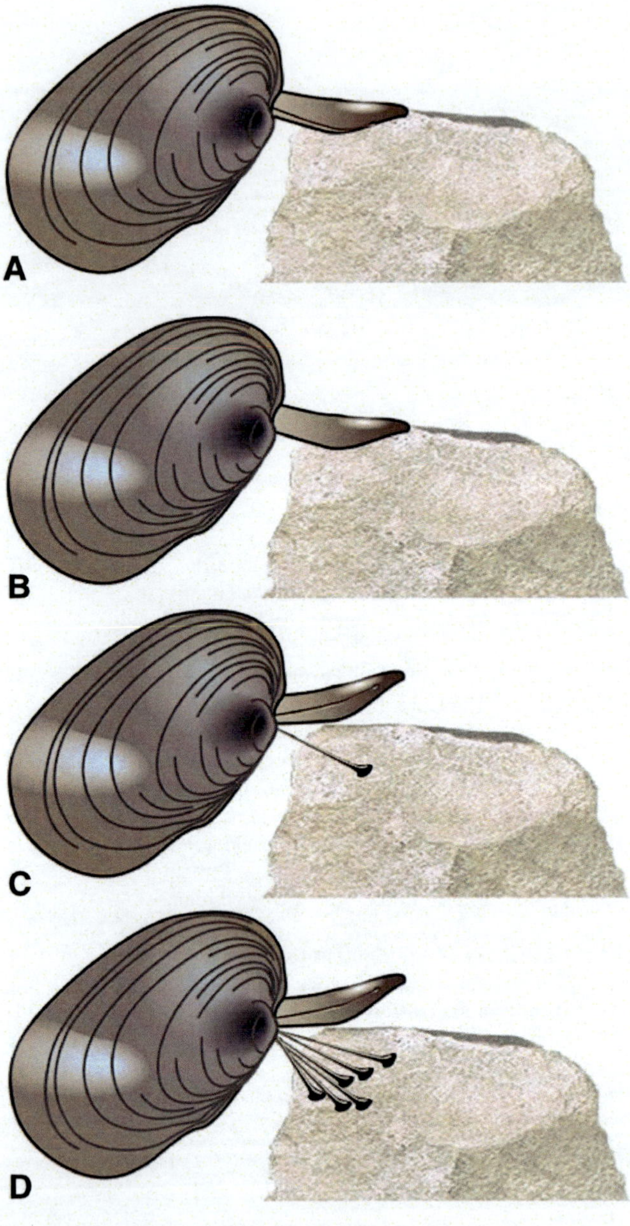

Fig. 18.2 Mussel byssogenesis in *Mytilus edulis*

Many factors have been proposed to trigger or modulate byssogenesis in *Mytilus* sp. Environmental variables include temperature, pH, season, salinity, exposure to air, or hydrodynamic factors, as well as natural surfaces for attachment. Individual characteristics of a mussel, such as size, age, metabolic state, and/or predatory defenses may also play a role in the frequency, size, strength, and morphology of mature byssi (van Winkle, 1970; Crisp et al., 1985; Bell and Gosline, 1997; Carrington, 2002; Selin and Vekhova, 2004; Moeser et al., 2006). Often these studies have appeared to contradict each other; however, differences in experimental design and the methodology used to quantify adhesive properties for specific cues or factors are plausible reasons for the disparity. Regardless of the environmental or biological impacts on byssogenesis, all byssi from *Mytilus* sp. are formed by the same process (Babarro et al., 2008).

18.2.1 Secretion of the Byssal Thread

For nearly 300 years (Brown, 1952), researchers have been intrigued with the remarkable speed with which *Mytilus* species are able to secrete individual threads. Waite et al. (2005) have commented on the similarity of this natural process with modern injection molding, in that the ventral pedal, or foot groove of the mussel serves as a mold in which the various foot proteins are secreted and polymerize in the complex composite holdfast of the organism. Within minutes of exploring a surface the mussel foot adds a new thread to the byssus, stabilizing the animal in turbulent seawater. Engel et al. (1971) described the process of thread formation as a "coordinated series of events involving muscular, sensory, and exocrine responses".

Early morphological studies differentiated the secretory processes of the foot into "white" and "purple" glands based on gross appearance and staining characteristics. The "purple" glands appear colorless until fixed in formalin. Phenolic proteins and "self-tanning" through the action of polyphenol oxidase had been associated with the overall thread formation process. Pyefinch is credited with identifying polyphenols in the thread in 1945 (Brown, 1952). Pujol (1967) conclusively identified collagen in secretory granules of the white glands, and recommended use of the term collagen gland in future work. Since the early 1970s, the key glands associated with the byssus have been termed the collagen, phenol, enzyme, and mucous glands, attributing each to the type of secretory granules produced in the respective tissues. Brown (1952) also described the "byssus gland" which is located at the base of the byssus stem, and whose function was connected with the rings connecting each thread to the extending stem and connective tissue rooting the byssus in the organism. No additional work has been published to further characterize this gland and its function, and our description will focus on the thread from the distal adhesive plaque to its attachment at the stem only.

18.2.2 Spatial Distribution of the Glands

Figures 18.1 and 18.2 provide a general idea of the orientation of the foot relative to shell and internal organs of the mussel. The pedal groove is the mold for thread protein assembly and curing, and a section of the foot tissues surrounding the groove is depicted in Fig. 18.3, with posterior, or groove surface up. The collagen glands make up the largest volume of non-muscular tissue within the organ. These cells are found adjacent to the base and lower walls of the pedal groove, with enzyme gland cells dispersed in smaller groups along the sides of the groove. It has been noted that the epithelial cells in close proximity to the collagen gland at the pedal groove are "ill-defined", and that the collagen granules appear to collect into the groove in a region lacking clearly defined cells. This is in contrast to the other glands of the foot, which are typical epithelial cells, where secretion of secretory granules is mediated by cilia. Phenol gland cells are located anterior to the collagen glands, and phenol granules secreted from these cells are collected and pass through the collagen gland tissue through ciliated secretory ducts, terminating in the pedal groove and distal depression. The greatest concentration of phenol gland cells is found associated with the distal depression where the adhesive plaque is formed, and along the sides of the pedal groove. Mucous glands have been described to concentrate in the posterior tissue around the distal depression as well, and these secretions are believed to have a role in preparing the surface or promoting release of the adhesive structure after thread formation (Tamarin et al., 1976; Wiegemann, 2005).

Ultrastructural examination of the collagen gland secretions (Pujol et al., 1972) revealed that collagen synthesis occurs in the extensive endoplasmic reticulum of the cells, and that through a maturation process, large, non-spherical secretory granules characteristic of the collagen gland are produced. The collagen within the granules is highly ordered, as evidenced by a fibrillar appearance and periodic shading of the material within the granules.

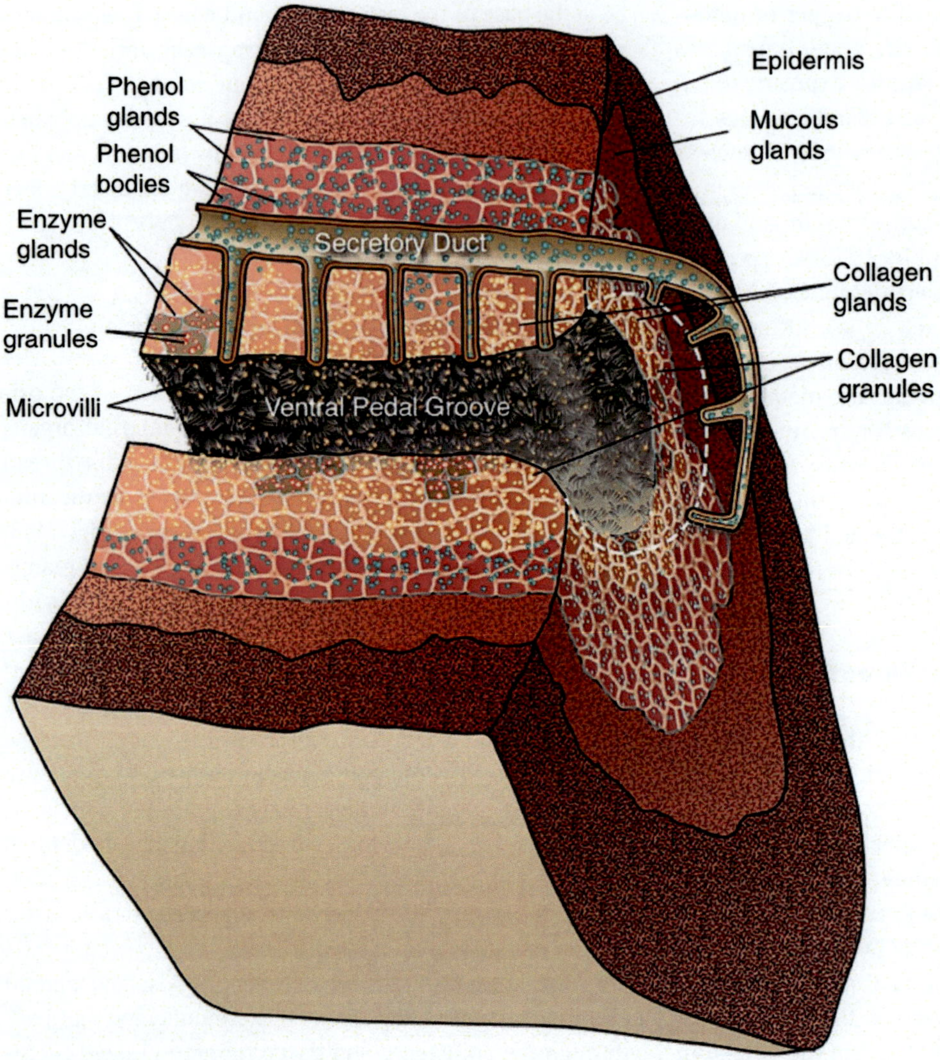

Fig. 18.3 Generalized glandular organization in the *Mytilus* foot organ

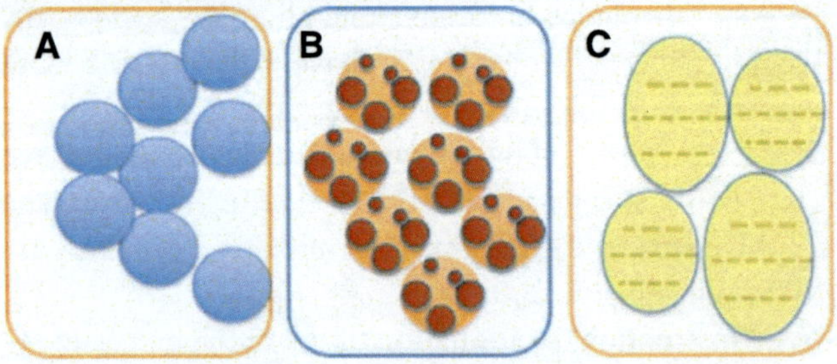

Fig. 18.4 Cartoon depiction of (A) phenol gland; (B) enzyme gland; and (C) collagen gland and their respective secretory granules. Colors are used consistent with the gland locations in Fig. 18.3

The ultrastructural study of Tamarin and Keller (1972) provided the first detailed contrast between the secretory granules of the three principal glands of the foot. The phenol and enzyme granules were noted to resemble other familiar granular secretions accumulated in spherical granules, while the atypical morphology of the collagen granules was again noted. In a follow up study (Tamarin et al., 1976) the authors also comment that unlike other secretory processes, intact collagen granules can be found in the distal depression. This may have a role in further maturation of the thread after the foot pulls away, or may reflect an incomplete release and fusion of collagen fibrils resulting from the physical processes associated with extrusion into the pedal groove. We provide a rough cartoon depiction of these granules in Fig. 18.4 as an accompaniment and guide to the gland cell localization shown in Fig. 18.3. The basis for the "mottled" appearance of the enzyme granules has not been further investigated, and is not always evident in other ultrastructural studies (e.g., Tamarin et al., 1976).

In a high resolution examination of the byssal plaque (Tamarin et al., 1976), it was shown that there are specialized, paddle-shaped cilia within the distal depression of the foot. It has been suggested that these cilia play a role in distributing the mucous and phenol gland secretions on the attachment substrate (Wiegemann, 2005).

18.2.3 Temporal Sequence of Adhesive Protein Secretion

The macroscopic process of byssal thread secretion has been described earlier (Fig. 18.2). Once the foot has settled in place on the attachment surface, over the course of a few minutes the coordinated secretion and polymerization of the products of the various glands occur. The thread is not uniform over its length from its connection to a surface at the byssal plaque, and its termination in the thread stem at the animal. Brown (1952) described the thread to be smooth and cylindrical at the plaque junction, and corrugated and flattened at the stem. The orientation of the glands relative to the thread mold (the pedal groove) places their secretions spatially where they are required: collageneous protein making up the core of the thread (preCOL-P towards the body of the mussel, preCOL-NG in the central region, and preCOL-D at the plaque end of the thread) is secreted in ellipsoid granules along the length of the thread, emanating from the base of the groove, phenolic proteins (fp-1 surrounding the thread core, fp-2 making up the foam-like material of the plaque, fp-3, -5 creating the interface between the attachment surface and the plaque base) secreted through cilia into the walls of the groove and into the distal depression, and mucoid proteins around the distal depression. Other proteins we now know to be associated with the byssal thread, including thread matrix proteins (TMPs; Sagert and Waite, 2009) appear to be included in the inventory of proteins secreted into the appropriate secretory granules with the primary thread components, such as collagens. Curing of the final composite protein assembly is likely to be catalyzed by one or more polyphenol oxidases whose presence has been detected in many studies, but remains to be purified and characterized.

A description of the specific adhesive proteins follows.

18.3 Proteins of the Byssal Thread

Byssal threads are protein fibers comprised of a flexible, collagenous core surrounded by a hardened sheath, currently referred to as the "cuticle". Threads vary in diameter (100–200 μM) and length (2–5 cm) and exhibit mechanical properties such as high stiffness, high ultimate strain, high ultimate stress, incredible toughness, and the ability to re-form following deformation (Bell and Gosline, 1996). Figure 18.5 illustrates the location of the individual proteins comprising the byssal thread and plaque as well as their morphological appearance in the mature bysuss. Note the rough-looking granular cuticle that encases the collagenous fibrous thread core. Scanning electron microscopy (SEM) has proven a valuable tool in elucidating the distinct components of byssal thread fibers (Waite et al., 2005).

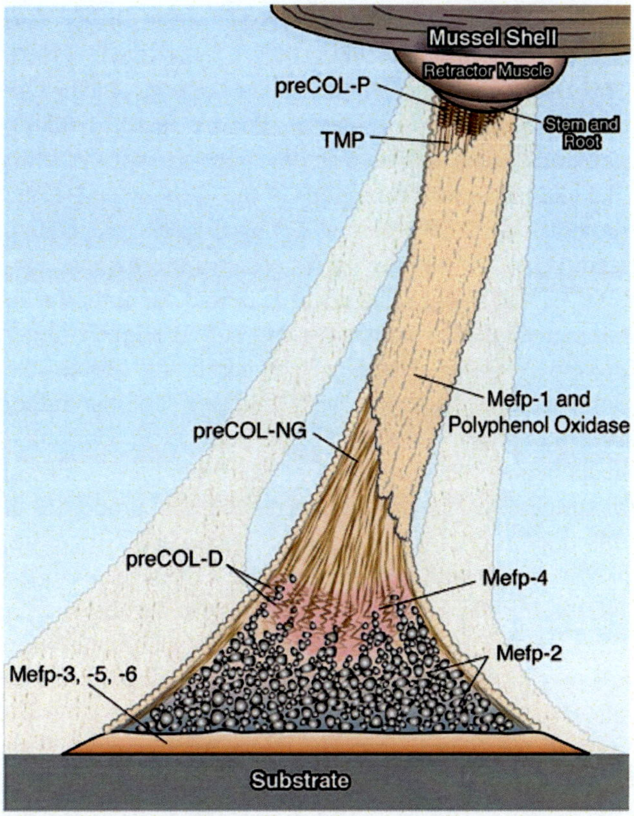

Fig. 18.5 Localization of adhesive proteins in the byssus

18.3.1 The Core: Precollagens

Three "bent core collagens", termed prepepsinized collagen (preCOL)-P, -D, and -NG, contain a common, triple-helical motif also found in other natural collagens: (Gly-X-Y)$_n$. These proteins are the major load-bearing macromolecules found in the byssus. The proximal region of the thread contains preCOL-P – the largest preCOL (95 kDa) with remarkable extensibility and toughness (Qin and Waite, 1995; Coyne et al., 1997). The central collagen domain is flanked by elastin-like domains, which in turn are flanked by small histidine-rich domains. A small acidic patch separates the elastin-like and histidine-rich regions at the carboxy end of the protein only. The proximal collagen fibers are coiled, lending to the pliable, "soft", elastic mechanical features of this region. The distal region of the thread contains preCOL-D – a considerably stiffer, slightly larger protein (97 kDa). PreCOL-D contains silk fibroin-like domains that flank the large collagen domain rather than the elastin-like domains of preCOL-P. The collagen fibers at the distal end of the thread are arranged in straight bundles, providing for the stiff properties at this region of the thread-plaque interface (Waite et al., 1998). Pepsin-resistant nongradient collagen, preCOL-NG, is distributed evenly or uniformly throughout the thread and is believed to mediate the function between the (elastic) proximal and (stiff) distal region by way of plant cell wall-like domains-$(X\text{-}Gly_n)_m$ – that flank the central collagen domain. In turn, silk fibroin-like domains flank these polyglycine-rich domains in preCOL-NG (Qin and Waite, 1998). Figure 18.5 illustrates the preCOL distribution throughout the thread. preCOL-P appears coiled and is most proximal to the stem and root; preCOL-D appears in a stiffer graphical form most distal from the animal at the thread-plaque interface; preCOL-NG is depicted with an indiscriminant graphical feature throughout the entire core. Elegant models depicting the distribution of the different preCOLs along a byssal thread appear in numerous publications (Waite et al., 2004, 2005). Transition metals directly bound to histidine have been detected in the preCOLs, supporting both a metal-binding role and a hardness property directly related to these domains (Coyne and Waite, 2000; Waite et al., 2004).

Harrington and Waite investigated fiber formation of manually-drawn preCOLs and determined that the fiber formation process for thread assembly was highly dependent upon pH (Harrington and Waite, 2008). Hagenau and Scheibel recently reported on recombinant collagen production and the feasibility of future applications in collagen engineering with the unique mussel preCOLs (Hagenau and Seibel, 2010). The self-assembly of the three collagen block polymers to form the core of the byssal thread is a very effective strategy to minimize stress and mechanical sudden changes in the thread where distinctive "stiff" and "elastic" regions are joined (Waite et al., 2004).

18.3.2 The Core: Thread Matrix Proteins

Thread matrix proteins, recently referred to as "TMPs", comprise a non-collagenous protein component of the byssal thread core (Sun et al., 2002; Sagert and Waite, 2009). The first TMP identified in *Mytilus edulis* was designated proximal TMP, or PTMP-1, by Sun et al. (2002). This non-collagenous, water-soluble protein was found to bind preCOL-P and aid in the stiffening effect in the proximal region of the thread. Sagert and Waite have introduced a family of TMPs isolated from the *Mytilus* species *Mytilus galloprovincialis* that is distributed throughout the thread (Sagert and Waite, 2009). The glycine-, tyrosine-, and asparagine-rich TMP family is highly prone to deamidation. The authors postulate that deamidation allows for smectic or lateral orientation of the preCOLs in byssal threads, providing a visco-elastic matrix around the collagen fibers. In essence, the TMPs "lubricate" the collagen microfibrils and aid in the reforming of byssal threads following deformation from tensile loads. The mix of collagenous preCOLs (~81% dry weight) and non-collagenous TMPs (~9% dry weight) make up the unique mussel byssal thread core.

18.3.3 The Cuticle: Foot Protein-1

The cuticle of the byssus – encompassing both the thread and plaque region – is comprised solely of a polyphenolic protein known by many names: *Mytilus edulis* foot protein-1 (Mefp-1), mussel adhesive protein (MAP), and currently foot protein-1 (fp-1). Over the past three decades, six (6) different families of mussel foot proteins (fps) have been classified from *Mytilus* mussel byssi, with designations based on their sequential discovery (fp-1, fp-2, fp-3, fp-4, fp-5, and fp-6 to date).

Foot protein-1 was the first polyphenolic protein isolated and identified in the mussel byssus of *Mytilus edulis* (Waite and Tanzer, 1981). Foot protein-1 is a ~115 kDa, acid-soluble protein containing numerous decapeptide (AKKPSYPPTYK) and hexapeptide (AKPTYK) repeats with approximately 60–70% of the proline and tyrosine residues undergoing posttranslational hydroxylation modifications. The open conformation of the protein, along with the numerous post-translational events, allows for cross-linking with other proteins, metals, and a variety of surfaces (Waite, 1983; Filupa et al., 1990; Haemers et al., 2005). *Mytilus edulis* fp-1 contains 10–15 mol% of 3,4-dihydroxyphenylalanine (DOPA). Numerous reviews and studies have demonstrated the importance of

DOPA (post-translationally modified tyrosine) as a binding mechanism of fp-1. The importance of iron chelate complexes in byssal cuticle strength and hardness and has also recently been confirmed by Harrington et al. (2010) using *in situ* resonance Raman spectroscopy. These authors (and Holten-Andersen and Waite, 2008) report that the density of metal complexation is varied at the submicron-scale, creating a hard as well as extensible outer compliant sheath.

18.3.4 The Cuticle: Polyphenol Oxidase

The *in situ* conversion of fp-1 tyrosine residues to reactive DOPA can be accomplished enzymatically with catechol oxidase or tyrosinase, or chemically *via* sodium periodate, pH>8.5, or dissolved oxygen. Polyphenol oxidases are a class of oxidoreductases that can act by catalyzing the hydroxylation of phenols to catechols and the dehydrogenation of catechols to orthoquinones. Their ability to act on diphenols (catecholase activity) and monophenols (cresolase activity) has led to speculations that a byssus-specific polyphenol oxidase enzyme is responsible for fp-1 cross-linking in the byssal cuticle. Although studies have identified phenol oxidase enzyme activity and localization in the foot organ (Zuccarello, 1981; Waite, 1985; Hellio et al., 2000), thread (Burzio, 1996) and plaque (Burzio, 1996), a byssus-specific polyphenol oxidase has not been definitively identified to date. The presence of catecholato-iron chelate complexes within the byssal cuticle supports the premise of fp-1 and polyphenol oxidase chemistry (Harrington et al., 2010). Figure 18.5 depicts fp-1 (depicted Mefp-1 in this image) cross-linked by a byssal-specific polyphenol oxidase in the outer cuticle of the thread extending along the length of the thread.

18.4 Proteins of the Byssal Plaque

The byssal structure culminates in a polyphasic plaque of varying size, dependent upon variables such as the size and species of the animal, as well as the age of the byssus. Plaques are commonly only ~0.15 mm in diameter where they meet the thread and ~2–3-mm diameter at the substrate interface. The plaque is a solid, structural foam-like region with a varying pore size gradient extending from the surface interface towards the interface of the thread and plaque – known as the thread-plaque junction (Waite et al., 2005). The sponge-like area contains voids and dense areas of protein material, which differs between *Mytilus* species (Laursen, 1992). The outer cuticle of the plaque is comprised of fp-1, presumably cross-linked *via* a byssal-specific polyphenol oxidase (as discussed above). Collagen fibers (preCOLs, also discussed earlier) extend into the thread-plaque interface. Five fps not previously mentioned in this chapter are arranged within the plaque. Biochemical properties dictate their specific localization and functional roles in adhesion. Similar to fp-1, post- or co-translational modifications to amino acids occur in fp-2, fp-3, fp-4, fp-5, and fp-6.

18.4.1 Thread-Plaque Junction: Foot Protein-4

Foot protein-4 has been reported as a 79 kDa protein, containing 4 mol% DOPA and high levels of the amino acids glycine, arginine, and histidine (Vreeland et al., 1998; Weaver, 1998; Warner and Waite, 1999). Foot protein-4 has been proposed to function as a coupling agent between the distal preCOLs of the thread and another plaque protein, fp-2 (see pink region of Fig. 18.5). Since 1999, the literature has been scarce regarding fp-4 from *Mytilus edulis*. However, two variants of fp-4 from *Mytilus californianus* were reported in 2006 (Zhao and Waite, 2006b).

18.4.2 Plaque Foam Matrix: Foot Protein-2

Foot protein-2 has been reported as a 42–47 kDa protein, containing just 2–3 mol% DOPA and 6–7 mol% cysteine. Tandem, repetitive amino acid repeats, substantial secondary structure and a resistance to proteases are properties that have resulted in the protein being functionally classified as a stabilizer in the byssal plaque (Rzepecki et al., 1992; Inoue et al., 1995). Figure 18.5 depicts the localization of fp-2 (in blue) between the thread-plaque junction containing fp-4 and the primer layer that contacts a substrate surface. A foam with protein properties similar to fp-2 has been reported in the marine polychaete, *Phragmatopoma californica* (Stewart et al., 2004).

18.4.3 Plaque Primer Layer: Foot Proteins-3, -5, and -6

The three fps that are localized to the primer layer of the plaque (fp-3, fp-5, and fp-6; see Fig. 18.5) are relatively small proteins in comparison to the other byssal adhesive

Table 18.1: Byssal adhesive proteins from *Mytilus edulis*

Adhesive protein	Mass (~kDa)	DOPA (~mol%)	Protein domain or other variables	+ Analogs within *Mytilus*
fp-1	115	10–15	Hexa- and deca-peptide repeats	ca, ch, co, g, j, t
fp-2	42–47	2–3	Tandem, repetitive repeats; 6–7 mol% cysteine	co, g
fp-3	5–7	20–25	Gene families	ca, g
fp-4	79	4	Tyrosine rich octa-peptide	ca
fp-5	9.5	27	Phosphoserine	ca, g
fp-6	11.6	4	Phosphoserine and tyrosine-rich	*Only* ca
preCOL-P	95	0	$[\text{Gly-X-Y}]_n$ Elastin domains	g
preCOL-D	97	0	$[\text{Gly-X-Y}]_n$ Silk-fibroin-like domains	g
preCOL-NG	76	0	$[\text{Gly-X-Y}]_n$ Plant cell wall-like domains	g
TMPs	50	0	Deamidation prone	g
Polyphenol oxidase	34–174	–	–	–

ca, *californianus*; ch, *chilensis*; co, *coruscus*; g, *galloprovincialis*; t, *trossulus*; j, JHX-2002; –, not determined

proteins presented thus far (fp-3 = 5–7 kDa; fp-5 = 9.5 kDa; fp-6 = 11.6 kDa). Foot proteins-3 and -5 contain substantially higher DOPA levels (~25 and ~27 mol%, respectively) than fp-1 or fp-2 (Papov et al., 1995; Warner and Waite, 1999; Waite and Qin, 2001). Foot protein-6 (so far identified only in *Mytilus californianus*) contains a small amount of DOPA, although tyrosine is present in large quantities (Zhao and Waite, 2006a). Some have hypothesized that the large number of gene variants for fp-3 may be related to specificity in substrate adhesion (Inoue et al., 1996; Floriolli et al., 2000). Further research is needed to determine the role, if any, fp-3 gene variants may play in byssogenesis and adhesion. Foot protein-5 contains variants of phosphorylated serine (pSer). Foot protein-5 may mediate binding to calcareous minerals – such as other mussel shells (Fant et al., 2000; Waite and Qin, 2001). To date, fp-6 has only been isolated from *Mytilus californianus* (Zhao and Waite, 2006a). Foot protein-6 also contains pSer residues like fp-5, and a relatively large amount of tyrosine (Zhao and Waite, 2006a; Flammang et al., 2009). A summary of protein properties of byssal adhesive proteins is presented in Table 18.1.

18.5 Chemistry of Adhesion at the Byssal Thread-substrate Interface

The unique attributes of the mussel byssal thread imparting its ability to attach to a wide range of surfaces and materials underwater have long intrigued scientists and industry. Recent developments in materials science have leveraged the association of DOPA (first established by Ravindranath and Ramalingam (1972)) with the byssal thread proteins and attachment, and will be described in the chapter by Lau and Messersmith in this book. Several excellent reviews (Waite et al., 2005; Wiegemann, 2005) have considered this topic, so our treatment of this topic will be brief. The abundance of modified (primarily hydroxylated) amino acids present in the plaque interface proteins fp-3 (Holten-Andersen and Waite, 2008) and fp-5, and the chemical interactions possible, particularly with the catechol moiety of DOPA, are clearly important to the tenacity and effectiveness of byssal thread attachment. Strong, stable complexes can be formed with transition metal ions and oxides, bonding proteins to surfaces as well as each other. In addition, hydrogen bonds with surfaces and other proteins, and covalent interactions between DOPA residues and other aromatic and amine species present in adjacent proteins may also stabilize the surface attachment of the byssal plaque. Dopamine itself has been shown to be an efficient bi-functional adhesive reagent (Lee et al., 2007), but there is still much to be elucidated regarding the mechanisms that contribute to adhesion in *Mytilus*, since simple chemical species have not been able to fully replicate the strength of mussel attachment.

Table 18.2 summarizes the biological classification of mussel proteins in the byssal plaque and thread, their functions in the adhesion process, and proposed bonding mechanisms.

Table 18.2: Summary of proteins involved in byssogenesis

Biological classification				Adhesion	
Byssus	**Region**	**Mussel proteins identified**		**Function**	**Bonding mechanism(s)**
Plaque	Primer layer	fp-3, fp-5, fp-6		Primer Link with other plaque proteins	Couple with inorganics (metals) and organics
	Plaque foam matrix	fp-2		Stabilization	Inter- and intra-molecular cross-linking
	Thread-plaque junction	fp-4			
	Plaque cuticle	fp-1		Protective coating	Inter- and intra-molecular cross-linking; couple with metals (i.e. Fe)
		Polyphenol oxidase		Cross-link fp-1	Oxidation
Thread	Thread cuticle	fp-1		Protective coating	Inter- and intra-molecular cross-linking; couple with metals (i.e. Fe)
		Polyphenol oxidase		Cross-link fp-1	Oxidation
	Thread core	Distal	PreCOL-D	Stiffness	Inter- and intra-molecular cross-linking
		Proximal	PreCOL-P	Elasticity	
		Nongradient	PreCOL-NG	Rigid and elastic	
		Entire region	TMPs	Collagen fibril organization	Metal independent deamidation influences?

18.6 Immunolocalization of Byssal Proteins

The primary technique to identify mussel adhesive proteins has been the use of classical biochemistry – where proteins are extracted (from mussel feet, threads or plaques), purified and analyzed for amino acid sequence and enzymatic responses. Molecular techniques such as hybridization of RNA (Northern Blot) or DNA (Southern Blot) have also been used to detect potential genetic transcripts of proteins in foot tissues. Immunochemical methods provide information on the localization of byssal adhesive proteins in both the glands of the foot organ and the exogenous byssus structure. A comparison of the location of specific byssal proteins stockpiled in the foot and their location in the thread and/or plaque help us to connect the association between the glandular and structural morphology of byssogenesis. To date, fp-1, the preCOLs and the TMP family have been localized to *both* glands within the foot and specific regions of the byssus using immunological techniques. Parallel documentation of the other byssal adhesive proteins has not been completed.

The first target protein for immunohistochemistry was the newly-discovered polyphenolic Mefp-1. Micrographs (using antibodies designed from the polyphenolic amino acid region of Mefp-1) revealed the phenol glands in the distal region of the foot near the distal depression fluoresced intensely (Waite and Tanzer, 1981). The thread sheath and substrate interface also fluoresced – at the time suggesting Mefp-1 was located in the thread and plaque. One of the first immunochemical studies for collagen was performed by Benedict and Waite (1986). Fluorescent antibodies for the collagen triplet amino acid motif were shown to bind only to the collagen vesicles in the proximal section of the foot. The gradient nature of preCOL-P and -D has been shown in both the ventral pedal groove of the foot (Qin and Waite, 1995) and byssal threads (Benedict and Waite, 1986; Qin and Waite, 1995). The smectic orientation and structural matrix role of the TMPs are also evident in the foot and byssal threads (Sun et al., 2002; Sagert and Waite, 2009).

18.7 Concluding Remarks

Our understanding of the basic morphology, ultrastructure and sequence of events leading to the secretion of the byssal thread has not changed dramatically in over 30 years. What has advanced tremendously over this time is our knowledge of the complex protein suite that comprises the mussel byssus and surprising parallels to non-biological materials. We now appreciate the ingenuity of nature in devising an extraorganismal attachment that imparts strength, flexibility, and resilience in rapidly formed threads synthesized from protein in freshwater and marine environments. There are still abundant questions to be asked as well. For example, the native polyphenol oxidase(s) responsible for the post-translational oxidation of amino acids in the mussel byssus remains elusive, even though enzyme glands and their secretory granules have been clearly evident in past ultrastructural studies. We continue to learn more about the association of metals with the polyphenolic proteins of the thread, yet a complete explanation for the tenacity of the byssal plaque on a wide range of substrates is only partially explained by our current understanding of the adhesive protein-surface interactions. Major innovations in materials science have been enabled by adapting the bi-functional chemistry of DOPA-containing amino acids to simple chemical mimetics, but we know that these biologically-inspired materials still fall short of their natural models. In spite of these challenging questions, continued study of the mussel byssus can be expected to yield new and exciting insights into nature, biochemistry, and strategies for attachment.

Acknowledgments

The authors thank Allen Haroldsen for his excellent graphic arts contributions to this paper. This work was supported by the U.S. Department of Energy under DOE Idaho Operations Office Contract DE-AC07-05ID14517.

References

Babarro JMF, Fernandez MJ, and Labarta U (2008) Secretion of byssal threads and attachment strength of *Mytilus galloprovincialis*: the influence of size and food availability. Journal of the Marine Biological Association of the United Kingdom 88(4): 783–791.

Bell EC and Gosline JM (1996) Mechanical design of mussel byssus: material yield enhances attachment strength. Journal of Experimental Biology 199(Pt 4): 1005–1017.

Bell EC and Gosline JM (1997) Strategies for life in flow: tenacity, morphometry, and probability of dislodgement of two *Mytilus* species. Marine Ecology Process Series 159: 197–208.

Benedict CV and Waite JH (1986) Composition and ultrastructure of the byssus of *Mytilus edulis*. Journal of Morphology 189(3): 261–270.

Brown CH (1952) Some structural proteins of *Mytilus edulis*. Quarterly Journal of Microscopical Science 93: 487–502.

Burzio LA (1996) Catechol oxidases associated with byssus formation in the blue mussel, *Mytilus edulis*. MS Thesis, University of Delaware, Newark, USA.

Carrington E (2002) Seasonal variation in the attachment strength of blue mussels: causes and consequences. Limnology and Oceanography 47(6): 1723–1733.

Coyne KJ, Qin X, and Waite JH (1997) Extensible collagen in mussel byssus: a natural block copolymer. Science 277(5333): 1830–1832.

Coyne KJ and Waite JH (2000) In search of molecular dovetails in mussel byssus: from the threads to the stem. Journal of Experimental Biology 203(Pt 9): 1425–1431.

Crisp DJ, Walker G, Young GA, and Yule AB (1985) Adhesion and substrate choice in mussels and barnacles. Journal of Colloid and Interface Science 104(1): 40–50.

Engel RH, Hillman RE, Neat MJ, and Quinby HL (1971) A study of the adhesive mechanisms of various species of the sea mussel. Report number NIH NIDR 70-2237, pp 1–35.

Fant C, Scott K, Elwing H, and Hook F (2000) Adsorption behavior and enzymatically or chemically induced cross-linking of a mussel adhesive protein. Biofouling 16(2–4): 119–132.

Filupa DR, Lee SM, Link RP, Strausberg SL, and Strausberg RL (1990) Structural and functional repetition in a marine mussel adhesive protein. Biotechnology Progress 6(3): 171–177.

Flammang P, Lambert A, Bailly P, and Hennebert E (2009) Polyphosphoprotein-containing marine adhesives. The Journal of Adhesion 85: 447–464.

Floriolli RY, von Langen J, and Waite JH (2000) Marine surfaces and the expression of specific byssal adhesive protein variants in *Mytilus*. Marine Biotechnology 2(4): 352–363.

Haemers S, van der Leeden MC, and Frens G (2005) Coil dimensions of the mussel adhesive protein Mefp-1. Biomaterials 26(11): 1231–1236.

Hagenau A and Seibel T (2010) Towards the recombinant production of mussel byssal collagens. Journal of Adhesion 86(1): 10–24.

Harrington MJ and Waite JH (2008) pH-Dependent locking of giant mesogens in fibers drawn from mussel byssal collagens. Biomacromolecules 9(5): 1480–1486.

Harrington MJ, Masic A, Holten-Andersen N, Waite JH, and Fratzl P (2010) Iron-clad fibers: a metal-based biological strategy for hard flexible coatings. Science 328(5975): 216–220.

Hellio C, Bourgougnon N, and Gal YL (2000) Phenoloxidase (E.C. 1.14.18.1) from the byssus gland of *Mytilus edulis*: purification, partial characterization, and application for screening products with potential antifouling activities. Biofouling 16(2–4): 235–244.

Holten-Andersen N and Waite JH (2008) Mussel-designed protective coatings for compliant substrates. Journal of Dental Research 87(8): 701.

Inoue K, Takeuchi Y, Miki D, and Odo S (1995) Mussel adhesive plaque protein gene is a novel member of epidermal growth factor-like gene family. Journal of Biological Chemistry 270(12): 6698–6701.

Inoue K, Takeuchi Y, Miki D, Odo S, Harayama S, and Waite JH (1996) Cloning, sequencing and sites of expression of genes for the hydroxyarginine-containing adhesive-plaque protein of the mussel *Mytilus galloprovincialis*. European Journal of Biochemistry 239(1): 172–176.

Laursen RA (1992) Reflections on the structure of mussel adhesive proteins. Results and Problems in Cell Differentiation 19: 55–74.

Lee H, Dellatore SM, Miller WM, and Messersmith PB (2007) Mussel-inspired surface chemistry for multifunctional coatings. Science 318: 426–430.

Moeser GM, Leba H, and Carrington E (2006) Seasonal influence of wave action on thread production in *Mytilus edulis*. Journal of Experimental Biology 209: 881–890.

Papov VV, Diamond TV, Biemann K, and Waite JH (1995) Hydroxyarginine-containing polyphenolic proteins in the adhesive plaques of the marine mussel *Mytilus edulis*. Journal of Biological Chemistry 270(34): 20183–20192.

Pujol JP (1967) Formation of the byssus in the common mussel (*Mytilus edulis* L.) Nature 214: 204–205.

Pujol JP, Houvenaghel G, and Bouillon J (1972) Le collagene du byssus de *Mytilus edulis* L. I. Ultrastructure des cellules secretrices. Archives de Zoologie Experimentale & Generale 113: 251–264.

Qin X and Waite JH (1995) Exotic collagen gradients in the byssus of the mussel *Mytilus edulis*. Journal of Experimental Biology 198(Pt 3): 633–644.

Qin X and Waite JH (1998) A potential mediator of collagenous block copolymer gradients in mussel byssal threads. Proceedings of the National Academy of Sciences 95: 10517–10522.

Ravindranath MH and Ramalingam K (1972) Histochemical identification of *Dopa*, *Dopamine* and *Catechol* in phenol gland and mode of tanning of byssus threads of *Mytilus edulis*. Acta Histochemica 42(1): 87–94.

Rzepecki LM, Hansen KM, and Waite JH (1992) Characterization of a cystine-rich polyphenolic protein family from the blue mussel, *Mytilus edulis* L. Biological Bulletin 183(1): 123–137.

Sagert J and Waite JH (2009) Hyperunstable matrix proteins in the byssus of *Mytilus galloprovincialis*. Journal of Experimental Biology 212(Pt 14): 2224–2236.

Selin NI and Vekhova EE (2004) Effects of environmental factors on byssal thread formation in some members of the family Mytilidae from the Sea of Japan. Russian Journal of Marine Biology 30(5): 306–313.

Stewart RJ, Weaver JC, Morse DE, and Waite JH (2004) The tube cement of *Phragmatopoma californica*: a solid foam. Journal of Experimental Biology 207(Pt 26): 4727–4734.

Sun C, Lucas JM, and Waite JH (2002) Collagen-binding matrix proteins from elastomeric extraorganismic byssal fibers. Biomacromolecules 3: 1240–1248.

Tamarin A and Keller PJ (1972) An ultrastructural study of the byssal thread forming system in *Mytilus*. Journal of Ultrastructure Research 40(3): 401–416.

Tamarin A, Lewis P, and Askey J (1976) The structure and formation of the byssus attachment plaque in *Mytilus*. Journal of Morphology 149(2): 199–221.

van Winkle W (1970) Effect of environmental factors on byssal thread formation. Marine Biology 7(2): 143–148.

Vitellaro-Zuccarello L, DeBiasi S, and Bairati A (1983a) The ultrastructure of the byssal apparatus of a mussel. V: Localization of collagenic and elastic components in the threads. Tissue and Cell 15(4): 547–554.

Vitellaro-Zuccarello L, DeBiasi S, and Blum I (1983b) Histochemical and ultrastructural study on the innervation of the byssus glands of *Mytilus galloprovincialis*. Cell Tissue Research 233(2): 403–413.

Vreeland V, Waite JH, and Epstein L (1998) Polyphenols and oxidase in substratum adhesion by marine algae and mussel. Journal of Phycology 34: 1–8.

Waite JH (1983) Evidence for a repeating 3,4-dihydroxyphenylalanine- and hydroxyproline-containing decapeptide in the adhesive protein of the mussel, *Mytilus edulis* L. Journal of Biological Chemistry 258(5): 2911–2915.

Waite JH (1985) Catechol oxidase in the byssus of the common mussel, *Mytilus edulis* L. Journal of the Marine Biological Association of the United Kingdom 65: 359–371.

Waite JH and Tanzer ML (1981) Polyphenolic substance of *Mytilus edulis* – novel adhesive containing L-DOPA and hydroxyproline. Science 212(4498): 1038–1040.

Waite JH and Qin X (2001) Polyphosphoprotein from the adhesive pads of *Mytilus edulis*. Biochemistry 40(9): 2887–2893.

Waite JH, Qin X, and Coyne KJ (1998) The peculiar collagens of mussel byssus. Matrix Biology 17(2): 93–106.

Waite JH, Lichtenegger HC, Stucky GD, and Hansma P (2004) Exploring molecular and mechanical gradients in structural bioscaffolds. Biochemistry 43(24): 7653–7662.

Waite JH, Anderson NH, Jewhurst SA, and Sun C (2005) Mussel adhesion: finding the tricks worth mimicking. Journal of Adhesion 81(3–4): 297–317.

Warner SC and Waite JH (1999) Expression of multiple forms of an adhesive plaque protein in an individual mussel, *Mytilus edulis*. Marine Biology 134(4): 729–734.

Weaver JK (1998) Isolation, purification, and partial characterization of a mussel byssal precursor protein, *Mytilus edulis* foot protein 4. MS Thesis, University of Delaware, Newark, USA.

Wiegemann M (2005) Adhesion in blue mussel (*Mytilus edulis*) and barnacles (genus *Balanus*): mechanisms and technical applications. Aquatic Science 67: 166–176.

Zhao H and Waite JH (2006a) Linking adhesive and structural proteins in the attachment plaque of *Mytilus californianus*. Journal of Biological Chemistry 281(36): 26150–26158.

Zhao H and Waite JH (2006b) Proteins in load-bearing junctions: the histidine-rich metal-binding protein of mussel byssus. Biochemistry 45(47): 14223–14231.

Zuccarello LV (1981) Ultrastructural and cytochemical study on the enzyme gland of the foot of a mollusk. Tissue and Cell 13(4): 701–713.

19 Wet Performance of Biomimetic Fibrillar Adhesives

K. H. Aaron Lau and Phillip B. Messersmith

Contents

19.1 Introduction

A number of legged organisms have evolved sophisticated, fibrillar attachment schemes that exhibit functional qualities highly desirable in synthetic reversible adhesives: substrate compliance, high adhesive strength, and sustained performance over many attach/release cycles (Creton and Gorb, 2007; Peattie, 2008). While a number of early synthetic mimics of fibrillar adhesives as well as the biological systems that inspired them are effective in ambient or low humidity environments, they are less effective in highly humid environments and function poorly in the presence of excess water. Yet, adhesives that function well under wet conditions are greatly desired for numerous industrial and consumer adhesive applications, as well as for biomedical uses (Yanik, 2009). This review chapter summarizes recent efforts in adapting or combining features of multiple biological adhesive strategies to develop biomimetic systems with enhanced wet adhesive performance. On-going research and development efforts are anticipated to lead to practical implementations of wet adhesives for a variety of uses.

19.2 Gecko Mimetic Fibrillar Wet Adhesives

19.2.1 Gecko: A Prototypical Biological Fibrillar Adhesive

The foot pad of the gecko is a celebrated example of a biological fibrillar reversible adhesive system, of which other examples include flies, beetles, and other insects (Autumn and Gravish, 2008). The gecko foot pad is carpeted with dense arrays of microstructured hair-like fibrils (setae) and illustrates some important characteristics of such fibrillar adhesive systems. The setae are thin keratinous fibers ~100 μm long, and are branched at the tips into numerous ~200 nm wide spatulae, which are tilted approximately normal to the long axes of the setae (Ruibal and Ernst, 1965; Autumn and Gravish, 2008). Although keratin has a high mechanical modulus (Autumn and Gravish, 2008), the high aspect ratios of the setae and the spatulae form a hairy surface microstructure that has a low effective modulus (Autumn and Gravish, 2008; Peattie, 2008; Jeong and Suh, 2009). This ensures intimate contact of the gecko foot over diverse length scales of roughness. Furthermore, contact mechanics arguments suggest that the hairy microstructure confers high adhesive strength *via* splitting a single large contact area into numerous small, independent contacts, the benefit of which is that the total adhesive force scales with the number of self-similar surface contacts (the power of the scaling being dependent on the tip geometry) (Autumn et al., 2002; Arzt et al., 2003; Spolenak et al., 2005a; Autumn and Gravish, 2008). In the case of the Tokay gecko (*Gekko gecko*), the high spatular density (~10^6 mm^{-2}) (Autumn et al., 2000) and van der Waals interactions between the tips of the spatulae and the opposing surface alone are sufficient to support more than the gecko's body weight (Autumn et al., 2000; Arzt et al., 2003), although capillarity arising from molecular water layers adsorbed from ambient moisture may also enhance the adhesion (Huber et al., 2005). In either case, reliance on such physical interactions enables the gecko to reach maximal adhesion almost instantaneously upon contact, and maintain adhesion over a high number of attach/release cycles. Finally, self-cleaning attributes have also been assigned to the fibrillar nature of the contact structure (Hansen and Autumn, 2005; Lee and Fearing, 2008), and the angled placement of the spatulae enables the gecko to engage adhesion with minimal shear loading and to detach at will *via* a peeling process (Autumn et al., 2000; Autumn and Gravish, 2008; Lee et al., 2008; Jeong et al., 2009). A number of reversible synthetic adhesives that mimic the setae/spatulae microstructure have been developed (Creton and Gorb, 2007), demonstrating adhesion on dry surfaces in the range of ~10 N/cm^2 (similar to the gecko) (Geim et al., 2003; Spolenak et al., 2005a; del Campo and Arzt, 2007; Schubert et al., 2007; Lee et al., 2008; Cutkosky and Kim, 2009; Jeong et al., 2009) up to 100 N/cm^2 (Ge et al., 2007; Qu et al., 2008).

19.2.2 Coated Gecko Mimetic Adhesives

For all of their remarkable capabilities, fibrillar adhesive systems in general and gecko adhesion in particular, are severely compromised under wet conditions. For example, van der Waals interactions decay in strength rapidly with distance in water (Israelachvili, 1992); correspondingly, gecko adhesion is observed to be greatly diminished (Huber et al., 2005; Sun et al., 2005). Nonetheless, if the gecko adhesive strategy can be adapted for wet conditions, the main features of gecko adhesion – contact splitting, compliant hairy structures – may inspire useful wet adhesives. Along this vein, several laboratories have employed the approach of adapting or combining features of multiple (biological) systems to enhance wet adhesive performance (Lee et al., 2007; Glass et al., 2008, 2009; Mahdavi et al., 2008). Recent advances in bioinspired fibrillar adhesives are the emphasis of this review. Ongoing efforts to identify, characterize and mimic biological adhesives are likely to produce more versatile biomimetic designs of reversible wet adhesives.

A simple mimic of gecko-type fibrillar adhesive designed for adhesion to wet tissue was described by Glass et al. (2008). They fabricated PDMS strips decorated with arrays of ~100 μm diameter cylindrical pegs for controlling the movement of a wireless capsule endoscope prototype (Fig. 19.1). The microstructured strips (Fig. 19.1B) were mounted on actuated "legs" (Fig. 19.1D) and were designed to be pressed against the intestine wall for attachment. The strips were test-

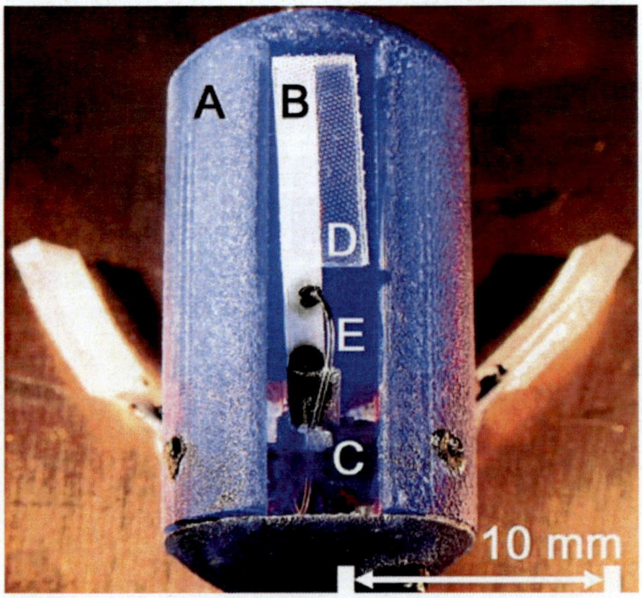

Fig. 19.1 Manufactured prototype of the three-legged capsule robot. (A) Capsule shell, (B) leg, (C) pulley, (D) adhesive pad, and (E) thin nylon cable providing the actuation force. Adapted from Glass et al. (2008) and reproduced with permission from IEEE

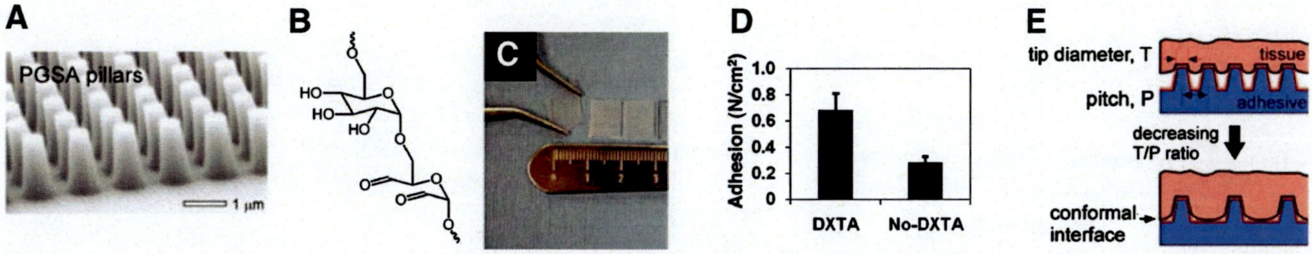

Fig. 19.2 Gecko-inspired biodegradable wet adhesive. (A) Scanning electron micrograph of synthetic gecko pattern. (B) Adhesive polymer: oxidized dextran (DXTA) possessing aldehyde groups. (C) 1-cm^2 patches of "gecko" tissue tape, which were used for *in vivo* experiments. (D) *In vivo* adhesion strength of coated and uncoated nanofibrillar samples after being implanted for 48 h. (E) Within a narrow range of diameter to pitch (T/P) ratio, patterned adhesives may exhibit enhanced surface area of contact with tissue when the pitch was sufficiently large and the tip diameter sufficiently low. Adapted from Mahdavi et al. (2008) and reproduced with permission

ed *in vitro* under shear loading on fresh porcine small intestines mounted flat on a mechanical testing device. Enhanced adhesion was demonstrated for the microstructured PDMS patterns, which could be further improved when coated with a silicone oil layer. However, due to complications related to the presence of silicone oil within an enclosed intestinal environment, enhanced adhesion of the capsule within intact small intestines was only tested for the uncoated patterned PDMS. Wet adhesives that may be activated and deployed *in-situ* would be useful for temporarily anchoring a swallowed capsule endoscope against the peristaltic action of the intestinal environment in order to facilitate detailed examination of a specific tissue location.

Mahdavi et al. (2008) created a wet adhesive coated nanofibrillar array composed entirely of biodegradable materials. The arrays were composed of a poly(glycerol sebacate acrylate) (PGSA) that were molded from a silicon template defined by photolithography (Fig. 19.2A). The adhesive layer was a thin layer of oxidized dextran (Fig. 19.2B) that was spin coated on the PGSA array. The PGSA used has been shown to be completely absorbed *in vivo* within 60 days (Wang et al., 2002), and the aldehyde groups of the oxidized dextran could covalently bond with amine groups present in proteins or the hydroxyl groups in PGSA (Mahdavi et al., 2008). The performance of the functionalized PGSA "bandages" (Fig. 19.2C) was tested under simple shear *in vitro* on porcine intestinal tissue, as well as on rat abdominal fascia tissue after 48 h implantations. The PGSA bandages were designed to function primarily as persistent rather than temporary/reversible adhesives.

In vitro shear tests of uncoated nanofibrillar arrays (4 mm disc samples) showed ~2 times enhanced adhesion over flat unpatterned PGSA. *In vivo* tests of the oxidized dextran coated nanofibrillar arrays (3 mm wide sections) showed more than twofold adhesion enhancement over uncoated pillars (Fig. 19.2D), but only a moderate enhancement was observed over a flat surface coated with the same dextran polymer. Interestingly, the best performance was found for arrays with the lowest densities of individual nanopillar protrusions (tip height = 2.4 μm pitch of array ~2.5 μm). Therefore the results are contrary to the predictions of contact-splitting theory, and the authors attributed this to the fact that the tissue substrate had a lower mechanical modulus than the PGSA microstructure, such that mechanical interlocking between the nanopillars and the soft tissue was likely to be an important mechanism responsible for the observed shear adhesion (Fig. 19.2E). Thus, the approach bears only limited resemblance to gecko adhesives, which do not achieve adhesive interactions through penetration into the substrate. Nonetheless, there may be future clinical applications for this type of adhesive.

19.2.3 Gecko/Mussel Mimetic Adhesives with poly(DMA-co-MEA) Coating

In a first demonstration of biomimetic reversible wet adhesives, Lee et al. (2007) united the design elements of gecko and mussel adhesives in a hybrid material that combined the reversibility of gecko-type adhesive with the wet performance of mussel adhesives. Proteins found within mussel adhesive plaques contain a high content of the catecholic amino acid 3,4-dihydroxy-L-phenylalanine (DOPA) (Waite and Tanzer, 1981; Papov et al., 1995; Waite and Qin, 2001). Strong interfacial adhesion has been demonstrated for both natural and synthetic polymers that contain DOPA, and single-molecule measurements in aqueous media show that DOPA forms extremely strong yet reversible bonds with surfaces (Lee et al., 2006).

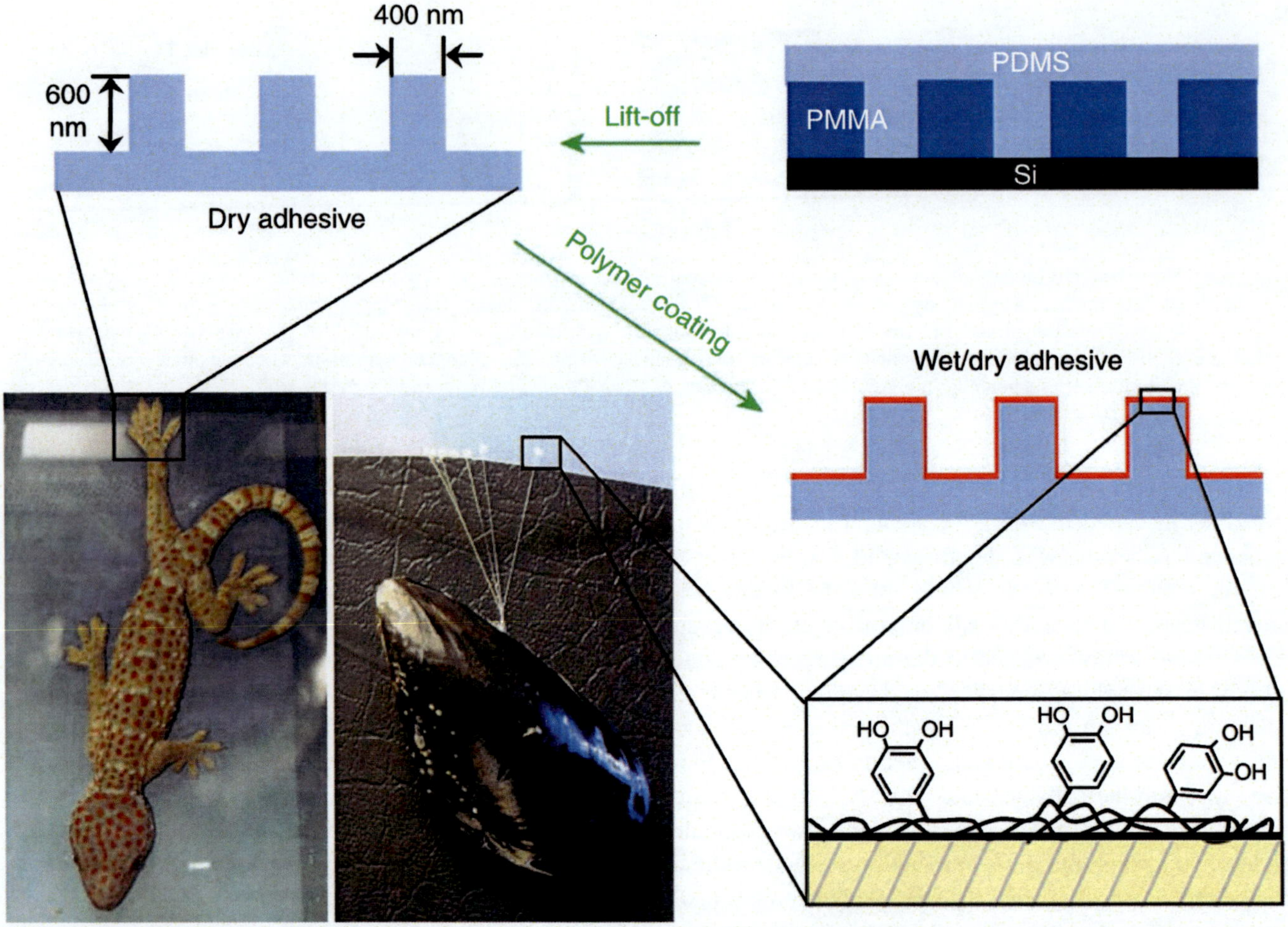

Fig. 19.3 Design and fabrication of gecko/mussel biomimetic nanopillar array adhesive. The PDMS nanopillar array was templated from an array of holes in a PMMA thin film supported on Si (defined by e-beam lithography). A thin layer of mussel-adhesive-protein-mimetic polymer (Fig. 19.4A) was adsorbed onto the fabricated nanopillars to confer wet adhesion properties. Adapted from Lee et al. (2007) and reproduced with permission

Therefore gecko-mimetic arrays of nanofibrils (Fig. 19.3) were coated with a physisorbed layer of poly(dopamine-co-methoxyethyl acrylate) (poly(DMA-co-MEA)) polymer with 17 wt.% DMA (Fig. 19.4A) to confer wet adhesive properties. Note that the adhesive layer is likely to be too thin (layer thickness $n = 10$ nm, Lee et al., 2007) to support the dissipative bulk viscoelastic deformations (Yamaguchi et al., 2009) that significantly contribute to the "tackiness" of conventional pressure-sensitive adhesives. The nanofibril arrays were composed of an elastomer – poly(dimethyl siloxane) (PDMS) – and were molded from a PMMA/Si template defined by electron-beam lithography. The diameter of the pillars (400 nm) was chosen to approximate the gecko's spatulae, while the pillar aspect ratio (1.5) was kept relatively small to prevent condensation (Geim et al., 2003; Spolenak et al., 2005a) of the elastomeric pillars.

The adhesive performance was characterized in detail using AFM force spectroscopy, whereby the pull-off force was measured during retraction of a tipless microcantilever from contact with a limited number of coated nanopillars (Fig. 19.4B). As expected, the adhesive force on uncoated pillars was greatly diminished (~85%) when immersed under water (5.9 ± 0.2 *vs.* 39.8 ± 2 nN/pillar in air). In comparison, the geckel coated pillars exhibited a higher adhesion than the uncoated ones in both air (120 ± 6 nN/pillar) and water (819.3 ± 5 nN/pillar). Although there was still a decrease in adhesion when immersed in water, the final adhesion force per coated pillar in water was still more than two times higher than the uncoated pillars in air (Fig. 19.4C). It was also demonstrated that the adhesive force scaled linearly with the number of contacting tips (Fig. 19.4D), but the predictions of contact-splitting scaling were not tested. Extra-

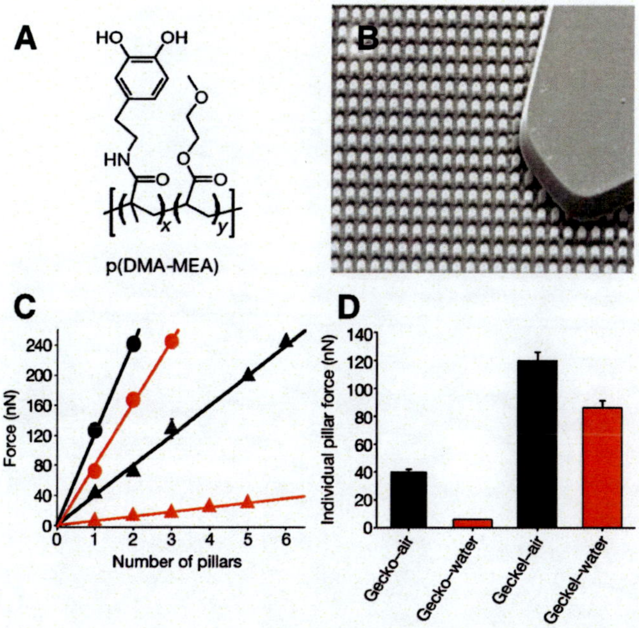

Fig. 19.4 Structure and performance of gecko/mussel mimetic wet adhesive. (A) Chemical structure of mussel-adhesive-protein-mimetic polymer: poly(dopamine-co-methoxyethyl acrylate) – p(DMA-MEA). (B) Scanning electron micrograph of an AFM cantilever contacting a pillar array (the number of pillars contacting the cantilever was controlled through the angle and tilt between the cantilever and the array). (C) Mean separation force *vs.* number of contacting pillars for uncoated (*triangle*) and coated (*circle*) pillars in water (*red*) and in air (*black*). (D) Separation force per pillar obtained from the slopes of the regression lines shown in C. Adapted from Lee et al. (2007) and reproduced with permission

polation of the AFM results to the macroscale (hypothetically at ~9 N/cm^2 in wet conditions) should be done with caution because of the challenges in maintaining equal load sharing of a large number of low aspect ratio nano-pillars. This level of adhesion was successfully observed over a thousand contact cycles, with little decay in adhesion strength.

The mussel-mimetic p(DMA-co-MEA) coating may also be applied to alternative structured adhesives in order to confer wet adhesive properties. For example, Glass et al. (2009) have coated a crack-trapping "membrane-over-pillar" microstructure and demonstrated more than 15 times enhanced wet adhesion over the uncoated microstructured surface (Glass et al., 2009). In particular, the adhesion was measured over a much larger contact area than studied by Lee et al. (2007), with a hemispherical glass probe 6 mm in diameter. As shown in Fig. 19.5A, the crack-trapping structure consisted of a thin, flat membrane suspended over an array of micropillars. The entire structure was composed of polyurethane. This microstructure has been shown to enhance adhesion because the membrane is too thin to efficiently transfer strain energy to the point where a crack between the membrane surface and the substrate could be initiated for detachment to occur (Glassmaker et al., 2007). The cracks are therefore "trapped" around membrane locations above the micropillars (Fig. 19.5B). Similar structures with

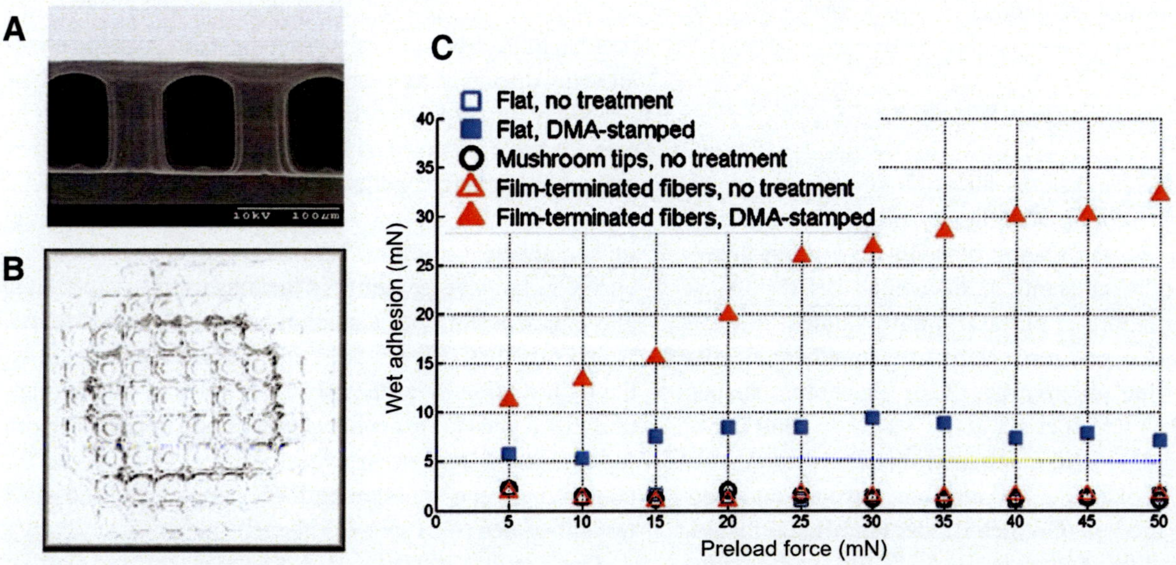

Fig. 19.5 Structure and performance of crack-trapping/mussel mimetic wet adhesive. (A) Scanning electron micrograph of the membrane over pillar microstructure. (B) Optical micrograph showing the adhesion interface as a preloaded glass hemisphere was pulled away from the pillar supported membrane submersed under water; the dark regions show that the contact line where detachment initiates was trapped along positions in between the micropillars. (C) Adhesion results under water as a function of preload. Adapted from Glass et al. (2009) and reproduced with permission

incompressible fluids filling the understructure has shown even higher levels of adhesion enhancement over a homogeneous flat surface (Majumder et al., 2007).

Figure 19.5C shows quantitatively the wet adhesive performance of the polymer coated crack-trapping structure. It shows a general feature of fibrillar adhesive systems in that the adhesion depends on the preload force. The coated structured-adhesive exhibited ~4.5 times adhesion enhancement over an unstructured (flat) surface coated with the mussel-mimetic polymer, whereas the enhancement was over 15 times compared to an uncoated microstructured surface. The mussel-mimetic polymer thus played a critical role in providing wet adhesion, while the microstructure brought additional enhancement of a magnitude in line with that observed under dry conditions (see reference Glassmaker et al., 2007). However, Fig. 19.5C also shows that the preload force was slightly higher than the final adhesion strength at pull-off, and the maximum wet adhesion was actually ~2.5 times less than the dry adhesion at the same preload. The extrapolated macroscopic wet adhesive strength (~1 N/cm^2) was also slightly lower than that extrapolated for the nanofibrillar array investigated by Lee et al. (2007) (Fig. 19.4). Nonetheless, due to the presence of a contiguous top membrane surface, the crack-trapping structure prevents pillar condensation, and the structure is also mechanically more robust than fibrillar structures. Furthermore, like the biomimetic hairy structures, the membrane-over-pillar design provides a relatively low effective modulus for compliant substrate contact.

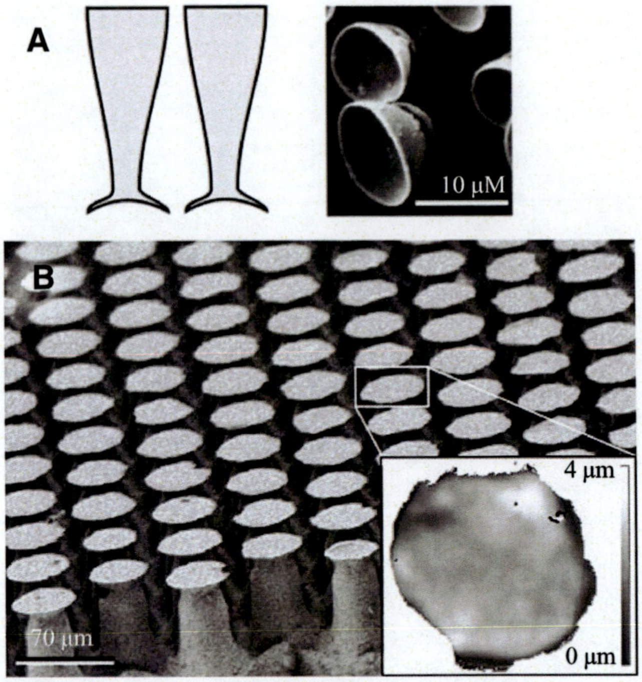

Fig. 19.6 Beetle-inspired mushroom-shaped fibrillar wet adhesive. (A) Schematic and micrograph of the "suction cups" on the vertical side of the foreleg tarsi of the male beetle *Dytiscus marginatus*. Adapted from Spolenak et al. (2005b) and reproduced with permission. (B) Scanning electron micrograph of beetle-inspired mushroom-shaped PVS microfibrillar adhesive microstructure. Inset displays the height image on top of a single "mushroom-cap" terminal contact plate. Adapted from Varenberg and Gorb (2008) and reproduced with permission

19.3 Beetle-inspired Fibril Design

In fibrillar adhesive systems, the tip shape has been shown to significantly modify both the individual fiber adhesive performance and the scaling behavior of fibrillar arrays with self-similar elements of different sizes (Spolenak et al., 2005b). Gorb et al. (2007) have studied in detail the wet adhesion performance of Chrysomelidae beetle-inspired fibrillar microarrays with mushroom-shaped tips (Fig. 19.6) (Gorb et al., 2007; Varenberg and Gorb, 2008). The biomimetic microarrays were prepared from a poly(vinyl siloxane) (PVS) elastomer using a commercial molding process (Gottlieb Binder GmbH, Germany). The fabricated tip diameters were ~50 μm, while native beetle tarsal seta tips are approximately 2–10 μm wide (Fig. 19.6).

Under dry conditions, the fibrillar system was assumed to adhere to flat surfaces *via* van der Waals forces. At a preload of 90 mN and a release rate of 100 μm/s, the fibrillar array displayed an average tensile adhesion strength of 3.8 N/cm^2 (extrapolated from a sample ~2 mm in diameter); when immersed in water, the glue-free array displayed an enhanced adhesion of 4.5 N/cm^2 (Fig. 19.7A). Optical interference microscopy characterization of the tip contact during pull-off showed that the detachment occurred initially at the center of the mushroom-shaped tips, and is consistent with a suction cup effect (Fig. 19.7B). Neither the dry nor wet adhesion appeared to depend on the hydrophobicity of the substrate (glass *vs.* silane functionalized glass). This enhancement stood in contrast with the dramatic decrease in adhesion strength observed for contact between unpatterned PVS samples and the glass test substrates (~2.1 *vs.* ~0.3 N/cm^2).

Gorb et al. (2007) have also shown that this fibrillar system has the ability to segregate contaminant particles in between the microfibrils. Thus the microstructured arrays are resistant to contamination of the actual contacting surface over repeated peel-and-attach cycles (Gorb et al.,

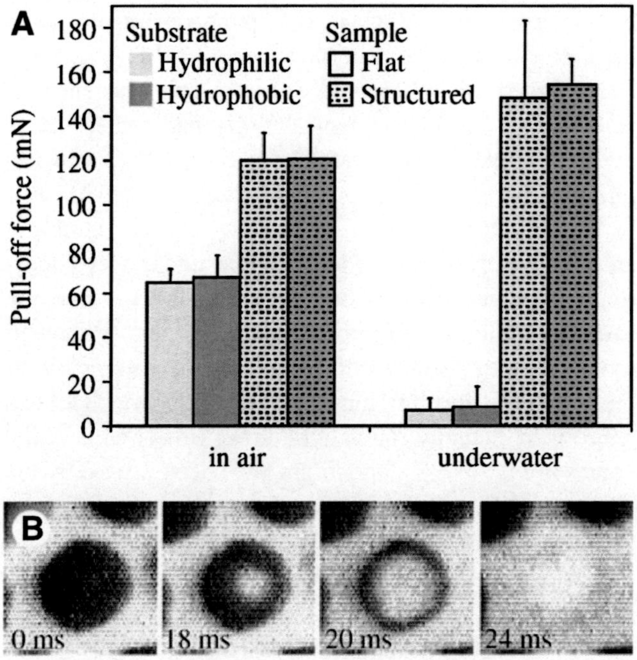

Fig. 19.7 Adhesive performance of mushroom-shaped fibrillar microstructure. (A) Mean pull-off force of flat and structured samples measured on hydrophilic and hydrophobic substrates in air and in deionized water. The error scales represent a standard deviation from mean values. (B) Behavior of a single terminal contact plate detaching from substrate in air. Dark areas resulted from destructive interference of reflected white light at the glass–PVS interface visualize the actual zones in contact. Adapted from Varenberg and Gorb (2008) and reproduced with permission

2007). Further studies on arrays of the mushroom tipped fibrils of different (smaller) sizes, and tests of non-ideal substrates with roughness on a scale larger than the tip diameter, may help elucidate the dimensional influence and practical limits of the suction-cup mechanism.

19.4 Tree Frog-inspired Wet Adhesives

Tree frogs display a remarkable capacity to cling to smooth and/or wet surfaces using their large toe pads. Barnes (2007) and Federle et al. (2006) have characterized their attachment capabilities in detail. Interestingly, it was found that the "flat" toe pads exhibited a hierarchically arrayed microstructure – the toe pads are composed of hexagonally arrayed epithelial cells several microns in diameter, and the surface of each cell is further divided into close-packed peg-like projections (Fig. 19.8). The projections and the epithelial cells are all flush with each other, and micro-indentation measurements have shown that the epithelial surface has a very low intrinsic mechanical modulus (<10 kPa in adult frogs) (Barnes, 2007). This stands in contrast to fibrillar systems, which derive their low effective modulus and substrate tolerance from a hairy structure composed of high aspect-ratio stiff fibers [stiffer fibrils are less likely to condense and collapse (Spolenak et al., 2005a)].

Tree frogs also secrete a watery epithelial mucus whose role in adhesion is not entirely clear. Federle et al. (2006) have shown by laser interferometry measurements that the liquid film between the centers of the toe pad epithelial cell surfaces and a glass test surface was only 0–5 nm thick in 40% of cases, and <25 nm thick on 95% of sampled cells. At such proximity, the role of viscous forces of the liquid secretion acting over the entire contact area for wet adhesion was likely to be small. Thus, the shear adhesion observed in single toe-pad measurements likely resulted from "dry" frictional contact, as also evidenced by the large shear force build-up and stick-slip events before sliding of a toe pad was initiated (Fig. 19.8D). Accordingly, Federle et al. (2006) also proposed that the

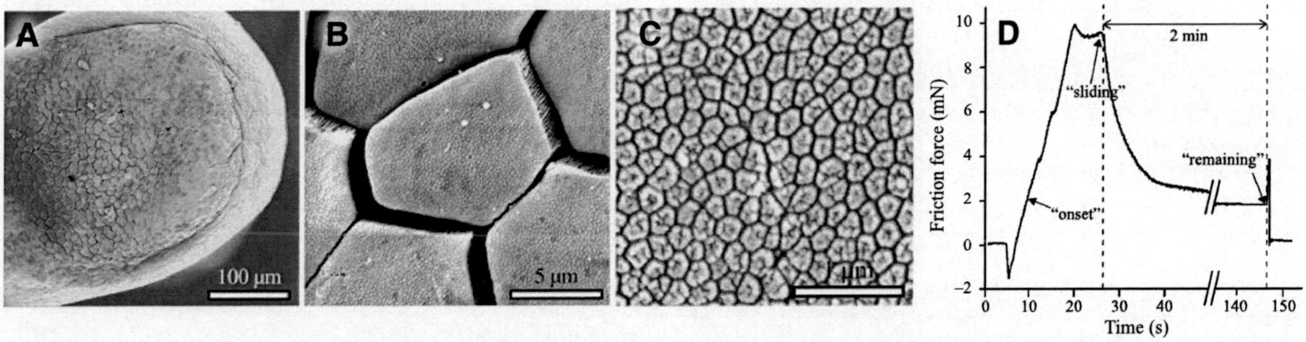

Fig. 19.8 Morphology and wet friction performance of tree frog toe pads. (A) Scanning electron micrograph of the toe pad of the tree frog *Litoria caerulea*. (B) Magnified view of the toe pad epidermis with hexagonal epithelial cells. (C) Further magnification showing the surface of a single hexagonal cell being composed of peg-like projections. (D) Shear stress measurement of a single toe pad during a sliding experiment consisting of 20 s sliding toward the body (500 µm/s) followed by 2 min standstill. Adapted from Federle et al. (2006) and reproduced with permission

channels in between the close-packed elements could facilitate drainage of the mucus to enable "dry" contact of the toe pad with a substrate. Furthermore, measurements of the (wet) tensile adhesion as a native toe pad in contact with a test surface was pulled apart showed that the increase in adhesion scaled approximately linearly with increasing toe pad area, which implied a capillary pressure force as the main contributing mechanism to the observed adhesion in tension (Barnes, 2007).

19.5 Cricket-inspired Wet Adhesives

It is interesting to note that the tarsal attachment pads of the bush cricket (*Tettigonia viridissima*) exhibit groove-separated hexagonal micro-elements (Fig. 19.9A) that are geometrically similar to the epithelial cell of the tree frog toe pad (Barnes, 2007). Accordingly, Varenberg and Gorb (2009) have investigated the wet frictional force between biomimetic micropatterns of flat hexagonal tiles and a flat substrate (Fig. 19.9B). However, unlike the tree frog toe pad, the hexagonal tiles were molded from a relatively stiff poly(vinyl siloxane) elastomer (Young's modulus = 3 MPa; also used for the microfibrillar arrays with mushroom-shaped tips prepared by the same authors – Fig. 19.6). The hexagonal micro-tiles were fabricated with lateral sizes ranging from 10 to 95 µm (all had a tile fraction = 75%) and aspect ratios = 0.01–0.1. The micropatterned samples (2 mm in diameter) were tested

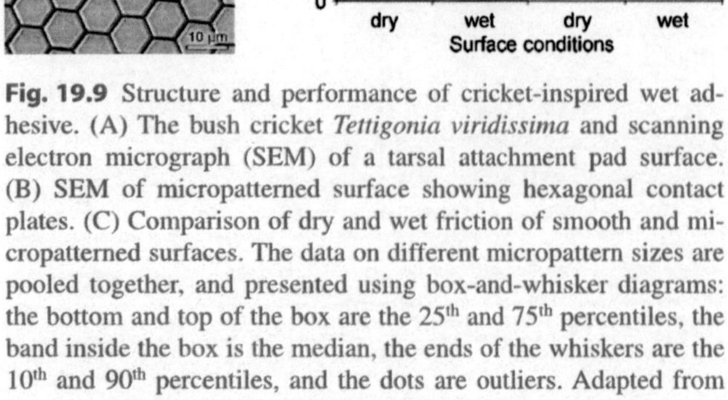

Fig. 19.9 Structure and performance of cricket-inspired wet adhesive. (A) The bush cricket *Tettigonia viridissima* and scanning electron micrograph (SEM) of a tarsal attachment pad surface. (B) SEM of micropatterned surface showing hexagonal contact plates. (C) Comparison of dry and wet friction of smooth and micropatterned surfaces. The data on different micropattern sizes are pooled together, and presented using box-and-whisker diagrams: the bottom and top of the box are the 25th and 75th percentiles, the band inside the box is the median, the ends of the whiskers are the 10th and 90th percentiles, and the dots are outliers. Adapted from Varenberg and Gorb (2009) and reproduced with permission

by placing them on glass substrates (average roughness, $R_a = 1$ nm) with a normal load of 60 mN, and the friction force was measured as the sample was sheared at 100 µm/s. To approximate the oily pad secretions of the bush cricket, a commercial mineral oil drop was applied on the samples before wet adhesion tests. As seen in Fig. 19.9C, the micropatterned surfaces exhibited an average of ~60 mN friction force under wet conditions (averaged over all micropattern sizes), while the force generated by an unpatterned PVS sample showed <10 mN. The "wet" friction exhibited by samples with the smallest tiles (10 µm wide) and grooves (1.4 µm wide) were actually found to be lower (lower ends of the values in Fig. 19.9C), and might have been caused by the increased difficulty in draining the interfacial fluid along the narrower and more tortuous groove channels.

19.6 Conclusions and Outlook

The biomimetic adhesives discussed in this review take inspiration from gecko and related fibrillar adhesive strategies, and are still at relatively early stages of development. Fortunately, the biological diversity of fibrillar attachment systems provides a great variety of adhesive strategies that can inspire novel biomimetic designs and lead to practical materials adapted for diverse wet environments (Arzt et al., 2003; Gorb, 2008; Peattie, 2008). Although high wet adhesive strength and high numbers of adhesion-release cycles have been demonstrated for some fibrillar microstructures under laboratory test conditions, large-area fabrication (>>1 cm^2) of (nano)fibrillar adhesive structures remains a significant technical challenge. Occasionally, unlikely combinations of highly contrasting biological adhesive concepts can be merged into novel biomimetic adhesives that exhibit desirable features inspired by more than one organism.

Acknowledgment

This work was supported by National Institutes of Health, USA, Grant R37 DE 014193.

References

Arzt E, Gorb SN, and Spolenak R (2003) From micro to nano contacts in biological attachment devices. Proceedings of the National Academy of Sciences of the United States of America 100(19): 10603–10606.

Autumn K and Gravish N (2008) Gecko adhesion: evolutionary nanotechnology. Philosophical Transactions of the Royal Society: Series A, Mathematical, Physical and Engineering Sciences 366(1870): 1575–1590.

Autumn K, Liang YA, Hsieh ST, Zesch W, Chan WP, Kenny TW, Fearing R, and Full RJ (2000) Adhesive force of a single gecko foot-hair. Nature 405(6787): 681–685.

Autumn K, Sitti M, Liang YA, Peattie AM, Hansen WR, Sponberg S, Kenny TW, Fearing R, Israelachvili JN, and Full RJ (2002) Evidence for van der Waals adhesion in gecko setae. Proceedings of the National Academy of Sciences of the United States of America 99(19): 12252–12256.

Barnes WJP (2007) Functional morphology and design constraints of smooth adhesive pads. MRS Bulletin 32(6): 479–485.

Gottlieb Binder GmbH, Germany, Fastening System. http://www.binder.biz/range.

Creton C and Gorb SN (2007) Issue theme: sticky feet: from animals to materials. MRS Bulletin 32(6): 466–508.

Cutkosky MR and Kim S (2009) Design and fabrication of multimaterial structures for bioinspired robots. Philosophical Transactions of the Royal Society: Series A, Mathematical, Physical and Engineering Sciences 367(1894): 1799–1813.

del Campo A and Arzt E (2007) Design parameters and current fabrication approaches for developing bioinspired dry adhesives. Macromolecular Bioscience 7(2): 118–127.

Federle W, Barnes WJ, Baumgartner W, Drechsler P, and Smith JM (2006) Wet but not slippery: boundary friction in tree frog adhesive toe pads. Journal of the Royal Society Interface 3(10): 689–697.

Ge L, Sethi S, Ci L, Ajayan PM, and Dhinojwala A (2007) Carbon nanotube-based synthetic gecko tapes. Proceedings of the National Academy of Sciences of the United States of America 104(26): 10792–10795.

Geim AK, Dubonos SV, Grigorieva IV, Novoselov KS, Zhukov AA, and Shapoval SY (2003) Microfabricated adhesive mimicking gecko foot-hair. Nature Materials 2(7): 461–463.

Glass P, Cheung E, and Sitti M (2008) A legged anchoring mechanism for capsule endoscopes using micropatterned adhesives. IEEE Transactions on Bio-medical Engineering 55(12): 2759–2767.

Glass P, Chung H, Washburn NR, and Sitti M (2009) Enhanced reversible adhesion of dopamine methacrylamide-coated elastomer microfibrillar structures under wet conditions. Langmuir 25(12): 6607–6612.

Glassmaker NJ, Jagota A, Hui CY, Noderer WL, and Chaudhury MK (2007) Biologically inspired crack trapping for enhanced adhesion. Proceedings of the National Academy of Sciences of the United States of America 104(26): 10786–10791.

Gorb SN (2008) Biological attachment devices: exploring nature's diversity for biomimetics. Philosophical Transactions of the Royal Society: Series A, Mathematical, Physical and Engineering Sciences 366: 1557–1574.

Gorb SN, Varenberg M, Peressadko A, and Tuma J (2007) Biomimetic mushroom-shaped fibrillar adhesive microstructure. Journal of the Royal Society Interface 4: 271–275.

Hansen WR and Autumn K (2005) Evidence for self-cleaning in gecko setae. Proceedings of the National Academy of Sciences of the United States of America 102(2): 385–389.

Huber G, Mantz H, Spolenak R, Mecke K, Jacobs K, Gorb SN, and Arzt E (2005) Evidence for capillarity contributions to gecko adhesion from single spatula nanomechanical measurements. Proceedings of the National Academy of Sciences of the United States of America 102(45): 16293–16296.

Israelachvili JN (1992) Intermolecular and Surface Forces: With Applications to Colloidal and Biological Systems (Colloid Science), 2nd Ed. Academic Press, San Diego.

Jeong HE, Lee JK, Kim HN, Moon SH, and Suh KY (2009) A nontransferring dry adhesive with hierarchical polymer nanohairs. Proceedings of the National Academy of Sciences of the United States of America 106(14): 5639–5644.

Jeong HE and Suh KY (2009) Nanohairs and nanotubes: efficient structural elements for gecko-inspired artificial dry adhesives. Nano Today 4: 335–346.

Lee H, Scherer NF, and Messersmith PB (2006) Single-molecule mechanics of mussel adhesion. Proceedings of the National Academy of Sciences 103: 12999–13003.

Lee H, Lee BP, and Messersmith PB (2007) A reversible wet/dry adhesive inspired by mussel and geckos. Nature Letters 448: 338–341.

Lee J and Fearing RS (2008) Contact self-cleaning of synthetic gecko adhesive from polymer microfibers. Langmuir 24(19): 10587–10591.

Lee J, Majidi C, Schubert B, and Fearing RS (2008) Sliding-induced adhesion of stiff polymer microfibre arrays. I. Macroscale behaviour. Journal of the Royal Society Interface 5(25): 835–844.

Mahdavi A, Ferreira L, Sundback C, Nichol JW, Chan EP, Carter DJ, Bettinger CJ, Patanavanich S, Chignozha L, Ben Joseph E, Galakatos A, Pryor H, Pomerantseva I, Masiakos PT, Faquin W, Zumbuehl A, Hong S, Borenstein J, Vacanti J, Langer R, and Karp JM (2008) A biodegradable and biocompatible gecko-inspired tissue adhesive. Proceedings of the National Academy of Sciences of the United States of America 105(7): 2307–2312.

Majumder A, Ghatak A, and Sharma A (2007) Microfluidic adhesion induced by subsurface microstructures. Science 318(5848): 258–261.

Papov VV, Diamond TV, Biemann K, and Waite JH (1995) Hydroxyarginine-containing polyphenolic proteins in the adhesive plaques of the marine mussel *Mytilus edulis*. Journal of Biological Chemistry 270(34): 20183–20192.

Peattie AM (2009) Functional demands of dynamic biological adhesion: an integrative approach. Journal of Comparative Physiology B 179: 231–239.

Qu L, Dai L, Stone M, Xia Z, and Wang ZL (2008) Carbon nanotube arrays with strong shear binding-on and easy normal lifting-off. Science 322(5899): 238–242.

Ruibal R and Ernst V (1965) The structure of the digital setae of lizards. Journal of Morphology 117(3): 271–293.

Schubert B, Majidi C, Groff RE, Baek S, Bush B, Maboudian R, and Fearing RS (2007) Towards friction and adhesion from high modulus microfiber arrays. Journal of Experimental Biology 21: 1297–1315.

Spolenak R, Gorb SN, and Arzt E (2005a) Adhesion design maps for bio-inspired attachment systems. Acta Biomaterialia 1: 5–13.

Spolenak R, Gorb SN, Gao H, and Arzt E (2005b) Effects of contact shape on the scaling of biological attachments. Proceedings of the Royal Society of London: Series A, Mathematical, Physical and Engineering Sciences 461: 305–319.

Sun W, Neuzil P, Kustandi TS, Oh S, and Samper VD (2005) The nature of the gecko lizard adhesive force. Biophysical Journal 89(2): L14–L17.

Varenberg M and Gorb SN (2008) A beetle-inspired solution for underwater adhesion. Journal of the Royal Society Interface 5(20): 383–385.

Varenberg M and Gorb SN (2009) Hexagonal surface micropattern for dry and wet friction. Advanced Materials 21: 483–486.

Waite JH and Qin X (2001) Polyphosphoprotein from the adhesive pads of *Mytilus edulis*. Biochemistry 40(9): 2887–2893.

Waite JH and Tanzer ML (1981) Polyphenolic substance of mytilus edulis: novel adhesive containing L-dopa and hydroxyproline. Science 212(4498): 1038–1040.

Wang Y, Ameer GA, Sheppard BJ, and Langer R (2002) A tough biodegradable elastomer. Nature Biotechnology 20(6): 602–606.

Yamaguchi T, Gravish N, Autumn K, and Creton C (2009) Microscopic modeling of the dynamics of frictional adhesion in the gecko attachment system. Journal of Physical Chemistry B 113(12): 3622–3628.

Yanik MF (2009) Towards gecko-feet-inspired bandages. Trends in Biotechnology 27(1): 1–2.

Subject Index

List of Contributors

Editors

Janek von Byern
Core Facility Cell Imaging and Ultrastructure Research
Faculty of Life Science
University of Vienna
Althanstrasse 14
1090 Vienna, Austria
E-mail: Janek.von.Byern@univie.ac.at

Ingo Grunwald
Fraunhofer Institute for Manufacturing Technology
and Advanced Materials (IFAM)
Department of Adhesive Bonding Technology
and Surfaces, Adhesives and Polymer Chemistry
Wiener Strasse 12
28359 Bremen, Germany
E-mail: ingo.grunwald@ifam.fraunhofer.de

Authors

Wolfram Adlassnig
Core Facility Cell Imaging and Ultrastructure Research
Division of Cell Physiology and Scientific Film
University of Vienna
Althanstrasse 14
1091 Vienna, Austria
E-mail: wolfram.adlassnig@univie.ac.at

Pierre T. Becker
Laboratoire de Biologie marine
Université de Mons
6 Avenue du Champ de Mars
7000 Mons, Belgium
E-mail: Pierre.Becker@umons.ac.be

Albrecht Berg
INNOVENT Technologieentwicklung e. V.
Biomaterialien
Prüssingstr. 27B
07745 Jena, Germany
E-mail: AB4@innovent-jena.de

Oliver Betz
Abteilung für Evolutionsbiologie der Invertebraten
Institut für Evolution und Ökologie
Universität Tübingen
Auf der Morgenstelle 28E
72076 Tübingen, Germany
E-mail: oliver.betz@uni-tuebingen.de

Michaela Bittner
Baxter Innovations GmbH
Industriestraße 131
1221 Vienna, Austria
E-mail: Michaela_Bittner@baxter.com

Jessica Blume
Nolax AG
Eichenstrasse 12
6203 Sempach-Station, Switzerland
E-mail: jessica.blume@nolax.com

Norbert Cyran
Core Facility Cell Imaging
and Ultrastructure Research
Faculty of Life Science
University of Vienna
Althanstrasse 14
1090 Vienna, Austria
E-mail: nbc555@gmx.net

Georg A. Feichtinger
Ludwig Boltzmann Institute for Experimental and Clinical Traumatology
Austrian Cluster for Tissue Regeneration
Donaueschingenstrasse 13
1200 Vienna, Austria
E-mail: Georg.Feichtinger@TRAUMA.LBG.AC.AT

James Ferguson
Ludwig Boltzmann Institute for Experimental and Clinical Traumatology
Austrian Cluster for Tissue Regeneration
Donaueschingenstrasse 13
1200 Vienna, Austria
E-mail: James.Ferguson@TRAUMA.LBG.AC.AT

Patrick Flammang
Laboratoire de Biologie Marine
Université de Mons-Hainaut
Pentagone 2B
6 Avenue du Champ de Mars
7000 Mons, Belgique
E-mail: Patrick.Flammang@umh.ac.be

Christian Gabriel
Red Cross Blood Transfusion Service of Upper Austria
Austrian Cluster for Tissue Regeneration
Krankenhausstrasse 7
4017 Linz, Austria
E-mail: Christian.Gabriel@o.roteskreuz.at

Charles Griffiths
Zoology Department
University of Cape Town
Private Bag X3
Rondebosch 7701, South Africa
E-mail: Charles.Griffiths@uct.ac.za

Ingo Grunwald
Fraunhofer Institute for Manufacturing Technology and Advanced Materials (IFAM)
Department of Adhesive Bonding Technology and Surfaces, Adhesives and Polymer Chemistry
Wiener Strasse 12
28359 Bremen, Germany
E-mail: ingo.grunwald@ifam.fraunhofer.de

Alfred Gugerell
Baxter Innovations GmbH
Industriestraße 131
1221 Vienna, Austria
E-mail: alfred_gugerell@baxter.com

Andreas Hartwig
Department of Adhesive Bonding Technology and Surfaces
Adhesives and Polymer Chemistry
Fraunhofer Institute for Manufacturing Technology and Advanced Materials (IFAM)
Wiener Strasse 12
28359 Bremen, Germany
E-mail: andreas.hartwig@ifam.fraunhofer.de

Elise Hennebert
Laboratoire de Biologie Marine
Université de Mons
6 Avenue du Champ de Mars
7000 Mons, Belgium
E-mail: Elise.Hennebert@umons.ac.be

Michael Hesse
Department für Palynologie und Strukturelle Botanik
Universität Wien
Rennweg 14
1030 Wien, Austria
E-mail: michael.hesse@univie.ac.at

Jaimie Jonker
Department of Zoology
Martin Ryan Institute
National University of Ireland
University Road
Galway, Ireland
E-mail: j.jonker1@nuigalway.ie

Waltraud Klepal
Core Facility Cell Imaging and Ultrastructure Research
Faculty of Life Science
University of Vienna
Althanstrasse 14
1090 Vienna, Austria
E-mail: Waltraud.Klepal@univie.ac.at

Lisa Klinger
Core Facility Cell Imaging and Ultrastructure Research
Faculty of Life Science
University of Vienna
Althanstrasse 14
1090 Vienna, Austria
E-mail: lisa.klinger@chello.at

Maria Kozielec
Fraunhofer Institute for Manufacturing Technology and Advanced Materials (IFAM)
Department of Adhesive Bonding Technology and Surfaces
Adhesives and Polymer Chemistry
Wiener Strasse 12
28359 Bremen, Germany
E-mail: maria.kozielec@ifam.fraunhofer.de

Krystyna Labuda
Ludwig Boltzmann Institute for Experimental and Clinical Traumatology
Austrian Cluster for Tissue Regeneration
Donaueschingenstrasse 13
1200 Vienna, Austria
E-mail: Krystyna.Labuda@TRAUMA.LBG.AC.AT

Ingeborg Lang
Core Facility Cell Imaging and Ultrastructure Research
Division of Cell Physiology and Scientific Film
University of Vienna
Althanstrasse 14
1091 Vienna, Austria
E-mail: ingeborg.lang@univie.ac.at

K.H. Aaron Lau
Messersmith Research Group
Biomedical Engineering Department
Northwestern University
2145 Sheridan Road, Tech E210
Evanston, 60208 Chicago, USA
E-mail: aaronlau@northwestern.edu

Thomas Lendl
Core Facility Cell Imaging and Ultrastructure Research
Division of Cell Physiology and Scientific Film
University of Vienna
Althanstrasse 14
1091 Vienna, Austria
E-mail: tom.lendl@gmx.at

Paul McEvilly
Department of Zoology
Martin Ryan Institute
National University of Ireland
University Road
Galway, Ireland
E-mail: p.mcevilly1@nuigalway.ie

Alexandra Meinl
Ludwig Boltzmann Institute for Experimental and Clinical Traumatology
Austrian Cluster for Tissue Regeneration
Donaueschingenstrasse 13
1200 Vienna, Austria
E-mail: Alexandra.Meinl@TRAUMA.LBG.AC.AT

Phillip B. Messersmith
Messersmith Research Group
Biomedical Engineering Department
Northwestern University
2145 Sheridan Road
Tech E210, Evanston
60208 Chicago, USA
E-mail: philm@northwestern.edu

Claudia E. Mills
Department of Biology
University of Washington
24 Kincaid Hall
Washington, 98195-1800 Seattle, USA
E-mail: cemills@u.washington.edu

Tatjana J. Morton
Ludwig Boltzmann Institute for Experimental and Clinical Traumatology
Austrian Cluster for Tissue Regeneration
Donaueschingenstrasse 13
1200 Vienna, Austria
E-mail: Tatjana.Morton@TRAUMA.LBG.AC.AT

Lila Nikkola
Department of Biomedical Engineering
Tampere University of Technology
P.O. Box 692, 331010 Tampere, Finland
E-mail: lila.nikkola@tut.fi

Sylvia Nürnberger
Department of Traumatology
Medical University of Vienna
Waehringer Guertel 18-20
1090 Vienna, Austria

Ludwig Boltzmann Institute for Experimental
and Clinical Traumatology
Austrian Cluster for Tissue Regeneration
Donaueschingenstrasse 13
1200 Vienna, Austria
E-mail: sylvia.nuernberger@meduniwien.ac.at

Hermann Onusseit
Henkel AG & Co. KGaA
Adhesive Technologies
Henkelstraße 67, 40191 Düsseldorf, Germany
E-mail: hermann.onusseit@henkel.com

Waltraud Pasteiner
Baxter Innovations GmbH
Industriestraße 131, 1221 Vienna, Austria
E-mail: Waltraud_Pasteiner@baxter.com

Marianne Peroutka
Core Facility Cell Imaging and Ultrastructure Research
Division of Cell Physiology and Scientific Film
University of Vienna
Althanstrasse 14, 1091 Vienna, Austria
E-mail: marianne.peroutka@univie.ac.at

Anja Peterbauer-Scherb
Red Cross Blood Transfusion Service of Upper Austria
Austrian Cluster for Tissue Regeneration
Krankenhausstrasse 7, 4017 Linz, Austria
E-mail: Anja.Peterbauer@o.roteskreuz.at

Fabian Peters
Curasan AG
Lindigstrasse 4
63801 Kleinostheim, Germany
E-mail: fabian.peters@curasan.de

Leon Ploszczanski
Department of Theoretical Biology
University of Vienna
Althanstrasse 14
1090 Vienna, Austria
E-mail: leon.ploszczanski@univie.ac.at

Anne Marie Power
Department of Zoology
Martin Ryan Institute
National University of Ireland
University Road
Galway, Ireland
E-mail: annemarie.power@nuigalway.ie

Heinz Redl
Ludwig Boltzmann Institute for Experimental
and Clinical Traumatology
Austrian Cluster for Tissue Regeneration
Donaueschingenstrasse 13
1200 Vienna, Austria
E-mail: Office@TRAUMA.LBG.AC.AT

Katharina Richter
Department of Adhesive Bonding Technology
and Surfaces
Adhesives and Polymer Chemistry
Fraunhofer Institute for Manufacturing Technology
and Advanced Materials (IFAM)
Wiener Strasse 12
28359 Bremen, Germany
E-mail: katharina.richter@ifam.fraunhofer.de

Klaus Rischka
Department of Adhesive Bonding Technology
and Surfaces
Adhesives and Polymer Chemistry
Fraunhofer Institute for Manufacturing Technology
and Advanced Materials (IFAM)
Wiener Strasse 12
28359 Bremen, Germany
E-mail: klaus.rischka@ifam.fraunhofer.de

Frank Roberto
Energy and Environment Department
Idaho National Laboratory
P.O. Box 1625
Idaho Falls 83415-2203, USA
E-mail: francisco.roberto@inl.gov

Robert Sader
Faculty of Medicine
Medical Centre of the Johann Wolfgang Goethe
University, Department of Oral, Craniomaxillofacial
and Facial Plastic Surgery
60596 Frankfurt, Germany
E-mail: R.Sader@em.uni-frankfurt.de

Thomas Schwaha
Institute of Zoology, Anatomy & Morphology
University of Vienna
Althanstrasse 14, 1090 Vienna, Austria
E-mail: thomas.schwaha@univie.ac.at

Willi Schwotzer
Nolax AG
Eichenstrasse 12
6203 Sempach-Station, Switzerland
E-mail: willi.schwotzer@nolax.com

Robyn Scott
Zoology Department
University of Cape Town
Rondebosch 7700, Cape Town, South Africa
E-mail: robynscott0@hotmail.com

Heather G. Silverman
Energy and Environment Department
Idaho National Laboratory
P.O. Box 1625
Idaho Falls 83415-2203, USA
E-mail: Heather.Silverman@inl.gov

Klaus Slenzka
OHB-System AG
Department Life Sciences
Universitaetsallee 27-29
28359 Bremen, Germany
E-mail: slenzka@ohb-system.de

Andrew M. Smith
Department of Biology
Ithaca College, Ithaca 14850, USA
E-mail: asmith@ithaca.edu

Russell J. Stewart
Department of Bioengineering
University of Utah
Salt Lake City 84112, USA
E-mail: rstewart@eng.utah.edu

Kelli K. Svendsen
Department of Bioengineering
University of Utah
Salt Lake City 84112, USA
E-mail: kelli.svendsen@utah.edu

Michael J. Tyler
School of Earth & Environmental Sciences
Ecology & Evolutionary Biology
Jordan Laboratories
The University of Adelaide
5005 Adelaide, Australia
E-mail: ellatyler@oze-mail.com.au

Martijn van Griensven
Ludwig Boltzmann Institute for Experimental
and Clinical Traumatology
Austrian Cluster for Tissue Regeneration
Donaueschingenstrasse 13
1200 Vienna, Austria
E-mail: Martijn.van.Griensven@TRAUMA.LBG.AC.AT

Janek von Byern
Core Facility Cell Imaging and Ultrastructure Research
Faculty of Life Science
University of Vienna
Althanstrasse 14
1090 Vienna, Austria
E-mail: Janek.von.Byern@univie.ac.at

Herbert Waite
Cell & Developmental Biology Department
University of California
Marine Biotechnology Building
Santa Barbara 93106, USA
E-mail: waite@lifesci.ucsb.edu

Ching Shuen Wang
Department of Bioengineering
University of Utah
Salt Lake City 84112, USA
E-mail: u0557755@utah.edu

Susanne Wolbank
Ludwig Boltzmann Institute for Experimental
and Clinical Traumatology
Austrian Cluster for Tissue Regeneration
Donaueschingenstrasse 13
1200 Vienna, Austria
E-mail: Susanne.Wolbank@TRAUMA.LBG.AC.AT

Vanessa Zheden
Core Facility Cell Imaging and Ultrastructure Research
Faculty of Life Science
University of Vienna
Althanstrasse 14, 1090 Vienna, Austria
E-mail: vanessa.zheden@univie.ac.at